P9-DTM-467

COMPUTING IN BIOLOGICAL SCIENCE

Computing in biological science

Edited by

MICHAEL J. GEISOW and ANTHONY N. BARRETT

Division of Biophysics, and Computing Laboratory,
National Institute for Medical Research, Mill Hill, London NW7 1AA, U.K.

1983

Elsevier Biomedical Press
Amsterdam · New York · Oxford

ISBN: 0-444-80435-8

PUBLISHED BY:

Elsevier Biomedical Press,
1 Molenwerf, P.O. Box 211, 1000 AE
Amsterdam, The Netherlands

SOLE DISTRIBUTORS FOR THE U.S.A. AND CANADA:

Elsevier Science Publishing Company
52 Vanderbilt Avenue,
New York, N.Y. 10017

Library of Congress Cataloging in Publication Data
Main entry under title:

Computing in biological science.

Includes index.
1. Biology--Data processing. 2. Biological chemistry --Data processing. I. Geisow, Michael J. II. Barrett, Anthony. [DNLM: 1. Biology--Methods. 2. Computers. 3. Models, Biological. QH 324.2 C738]
QH324.2.C665 1983 574'.028'54 82-18165
ISBN 0-444-80435-8

Printed in The Netherlands

Contents

IMAGE PROCESSING AND ANALYSIS

NUMERICAL ANALYSIS

MICROPROCESSING

List of contributors

Anthony N. Barrett	Computing Laboratory, National Institute for Medical Research, Mill Hill, London NW7 1AA, U.K.
Robert J. Beynon	Department of Biochemistry, University of Liverpool, P.O. Box 147, Liverpool, L69 3BX, U.K.
Ian Burdett	Division of Microbiology, National Institute for Medical Research, Mill Hill, London NW7 1AA, U.K.
G. Marius Clore	Division of Molecular Pharmacology, National Institute for Medical Research, Mill Hill, London NW7 1AA, U.K.
Michael J. Geisow	Division of Biophysics, National Institute for Medical Research, Mill Hill, London NW7 1AA, U.K.
Richard Gordon	Faculty of Medicine, University of Manitoba, 770 Bannatyne Ave., Winnipeg, Manitoba, Canada, R3E 0W3
Peter F. Lemkin	Image Processing Section, Division of Cancer Biology and Diagnosis, National Cancer Institute, National Institutes of Health, Bethesda, Md. 20205, U.S.A.
Lewis E. Lipkin	Image Processing Section, Division of Cancer Biology and Diagnosis, National Cancer Institute, National Institute of Health, Bethesda, Md. 20205, U.S.A.
W. Jack Perkins	Computing Laboratory, National Institute for Medical Research, Mill Hill, London NW7 1AA, U.K.

ERIC PLATT — Department of Biochemistry, University of Manchester, Oxford Road, Manchester, M13 9PL, U.K.

BARRY ROBSON — Department of Biochemistry, University of Manchester, Oxford Road, Manchester, M13 9PL, U.K.

BRUCE A. SHAPIRO — Image Processing Section, Division of Cancer Biology and Diagnosis, National Cancer Institute, National Institutes of Health, Bethesda, Md. 20205, U.S.A.

PETER SHAW — Department of Ultrastructural Studies, John Innes, Colney Lane, Norwich, NR4 7UH, U.K.

MICHAEL J. STERNBERG — Laboratory of Molecular Biophysics, Department of Zoology, South Parks Road, Oxford OX1 3PS, U.K.

ALAN MACKAY — Department of Crystallography, Birkbeck College, Mallet Street, London WC1E 7HX

Geisow & Barrett (eds.) Computing in biological science
© Elsevier Biomedical Press, 1983

Introduction

This book presents a set of chapters illustrating various applications of computers to selected areas of biological science. It would not be possible to include a comprehensive account of all applications covering the whole of biology in a single volume; instead, a number of individual scientists actively engaged in biological research give personal assessments of their particular problems and the contribution that computers make towards solving them.

Although the basis of the scientific work is biological, the applications themselves can be divided into several distinct but not necessarily exclusive areas of computing already familiar to scientists working in other disciplines such as physics and engineering. The subject matter therefore has been arranged according to application type, viz. image processing, modelling, numerical analysis and microprocessing, thus demonstrating a relatively wide coverage of applications in both computing and biology. Most of the chapters give descriptions of the software and related algorithms, together with details of the biological problem and its significance. Discussion relating to hardware has been kept to a minimum except in the chapter dealing with microprocessors.

Due to the prohibitively large amount of space that would be required to give the programs in coded form, we have instead included an appendix indicating the relevant sources from which further information can be obtained.

MODELLING

A problem frequently encountered in biochemical research is the need to produce a descriptive model of some biological system. This might be a structural model of a protein or virus particle or it might be a mechanistic construct, such as a series of relational equations describing an enzyme reaction or information flow in neurones. The data available for the solution of these problems can rarely be handled by analytical formulae, because these are usually unknown or have no unique solution. Furthermore, the data may consist not only of measurements, but of qualitative observations and empirical considerations, which cannot be combined in a programme which leads to a solution independent of the experimenter. However, the power of the computer in these circumstances lies not in its speed but its interactive capability. The human operator must make decisions in the face of information supplied by the computer; decisions which require the integrative abilities of the experienced experimenter and which are still beyond our ingenuity to reproduce by programming. A common factor in these computer applications is the high probability of more than one solution to the problem. This need not be a drawback, however, since the possession of a small number of possibilities may represent a sufficient advance in understanding to allow design of experiments which do provide a unique solution.

Interactive computer models of biological systems. This chapter gives the basic reasons for the modelling approach and assesses the relative merits of different types of models in relation to a number of applications based on the author's experience. A high value is attached to user interaction via extensive input/output facilities and visual display systems.

Computational embryology of the vertebrate nervous system. Simulation of the development of the shape of an embryo and its component structures in terms of cellular interactions is described. This contribution illustrates the increasing importance of modelling techniques for testing and improving concepts relating to morphogenesis.

Computer analysis of bacterial cell walls and modelling of the surface growth process. The application of mathematical modelling and image processing techniques to electron micrographs of bacterial cell walls is discussed. The computer is used as a data acquisition device for the accurate digital analysis of shapes and then as an image processing and modelling tool for studying mechanisms controlling the processes of bacterial cell growth and division.

The above chapters represent the macroscopic modelling approach in which simulations of changes in shape of complete organisms can be examined in relation to ideas about components of those organisms. By contrast, modelling at the molecular level is examined in the last two chapters. Here the aim is to predict the conformation of molecules, often of known chemical structure, but unknown geometry.

Calculation of biomolecular conformation. The complete representation of molecular structures of the size of biopolymers is computationally expensive using ab initio quantum mechanical techniques and requires very large, fast machines. This chapter examines alternative methods for the calculation of molecular conformations.

Analysis and prediction of protein structure. Descriptions of the conformation of proteins using quantum mechanical methods or approximations based on minimizing the energy of trial structures is beyond the present capability of computers. Very promising alternatives have been developed in recent years using a combined approach of information available from known protein structures, constraints provided by well-explored protein atomic stereochemistry and aspects of biological function.

IMAGE ANALYSIS

Database techniques for two-dimensional electrophoretic gel analysis. A powerful application of computers in bioscience is the development of data bases familiar to all biologists in the form of keyword searches of the scientific literature. Information filed in such bases can be searched with high efficiency to produce previously hidden relationships between the entries in the data base. One such ambitious application is the concept of a total human protein map (Clark, B.F.C., [1981] *Nature, 292,* 491). The resolution of two-dimensional polyacrylamide gels is now such that it is reasonable to attempt the resolution of all the proteins in a mammalian cell. This is certainly possible when given cell types are fractionated into two well-defined sets of proteins, for example, nuclear and cytoplasmic cell membrane pools. Automatic identification and quantitation of polypeptides in two-dimensional gels is described as well as the construction of a multiple gel data base. The emergence of such a data base would be a world resource of exceptional value to cell biologists. Qualitative and quantitative variation in the protein composition of a given cell type could be linked to function or pathology in a manner presently impossible.

Nucleic acid morphology: analysis and synthesis. This chapter illustrates the complementary overlap of image analysis and modelling in most approaches to understanding the structure of biological material. The maximum information is extracted from high-resolution electron microscope images of spread nucleic acid molecules, in order to map base-paired regions. In parallel with this effort an interactive modelling approach, similar to the protein structure folding described in the chapters by Sternberg and Platt and Robson, is employed to predict based-paired regions from sequence.

Two-dimensional and three-dimensional reconstruction in electron microscopy. Crystallographers and an increasing number of electron microscope users have successfully determined biopolymer structures ranging from small peptides to whole viruses or ribosomes. The degree of success (and information) obtained is measured as a resolution in Å units, where 3 Å resolution allows an approach to atomic detail, but even 10 Å resolution would reveal much of interest in a particle the size of a ribosome.

These relatively high resolutions are made possible by the presence of two or three-dimensional order in the diffraction patterns (crystallography) or imaged structures (electron microscopy). Well-developed methods are now available for crystallographers and are not described in this volume (a recent practical description is given by Blundell and Johnson). However, very exciting studies of two-dimensional protein assemblies have recently been reported by electron microscopy at 6–3 Å resolution. The state of the art of this emerging technique, which is almost wholly dependent on the computer for performing the large number of calculations involved, is described.

NUMERICAL ANALYSIS

Computer analysis of transient kinetic data. Biological processes are dynamic and, at the molecular level, extremely fast. Important understanding of underlying mechanisms has emerged from the careful analysis of individual components contributing to a fast biological response (for example, the action potential of myelinated nerves). This chapter describes a new computational approach to analysing transient kinetic information, with the aim of establishing a very powerful, flexible tool of wide utility.

The numerical geometry of biological structures. A large component of the total computer usage in biology involves the elaboration of models, as described at the beginning of this introduction. Most of the programmes at work on these applications make use of three-dimensional geometry (we live in 3D space). The chapter by Alan Mackay provides a valuable and comprehensive reference collection of geometrical properties and operations in space, with emphasis on implementation on small computers.

MICROPROCESSING

At the other end of the spectrum of computing power, the invasion of labs by mini and microcomputers is considered by Robert Benyon. Although programmers used to the facilities provided by a large central computer tend to be critical of the speed and limited memory of microprocessor-supported machines, the basis of even this criticism is being eroded with the emergence of cheap micros with 16 bit bytes, RAM starting at 65 K and integral hard discs (high density storage) which support a variety of high-level language compilers. However, speed of computation is frequently unnecessary in many online control applications in the biological laboratory, and for this type of application, the microcomputer can be confidently predicted to achieve considerable importance in the coming years.

ENVOI

The book has two main aims: one is to make biologists more aware of the scope of computer applications within their field, particularly as this is still a rapidly developing area, thus encouraging a wider use of computers and computational techniques. The second is to inform mathematicians and computer specialists of the existence of biological problems in which computers are playing an increasing part in their eventual solution. This, it is hoped, will stimulate further developments relevant to biology from scientists working in other disciplines.

Modelling

Geisow & Barrett (eds.) Computing in biological science

1

Interactive computer models of biological systems

W. J. PERKINS

1. INTRODUCTION

The commercial development of digital computers from 1949, provided methods for the rapid sorting and rearrangement of large quantities of data, and for complex calculations. In these respects they soon became widely used in the business world and in science and engineering. Those concerned with medicine and biology were much slower to react to the introduction of this new technology, clinicians generally being reluctant to put their faith in a complicated machine with which they were unfamiliar. A big problem was the lack of understanding, between computer experts and potential medical users, of each others subject. Searching, sorting, rearrangement and calculation were all features that could be used for computer diagnosis [1] but clinicians, not being familiar with how this would be done and what errors might be introduced, were reluctant to countenance decision making by the computer. Computer aided diagnosis was an obvious compromise. An earlier and more effective use of computers in medicine could have been achieved if, instead of relying upon experts in computing and medicine to communicate, the development of biomedical computing groups, with some competence in both subjects, had been encouraged. This approach had already been established successfully in bio-engineering, a closely related subject.

In biomedical research, administrators seemed to be aware of only one use – statistics – no doubt to try and glean more information from the relatively poor data normally obtainable. However, developments in other directions, such as biochemistry for the analysis of blood samples and X-ray diffraction pattern analysis, were able to be pursued, and are mentioned by Ledley [2].

A major advance came with the development of the small laboratory computer, particularly the PDP8, introduced in 1964, which was taken up by biomedical research workers all over the world, with physiologists being well to the fore in using them for real-time analysis of experiments to provide an indication of how the next stage might be directed. This paved the way for a more effective exploitation of computers generally as the limited memory of the mini meant that more detailed analysis of large quantities of data had to be carried out on the larger computer systems.

2. BASIC RESEARCH REQUIREMENTS

In basic biomedical research, data is usually poor and scarce. As a bio-engineer I concentrated first upon improving measurement procedures and then upon processing the data by electronic hardware techniques [3]. It was therefore a natural step into biomedical computing to take advantage of the more flexible software techniques. There was also another reason. One's understanding of the behaviour of a particular biological system is comprised of available data and ideas. A distinction between basic and applied research is that in the former it is the understanding from ideas which predominates. It can thus be more rewarding to assess ideas than to attempt to acquire further data [4]. A standard scientific procedure is to formulate an idea into an hypothesis and to test this by experiment. An alternative approach is to model the hypothesis in a computer and to test it theoretically with respect to the system being considered. If it is shown to be untenable, there is little point in establishing experiments to test it biologically. The main value of such models is in providing a guide to the design of experiments which might produce further information.

3. MODELLING AS A RESEARCH TECHNIQUE

Where it is difficult to study a system directly it is possible to consider another system, similar in certain respects, which may be more amenable to study. This similar system, be it actual or conceptual, becomes a model and it should be possible to relate results obtained from it to the system of interest. Modelling should not be just an academic exercise and should have a purpose. A model should either have some practical value in providing further information about how a particular system behaves, or act as a conceptual tool in providing a better understanding of a systems behaviour [5].

Biological systems differ from man made systems in that they have been created, making it necessary to establish their basic design as well as to understand their behaviour. This might suggest that models of biological systems are unlikely to have much practical value until they have first been used as a conceptual tool in research. Fortunately, it is not essential for a model to be exact in all respects for it to be used in a practical way. A model may be sensibly correct when confined to a specific situation and can be of value when it represents the understanding of the user for that situation.

Thus a number of biological models [6], as for example in lung function and renal activity, are used in clinical practice, where the behaviour of a model to data fed from a patient may be compared with established normals. A practical application demonstrating the value of simplified models is in education [7] to enable students to obtain an understanding of behaviour for extreme conditions that could be difficult to assess by experiment.

4. MODELLING METHODS AND TECHNIQUES

4.1. Physical models

Physical models are very useful in presenting a final situation, especially in three dimensions, but have very limited value in representing dynamic situations. My first excursion into modelling was to develop a system of resistance/capacitance networks to simulate protein metabolism which appeared to behave as a series of exponential and decay functions. This behaved reasonably well for parameter changes but any reorganisation of the model itself required a major modification. It also had the attendant disadvantage that network theory, though familiar to electrical engineers, was a totally new concept to most biologists and it is rather pointless for biologists to simulate a system with whose concepts they are familiar, by a model utilising concepts with which they have little knowledge. Better therefore to use the mathematical representation of the exponential functions, though unfortunately, mathematics is not the well-understood common language it should be.

4.2. Mathematical models

Mathematical models still demand an understanding of the relevant mathematics and can only be applied to systems that are amenable to mathematical description and where the quality of available data is adequate to provide some comparison of the model behaviour with that of the system. Models of biomedical systems may be formulated from a mathematical basis which considers the transfer function from input to output. It is usually possible to derive an expression, however complex, to fit a data curve but this may have no biomedical significance. The real value of this approach is in suggesting to users how a system might be behaving, as a guide to further studies. As it is the biomedical worker who seeks a better understanding of behaviour, a formulation from biomedical reasoning is more realistic. This can be done using compartmental analysis which represents the behaviour of a system by interconnected biological compartments having variable rate constants between them [8]. The equations representing the dynamic behaviour of the system can then be solved for the varying values of the rate constants. An inherent danger of mathematical models is that mathematical assumptions, necessary to simplify the equations, may also result in biological assumptions being made.

4.3. The analogue computer

The analogue computer was most effective for solving the differential equations describing the behaviour of biological systems. Having a rapid response, it was possible for a user to adjust the manual controls representing biomedical rate constants and to observe instantaneously their effect upon the outputs of the various biomedical compartments.

Numerous biological systems were studied in this way using an EAL, TR48 computer. These included protein metabolism, neural functions, erythrocyte hydrodynamics, and respiratory dynamics [9]. Though suitable for allowing operator interaction with a model, the analogue computer was inferior to the digital in most other respects, such as accuracy, comparison, and function generation.

4.4. A hybrid system

A hybrid system would seem to contain the advantages of both. Thus when a Honeywell DDP516, 8 Kwords, digital computer was obtained, the two machines were interfaced to each other [10]. This system was used to solve the differential equations representing the metabolism of γ-aminobutyric acid (GABA) [11] (Eq. 1) which is found exclusively in brain tissue and affects the transfer of electrical pulses between neurons.

$$s_1 = \int k\mathrm{d}t - c\int s_1\mathrm{d}t - s_2 \qquad (1)$$

where s_1 is the amount of GABA in the cells
s_2 is the amount of GABA in the medium
k and c are unknown constants

The values for s_2 were obtained from a function generator which produced an interpolated curve through the available s_2 data points. Each combination of specified constants gave solutions for s_1 which were compared with the experimental data for s_1 to give the sum of the errors squared ($\Sigma\varepsilon^2$). Here, the digital computer took over the cycling of parameters, multiplication, generation of the functions obtained from experimental data (s_1 and s_2), comparison and summation of mean error square, whilst the analogue computer was used for addition and integration. The hybrid system, though effective for a dedicated task, such as monitoring the effect of parameter changes in a known system, was not suitable in a research environment where problems were continually changing. The analogue computer though was adequate at the early stage of evaluating a model which could then be transferred to the digital computer for more detailed analysis.

5. USER INTERACTION

In basic biomedical research, problems are rarely clearly defined at the outset but are iteratively refined as each proposed solution is tested. The large remote computer, with the capability of providing solutions to the problems presented, does not satisfy this requirement. I thus became interested in interactive computing to allow researchers to use computers as an extension of their own thought processes [12]; to combine their own intuitive and subjective reasoning with the high-speed calculations and display facilities of the computer; to assess ideas as well as to analyse the data. Suitable input devices and output displays were interfaced to the DDP516 and a philosophy of programming adopted to allow easy intervention with the computation. This was achieved by incorporating pre-determined options that could be selected by simple keyboard instructions and by providing control devices to change parameters. [13]

5.1. *Computer graphics*

Detailed analysis and print out of results was always possible but for a rapid understanding of the effect of any change in the computation, and thus in the models' behaviour, a graphic display was essential. To determine which sets of neurons in the visual cortex respond to a given angle of visual stimuli, their activity is recorded as an illuminated line is rotated. By doing the analysis and display online with the experiment, the display can be seen building up then, if no directional pattern is observed, the electrode positions can be changed.

For conventional two-dimensional graphics, it is a simple procedure to quickly change the axes, their labels and the scale. Smoothing and curve fitting procedures can be brought in as required and the user may test numerous graphs before deciding which to select for hard copy. Three-dimensional displays can be obtained by adding perspective. This does distort the image in terms of measurement but the amount is known and can be dealt with in the calculations. Rotation about the three orthogonal axes X,Y,Z, either manually or continuously, gives a clear indication of the three-dimensional nature of a displayed structure. For fixed displays, a stereoscopic pair can indicate depth when seen through a suitable viewer, or red/green anaglyphs provide an alternative stereo display when viewed through appropriate colour filters. Colour displays provide another dimension in pattern discrimination.

6. STRUCTURAL MODELS

The interactive system, developed so far, was thus capable of displaying models of biological structures built up from available data. These models, which I term descriptive models, may be used for visual qualitative assessment to obtain a better appreciation of a structure, and for more detailed analysis.

6.1. *Molecular models*

The first models of this type to be investigated were of molecular structures. Data from X-ray crystallography provides information on the number and position of atoms of a molecule in the crystal state, their connectivities and the bond lengths. These can be plotted as a stick diagram (Fig. 1) which can be rotated about the

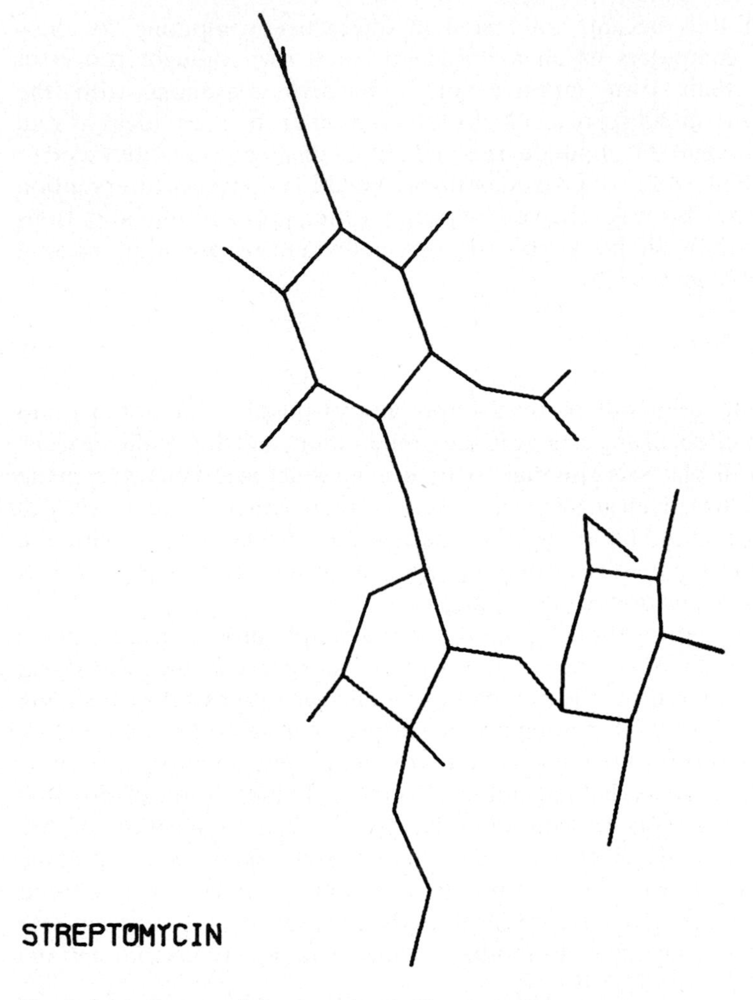

Fig. 1. Stick diagram of simple molecule (Streptomycin).

orthogonal axes and then produced as a stereo display [14]. Perspective can be added to both bonds and atoms to produce a more realistic display [15] (Fig. 2). The normal state of a molecule in solution, though having the same atoms and bonds, can take up different conformations depending upon the degrees of freedom provided by the rotation of certain bonds about their own length axes [16]. The normal conformation can be determined by calculating the minimum energy condition for the molecule as the free bonds are rotated sequentially [17]. For even small molecules with some degrees of freedom, there will be numerous changes of energy for each finite change. The energy distribution for changes in position of the atoms due to bond rotation are then plotted as in Fig. 3 which shows the distribution for a base component. Interaction with the display may help to reduce the search time as information from other sources, such as circular dichroism,

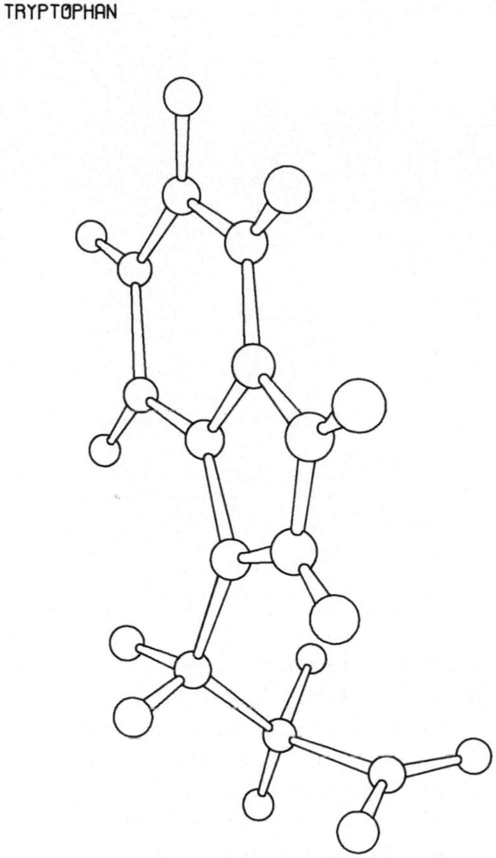

TRYPTØPHAN

Fig. 2. Perspective plot of TRYPTOPHAN.

spectroscopy, NMR, can be incorporated with the experience of the operator, to vary the intensity of the search in particular areas. The points of bond rotation may also be identified and a separate control used to rotate a given section of the molecule about the bond. A structure in the crystal and normal states, is shown in Fig. 4.

6.2. Serial sections

The growth and development of biological systems may be studied by cutting a number of thin sections sequentially for examination under a light or electron microscope. For a more detailed analysis, relevant features of each section image can be entered interactively into a computer via a drawing tablet [18]. Images may be entered directly from a light microscope by using a camera lucida attachment which allows the operator to see the image superimposed upon the tablet. Negatives can be projected on to the tablet from below or photographs placed on the tablet. A program, PROF, for entering features of the image, allows the operator to edit sections as they are entered. To allow flexibility in any future analysis or display, the

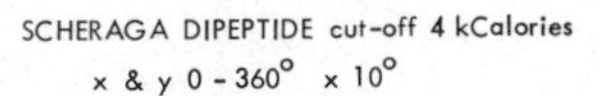

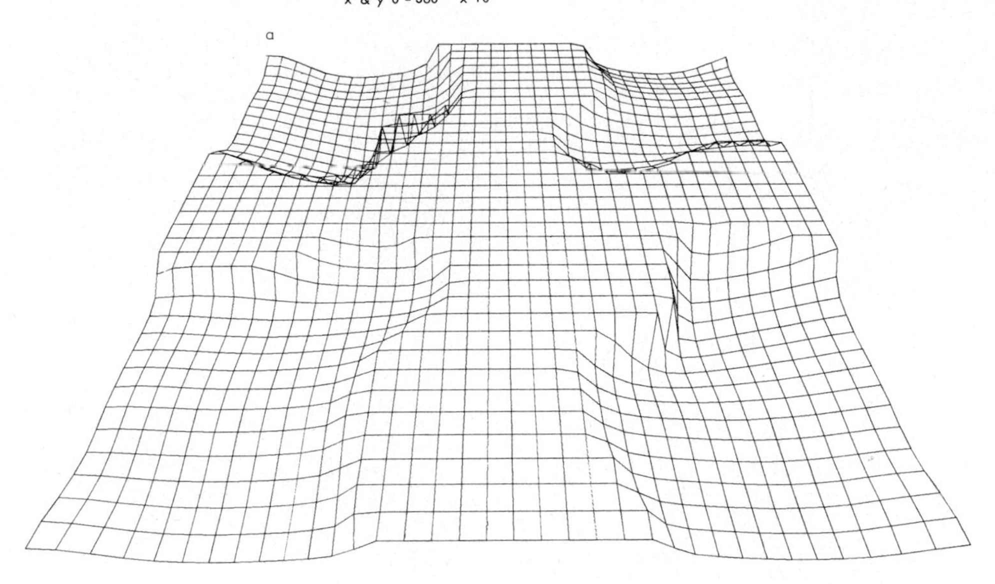

operator categorises the sub-sections defining the internal structures, as they are entered. Options, provided as a menu area on the tablet, guide the operator in entering sections and sub-sections from the tablet (Table 1). These also allow the sections to be edited. On EXIT, a file is produced for the total number of sections entered. The sections can be recombined and displayed by calling the file into a program, RECON, which contains numerous display options (Table 2). First the complete structure is displayed in the three orthogonal axes but can be stopped at any time by touching the space bar. A central image can be selected and displayed, then rotated about any selected orthogonal axis. Scale factor and section separation may be changed and different categories given different hatch patterns, or colour if a colour monitor is available. Although complex structures can be completely displayed they become too detailed for visual comprehension. The categorisation procedure allows the operator to remove and replace sections or subsections by the simple keyboard instructions. This produces displays of selected features which can be separately filed and called into an appropriate analysis program (ANSIS); for example, to calculate the rate of change of volume of a sub-structure between defined sections of the structure. Stereo displays can be obtained on the screen or as hard copy, for viewing by mirrors or as two-colour anaglyphs. If any of the sections are misaligned, they can be corrected in an ALIGN program, and individual sections can be deleted from or inserted into the structure by means of an EDIT program.

This system has been used for neurobiological sections to study developments of the visual system in tadpoles and frogs (Fig. 5), also neurobiological sections to consider growth patterns of neuronal structures, and for the study of nemotode sections. Sections obtained by X-ray, ultrasonic or NMR tomography can be studied in the same way [19].

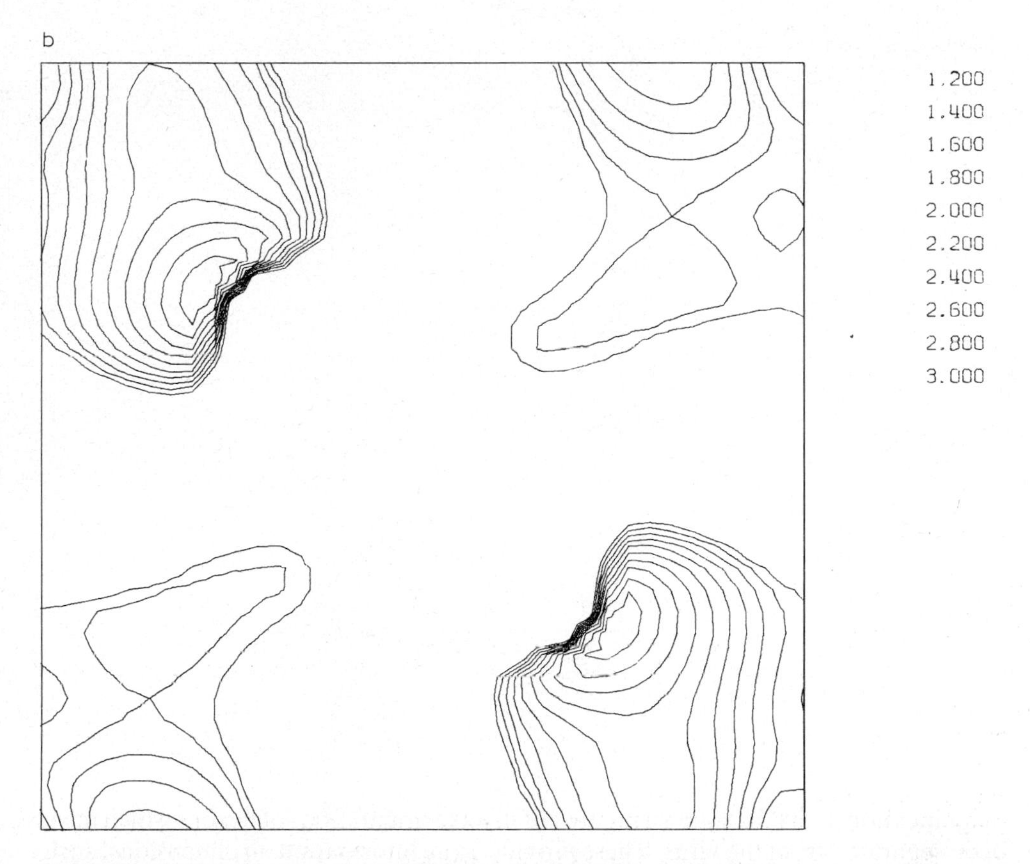

Fig. 3. Energy plot of base component (Scheraga): (a) Graphic; (b) Contour

This system was initially developed on the DDP516, with 16 Kwords of memory plus disc, and is now implemented on a PDP 11/34, with 64 Kwords plus disc. A schematic arrangement of the reconstruction system is shown in Fig. 6.

6.3. *Deductive models*

Where an understanding of a system is limited by a lack of data, a model can be produced from ideas about the system's behaviour, and used in a deductive manner by modification of the model to fit the known facts given by the data. The shape of the structure of a bacterium or virus can provide an indication of structure. The adenovirus was known to be an icosahedron and so could easily be modelled mathematically in a computer and displayed as a three-dimensional structure. Attached to its outer surface are protein structures called hexons, whose three dimensional structure was not known. The electron micrograph of Fig. 7, with a

a

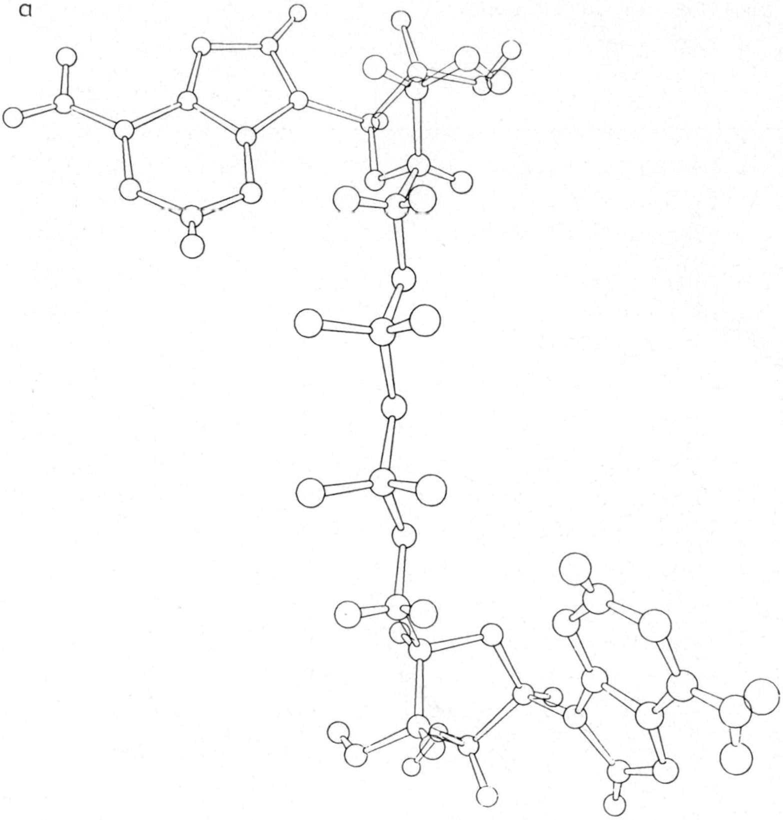

magnification of 500 K, shows a number of diverse orientations of hexons which have been separated from the virus. These provide some information on dimensions, such as a maximum observed length of about 11 nm, and widths of 7.5 nm and 8.5 nm at each end. The electron micrographs (EMs) and other sources, indicated a triple polypeptide structure; EMS of the adenovirus suggested that the hexon was attached to the virus with its length axis perpendicular and the wider section at the bottom. The hexons are positioned on to the icosahedral surfaces of the virus, in groups of nine (GONS). EMs of GONS isolated from the virus, showing normal and reversed orientations, produced an image in the shape of a Y for one end (Fig. 8.1), and an image in the shape of an O for the other (Fig. 8.2). Measurements indicated that the O shaped end was smaller than the Y shaped end which would suggest that the O shape should be at the narrower top of the hexon. That was the extent of the data.

As analytic methods of reconstruction were not feasible, due to radiation damage to a specimen for the larger number of images required, I decided to build a computer model that could be manipulated for comparison of its projections with the diverse EM images of the hexons [20]. This was built up in sections from spheres as these, when filled with random dots, could indicate the different densities of the model as it was rotated. The model was then successively modified to try and match

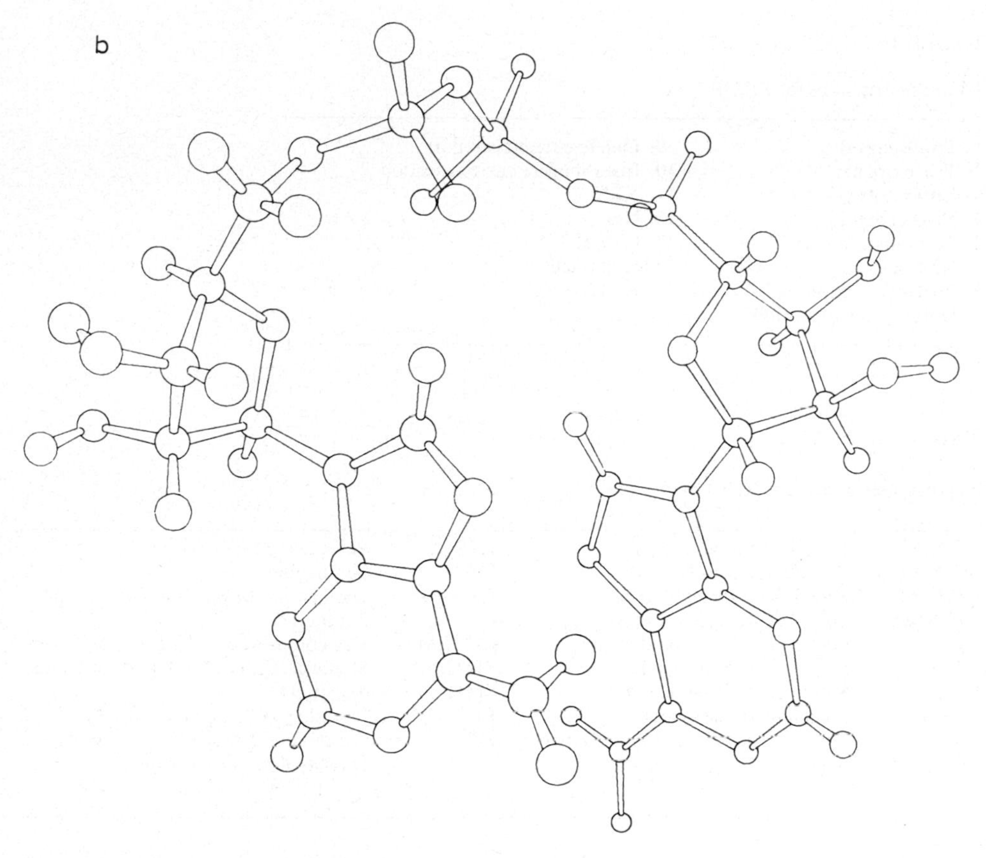

Fig. 4. Molecular structure APPA: (a) Crystal; (b) Normal state.

the diverse EM images observed (Fig. 9). These could be roughly categorised as in Fig. 10. In order to obtain the X patterns, I found it necessary to associate the O shape with the larger base, which made the O shaped end of the model large than the Y shape, contrary to the previous assumption obtained from measurements on the GONS.

Although at this stage the model images were approaching those of the EM images of the hexon, these could not match exactly as, in order to observe the protein structures in the EM, the specimens had to be stained and this could distort the resultant images. It was therefore necessary to simulate the effect of stain upon the different orientations of the model and this was another unknown factor.

Only two of the EM patterns could be identified with a known orientation, the normal and reversed positions of the GONS (O and Y), so these two positions were chosen to assess the effect of stain. At this stage, with two unknown variables, I was moving further away from any match so I decided that the O shape had to be smaller than the Y, as measured from the EMs. In order to satisfy the other criteria of obtaining the X patterns, it also had to be associated with the larger base – two incompatibles. However, the introduction of stain into the model, which had

TABLE 1

Tablet instructions for PROF

1. Trace closed	9. Delete current category
2. Trace open	10. Insert before current section
3. Enter dots	11. YES
4. Next category	12. NO
5. Previous category	13. Cancel
6. Next section	14. End trace
7. Previous section	15. EXIT
8. Delete current section	

TABLE 2

Typical instructions for RECON

SE N,M	Enable sections N to M	ZI S	Set spacing to S microns
SD N,M	Disable sections N to M	DF N	Set display factor to N
CE N,M	Enable categories N to M	CS N,M	Use symbol M with category N (for
CD NM,	Disable categories N to M		dot mode)
3V D	Orthogonal views on device D	CC N,M	Use colour M with category N
DI D	Single view on device D	SH C,P,D	Shade category C with pattern P and
ST D	Stereo view on device D	SH C	density D
RG D	Anaglyph on device D	FI	Disable shading on category C
RX N	Rotate about *X* axis (*Y* or *Z*) by N	HE	Output current display to new file
	degrees		Display these instructions

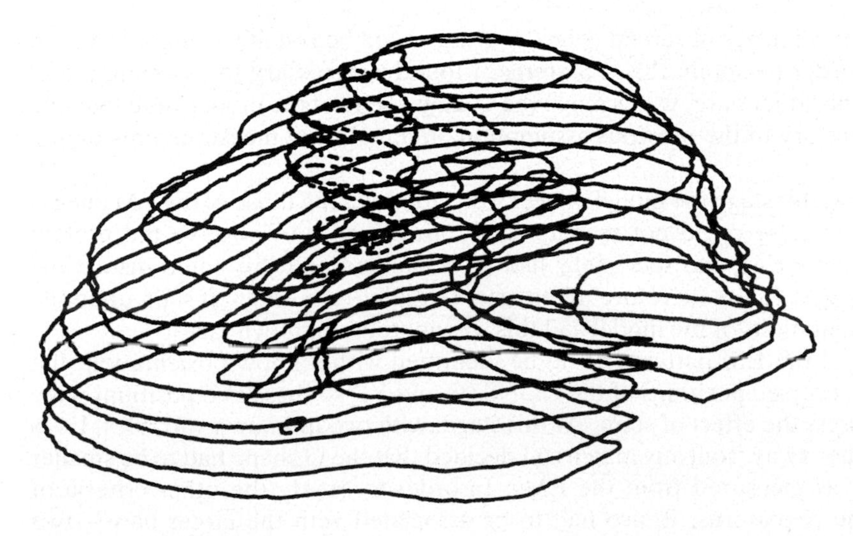

Fig. 5. Frog brain sections indicating surgical intervention (dotted).

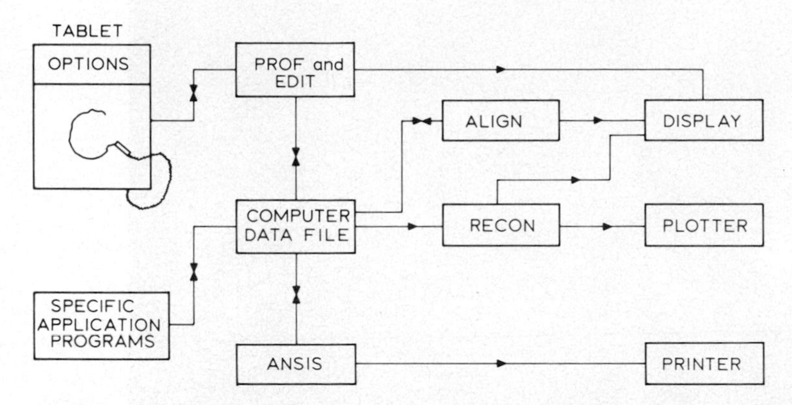

Fig. 6. RECON system.

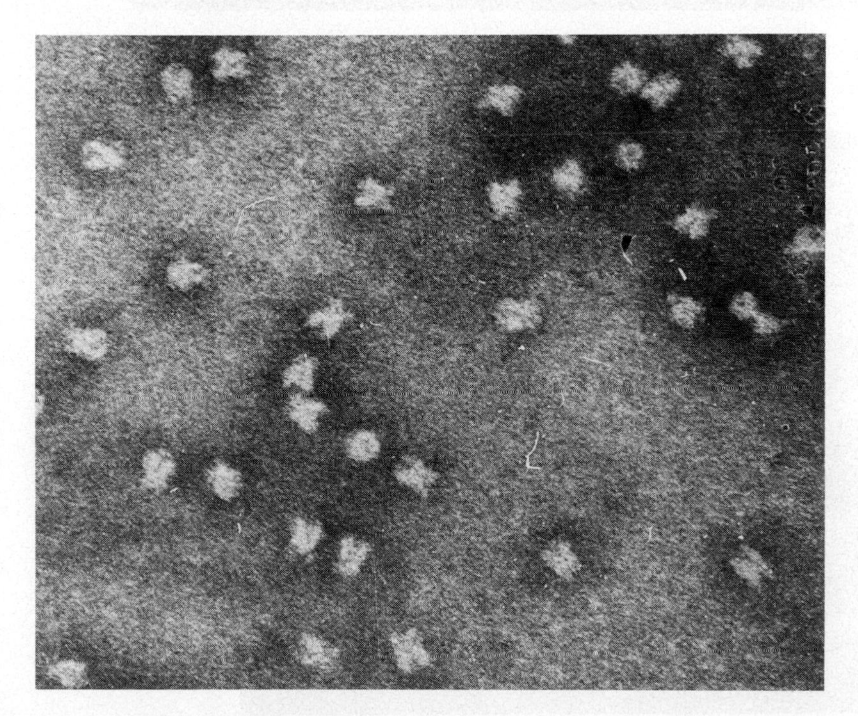

Fig. 7. Diverse orientations of adenovirus hexons.

complicated the problem, now provided the solution. These two features could be reconciled by reducing the base section at the very bottom and only allowing the stain for the normal upright position to reach a low level of about 1.4 nm (Fig. 11). EM tilting experiments on specimens containing hexons in the upright position, confirmed this low level for the stain.

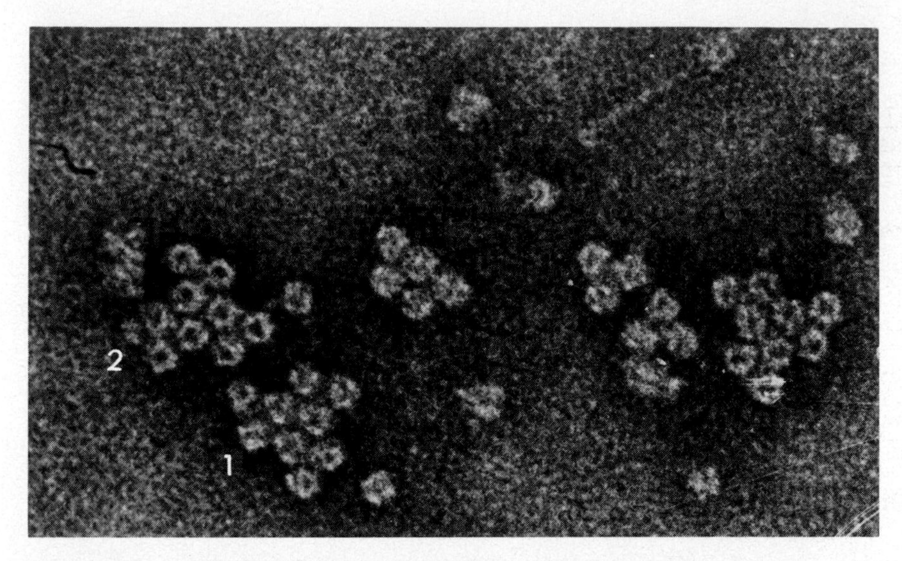

Fig. 8. Groups of nine hexons (GONS): (1) Y shape; (2) O shape.

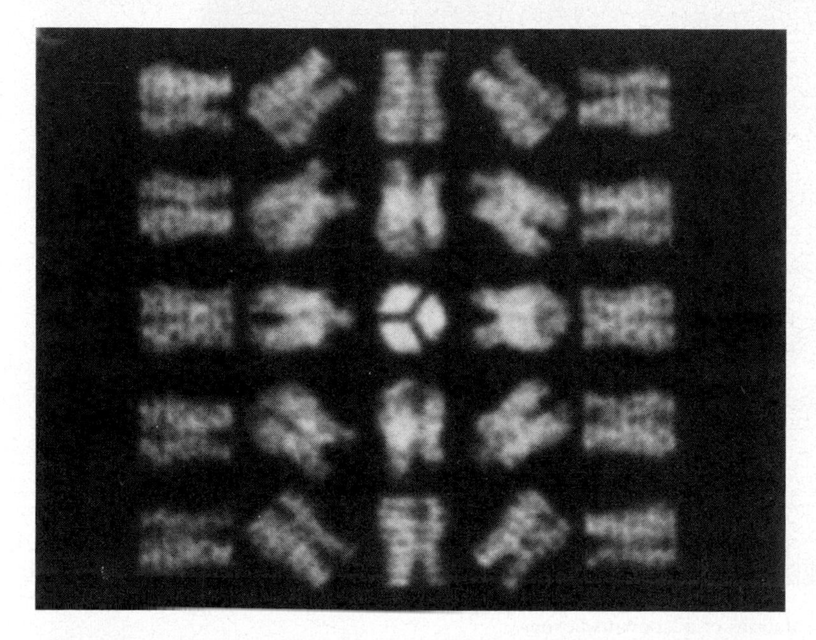

Fig. 9. Successive orientations of model.

Another possibility to be tested was whether the hexon might be twisted. No difference could be ascertained from their respective rotational patterns. However, gaps left in the bottom section of the model to obtain the X shapes, also let through stain to superimpose a Y pattern on the O. This could be prevented at the lower level

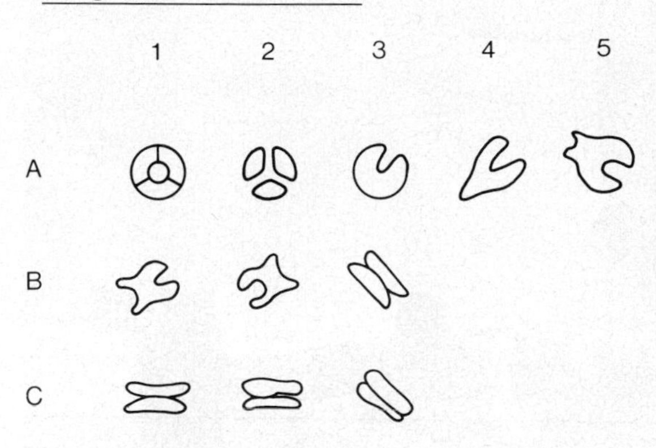

Fig. 10. Categories of main patterns.

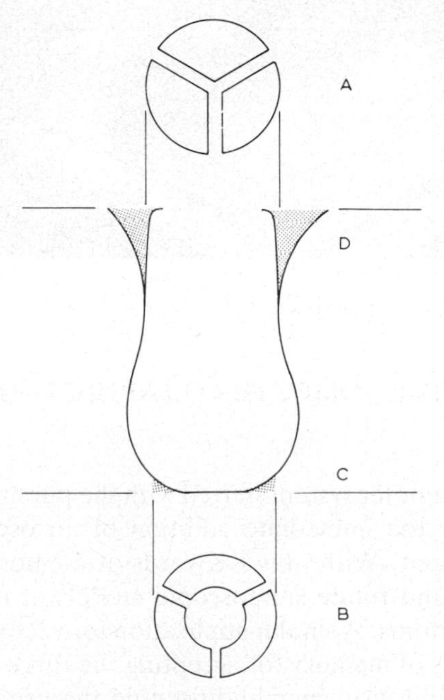

Fig. 11. Staining of model in upright position.

of stain by twisting the model, which also produced a required narrowing of the Y pattern for the other end. Techniques of interactive computing thus proved effective in allowing ideas to be incorporated into a three-dimensional model of the adeno virus hexon, and for testing this model against the available data [21].

Fig. 12. Interactive computer system.

7. DESCRIPTION OF INTERACTIVE COMPUTER GRAPHICS AND DISPLAY SYSTEM (Fig. 12)

The development of the interactive computer system started with the purchase of the Honeywell DDP516 in 1968, and by the immediate addition of an oscilloscope display and A/D conversion at the input. With only 8 Kwords of memory and no disc, the system was able to display and rotate stereoscopic models of molecular structures containing up to a hundred atoms. A similar application for vectorcardiography used a PDP8 with only 4 Kwords of memory for accepting the three weighted values of the analogue ECG signals and displaying and rotating the vectorcardiogram [22]. The addition of a disc to the DDP516 helped considerably in being able to file and call in different structures. In building up virus models, it was also convenient to store sub-structures, used in building the main structure. An additional 8 Kwords of memory allowed more complex structures to be dealt with. Once this was added we were less restricted in our program development which had become a limitation that had been overcome by overlaying programs. The philosophy adopted for developing interactive programs was to prepare the main program in FORTRAN and for this to call in various sub-routines for control, calculation and display. The systems type of sub-routines, such as PLOT, are written in ASSEMBLER and those sub-routines written specifically for the application are normally prepared in FORTRAN. Users are presented with a set of operating instructions for a particular program but there is also a HELP instruction which displays all the available options on the VDU. These are in the form of two letter instructions followed by the control parameters, e.g., RX θ, which can be entered from the keyboard. The system now uses a PDP 11/34 with 64 Kwords of 16 bits and two discs. A floppy disc has also been added. Rotations of the display are also achieved by the addition of potentiometer controls for X,Y,Z and separation. An alternative joystick control is also provided.

Complete images are entered by a Joyce–Loebl scanner modified to operate under computer control. A separate image processing colour display unit uses a 6800 microprocessor to control images which can be entered from a TV camera or transferred from a host computer. In this unit, features of an image can be selected by the operator and controlled by means of a light pen. The other method for feature selection is to use a graphic tablet. Initially we developed our own tablet which used a conductive transparent glass plate under which the image was placed, a soft pencil probe being used to transfer the X,Y coordinates to the computer. The present commercial tablet, which uses a magnetic probe, is incorporated into a desk unit and allows an image to be traced directly from a photograph, or projected on to the tablet, from either 35 mm negatives or slides.

A large screen oscilloscope (20″ × 15″) was used initially to produce a linear refresh display of rotating images. This could display molecular structures containing up to a hundred atoms with acceptable flicker but beyond that a storage display was necessary. For this a 611 storage oscilloscope was used, rotations being effected in steps by deleting then redisplaying the new image. A charge coupled memory was later added to the large screen oscilloscope to provide alternative refresh or store modes. Any display could be photographed or transferred to a Calcomp plotter for hard copy. A colour display has also been added which operates in a scanning mode, as distinct from the X–Y mode, of the oscilloscope displays. In terms of accuracy, this produces a 512 × 512 bit display, interlaced on a 9″ × 8″ screen, compared with the 256 × 256 bit display on the oscilloscope and 4096 × 4096 bits on the 9″ × 8″ storage oscilloscope.

8. COMMENTS

The dedicated interactive computer system has proved invaluable in basic biomedical research, being capable of operations that could not be undertaken with a larger time-shared computer system. This has been used as a general development system for devising new interactive techiques, producing new methods of display and looking at biomedical research problems amenable to this type of computing. A major difficulty has been that as a particular procedure demonstrates its value, biomedical staff wish to use it in a routine manner. This of course was an objective, to demonstrate the value of interactive computing in biomedical research but, being a dedicated system, routine use inhibits further development, both of the system and for other applications. One possibility is to transfer the separate features of data entry, display and calculation, to other computers. Data acquisition from the tablet is the most time consuming activity so this feature has been implemented on a MINC microcomputer for use in the individual laboratories. It could also be implemented on a DEC 1123 or an LSI11, all of which are compatible with the 11/34. Data files can then be transferred to the 11/34 for display or to the central time shared DEC 2060 computer for analysis. Alternatively, the local system may be expanded to deal with all features. The separate interactive display system, based at present on a 6800 microprocessor, can be linked to the DEC 2060 for use as an interactive display terminal in the user area.

The next step being considered is to minimise the tedium, for the user, of data selection and entry of features. Automatic pattern recognition of biological images poses difficult problems but these may be simplified by introducing interactive pattern recognition techniques. The continuing advances being made in microcomputers and displays now brings the techniques described here, within the budgets of individual laboratories and should have a major impact on the biological research.

ACKNOWLEDGEMENTS

I would like to acknowledge the following persons for providing the original pictures: Dr. J. Thornton (Molecular Structure APPA); Dr. S. Udin (Frog brain sections); Dr. N. Wrigley (Orientations of adenovirus hexons).

REFERENCES

1. Ledley, R.S. and Lusted, L.B. (1959) Science 130, 9–21.
2. Ledley, R.S. (1965) Use of Computers in Biology and Medicine. McGraw-Hill
3. Perkins, W.J. (1962) In: K. Enslein (Ed.) Data Acquisition and Processing. Pergamon, pp. 117–122.
4. Perkins, W.J. (1982) In: J. Paul and M. Jordan and M. Ferguson-Pell (Eds.), Computing in Medicine. MacMillan.
5. Perkins, W.J. (1977) Jap. J. Med. Elec. 15, 1–5.
6. Hammond, B.J. (1967) Proc. Inst. Elec. Rad. Eng. 5, 109–115.
7. Dickinson, C.J. (1977) Computer Model of Human Respiration. MTP Press.
8. Berman, M. (1965) In: R.W. Stacy and B.D. Waxman (Eds.) Computers in Biomedical Research, Vol. 11 Academic Press p. 173.
9. Perkins, W.J. and Hammond, B.J. (1968) Bio-Med Eng. 2, 113–118.
10. Perkins, W.J. (1969) Computer Science and Technology, IEE No. 155, 9–12.
11. Machimiya, Y., Balacz, R. and Hammond, B.J. (1970) Biochem. 116, 419–481.
12. Perkins, W.J. and Hammond, B.J. (1975) Nature 171, 256.
13. Perkins, W.J. (1975) Modular Programs to Facilitate Computer Interaction. XII, Rassegna Internazionale Elettronica, pp. 136–141.
14. Perkins, W.J., Piper, E.A., Tattam, F.G. and White, J.G. (1971) Comp. Biomed. Res. 4, 1–3.
15. Perkins, W.J., Piper, E.A. and Polihroniadis, P. (1973) Comp. Biomed. Res. 6, 509–521.
16. Perkins, W.J., Piper, E.A. and Thornton, J. (1976) Comp. Biol. Med. 6, 23–31.
17. Nelder, J.A. and Mead, R. (1965) Comput. J. 7, 308.

18. Green, R.J., Perkins, W.J., Piper, E.A. and Stenning, B.F. (1979) J. Biomed. Eng. 1, 240–246.
19. Perkins, W.J. and Green, R.J. (1982) J. Biomed. Eng. 4, 37–43.
20. Perkins, W.J., Polihroniadis, P., Piper, E.A. and Smart, P. (1976) J. Biomed. Eng. 14, 274–281.
21. Nermut, M.V. and Perkins, W.J. (1979) Micron 10, 247–266.
22. Kalff, V., Perkins, W.J. and Dewhurst, D. (1974) Comp. Biol. Med. 4, 137–144.

Geisow & Barrett (eds.) Computing in biological science

2

Computational embryology of the vertebrate nervous system

RICHARD GORDON

Dedicated to the memory of Wallace Kirkland, photographer for *Life* magazine and author of *The Lure of the Pond* [1].

1. INTRODUCTION

The central nervous system (brain and spinal cord) is the first tissue to differentiate in the vertebrate embryo. Computers are essential to the analysis of every stage of its development. This chapter will demonstrate the central role of computers in designing models for early nervous system morphogenesis. Computers are also important for evaluating these models by comparing their predictions with observed morphology and cell trajectories. They are furthermore useful for collecting the basic observational data on cell trajectories and cell behavior.

A distinction has recently been made between experimental, theoretical, and *computational* physicists [2]. The same distinction can be made in biology. Although they are yet few in number, some scientists could be called *computational biologists*. Of these, a small number are *computational embryologists* who try to extract the essentials of cellular behavior in an embryo, and then *prove*, by computer simulation, that this behavior is sufficient to explain the observed morphogenesis.

In an embryo the forces generated by each cell may be transmitted through the whole embryo. Thus each cell can contribute to the motion of the whole embryo. The result is a complex relationship between morphogenesis of the embryo and the behavior of its component cells. This relationship is too complex to work out by existing mathematical methods. It is for this reason that we turn to computer simulation. Computations for simulating real, three-dimensional embryos are therefore often as complex as those needed to predict the weather or the motion of individual molecules in a gas or liquid.

When a computer simulation accurately portrays our understanding of cell behavior, and yet does not generate patterns or shapes corresponding in detail to the embryo, then we may say with some certainty that our understanding is insufficient. Erroneous simulated shapes or shape transformations may give us clues about the sort of information we lack. We are sent back to our data, our time-lapse movies, or our measurements to see what we had not seen before, or to the embryo to carry out new experiments. Thus, because of the precision and attention to detail computer simulation requires, it becomes an incisive tool for testing and shaping our ideas on morphogenesis.

There may be alternative models that are plausible explanations of the observed morphogenesis. The simplest models, usually being 'first approximations', may leave out something essential or central to the real biology. Thus the best approach is to use the computer to test all plausible models, subjecting each to the same criterion that it quantitatively predict the correct shape and the course of development over time. Often even just the attempt to make them precise enough for computer simulation clearly shows just how much many models are mere handwaving. But the important thing to keep in mind is that *I am not speaking of computer modeling, but rather the use of computers to test and refine models*. When we are done, we are left with a quantitative model and a set of concepts which we have found adequate to explain the phenomenon at hand. At this point the computer simulation itself may be safely discarded. It is not a crutch substituting for understanding.

I have reviewed the many models for early development of the nervous system elsewhere [3]. That there are so many seemingly plausible models of neurulation may be due to the lack of observation of cell movement, cell division, neighbor changes, cell death, etc. There is such a wealth of data in the morphological transformations of tissues (if they are followed at the cellular level), that one is, on the other hand, both hard put to come up with an accurate explanation, or having one, to suggest plausible alternatives. Computer-aided time-lapse work, coupled with careful simulations, may be the key to unraveling the morphogenesis of each organism. For this reason I have included a summary of potential developments in computer aided microscopy that should make these observations possible and practical.

Everything described in Sections 4.2.2 through 4.2.5 has been part of a quantitatively successful simulation of the shaping of the neural plate [4,5]. The rest of this chapter concerns work in progress.

2. OUTLINE OF THE EARLY DEVELOPMENT OF THE NERVOUS SYSTEM

In the first stages of its development the successive divisions of the fertilized vertebrate egg give rise to a hollow, single-layered ball of cells called the *blastula* (Fig. 1f). (I will always be referring to the California newt, *Taricha torosa*, unless otherwise specified.) Half of the tissue then tucks itself through a slit in the blastula, called the *blastopore*, forming two new layers inside the spherical embryo: the

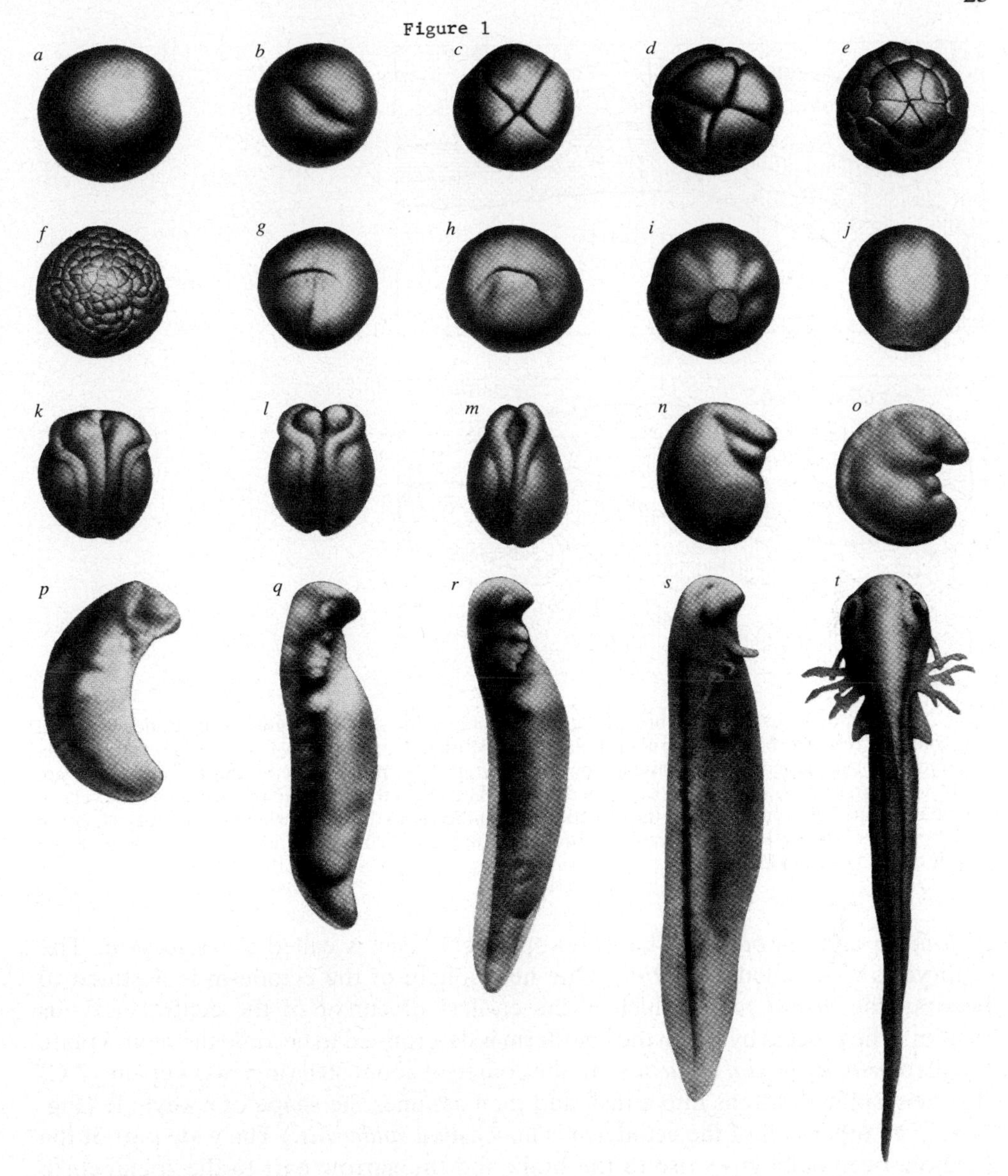

Fig. 1. Developmental stages of the California newt *Taricha torosa* abbreviated from Twitty and Bodenstein in Rugh [19]. First the fertilized egg divides many times over (a–f, stages 1,2,3,4,6,8), giving rise to a hollow sphere of cells, or blastula. Then a groove (the blastopore) appears (g, stage 10) and gradually deepens to form an internal cavity as more and more surface tissue moves into the interior (h,i, stages 11,13). This process, termed gastrulation, gives rise to two internal tissue layers: the endoderm and the mesoderm. The mesoderm later forms the main body musculature and the (usually transient) notochord. One hemisphere of the outer layer of the embryo then flattens into a disk (j, stage 14) and forms itself into the keyhole-shaped neural plate (k, stage 15). At the same time the notoplate and the underlying notochord undergo a considerable elongation and narrowing along the midline (see Fig. 3). In the succeeding stages the neural plate rolls up to form the neural tube (l–n, stages 17,19,22), the eyes develop and the embryo elongates into the larval form (o–t, stages 24,27,33,36,37,40). Figures g–i are shown in posterior view, j–m and t in dorsal view, and n–s are seen in lateral view. From Gordon and Jacobson [5].

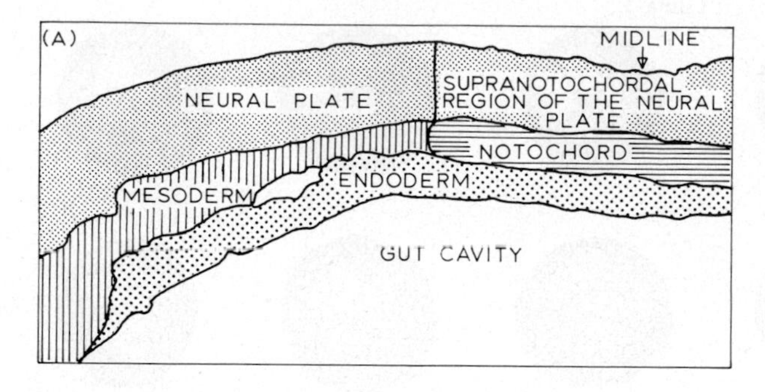

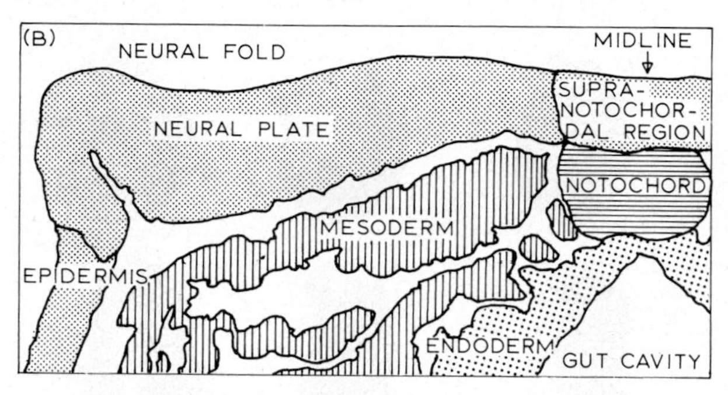

Fig. 2. Schematic cross-sections through the newt embryo at stages 13 and 15 of its development illustrate the relations between the neural plate and the underlying tissue layers. (A) The tissues of an embryo that has just completed gastrulation; the neural plate, the mesoderm and the endoderm are visible. (B) An embryo with a keyhole-shaped neural plate. Note that between the two stages the cells of the neural plate have lengthened in the direction perpendicular to the plate, whereas those of notochord and notoplate (here labeled 'supranotochordal region') have converged along the midline. From Gordon and Jacobson [5].

endoderm and *mesoderm*. The outer spherical layer is called the *ectoderm*. The embryo is now called a *gastrula*. One hemisphere of the ectoderm is destined to become the *neural plate*, which is the earliest precursor of the central nervous system. The process by which the ectoderm is determined to become the neural plate is called *primary neural induction*. In the course of about 30 h (in newts kept at 17°C) this hemisphere flattens into a disk and then assumes the shape of a keyhole (Fig. 1k). (The other half of the ectoderm is now called *epidermis*.) The wide part of the keyhole eventually gives rise to the brain and the narrow part to the spinal cord. (They are thus called *presumptive* tissues.) Once the flat neural plate is formed it rolls up into the *neural tube* (Fig. 1 l,m and Fig. 2). (In human embryos, what I have called the epidermis is frequently referred to as the *non-neural surface ectoderm*. Neural plate is called *neuroectoderm* [6]. Other names for the neural plate are *medullary plate*, *neuroepithelium* and *neural primordium*.)

Neurulation may be divided into three steps: (1) primary induction of the neural plate, which sets the presumptive neural plate cells apart from the rest of the ectoderm; (2) shaping of the neural plate from a hemisphere into a keyhole shape; followed by (3) rolling up (or folding) of the neural plate to form the neural tube.

(Neural tube formation occurs by a quite different mechanism in fish [7].) These phases and steps in reality form a continuum with events starting with oogenesis (Fig. 1). The divisions of time are really ours, not the embryo's. Cf. Ballard (1976) [8].

The neural plate is a sheet of columnar cells, one cell thick in newts and mammals. (The *apical* end of the cells is external at the neural plate stage. The other end of the cells is called the *basal* end. The apical ends of the cells become the inner lining of the neural tube.) This monolayer structure makes newts (urodeles) more amenable to analysis than, say, frogs (anurans), which are covered by an extra *superficial cell layer* [9,10]. The newt embryo is hardy and tolerant of experimental surgery. The embryonic newt cells are large enough for examination with a low-magnification dissecting microscope. The individual cells contain varying amounts of dark pigment and are readily identified and followed during development without the need for stains or other markers. Moreover, each embryonic cell contains its own supply of stored food (yolk), so that it is possible to culture groups of embryonic cells or even single cells in simple salt solutions. Because the yolk is inside each cell, the embryo does not increase in size until it can eat. This contrasts with chick and human embryos, for instance, which obtain nutrients from outside the embryonic cells and are actually growing in volume during stages that roughly correspond to those in early newt development. (Culture of mammalian embryos may soon make them amenable to similar investigations.)

In the course of our early attempts at simulation we found that part of the mesoderm, called the *notochord*, seemed to play an essential role in neural plate shaping [4,5]. The keyhole shape was not obtained unless the notochord remains attached to the neural plate. The notochord is a broad area under the neural plate, which changes shape to a long, narrow structure (Fig. 3a). The overlying part of the neural plate follows exactly the same change of shape. In order to distinguish it from the rest of the neural plate, I call it the *notoplate*. (The rest of the neural plate may be called *mesoplate*.) Like the notochord (in most chordates), the notoplate is thus a transient structure. It is intimately associated with the notochord until neural tube closure, and afterwards, at least, supports its own elongation. It eventually becomes so narrow that its cells string out amongst the rest of the neural plate cells (A. G. Jacobson, personal communication). Malacinski and Yoon [11] have recently observed that embryos lacking notochords undergo normal neurulation. This suggests that the notoplate is responsible for its own elongation and for the keyhole shape of the neural plate.

We discovered a temporary ten-fold spurt in the rate of elongation of the neural plate just at the time of neural tube closure (Fig. 4). This may be due to the *sheath* which forms around the notochord. A similar phenomenon has been found in chick neurulation [12]. It has not been looked for in embryos lacking a notochord.

The neural plate of the newt increases from 6,600 to 10,000 cells from neural plate formation to neural tube closure. If we assume that the two daughter cells of a divided cell have a total volume equal to the volume of the cell from which they came, then there will be little or no effect of this cell division on the rest of the neural plate. Thus, to a first approximation, we may ignore cell division in our

Figure 3

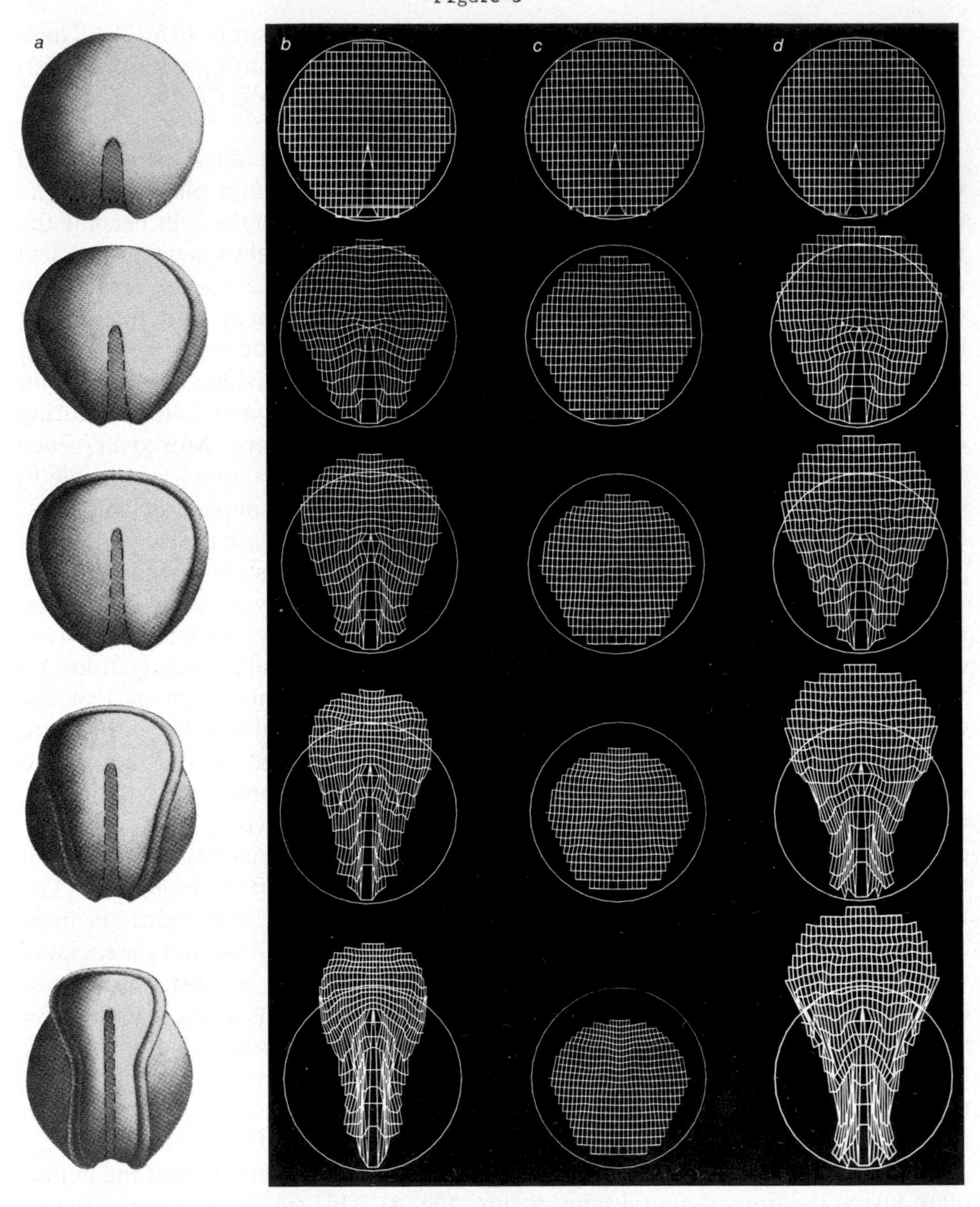

Fig. 3. Computer simulations of the formation of the neural plate. The shaping of the tissue is represented as the distortion of a grid placed over the embryo. The grid is obtained from the underlying representation as shrinkage units (Fig. 19) by keeping a record of the initial left and right nearest neighbors of each unit. Lines are drawn connecting the centre of a unit to the centres of its initial left and right neighbors and to the midpoints between the pairs of units which were initially above or below it. Since the shrinkage units are initially hexagonally close-packed, the grid starts out with rectangles whose height to width radio is $\sqrt{3}$. Column *a* shows a sequence of schematic diagrams of the developing neural plate from stages 13 to 15, derived from a time-lapse motion picture. The dashed line encloses the notoplate and its underlying notochord. Column *b* shows the results of the computer simulation with

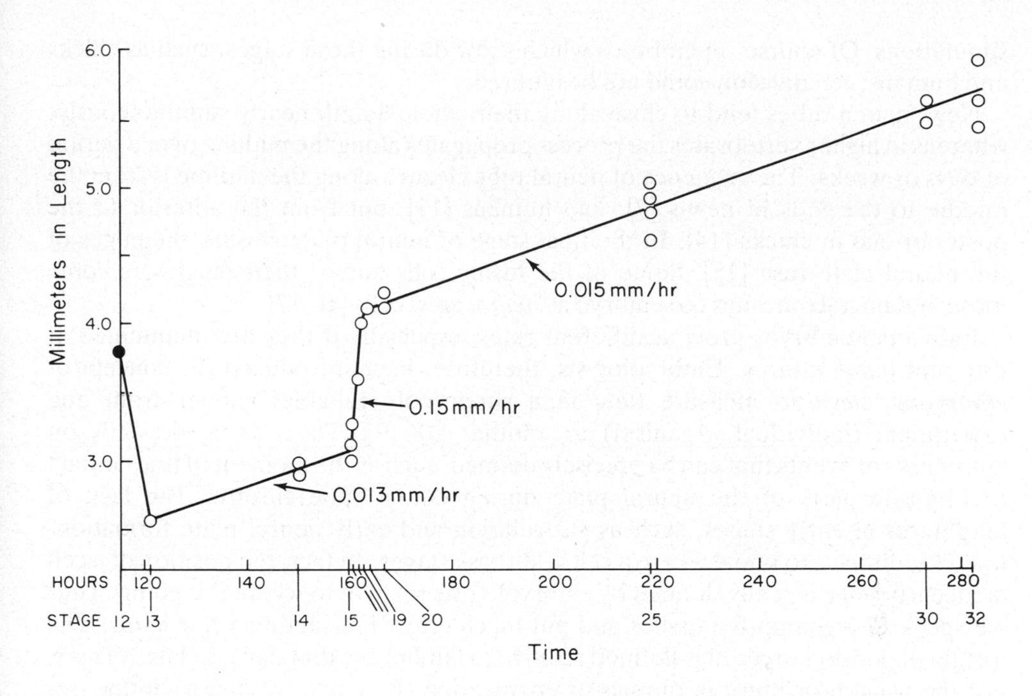

Fig. 4. Length of the excised newt nervous system versus stage of development. The point at stage 12 is the calculated hemicircumference of the spherical embryo (radius 1.2 mm). Between stages 12 and 13 the length of the nervous system decreases as the hemisphere converts to a disc. Thereafter the length increases. A tenfold increase in the rate of elongation is measured from mid-neurula stage 15 to neural tube closure at stage 19. From Jacobson and Gordon [4].

---

linear interpolation of the height programs (Fig. 14). This incorporates two forces: non-uniform apical shrinkage of the neural plate cells and elongation of the notoplate along the midline. Except for foreshortening of the neural plate in the three dimensional embryo, the resulting shape of the transformed grid is virtually identical. In column *c* the notoplate elongation has been turned off so that the only force is that of cell shrinkage; the resulting grid is reduced in size but does not attain the keyhole shape. In column *d* cell shrinkage has been turned off so that the only driving force is notoplate elongation; this experiment could not be done with the living embryo, only by computer simulation. Although the grid does attain a keyhole shape, the anterior end is overlarge. Thus both apical shrinkage and notoplate elongation appear necessary for neural plate shaping. From Gordon and Jacobson [5].

simulations. Of course, in embryos which grow during these stages, such as chicks and humans, cell division could not be ignored.

Newt neural tubes tend to close along their whole length nearly simultaneously, whereas in higher vertebrates the process propagates along the midline over a period of days or weeks. The sequence of neural tube closure along the midline is from the middle to the ends in newts [7], and humans [13], but from the anterior to the posterior end in chicks [14]. In the final stage of neural plate closure the edges of the neural plate fuse [15]. Some of the fusing cells and/or their neighbors come loose and migrate around the embryo as *neural crest* cells [16,17].

Individual embryos grow at different rates, especially if they are maintained at different temperatures. Embryologists, therefore, have introduced the concept of *embryonic stage* to measure time in a supposedly reliable manner from one experiment (individual organism) to another [18,19]. The staging depends on landmarks or events that can be precisely defined, such as the moment of first contact of opposite parts of the neural plate during neural tube closure. The lack of landmarks at early stages, such as gastrulation and early neural plate formation, makes it difficult to know where a cell is at these stages. In fact, the position of a cell at an early stage is really defined by embryologists relative to where it is going. Thus we speak of presumptive tissues and put much effort into defining *fate maps*. The 'original' position of a cell is defined relative to landmarks that develop later. This is not the usual procedure in physics or engineering. It is as if we had to define our coordinate system on a bridge by the location of its twisted parts after it collapsed. To do this, we would need a record of the trajectory of each portion, and an ability to trace it backwards in space and time. The record must be a movie that misses nothing, and the parts must be clearly distinguishable to be able to follow them backwards through each frame of the movie, without mixing them up. Fate maps of embryos are even more complicated than this, since cell division and changes of neighbors keep changing the number and relative positions of the parts.

3. THE ASSUMPTION OF A CELLULAR BASIS FOR MORPHOGENESIS

Most organisms are divided into cells. This natural subdivision permits us to think about embryonic development in terms of the interactions between cells [20]. We can then abstract the concept of the behavior of an individual cell. What does it do in a particular situation? When does it divide? What sort of information does it receive from other cells, near or far? What information about its own history can it bring to bear? What kind of responses can it make? What information can it transmit to other cells? A presumption of modern embryology is that the shaping and differentiation of tissues in embryos can be explained in terms of answers to such questions. As purposive as the development of an embryo may seem, we stringently attempt to leave out all goal directed behavior when speaking of cells. Since computers are neutral on this point, it is sometimes difficult to eliminate such anthropomorphic thoughts from one's own description (the computer program) for cell behavior. The smallest change in the order of operations carried out in a computer program can change the simulation into one imitating goal directed behavior.

The hierarchical division of an organism into cells and tissues may be inadequate for understanding its mechanical properties. Kastelic and Baer [21] consider a number of hierarchical levels for the mechanical analysis of tendons. When present, the cellular level is but one of these hierarchical levels. On the other hand, the living cell is not a mere passive mechanical element, but can itself exert the forces that lead to development. Thus we will proceed on the assumption that the cell is the important level at which the forces of development are organized and generated.

4. THE REPRESENTATION OF EMBRYONIC EPITHELIA IN COMPUTERS

A computer simulation of an embryo is an embodiment of our understanding of how an embryo works. If we approach a simulation from the cellular level, we must decide how we will simplify each cell enough to represent it in the computer, how the cells can influence one another, and how the cells will respond to external conditions or forces. We usually do not actually know these things, and so must make educated guesses. Honing in on the actual parameters and forces used by a given embryo takes an alternation between computer simulation and experiments or observations. Here I will concentrate on the choices for describing cells in computers that I and others have found useful and plausible. Choices must also be made about how the flow of time is to be represented. Before getting into the representations of cells and tissues, a few general remarks on the representation of time are necessary.

Ordinarily I represent the flow of time by carrying out embryonic simulations in a *quasi-static* way [22]. This means that the tissue is allowed to relax after each increment of time. All forces are allowed to come into balance by iterating the computation until the residual movement of all cells is insignificant. In real life, this is equivalent to assuming that if the time-dependent parameters of the cells suddenly stopped changing, then the whole embryo would come to rest in a time short compared to the time increment. We may express this in terms of concepts from hydrodynamics or continuum mechanics: we are dealing with very *low Reynold's number* hydrodynamics or *creeping motion* in which momentum is unimportant [23]. I have estimated, for instance, that in the case of sorting out of embryonic cells, the momentum is 10^{-8} less than the momentum needed to give a significant departure from the low Reynold's number limit [24,25].

Fast relaxation relative to the time of a morphogenetic process may be an important general embryological principle: the rate of growth could vary with fluctuating temperature without altering the sequence of forms. If relaxation were slow, as postulated by Odell et al. [26], different growth rates might produce different forms. It takes a *Taricha torosa* embryo 25 days to go from stage 1 to stage 16 at 5 °C, but only 3 days at 25 °C, with no apparent difference in shape of the neural plate [27].

I will now consider each representation for neural plate cells and their

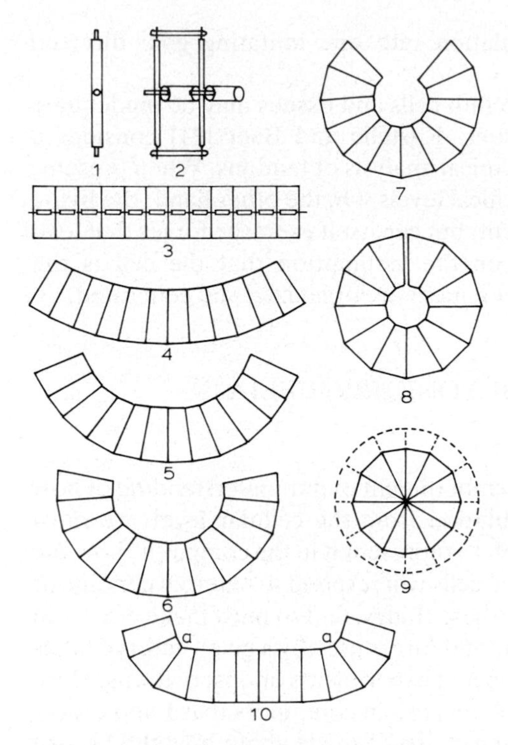

Fig. 5. The mechanical model of Lewis [33] for a cross-sectional representation of neurulation. '(1) Flat bar with a peg at each end and on each side. (2) Two bars in position. A rubber band at each end, a rubber tube spacer in the middle. (3) Model with the tension of the rubber bands equal on the 2 sides. (4–8) Equilibrium curves with different numbers of bands on the 2 sides. (9) Idealized condition with lumen obliterated by still greater increase of tension. Dotted line for no decrease in size of segments. (10) Local increase of tension at *a* and *a*, a neural fold imitation'.

connections. The representations each have their advantages and disadvantages. I have arranged them in order of increasing realism in their portrayal of the geometry and forces of the cells.

4.1. Representation as a cross-section

When the neural plate of an embryo is fixed and sectioned, the cells appear approximately as trapezoids in cross-section. This view, in its tantalizing simplicity, has led many to regard the cells in an epithelium as a set of connected trapezoids (Fig. 5). This can be called a *one-dimensional* representation, because the cells are strung out in a line. (The line may become a closed curve, like a circle.) All models of this kind are intended to represent a cross-section of the embryo. No actual epithelium looks like this. Any number of forces applied to such a representation will cause folding and rolling up, so that one gets the impression that one has uncovered the mechanism for neural tube closure. However, as I will show below, because they are one dimensional, all such models cannot explain neurulation.

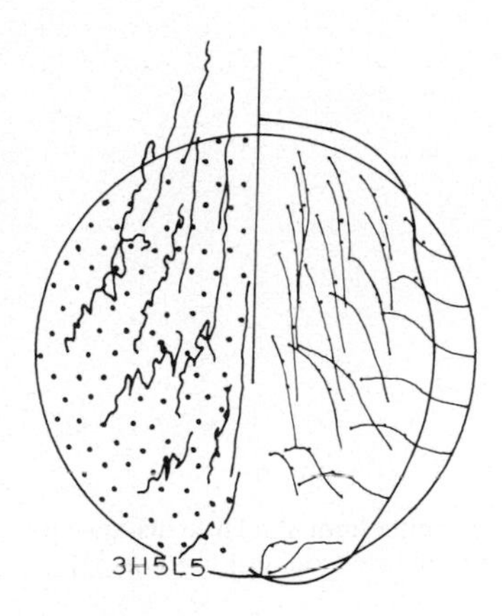

Fig. 6. Trajectories of simulated shrinkage units (left half of figure) are compared to trajectories of cells in the actual embryo (right half of figure) through the same time period, stages 13 to 15. The cell trajectories are from Burnside and Jacobson [39]. Along the midline, in both the simulated and normal embryo, units (cells) move directly anteriorly (toward top of figure).

The history of cross-sectional models goes back quite far. For instance, Glaser [28] accepted his own evidence that the neural plate cells must swell up, more at the basal end than at the apical end, changing the shape of each trapezoid, and thus causing rolling up. He suggested that the basal ends of the cells 'weakened' in response to changing acidity conditions within the embryo [29]. Similar models were considered by Rhumbler [30], Moore [31,32], Lewis [33,34], and Odell et al. [26,35].

Observations are also often made only with cross-sections in mind. Cell shapes are often compared at different stages without taking into account the motion of the cells over the embryo [36–38]. The cells compared are thus not necessarily the same cells.

In effect, theories and observations that deal with cross-sections carry the assumption that either cells remain approximately in the same cross-section during neurulation, or that the lateral and anterior/posterior movements of the cells are independent, and therefore separable in the mathematical sense (G. Oster, personal communication). These assumptions may not be valid, given the enormous distances neural plate cells move over the embryo [39]. Some of these distances are two-thirds of the radius of the embryo. Moreover, most of this motion is in the anterior/posterior direction (Fig. 6), perpendicular to a typical cross-section showing the folding process. The motion is also not the same for cells at different distances from the sagittal midline. Thus any cross-sectional representation does not represent the same cells from one moment to another. Further, since the bulk of the movement is in the anterior/posterior direction (Fig. 6), there are likely to be considerable force components in this direction, which are not included in any cross-sectional representation.

Another difficulty with one dimensional calculations is that they can lead to a false impression of the rigidity of an embryo. G. Oster (personal communication and computer movie) has emphasized the rigidity with which his computer cells communicate by mechanical means in transmitting 'triggering' events in one-dimensional simulations. Often a number of cells will be skipped over and a cell even on the far side of an embryo will be triggered by a local event. But consider the effect of one

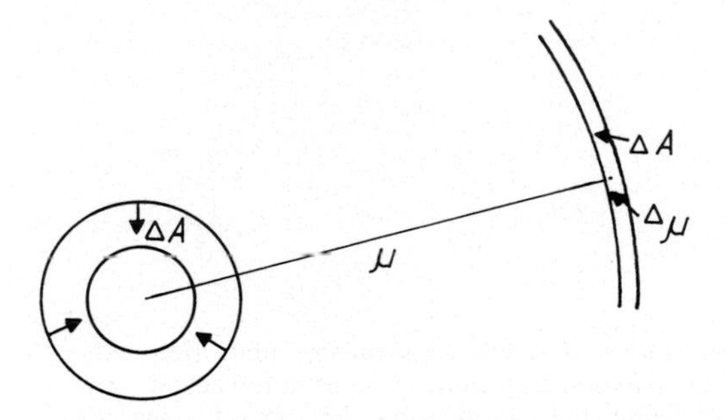

Fig. 7. When a circular cell decreases its area by ΔA, a point in a planar epithelium at a large distance μ from its centre will move towards it by a distance $\Delta\mu = -A/(2\pi\mu)$. From Jacobson and Gordon [4].

cell on another in two dimensions (Fig. 7). The tug of one cell on another falls off as $1/\mu$, where μ is their distance apart. In a three-dimensional tissue, the fall-off would go as $1/\mu^2$. But in one-dimension, the corresponding fall-off is $1/\mu^0 = 1$, meaning no fall-off at all. Thus a one-dimensional tissue is especially rigid, transmitting mechanical effects along the whole chain of cells without damping. (Ziman [40] has similarly emphasized the lack of physical reality in one-dimensional models of two- or three-dimensional systems.) We therefore see that the process of neurulation is fundamentally two-dimensional and must be simulated as such. (By this I do not necessarily mean a two-dimensional plane, but rather a *surface* which exists in three dimensional space.)

4.2. *Representation as a planar graph*

The simplest representation of a cell is as a point in space, which can be called its *centre*. In such a representation a neighbor of a cell may be indicated by a line or *bond* from centre to centre. A set of points connected by bonds is called a *graph* [41]. A graph in a plane, which has no bonds crossing one another, is called a *planar graph*. This concept of no crossings has a simple basis in the geometry of epithelia. If cells ABCD are connected, so their centres form a quadrilateral, and if A and C are connected, then B and D ordinarily cannot be connected. The only exception is if all the interfaces between the four cells meet at a single point (Fig. 8). This is likely to occur only transitorily after a cell division or when cells change neighbors. Ordinarily only three cells meet at a single point, which I will call a *vertex*.

4.2.1. *Representation of a cell*

We can also represent a cell by its area, as seen looking down on it in an epithelium. The simplest representation of this area we could adopt is a circle. Considered in its depth, the cell then looks like a cylinder, if the epithelium is flat (Fig. 9). For now I will assume that it is flat and remains so. (This assumption is plausible up until neural tube formation begins. It is, nevertheless, a severe constraint to keep a tissue flat that

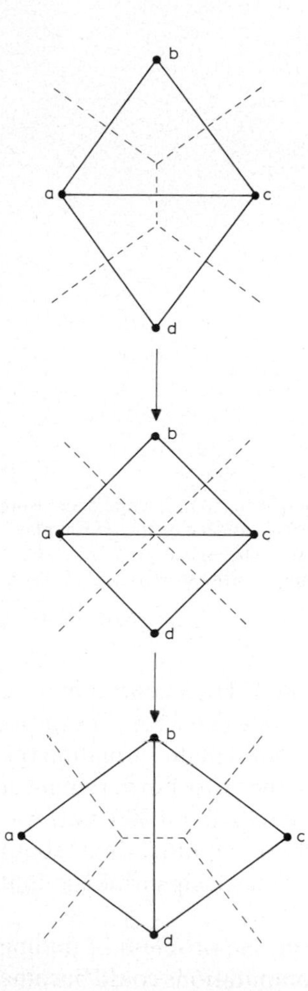

Fig. 8. Representation of an epithelium as a planar graph. The solid lines indicate the bonds between the centres of the cells at a,b,c, and d. The dashed lines are the cell–cell interfaces. Ordinarily three cells meet at a vertex. However, when there is a change of neighbors, as shown, there is a transitory configuration in which all four cells meet at a single point.

on its own might buckle into the third dimension; cf. Section 4.2.6.) Each cell is then represented as its centre, its apical radius, and its height. In the spirit of only using local interactions, we shall assume that the centre of the cell can only be moved by forces from its neighbors. This may not seem an obvious restriction, since many cells can migrate reasonably freely, apparently on their own, through embryos [42]. But experiments going back to Roux (1888) [43] indicate that the neural plate can change shape and roll up into a neural tube when isolated from the rest of the embryo (cf. Jacobson and Gordon [4] and Gordon and Jacobson [5]). Thus the cells have no substratum against which to push in these experiments. Since morphogenesis in this situation appears normal, I assume that there is no need for, nor presence of, pushing of neural plate cells against the mesoderm in intact embryos.

Most computers are too slow and small to handle every cell in the neural plate (about 10,000 in newts). For this reason, each cylinder must actually be taken to represent a group of cells. Because of the shrinkage in apical area, I call each group a *shrinkage unit*. At the beginning of a simulation, it is convenient to give all shrinkage

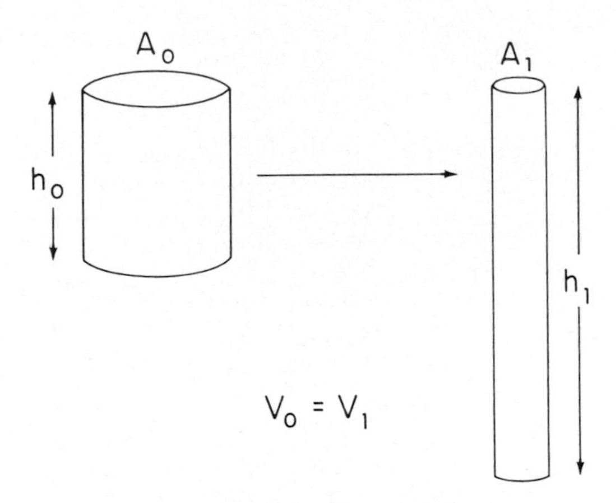

Fig. 9. Keeping the same volume, a cell or shrinkage unit increases its height (h_0 to h_1), while decreasing its apical surface (A_0 to A_1). The apical surfaces of the cells form the dorsal surface of the neural plate. The long axes of the cells are perpendicular to the surface of the plate. The apical area is inversely proportional to the height, $A = V/h$, since the cells retain a columnar shape during stages 13 to 15. From Jacobson and Gordon [4].

units the same radius and have them hexagonally close-packed. Their centres may be chosen to fall within a disc, to represent the stage 13 neural plate (Fig. 10). The initial radius of the units should be set small enough so that *after* neural plate formation the narrowest regions of the neural plate, to the left and right of the notochord, remain at least a few units wide (Fig. 11). Division of the neural plate into about 300 units was just sufficient to give both a fine spatial division and plausible computing time (about 15 min cpu time per run on a Digital Corp. PDP10 [4]). Thus each shrinkage unit represented about 30 cells.

If one were to attempt a longer time span in a simulation, the problem of finding the minimal division of the tissue into units adequate for computations could become severe. All the cells within a unit are, by definition, forced to move together. Thus phenomena attributable to cell division or cell rearrangement may become poorly approximated. As it is, one must make some presumptions about how the relationships between cells are reflected in relationships *between units*. Careful dimensional analysis [44,45] should be done to handle any scaling problems between cells and shrinkage units. With the increasing memory becoming available on computers, it may be wisest to avoid this problem by making every unit represent just one cell. Thus I will use the term ‘cell’ to mean a shrinkage unit from here on.

A cell’s height versus time is called its *height program*. There is some controversy about whether or not the height programs are intrinsic to each cell in neural plate development, or are the result of dynamic interactions with neighbouring cells [3]. In our simulation [4], we assigned each cell a digit from 1 to 9, representing one of the nine height programs it could follow (Fig. 10). For an individual cell, then, we have

$$V_i = \pi r_i(0)^2 = \pi r_i(t)^2 h_i(t) \tag{1}$$

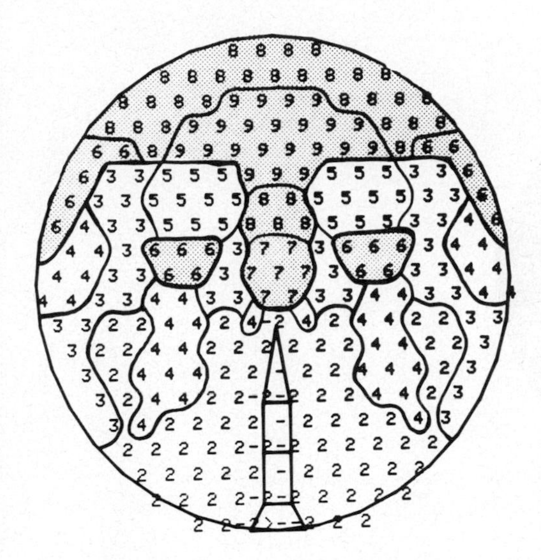

Fig. 10. A shrinkage pattern or spatial distribution of height programs on a starting disc representing a stage 13 neural plate. Shrinkage units with higher numbers shrink more over stages 13 to 15 than units with lower numbers, as specified by the height programs (e.g., Figs. 14 to 15). Minus signs indicate units on the perimeter of the notoplate. Contour lines have been added to make the pattern more visible. The area that will shrink most (height programs 6–9) is shaded. The shaded area is approximately shaped like an Y. Compare to Fig. 11. From Jacobson and Gordon [4].

or

$$r_i(t) = r_i(0)\sqrt{h_i(0)/h_i(t)} \tag{2}$$

where r_i and h_i are the apical radius and height of cell i and V_i is its volume, presumed constant (Appendix 3, Assumption 1, in Jacobson and Gordon [4]). The time is represented by t.

The spatial pattern of height programs is called the *shrinkage pattern.* (This is, in effect, the very pattern formed by primary neural induction.) It may be entered into the computer by hand using interactive *computer surgery,* which allows us to alter the shrinkage pattern in any way desired: an initial pattern of labels of the cells is displayed on a graphics terminal, and a pointer moved from cell to cell using keyboard commands. The height program of the indicated cell can be changed to any other. If the label is set to 0, then the cell will be removed before proceding with the simulation. If the pointer is moved to a space not yet containing a cell, a new cell can be inserted at that position. Thus the program allows us to imitate any surgical operation involving excisions, transplantations, or inversions. Since initially the cell centres are located on a hexagonal lattice, the surgery procedure is particularly simple to write.

4.2.2. Representation of the connections between cells

When the cells move relative to one another within the epithelium, if the computer simulation is to continue to represent a confluent monolayer of cells, it is essential to restrict the motion of each cell so that crossed bonds do not develop. Crossed bonds

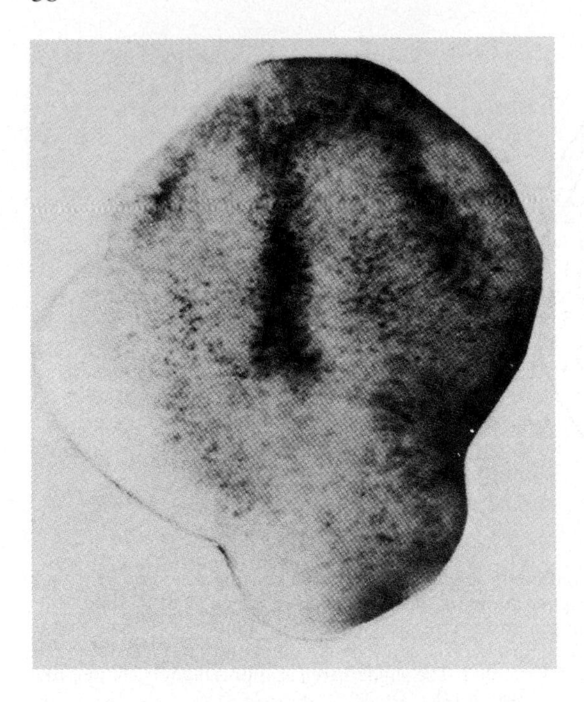

Fig. 11. A neural plate with no other tissue was explanted onto neutral agar at stage 13 and held flat by a piece of beaded cover slip. The pigment concentration that results by the time controls have reached stage 15 is a distorted version of the stage 13 shrinkage pattern (cf. Fig. 10). The pattern resembles the Greek letter ϒ. Posterior is at the bottom. ×33. From Jacobson and Gordon [4].

would be topologically equivalent to folds, wrinkles, removal of cells from the sheet, or just geometric impossibilities. The computer, of course, knows nothing of topology. Thus restrictions must be put in the computer program to prevent bonds from crossing.

The topology may be taken into account as follows. Consider a cell about to be moved. Its centre lies inside the polygon formed by the bonds between its nearest neighbors. If the motion which we calculate would carry it outside this polygon, that motion must be restricted (Fig. 12). When a need for such restriction arises, it indicates a need for a smaller time step. A refined simulation would back up all the cells to their positions at the previous time step and resume with a smaller time increment. Such a dynamic adjustment is analogous to the 'step-size control' numerical methods for integrating ordinary differential equations [46].

If two cells are in contact, the distance between their centres is the sum of their radii, which I call the *relaxed distance*, because we may asume that at this distance these cells will neither push nor pull on one another. The simplest way of modeling the force between two cells is to assume there is an 'equivalent spring' whose relaxed position corresponds to this relaxed distance. If they are at a greater distance, the 'spring' pulls them back. If they are too close, it pushes them apart. I used a stress/strain relationship (Eq. 1.9 in [41]) for this equivalent spring that reduces

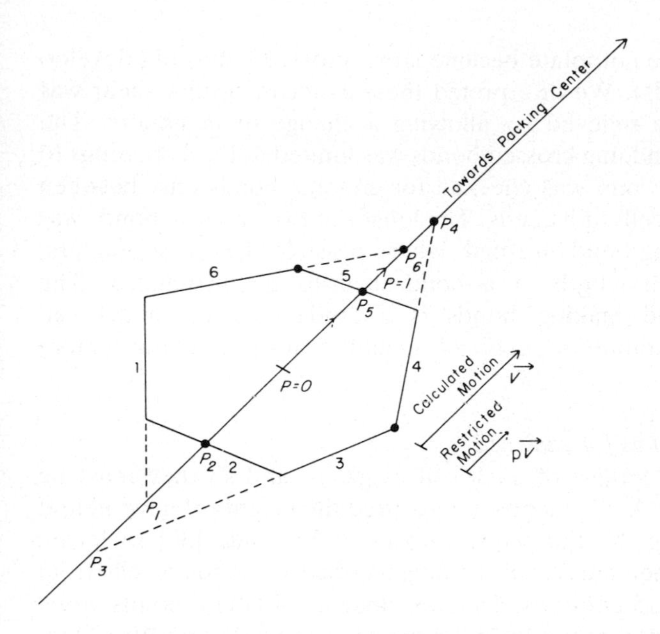

Fig. 12. The motion of a cell located at $p = 0$ is restricted to remain within the polygon formed by the bonds, numbered 1 to 6 here, between its nearest neighbors. (The centres of the neighbors are at the vertices of the polygon.) The vector of motion is parametrized so that it lies along the line of motion from $p = 0$ to $p = 1$. Each bond intersects this line at a given value of p. The direction of motion is either towards a packing centre, or better, perpendicular to the local direction of propagation of the current repacking. From Jacobson and Gordon [4].

to the simple linear relationship:

$$\Delta d = d_{ij} - (r_i + r_j) \tag{3}$$

where d_{ij} is the distance between two cells of radii r_i and r_j, and Δd is proportional to the change in their distance, along the line between their centres. This is equivalent to Eq. (1) in [26], for 'a simple linear viscoelistic element'.

An advantage of regarding the computation as quasi-static is that it reduces our need for detailed knowledge of the stress/strain relationship. The repeated iterations bring each pair of bonds close to its relaxed distance, which in terms of the equation of motion is the position of minimum potential energy. The elastic moduli of the cells will thus have little effect on the location of their relaxed positions and need not be measured. The viscosity need not be determined either, since it has no effect once motion has stopped. This is accomplished by reiterating Eq. (3) after each time increment until the Δd's are all small (I will call this viscosity the *bulk viscosity*, since it has to do with the response of the whole cell to distortion. *Tissue viscosity* will refer to the sliding friction between cell membranes [25]. A typical bulk viscosity for protoplasm is 10^{-1} to 10^3 poise [47]. Tissue viscosities, measured for embryonic tissues in cell sorting experiments, are in the range of 10^6 to 10^8 poise.)

Despite our attempts to retain the bonds in a planar graph, when the distortions produced by elongation of the notoplate became large, crossed bonds did develop (as did excessively long bonds). We interpreted these as meaning that shear was occurring and could only be relieved by allowing a change of neighbors. The algorithm for detecting and undoing crossed bonds was limited to local crossings to reduce computer time: each bond was checked for crossing bonds only between nearest neighbors of the two cells at its ends. The longer of two crossing bonds was broken and new, non-crossing bonds formed, where possible. Extra parameters, the minimum and maximum length of a bond, had to be introduced. The operations for breaking and making bonds is a crude way to model cell rearrangement. The representation in Section 4.3 handles this problem in a more natural way.

4.2.3. Obtaining the parameters for each cell

We are taking the height programs of each cell as given, and so they must be measured from real embryos. A. G. Jacobson measured the heights of newt neural plate cells at stage 13, using the time-lapse movies of Burnside [39] to locate those cells at stage 15, and then measured the heights of corresponding cells from serial sections of other stage 15 embryos. This was done at 24 lateral points and 9 points along the midline, mostly at the intersections of coordinate grid lines (Fig. 13). The heights for the remaining areas were interpolated from observed cell trajectories (Fig. 6). Nearly every cell differed from every other cell in its starting and final heights. Height data were normalized to a uniform starting height of 100 microns. The dynamics of the shape depends on the *ratio* of the height of a cell at a given time to its initial height, and, therefore, is unaffected by this normalization.

These measurements gave us the starting and final heights for each height program, but not the intermediate values. To estimate the temporal course of each height program, we used two interpolation methods. In the first, we assumed that the height of a cell varies linearly with time (Fig. 14). In the second, we made the heights inversely proportional to the area of the entire neural plate at a given time (Fig. 15). The latter we could measure. Our inability to accurately stage different embryos at stages intermediate between 13 and 15 made it undesirable for us to attempt to use sectioned material for the intermediate parts of the height programs.

We were fortunate that the neural plate is approximately flat, so that the sections were, for the most part, perpendicular to the long axes of the cells. (Sagittal sections were used when necessary.) Measurement of height programs could be made more accurately using the following combination of computational tools:

(1) Preparation of time lapse movies under computer control (see Section 5.)
(2) Entry of serial sections of staged embryos, covering the same time period, into a computer using a graphics tablet [48]. For a given epithelial tissue, the cross-sectional lines of both the apical and basal surfaces would be traced in.
(3) Three-dimensional surface smoothing and fitting to remove irregularities in the hand-tracing contours [49]. The smoothed coordinates can be used to prepare and display three dimensional reconstructions [50,51].

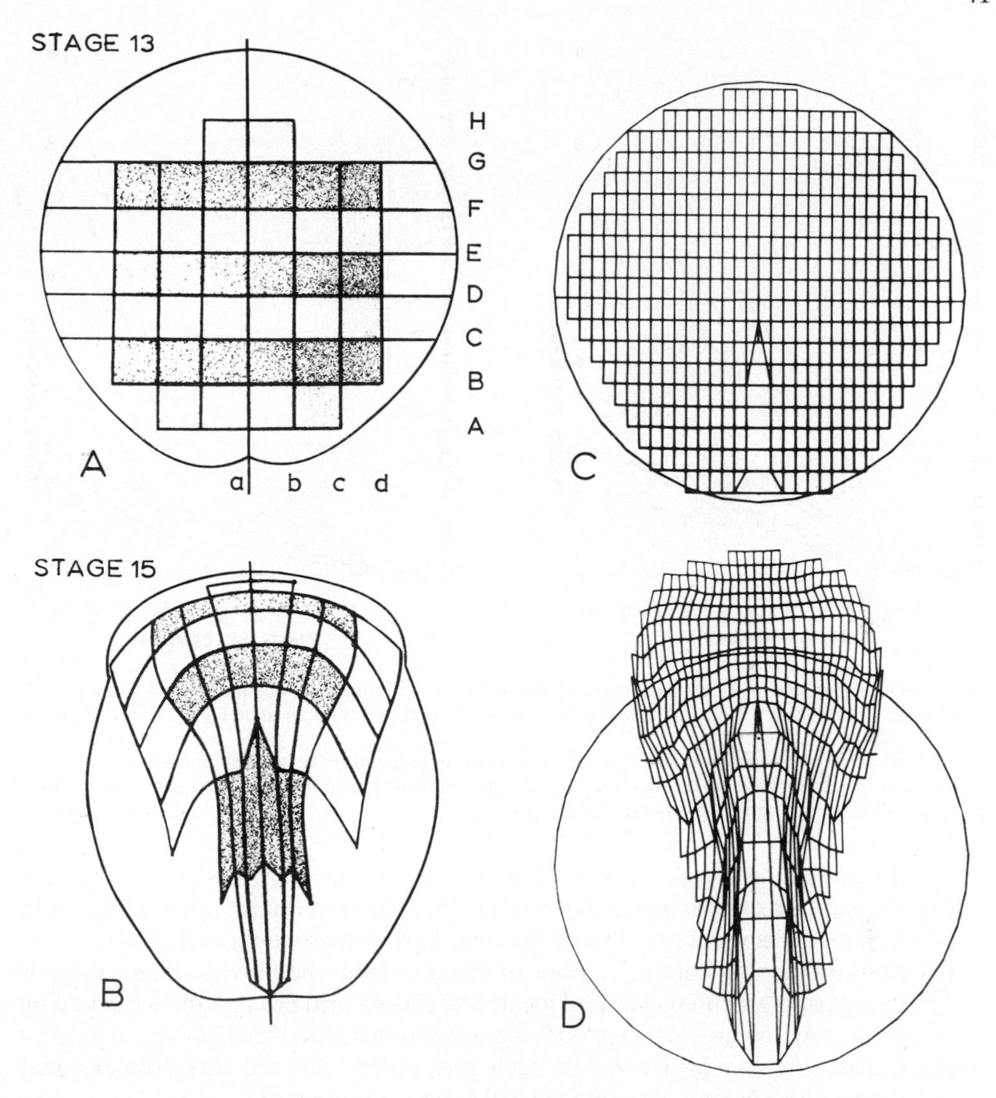

Fig. 13. A coordinate grid is placed on the outline of a stage 13 embryo (A) and the distortion of that grid shown at stage 15 (B). These are compared to the starting coordinate grid of the computer simulation (C) and its distorted grid (D). A and B are from Burnside and Jacobson [39]. Note that the anterior end of the neural plate in B is seen in perspective, wrapped over a flattened spherical surface. From Jacobson and Gordon [4].

(4) Calculation of the distance between the best-fit tangent planes to both the apical and basal surfaces. The cell height is then approximated by the distance between these tangent planes. (Of course, the cells could be twisted or inclined from their basal to apical ends, giving them far from simple columnar shapes. Cf. Fig. 53 in [4]. A. G. Jacobson [personal communication], also has evidence for massive, spaghetti-like twisting of

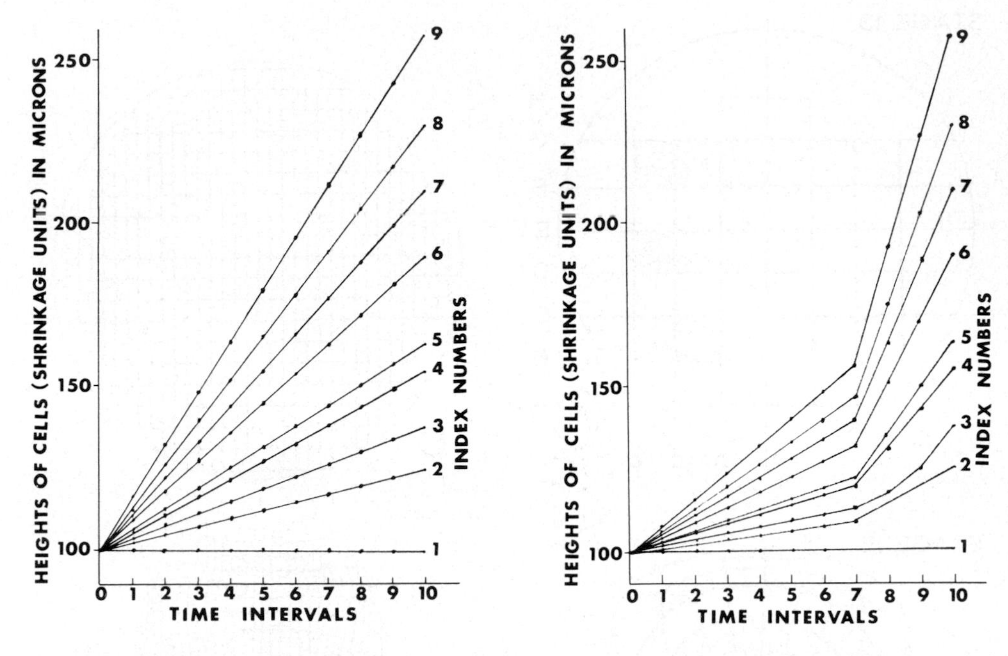

Fig. 14. Height programs for shrinkage units labeled 1 to 9. The time intervals evenly divide the period from stage 13 to stage 15. Here the interpolation between the stages is taken to be linear. From Jacobson and Gordon [4].

Fig. 15. This set of height programs has the same starting and ending heights as those in Fig. 14, but rather than increasing linearly through time the height is inversely proportional to the area of the neural plate (nonlinear). From Jacobson and Gordon [4].

the cells of the notoplate, where considerable cell rearrangement is occurring.)

(5) Preparation of a perspective view of the three dimensional reconstruction from a view corresponding to that of the time-lapse movie camera (cf. [52]).

(6) Measurement of the trajectories of the cells from the movies. Single cells or groups of pigmented cells [39] located at chosen grid points can be tracked by projecting the movie image onto a graphic tablet (see Fig. 13).

(7) Calculating the projection of each grid point onto the three-dimensional reconstruction and recording the cell height at that point.

(8) Following the trajectory of the grid point to the later stage, calculating the projection of its new location onto the three dimensional reconstruction of the later stage, and recording the cell height at the new location.

The two cell heights measured in steps (7) and (8) should then be of corresponding cells at the two stages, if all alignments, magnification factors, etc., have been properly accounted for. (Variability in the size of the embryos should be taken into account. The normal embryo volume can easily vary by 30% cf. [4]).

The existence of the shrinkage pattern was anticipated by Lewis [33]: 'It is evident ... that there is a lack of uniformity in timing, strength and location of the contractions of the superficial [apical] ends of cells in different regions at different stages. This lack of simultaneous behavior is responsible in part for the many different contours along the length of the organ during neurulation'. The shrinkage pattern is the first nontrivial pattern to appear amongst the cells of the vertebrate embryo. Our current characterization of it, on some 300 units, is rather crude (Fig. 10). It would be well worth knowing if there are finer details to the pattern. It also deserves confirmation and a check for universality, by measuring the shrinkage pattern for other embryos.

The above eight-point procedure for measuring cell heights does not, unfortunately, solve the problem of obtaining the actual height programs. We would still have to make interpolations of dubious validity. We were saved from this work by the fortunate circumstance that the course of the simulated neural plate formation was reasonably independent of the interpolation scheme chosen (Fig. 16 in [4]). But there are good reasons for wanting direct measurement of the height programs of individual cells. A cell's height program may depend on the initial height of the cell. Details of the height program may correlate with its developmental fate, such as formation of the eye placodes at a later stage. We are thus left with a need to measure the heights of the cells on living embryos. Computer microscopy tools that could be used to attempt this feat are discussed in Section 5.

4.2.4. *Repacking of the cells as they change shape*

Although the simulated cells are initially hexagonally close-packed, as soon as their radii change by different amounts in accordance with their height programs (Eq. 2) this arrangement is disrupted. Two features of hexagonal close-packing are lost. First, it is no longer possible for each cell to just contact each of its neighboring cells (Fig. 16). Second, as a consequence, the position of a cell is no longer uniquely specified by the positions of its neighbors. Thus any *algorithm* (set of instructions) which moves or *repacks* the cells in any attempt to retain them in a contiguous epithelial sheet must make compromises about the positioning of the cells and will be arbitrary to some degree.

Care must be taken in designing a repacking algorithm. Any systematic method for adjusting the distances between cells can lead to unrealistic, skewed repackings. Yet, of course, some particular order must be chosen. (Parallel computers may someday permit us to do the physically more realistic operation of permitting each cell to pull on all the others simultaneously.)

Let us temporarily designate one cell as a *packing centre* and move all the other cells towards it. This can be done by using consecutive rings of cells. All the nearest neighbors are first adjusted to the packing centre. Then *their* nearest neighbors, the next ring, are adjusted to them, etc., so the adjustments propagate outwards from the packing centre (Fig. 17). Within a given ring, the cells can be adjusted in random order, to avoid any systematic effects. All of the cells may be shrunk slightly (Eq. 2) before starting a repacking.

Since hexagonal close-packing is not in general possible, if we were to move any pair of neighbors into perfect contact we would disrupt other pairs. Thus only a small step towards the relaxed distance between each pair of cells should be made. This step may be taken as a fixed fraction of the current Δd. I found that if this

Fig.16. A configuration of dimes, pennies, nickels and quarters put together so as to approximate close packing. Note the curvature and the inability to obtain close packing everywhere.

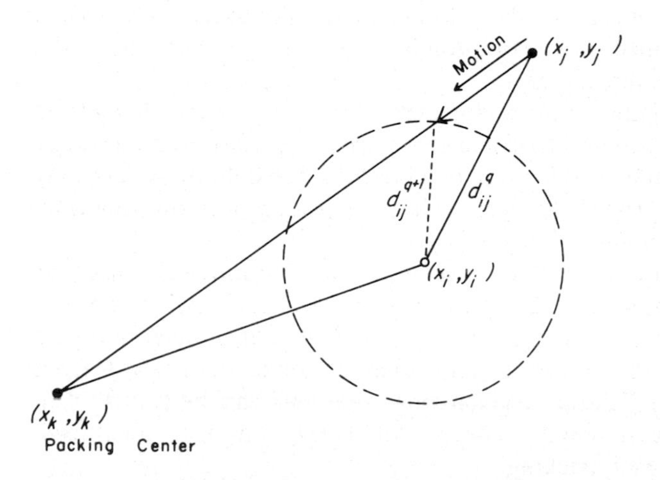

Fig. 17. If a cell j is being moved towards a packing centre, as in Figure 12, there is a circle of positions around its neighbor i on which the centre of j could be placed to bring i and j closer to their ideal distance apart. The line of motion (adjustment line) intersects this circle in at most two points. The closer one to the previous position of j is chosen. From Jacobson and Gordon [4].

damping factor was too large, cracks propagated through the disc, as though it were brittle. The maximum usable damping factor can be found by trial and error.

When one cell is moved to a given distance away from another, there is a whole circle of radius equal to that distance on which its centre may be placed. A unique point on this circle may be chosen by moving the cell along an *adjustment line* towards the packing centre (Fig. 17).

In order to retain continguity of the sheet of cells, I found it important to utilize three consecutive rings, each one bond further from the packing centre. Initially the first ring contains only the packing centre; the second contains its nearest neighbors. The third ring is built from nearest neighbors of cells in the second ring which are not in the first two rings. These rings propagate by deriving the first ring from the second, the second ring from the third, and calculating a new third ring (Fig. 18). As mentioned, the cells in the first ring are considered in random order. When it is considered, a cell is *anchored,* and will no longer be moved during the current repacking. Each of its unanchored first nearest neighbors, belonging to the first or second ring, is adjusted to it.

If two cells being adjusted to one another are in the same ring, then the adjustment is along the line between their centres. This can be justified in terms of the orthogonality between radial and tangential adjustments. Inclusion of this nuance leads to considerable improvement in the results of repacking.

When a cell is anchored, its net vector displacement since the last repacking is calculated. This may be considered the *local displacement* of the sheet. When a cell is anchored, its second nearest neighbors (neighbors of the neighbors) which are in the third ring are moved according to this local displacement. (A flag is set for the cell in the third ring, so that it undergoes only one local displacement per repacking.) In this way the relative positioning of the cells, i.e., the *local structure* of the sheet, is preserved. If local displacements were not taken into account, cells of consecutive rings could overlap more and more severely with increasing distance from the packing centre, scrambling the bonds so that they cross and no longer represent a planar graph.

To make the simulation quasistatic, repacking has to be done iteratively, as many times as necessary to bring the cells to rest. It is difficult to design a criterion for convergence, apparently for the following two reasons. First, the repacking algorithm adjusts cells to each other one pair at a time with some degree of random choice of the pairs. Thus in consecutive repackings a cell may be moved first towards one neighbor, then towards another. Second, the packing centre exerts a systematic pull. If the packing centre is continually changed, as described in Section 4.2.5, all cells are tugged first one way, then another. A convergence criterion must partially absorb these factors to be useful.

One such criterion may be obtained as follows. A vector is calculated, equal to the mean displacement of all the cells since the last repacking. If, as in the neural plate, bilateral symmetry would cause some of the components of the individual cell displacement to cancel out in this mean, a sign change is made to force them to add instead. The running average of this mean vector, over consecutive repackings, is also calculated. It tends to smooth out any overall oscillatory movements. When the

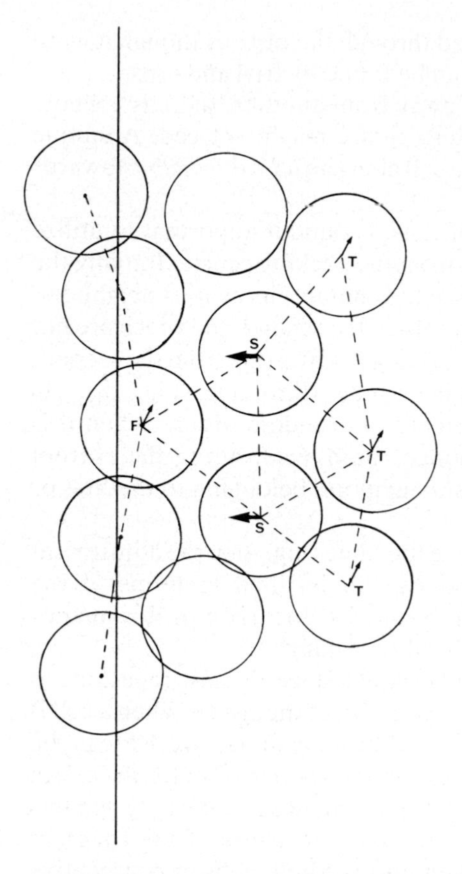

Fig. 18. The three rings of cells used for repacking. The best fit straight line is calculated for a randomly selected cell F in the first ring and its 2–4 consecutive neighbors in that ring. The cell is anchored, and its second nearest neighbors in the third ring (T) are moved according to the displacement of cell F since its last repacking (small arrows) to preserve the 'local structure'. Its nearest neighbors in the second ring (S) are then adjusted towards their ideal distances from cell F (thick arrows). The dashed lines indicate some of the bonds between the cells.

length of the current vector, or the running average, falls below some minimal length, the repackings are stopped and we say that the simulation has relaxed to its quasi-static equilibrium.

4.2.5. External moving boundaries

We cannot yet attempt to simulate the whole embryo at once. Thus if a tissue has an effect on the tissue we choose to simulate, we must regard the former as setting 'external' boundary conditions or constraints on the latter. For instance, we may regard the perimeter of the notoplate as a moving boundary 'external' to the rest of the neural plate. The shape of the boundary may be approximated by an inverted parabola along which cells are evenly spaced (Fig. 19). During each time increment, the simulated notoplate is made taller and narrower, to retain constant area, as

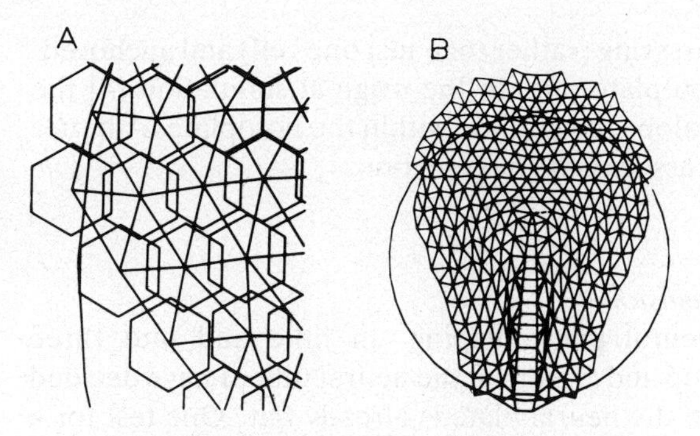

Fig. 19. (A) Enlargement of a portion of the neural plate simulation. Each shrinkage unit is indicated by a hexagon rather than a circle, since this reduces the plotting time substantially. The lines between their centres are the bonds. Note the overlaps and gaps between the units in this representation. (The curve on the left is the original disc within which the original coordinates of the units were chosen.) (B) Display of the bonds only. From Jacobson and Gordon [4].

observed [4]. As the notoplate elongates, its perimeter increases. Thus new cells have to be intercalated between the old ones along the perimeter. (These cells may be considered to have come from cells within the perimeter). This simulates our observation that the notoplate elongates uniformly along its length (Fig. 20).

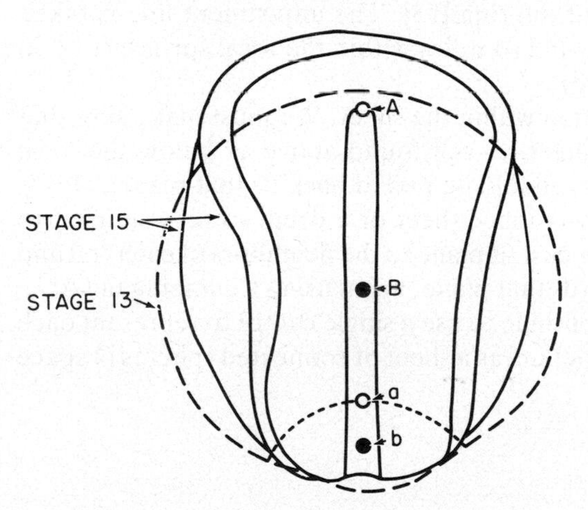

Fig. 20. The outline of the embryo and the notoplate are shown at stage 13 (dashes) and at stage 15 (solid lines). On a time-lapse movie of development of the embryo, cells were located at stage 15 at the anterior end of the notoplate and halfway to the blastopore (A and B). These cells were then followed as the movie was run backwards to stage 13. The two cells then occupied positions a and b. Point a was still at the top of the notoplate and point b half way to the blastopore. Thus the regions between a and b and between b and the blastopore appear to elongate proportionately. From Jacobson and Gordon [4].

When a spatial process has a localized energy input, it is best to use an order for considering the cells which starts at the energy source and works out from it (M. Klee, personal communication). For instance, the computation of the wake of a ship should start from the ship, rather than from the shoreline. By analogy with this situation, we started the repacking from the notoplate. All the cells of the notoplate perimeter are initially placed in the first ring (rather than just one cell) and anchored. Repacking thus occurs from the notoplate out. In the original simulation [4] the packing centre was taken at random along the midline within the notoplate. A better choice for the adjustment line is discussed in the next section.

4.2.6. *How to include the third dimension*

Extension of the simulation of neural plate shaping, in time and into three dimensions, may explain the rolling up and closing of the neural tube. I have decided to start, as before, at stage 13, when the neural plate is already flat. One test for a successful three-dimensional simulation will be that it repeat the essentially planar neural plate formation before going on to neural tube formation. (An earlier starting point, such as that attempted by Odell et al. [26] might be more appropriate, but time-lapse analysis to provide empirical comparison is lacking at these stages.)

The notoplate can again be represented as a parabola whose shape changes with time in a predetermined way. The cells in a propagating ring are listed in order around the ring. To get the local orientation of the propagating ring, we calculate the least squares best fit of a straight line in space to the coordinates of the cell and its two to four neighboring cells in the same ring. This requires that the cells in a ring be ordered in a doubly linked list around the ring [53]. The adjustment line is taken perpendicular to the propagating ring and so it lies within the local surface (i.e., in the best fit tangent plane to the surface).

These calculations take care of motion within the sheet. We must make sure that each cell does not depart from the sheet. A cell found above or below the local tangent plane defined by its neighbors should be pulled back to that plane. This is analogous to the force that causes the rubber sheet of a drum to restore itself to planarity. To simulate this, we find the best fit plane to the neighbors of each cell and move the coordinates of the cell towards that plane, again using a damping factor.

Note that in three dimensions we continue to use a single centre to represent each cell. In effect, we are treating the epithelium as a sheet of connected spheres in space (Fig. 21).

4.2.7. *Bottle-shaped cells*

An epithelium may alternatively be represented in three-dimensions as a sheet of *bottle-shaped* cells. What is the minimum number of parameters needed to describe such a cell? Consider the following. One could draw a bottle-shaped cell as a truncated cone (Fig. 22). The parameters describing such a cone would be the major and minor axes of the ellipses at either end, their orientations in space, plus the coordinates of their centres, amounting to 16 parameters in all. A simpler model can be formulated by specifying only the coordinates at the ends of the long axis (6 parameters) and the radii of spherical caps at either end, for a total of just 8 parameters (Fig. 23). Disregarding the orientation and placement of the cell in

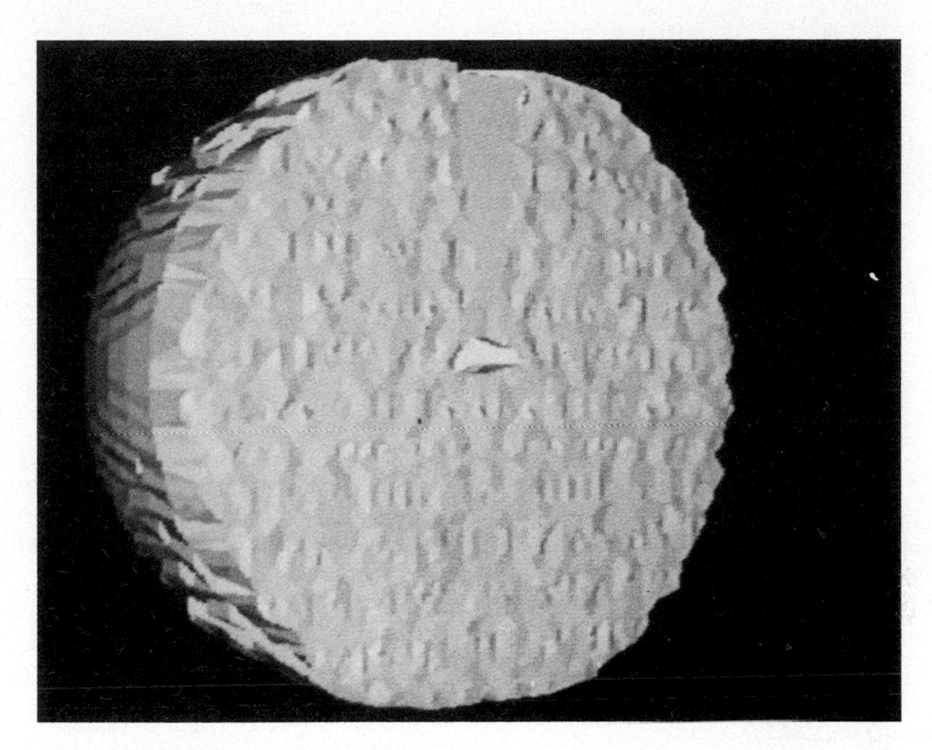

Fig. 21. A hidden surface, halftone, computer drawing of the initial configuration for a three dimensional simulation of neural plate shaping and neural tube closure. The neural plate, at stage 13, is the flat portion, with the blastopore in the foreground. The notoplate is the smooth portion of the neural plate shaped as an inverted parabola. The curved portion is the epidermis. (Prepared by A. Stein and Y. Cheng.)

space, its intrinsic parameters are its height h and its apical and basal radii, a and b. The volume, less the spherical caps, is just $V = \pi h(a^2 + ab + b^2)/3$. If the volume of a cell is constant, as it is during the newt neural plate formation, then there are two independent parameters, say h and a. This means that there is no a priori relationship between the height and apical radius of a bottle shaped cell. This is in contrast to the situation for the flat neural plate, for which $V = \pi a^2 h$ (cf. Eq. 1), in which case a and h are inversely related.

A bottle-shaped cell can change shape in a controlled fashion then, only if both the height and apical programs, $h(t)$ and $a(t)$, are specified. These could be controlled by microtubules and microfilaments, respectively. In the neural plate the microtubules and microfilaments would have to act in concert, so that the shape transformation occurs in a plane. However, they must then deliberately go out of register to change each cell from a columnar to a bottle shape. Any lack of coordination could lead to too much curling or curling the wrong way (and could thus be the basis for neural tube defects). Such curling could constitute the bending movements discussed by Gierer [34].

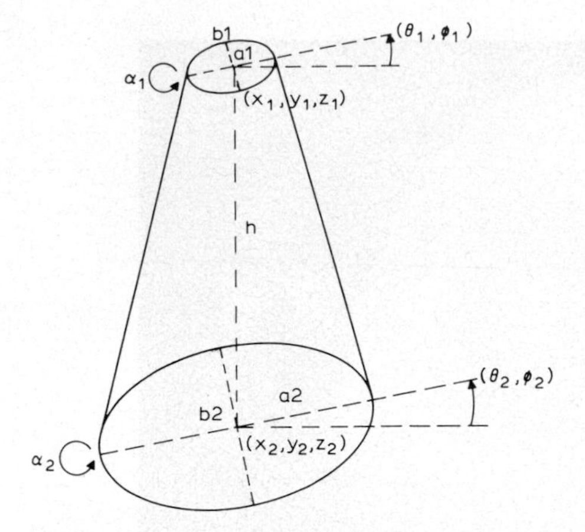

Fig. 22. Truncated cone model for a bottle cell. The sixteen parameters needed to describe a three dimensional cell in this representation are indicated. Three angles are needed to describe the orientation of the major axis of each ellipse at the apical and basal ends.

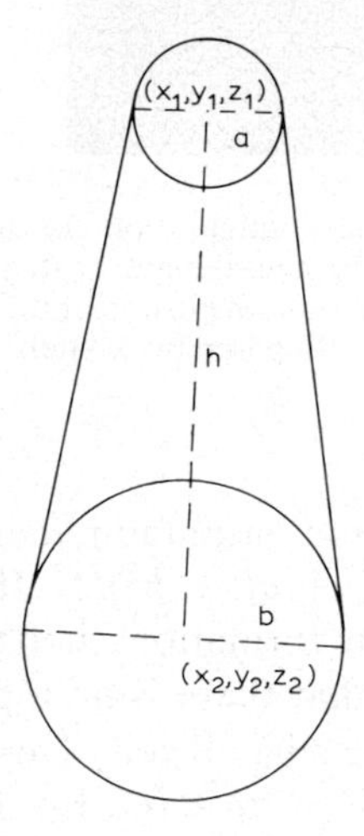

Fig. 23. Spherical cap model for a bottle cell. The number of independent parameters is reduced to eight, compared to the sixteen parameters in the representation in Fig. 22. The lateral surface of the cell may be taken as the conical surface tangential to both spheres, or connecting parallel equators, as shown.

This model, then, gives a plausible explanation for the existence of two shape changing macromolecules in each cell: they are both necessary to give the cells explicit three-dimensional shapes versus time which can lead to controlled folding and rolling up of the neural tube.

4.3 *Representation as a polygonal array*

The above representations of epithelia are inadequate to describe certain models for morphogenesis during neurulation. For instance, the notoplate is a fascinating part of the neural plate that changes shape from a broad structure to a long, narrow one, at constant area (Fig. 3a), eventually becoming so narrow that its cells string out amongst the rest of the neural plate cells (A. G. Jacobson, personal communication). Like the notochord (in most chordates), the notoplate is thus a transient structure. It is intimately associated with the notochord until neural tube closure, and afterwards supports its own elongation. Recent work by Malacinski and Yoon [11] would suggest that the elongation of the notoplate may occur without help from the notochord.

The mechanism of notoplate elongation, long an enigma [5], could possibly be explained by assuming the existence of a negative surface tension. When two ordinary embryonic tissues are placed in contact, one tissue envelops the other. This phenomenon may be accounted for by *differential adhesion* or, equivalently, a surface tension between the two cell types [54]. The motion is resisted by a tissue viscosity, defined in Section 4.2.2. (The tissue viscosity is probably caused by the friction between cell membranes [24,25] as they slide past one another.) There is a tremendous rearrangement of notoplate cells during early neurulation. Let us suppose that this rearrangement is driven by a surface tension between notoplate and mesoplate cells. Since the perimeter of the notoplate increases with time, this surface tension must be negative.

There is nothing nonphysical about a negative surface tension. If one tries to layer ethanol on water, the interface increases indefinitely until the ethanol and water are thoroughly mixed at the molecular level. The process is driven by what amounts to a negative surface tension [55]. This is just the 'checkerboard' pattern of cell sorting [56,57].

If a fluid object is increasing its perimeter or surface area due to a negative surface tension of its interface, its changing morphology is likely to be quite sensitive to the slightest external force. This is because a negative surface tension exerts no restorative effect on the shape. Now, in the normal newt embryo there is a 'tongue' of mesoplate cells anterior to the notoplate which undergo the maximum shrinkage (Fig. 10). Since these cells are arranged in a line, they should pull on the notoplate. This pull may bias the lengthening of the notoplate into the anterior/posterior direction.

For a more refined model, let us suppose that instead of a sharp boundary of negative surface tension there were a gradient of this property. The cells have the property that the further apart they are, the more strongly they adhere. *Provided* they had an opportunity to touch, they would pull up to one another, squeezing the other cells out of the way. If there were a bilaterally symmetrical, lateral gradient of such cell–cell adhesions, cells would squeeze together in the lateral directions and squeeze the others out of the way in the anterior/posterior direction. The result would be a narrowing and lengthening of the notoplate. As with a sharp boundary, the stringing out into a single row of cells would still be explainable. Such a gradient

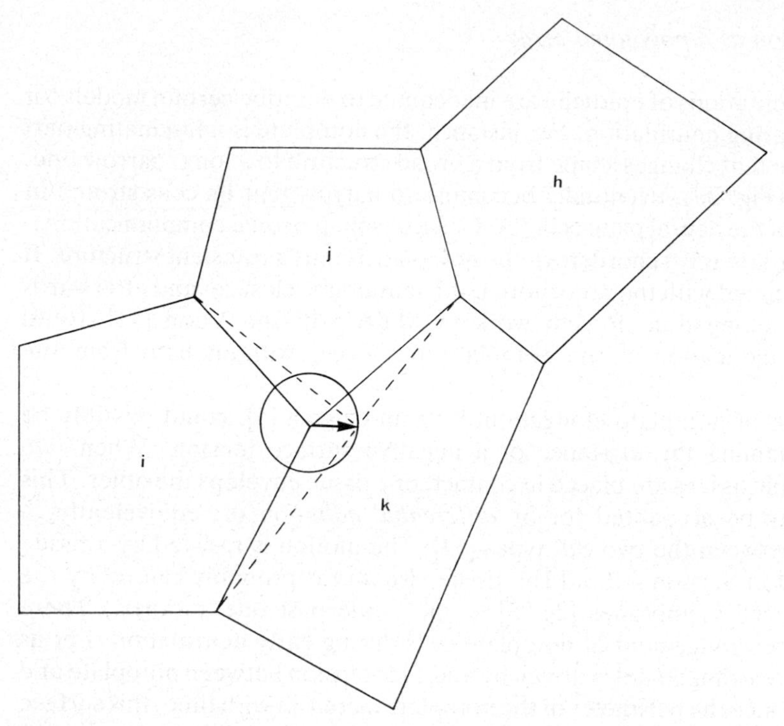

Fig. 24. Four cells in a polygonal representation of an epithelium. The cells cover the plane, with no gaps or overlaps between them. One has to calculate the motion of the vertices. A circle of radius ε is taken around a vertex. The centre of this circle is taken as the origin of a polar coordinate system. Each of the three interfacial edges would change in length in a different way for each orientation of the radius vector. However, there is a unique minimum of E (Eq. 5) at the angle given by Eq. 6. If the length of the interface between cells j and k were to get too small, a switch of neighbors would occur, as in Fig. 8, so that h and i would share an interface.

mechanism would also lead to the same change of shape, independent of exactly where cuts excising the notoplate/notochord fragment were made (cf. the experiment in Jacobson and Gordon [4]). The shape change would be intrinsic to the tissue, and not depend on the presence of the mesoplate.

If we adopt this representation, we are faced with the problem of simulating cell sorting in a confluent monolayer. Every cell in the epithelium can be represented as a convex polygon, with no spaces between polygons (Fig. 24). Each vertex of the cell is shared by exactly three cells (unless the vertex is at the free boundary of the tissue). For each pair of cells i and j, we assign an interfacial tension γ_{ij}. (The units of surface or interfacial tension, [dynes/cm], may also be expressed as work per unit area, or [ergs/cm^2].) In the above model, if the x-axis has its origin on the midline and measures lateral distance, we would be taking, for example:

$$\gamma_{ij} = c|x_i - x_j| \tag{4}$$

where c is an adjustable parameter. Let l_{ij} be the length of the interface between cells i and j and let h_{ij} be its height (essentially the local cell height). Then we may write an equation for the total interfacial energy of the tissue:

$$E = \Sigma\, l_{ij}\, h_{ij}\, \gamma_{ij} \tag{5}$$

Our computing problem is then to stimulate the dynamics of an epithelium changing at each small time interval in such a way as to reduce E. However, we cannot simply go for a global reduction in E, since the real embryo, and thus our computer simulation, may and should get trapped in metastable states, if they are encountered [56,57]. The answer to this difficulty is found in taking the fluid nature of the cell mixtures literally [24,25]. For the case in hand, it means that our simulation of sorting, confluent epithelia must have appropriate fluid-like behavior. We must therefore take especial care in designing the rules by which cells change their relative positions and change neighbors. Unfortunately, we must work partially in the dark, because observations of epithelial tissues known to be acting this way have not yet been made. Observations of cells changing neighbors either lack sufficient resolution to see individual cells [4] or are of very few cells [9].

We have found two rules of motion of the vertices that may prove adequate for fluid-like behavior. Each vertex is moved in the direction of maximal decrease of the local value of E or E_v. The formula for E_v is identical to that for E (Eq. 5), except that the summation is taken only over the three interfaces (edges) emmanating from the vertex v. Consider a polar coordinate system centred at the vertex, so that the three edges, with specific interfacial energies γ_{ij}, γ_{jk}, γ_{ki}, extend to the points (l_{ij}, π_{ij}), (l_{jk}, π_{jk}),(l_{ki}, π_{ki}) (Fig. 24). If we take a circle of infinitesimal radius ε around the vertex, and consider moving the vertex to a point at angle θ on the circle, then we can calculate $E_v(\theta)$. We find that $E_v(\theta) - E_v$ has its maximum at:

$$\tan \theta_{max} = \frac{\gamma_{ij} \sin \pi_{ij} + \gamma_{jk} \sin \pi_{jk} + \gamma_{ki} \sin \pi_{ki}}{\gamma_{ij} \cos \pi_{ij} + \gamma_{jk} \cos \pi_{jk} + \gamma_{ki} \cos \pi_{ki}} \tag{6}$$

We thus calculate this direction for each vertex, and then take a small step in that direction. To keep the proportions of the steps commensurate, the length of each step is taken as a monotonically increasing function of $E_v(\theta_{max}) - E_v$. The steps must be small enough that a vertex never crosses over a cell–cell interface, disrupting the topology.

Cells are given an opportunity to change neighbors when two vertices come within a small distance of one another (Fig. 8 and 24). The criterion for exchange of neighbors is particularly simple: we replace the interface between cells A and B by a perpendicular interface of equal length between cells C and D. The interchange is made if $\gamma_{CD} > \gamma_{AB}$, otherwise it is not.

It will still be necessary to restore the apical area of each cell after its vertices are moved. We again use a repacking algorithm that propagates from a packing centre. Two of the vertices of a cell are anchored, and the rest are extended (or withdrawn) radially along straight lines through the centroid of these anchored vertices, until the desired (shrunk) area of the cell is matched (Fig. 25).

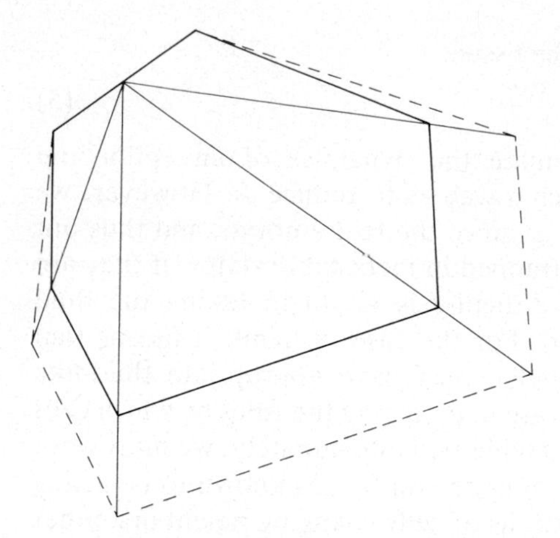

Fig. 25. In the polygonal representation of an epithelium, each cell may often have to have its area adjusted to its current value, according to the cell's apical shrinkage program. This can be done by propagating a repacking of the cells which does not disrupt their topology. Here a cell's area is adjusted by anchoring two of its vertices and expanding the rest radially from the centroid of the anchored vertices.

Cell division may be included, but the manner in which it occurs must be carefully designed to match real cell behavior [58,59].

4.4. *Representation as a bubble raft*

Cells have been treated as bubbles for well over a century [58]. I would like to show how, if we regard an embryonic epithelium as a bubble raft, the cell–cell adhesivenesses postulated in Section 4.3 may be directly estimated from the radii of curvature of the cell–cell interfaces [60]. In all artificial bubble rafts, the surface tension is the same for all bubbles. But since we anticipate adhesion gradients in embryonic tissues, let us assign every cell interface a (possibly) different surface tension, γ_{ij}, as in the previous section. Let us also suppose that each cell can support a hydrostatic pressure P_i, say just enough to keep the cell from being flaccid. (We are assuming that the pressure is a scalar quantity, and that it is uniform everywhere in a cell.) Under these assumptions, for every cell–cell interface we may write a Laplace equation [61]:

$$P_i - P_j = 2\gamma_{ij}/R_{ij} \tag{7}$$

where R_{ij} is the *radius of curvature* of the interface. An epithelium is now represented as a confluent layer of *poly-arcs* instead of polygons. For convenience let us use the *curvature,* defined as:

$$K_{ij} = 1/R_{ij} \tag{8}$$

or

$$P_i - P_j = 2\gamma_{ij} K_{ij} \tag{9}$$

For a flat interface between two cells the curvature is zero, making it easier to deal with than the corresponding infinite radius of curvature.

From a micrograph we can measure each K_{ij} [60]. We do this by using a graphics tablet, pointing to two attached vertices, and then recording the coordinates of a few points along the membrane between the vertices. A best fit to these points of a circular arc through the vertices is then calculated. The best fit circle is defined as the one for which the sum of the distances to the measured points is a minimum. It is found by doing a binary search [62] to find the curvature which minimizes this function.

We have one equation of the form of Eq. (9) for every cell–cell interface. The equations are linked through the pressures in the cells. Consider the three cells i,j,k sharing a single vertex. Then we have:

$$\begin{aligned} P_i - P_j &= 2\gamma_{ij} K_{ij} \\ P_j - P_k &= 2\gamma_{jk} K_{jk} \\ P_k - P_i &= 2\gamma_{ki} K_{ki} \end{aligned} \tag{10}$$

or, adding these together and dividing by 2:

$$0 = \gamma_{ij} K_{ij} + \gamma_{jk} K_{jk} + \gamma_{ki} K_{ki} \tag{11}$$

Let us make one further assumption, namely that all the γ's cluster around a mean value Γ, and define ε_{ij} by

$$\gamma_{ij} = \Gamma(1 + \varepsilon_{ij}) \tag{12}$$

Equation (11) then becomes:

$$-(K_{ij} + K_{jk} + K_{ki}) = \varepsilon_{ij} K_{ij} + \varepsilon_{jk} K_{jk} + \varepsilon_{ki} K_{ki} \tag{13}$$

While Eq. (13) is one equation in three unknowns, it has considerable structure and can give us some information about the ε's. In particular, we know they cannot all be zero, because the left hand side of Eq. (13) is rarely zero in the tissues we have measured.

For every vertex between three cells, we obtain an equation of the form of Eq. (13). Let us define

$$\varepsilon^2 = \Sigma \varepsilon^2_{ij} \tag{14}$$

We can solve the simultaneous Eqs. (13) subject to the constraint that ε^2 be minimized, using Lagrangian or undetermined multipliers [63,64] (see Fig. 26). As more vertices are considered, more interdependence of the unknown ε's is forced into the equations. For a group of bubbles the ε's cluster around zero, as expected. The cells have a wider range. (Values below -1 may either be part of the numerical error of this estimation procedure, or may be indicative of a role of other physical processes, such as elasticity, in determining the shape of a cell–cell interface. We take the bubble model as only a first approximation.)

By puncturing and cannulating an individual cell we may be able to drop its internal pressure to zero, establishing the value for one of the unknowns experimentally. Movies of the changing shapes of the nearby cell–cell interfaces could then provide more information on the ε's.

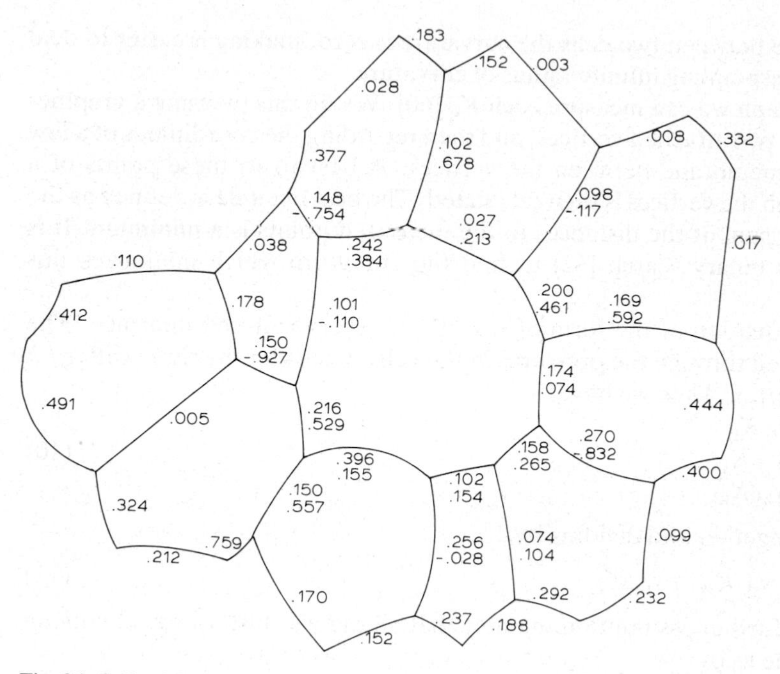

Fig. 26. Cells of the superficial cell layer of a frog embryo at gastrulation, from a time-lapse movie kindly provided by R. Keller. The interfaces between the cells are recorded using a computer graphics tablet with a computer program that yields the best fit of a circular arc for each cell–cell boundary. Those circular arcs are plotted by computer here. The top number is the relative curvature K_{ij}. The bottom number, where available, is ε_{ij}, the estimated deviation from the mean surface tension. From Stein and Gordon [60].

The additional parameters needed for a poly-arc representation of the neural plate would be the curvatures of each interface and the pressures within the cells. This detailed shape and pressure of each cell may be important in testing models of the type considered in the next section.

4.5. *Representation of intercellular transport*

One observation suggested by the above analysis is that the internal pressures of embryonic epithelial cells differ from one cell to the next. The pressure differences between adjacent cells may lead to the flow of molecules between them (R. Webber, personal communication). These flows could become organized through a cooperative process that would explain the pattern of primary neural induction (Fig. 10). (The model to be presented is clearly a first order approximation, since it cannot explain the qualitative differences between various inductive events that occur later in embryogenesis.)

Each neuroepithelial cell may receive and release molecules due to pressure differences with its neighbors. At a given stage, there may be an imbalance in the inward and outward flows. This imbalance would alter the pressure within the cell, and thereby the flow pattern into and out from the cell. If we were to assume completely passive diffusion, these flows would settle down to a spatially uniform equilibrium. However, if the diffusion were activated, i.e., oriented, other results would be possible. A cell might respond to an imbalance by reversing the flow of one or more of its interfaces, so that the diffusing molecules would come in one side and go out another. Such a property would constitute a *polarity* to the cell (cf. [65]).

The flow through a cell may change by redistribution of the 'pumps', such as sodium pumps, floating in the fluid cell membrane. This could happen in response to net electric fields set up by the flow of the pumped ions, for example. A model of this kind has been proposed for the development of polarity in undivided *Fucus* eggs by Jaffe [66]. The breakdown of the symmetry of the *Fucus* egg has been analyzed theoretically by Larter and Ortoleva [67]. Models for electrophoresis of protein molecules floating in bilayer membranes that act as two dimensional fluids have been analyzed by Poo [68].

While no one has looked for molecular pumps between neural plate cells, there are many micrographic and physiological observations of gap junctions, etc., which undergo dynamic changes during neurulation (cf. Decker and Friend [69]). (Their role remains unknown. See Schoenwolf and Kelley [70].) The backflow through the gap junctions may be what is reflected by the cell–cell interface curvatures. Sheridan [71] remarked that these low-resistance intercellular junctions 'clearly permit cells to share the pumping or distributing of small ions.... Another possibility is that the coupled cells can control each other's activity by exchanging larger molecules across the junctions used by small ions... the possibility of direct exchange of molecules of this size between cytoplasms obviously must be kept in mind in considering induction and morphogenetic movements'. Moreover, although gap junctions may be without polarity, they need not be passive, since hormones can alter the size of their openings [72].

If a cell were to respond to a flow imbalance, it may then create a similar imbalance for its neighbors. Thus we can imagine that such an effect could propagate across the whole epithelium. (The intercellular spaces may be involved in this propagation, since they would otherwise be drained or swelled by an imbalance of pumps on their facing membranes.) If the epithelium is topologically equivalent to a spherical shell, the overall flow pattern may be confined to a relatively few geometric patterns (c.f. Sadourny [73].) It is clear that one of these patterns is similar to a Greek capital upsilon (Fig. 10). The cooperative effect of flow orientation between neighboring cells could sharpen this pattern. The pattern of electric current over the primitive streak of chick embryos [74] may be 'leakage' from these cell to cell currents, or otherwise related to them.

This proposed mechanism for primary neural induction may also explain the ability of the presumptive neural plate to regulate. If a patch of the epithelium is experimentally rotated in place, a number of things will happen. Until the wound boundaries reunite, all the flows along the boundaries may cease. During this interval, new flow patterns may be established. (This depends on the rapidity of flow reorientation compared with wound healing.) Once the gates are, so to speak, open, two possibly incompatible flow patterns will be running into one another. We may anticipate that the subsequent reorientation of the flows would normally be dominated by the larger portion of the tissue. Thus, eventually, the normal flow pattern may be restored. We thereby may have an explanation for *regulation*.

The role of the mesoderm in inducing the upsilon pattern may be to start the flow in the right direction. The presumptive notochord, which is the leading edge of the underlying mesoderm, may somehow organize those neural plate cells over it (the notoplate) into a common flow pattern along the anterior/posterior axis, either towards or away from the *dorsal lip of the blastopore* (Fig. 1). This region could, indeed, be regarded as an organizer [75] since if it were transplanted to an area with no flow pattern, or a weakly established one, it could induce a flow into the whole new area. (The *organizer* is a piece of tissue which, when transplanted to other parts of a vertebrate embryo, causes development of a whole additional nervous system.) The 'strength' of the organizer comes from two plausible properties: (1) the flows have a sharply defined common orientation, compared to the rest of the epithelium; and (2) the speed of flow may stabilize the direction. Due to the feedback nature of this mechanism, the organizer may, effectively, be regarded as a region in which a cooperative phase transition has locked the cells into a given orientation of flow.

This cooperative effect may, then, propagate over the whole tissue, locking it into a particular flow pattern. When the pattern is fully established, each region may be similar to the organizer in resisting change. This property corresponds to what is called *determination*.

The mesoderm has tissue movements similar to those of the overyling neural plate [76]. This may reflect a similar upsilon pattern of cell behavior and flow patterns. In fact, we may regard both patterns of a topologically continuous flow pattern.

In this model the maximum flow occurs along the cells of the notoplate. It may be this maximum which sets these cells apart from the rest of the neural plate. There will also be a sharp, bilateral gradient of flow rate across the notoplate. This flow gradient may be responsible for the adhesiveness gradient amongst these cells postulated in Section 4.3. A similar effect may separate the notochord from the rest of the mesoderm.

My first attempts at modelling interactions of this type between polar cells were attempts to model the morphogenesis of hydra with L. Wolpert in 1970. In these models, only a single line of cells was considered. Each cell in this one dimensional representation had to make a 180° flip in orientation to change polarity. Due to this coarse behavior, computer simulations of the model had no predictive power. However, if we were to now model the hydra as a two-dimensional (cylindrical) epithelium, with cooperative cell–cell polarity interactions, the orientations of the polarities could change more gradually from cell to cell. The mechanism proposed here for primary neural induction may thus also be applicable to the morphogenesis of other epithelia. A factor in favor of this view is that hydra containing only epithelial cells undergo normal morphogenesis [77]. Vegetative budding of hydra may correspond to the development of alternative flow patterns, rather than peaks in a reaction/diffusion mechanism [78,79].

Certain major embryonic malformations, such as Siamese twinning, two heads, etc., may be due to alternate, metastable flow patterns. Small perturbations may be all that is necessary to get the cooperative flow pattern into a metastable state. The effect of elevated temperature in producing gross abnormalities may just be to influence the strength of the cooperative effect between polar cells. Thus we see that this model may give a rational basis for understanding some of the major malformations of embryos and fetuses.

The computer simulation of intercellular transport requires a few additional parameters beyond those in the previous sections. Each cell must be allotted a population of molecular pumps, which distribute themselves amongst the cell–cell interfaces with orientations either into or out from their cell. (Some may, of course, be on the apical or basal surfaces, and not be involved in cell to cell pumping.) Each cell and intercellular space must contain some concentration of the pumped chemical species. A mechanism has yet to be proposed that would couple a high or low concentration with the electrophoretic mobility or flipping of the pumps. Some such mechanism is needed to move the pumps from one cell–cell interface to another within the same cell. The work of Kell, Clarke and Morris [80] and Kell and Morris [81] may suggest a mechanism. (The exchange of pumps between cells, through exocytosis and endocytosis, etc., is not out of the question.) Since we are dealing with molecular events, Monte Carlo computing methods, depicting their stochastic nature, may be in order [82].

As the number of parameters describing a cell increases, so does the potential for interesting interactions amongst the parameters. For instance, the sliding of membranes must somehow alter the distribution of the pumps they contain. Conceiving and unraveling these interactions may be one of the important future challenges in uncovering the molecular basis for morphogenesis. It is clear that computational embryology will play a large role in evaluating such models.

5. COMPUTER-AIDED OBSERVATION OF CELLS IN LIVING EMBRYOS

Many of our attempts to model vertebrate neurogenesis are limited by a lack of information about what the cells of the embryo are actually doing. In addition, attempts at simulation often lead to a requirement for new information. In this section I will show a few ways in which microscopes and computers may be used in an integrated way to build new, powerful tools for embryological investigation at the cellular level. I will emphasize those methods that permit enhanced in vivo observation, since movements, changes of neighbors, mechanical responses, etc., are only observable on living material.

5.1 Computer-controlled focusing

Any reader who has attempted to take time-lapse movies of developing embryos has experienced the frustration of drifting focus. The causes can be thermal heaving of the microscope, hysteresis in its moving parts, changes in humidity affecting a gel supporting the embryo, or even growth of the embryo out of the focal plane. The higher the magnification, the more rapidly the image goes out of focus. Thus a major improvement in time-lapse microscopy of embryos would come about if we could keep the embryo in focus automatically.

It is possible to bring an image into focus using image processing methods before recording it permanently on film, videotape, or in a computer's memory. Thus I will only briefly mention some of the problems unique to embryological work, rather than the special purpose hardware or software (computer programs) needed to get a picture into focus.

There are two types of embryos we need to consider: transparent and opaque. The newt embryo is essentially opaque, so that we are restricted to viewing its surface. High resolution movies have so far only been made of the part which is uppermost in the floating embryo [39]. Since the cells change positions relatively slowly, and the variegated pigment can be used to obtain a high contrast image, it should be possible to keep ectodermal newt cells in focus by simple Fourier analysis of the image. One would acquire the image into a computer via a digital television camera, calculate the Fourier components of a few representative scan lines across the image, and repeat at different focal levels. The level giving the most high-frequency components should correspond to the sharpest focus. A 'binary search' [62] on the focal levels could be done, to home in on the level of sharpest focus in the smallest amount of computing time, and with the fewest trial pictures.

With transparent embryos one must face the problem of multiple levels of focus [83,84,42]. That is, each level in the embryos is likely to have some cell or parts of cells in focus. Here we might best acquire an image of the cell(s) we want to follow in the computer, setting the initial focus carefully ourselves. It should then be possible to use 'template matching' [85] to find the focus level which gives an image best corresponding to the image we started with. Of course, after a while the cell(s) will change shape or configuration so much that they will not match the original template image at all. Thus we may consider a 'running template' method, in which each image, after it has been matched as accurately as possible to the previous template, itself becomes the template for the next round of refocusing. This is, in essence, an extension of cell tracking methods for tissue culture into three dimensions [86].

Cell tracking has the unsatisfactory nature of being a pattern recognition problem. Many difficulties can arise. The cell can divide. Its image can overlap that of another cell. Or it might die and disappear altogether. In the meantime, other interesting cells and events could be missed. Pattern recognition techniques thus presume a considerable foreknowledge of the phenomenon being observed. For this reason, I prefer to emphasize image processing methods that are not so dependent on knowing the nature of the picture in advance.

5.2. The through-focus mosaic

In any embryo, transparent or opaque, we only see one plane in focus at a time. Suppose that we took a through-focus series of pictures and 'cut-out' the parts of each picture which are in focus. These parts could be combined into a single picture, which I call the 'through-focus mosaic'. A hand made through-focus mosaic, using only two focal levels, is shown in (Fig. 27). The equivalent operation could be done by

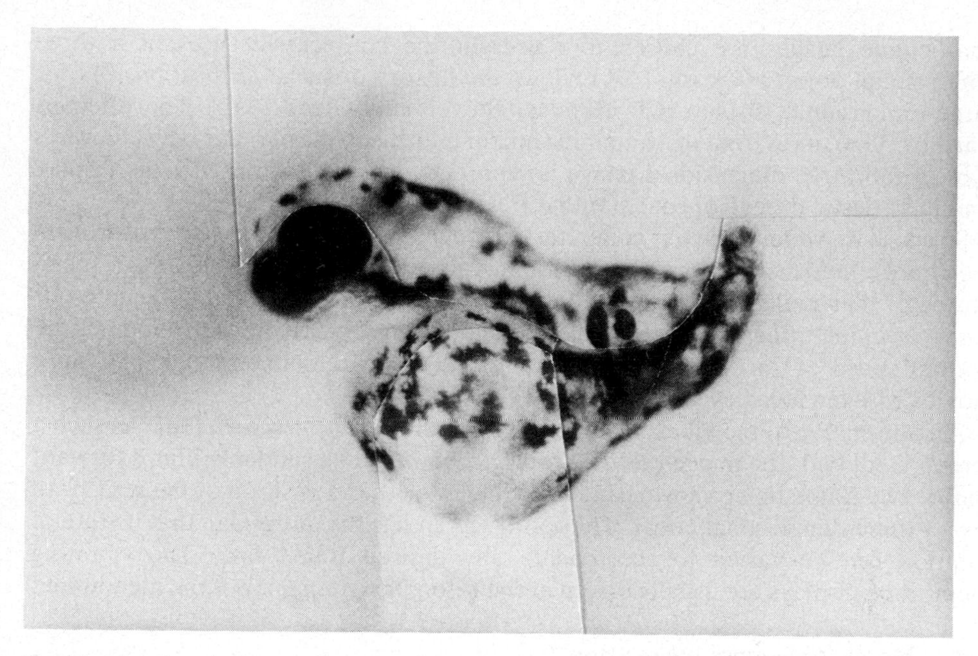

Fig. 27. A hand-made through-focus mosaic picture. This 'tailless' fish embryo was photographed at two focal levels. The in-focus parts of each print were cut out and taped together.

computer, by using what is called a 'block transform' method [87]. Each focal level picture would be divided into small squares, each of which would be Fourier analyzed. Those blocks which had fine detail, reflected in the presence of high-frequency Fourier components, would be copied to a new picture. If two blocks in the same part of pictures at different levels were both found to be in focus, they could either be averaged to give a 'see through' effect, or a choice could be made of the one in front or in the back. The latter approach allows us to separate the front from the back layers of the embryo. The through-focus mosaic effectively converts our microscope to a telescopic lens.

5.3 Three-dimensional time-lapse movies

A device has recently been invented which produces three-dimensional computer pictures in a rather unique way, and which can be adapted to produce three-dimensional time-lapse movies. Such movies would permit us to keep a record of the whole thickness of the living embryo at once. In this new three-dimensional display, a membrane is set vibrating at 30 cycles per second using a low-frequency loudspeaker. Its surface is reflective, so that as the membrane vibrates it changes from a convex to a concave mirror and back. A cathode ray screen is viewed by reflection off the vibrating mirror. Whatever is displayed on the screen thus appears near or far, depending on the momentary shape of the mirror. The display is synchronized with the vibrations of the mirror, so that the image changes as the mirror moves. The consecutive images are chosen to represent the consecutive slices of a three dimensional object. (Okoshi [88] reviews the history of such *varifocal mirrors* for three-dimensional display. Cf. Rhodes, Stover and Glenn [89].) The effect is startling. One sits in front of a humming mirror and due to the parallax between one's eyes, a full three-dimensional image is seen. Two people can observe the display simultaneously, though of course with different viewing angles.

Suppose we were to use our computer controlled microscope focus to synchronize the change in focus with the taking of a movie frame. If we then had a high speed projector that could be synchronized with a vibrating mirror, we could see into our embryos in full three dimensions. The projections must be run at, say, ten times the normal speed of 24 frames per second, to give us three-dimensional time-lapse movies with ten focal levels.

An alternative to the vibrating mirror is a spiral projection screen [90]. As such a screen is rotated, the image recedes from the viewer, then suddenly jumps forward again. The frame being shown can be correlated with the position of the screen, to give a three-dimensional effect. The spiral screen has the advantage that its rate of rotation can be varied to accomodate any desired frame rate. The vibrating membrane displays are usually driven at the resonance frequency of the membrane.

5.4. *Mosaic time-lapse microscopy*

When we look through a microscope at high magnification, our field of view is rather restricted. If we were to use a computer to drive stage motors in x and y, we could acquire a set of images at high magnification covering a whole embryo. If this were done at regular intervals, we would have time-lapse mosaic microscopy. As I have noted, single cells can move considerable distances over an embryo. Thus we often could use a record of the whole surface of the embryo at high magnification.

The amount of data stored in time-lapse mosaic microscopy is immense. A typical computer image contains 512×512 pixels, each of which maybe digitized to 64 grey levels (6 bits). If we want each pixel to correspond to half a resolution element, then our optics should be arranged so that each pixel is, say, 0.1 μm wide. A small embryo, 1 mm in diameter, would then require an array of $10{,}000 \times 10{,}000$ pixels or a mosaic of 20×20 or 400 pictures. The total digital storage needed for one mosaic snapshot would be 6×10^8 bits. A single 9-track computer tape 2,400 feet in length, at 6,400 bits per inch, holds 1.4×10^9 bits, or a little more than two mosaic pictures. For this reason we are likely, for the time being, to use the computer mostly to keep the mosaic images in focus, but to record them on movie film or video tape or disc, with a camera triggered by the computer.

The film we obtain, however, will be rather peculiar. It could not be shown as a continuous motion picture. Rather, we will have to turn to the computer even to view it. Using a computer controlled single frame (stop motion) movie projector, we may follow a single cell, letting the computer locate the consecutive frames we should find it on. Stevens et al. [91] have designed a special, high-speed film transport which could be used for rapidly jumping to appropriate mosaic serial section images under computer control. If the cell being followed goes off the edge of the frame, the computer could calculate which frame number to skip to for continued tracking. One could envisage returning such frames to the computer, to produce a time-lapse movie of the change of shape or neighbors of a given cell over a major segment of embryonic time. It would be as if we have trained our camera on that very cell. Thus cell tracking can be done retrospectively, using an interactive system, rather than a fully automated one.

Mosaic microscopy may have some of the same difficulties as mosaic methods for earth satellite surveys and planetary exploration [92]. Lens aberrations may have to be corrected, seams due to non-uniform illumination may have to be removed, and most importantly, discrepancies due to cell motion during acquisition of a mosaic group of frames will have to be accounted for. This approach may nevertheless be worth the immense effort, since it permits us to follow every visible cell during morphogenetic processes.

On a nearly spherical embryo, such as the newt, we have the additional problem that due to its opacity we can only see one side at a time. To solve this problem I am designing a rotating microscope. The microscope will rotate around the embryo, which is carried by a stage that can be moved in x, y and z in order to bring mosaic patches into focus. Mosaic methods more nearly equivalent to image processing for planetary exploration will be needed to handle the spherically arranged data.

5.5. *Computed tomography with light and X-ray microscopy*

Light microscopy and classical medical tomography, an X-ray technique [93, 94] share the characteristic that they produce images of a plane within an object by passing radiation through the volume of material above and below that plane. Classical tomography is rapidly being replaced by what is called *computed tomography* or *reconstruction from projections* [94] which has three major characteristics:

(1) the image of the desired plane is computed from measurements of the transmitted radiation;
(2) the radiation traverses only the plane of interest;
(3) the images are superior in density discrimination to those produced by classical tomography.

One may attempt to imitate computed tomography in light microscopy by placing a live embryo in a capillary tube and rotating it to obtain views from many angles. Each plane perpendicular to the axis of the tube, which is also the axis of rotation, could then be reconstructed.

Computed tomography is ordinarily accomplished with a narrow X-ray beam at resolutions orders of magnitude worse than the diffraction limit for X-rays. However, if we were to try to form a narrow beam of light for these purposes in a light microscope, diffraction effects would severely limit the attainable resolution. A pencil beam of light would have to be formed by an objective and condensor of nearly zero numerical aperture. If we were to increase the numerical aperture, then the form of the beam would be a cone. The cone of light would give us a measure of

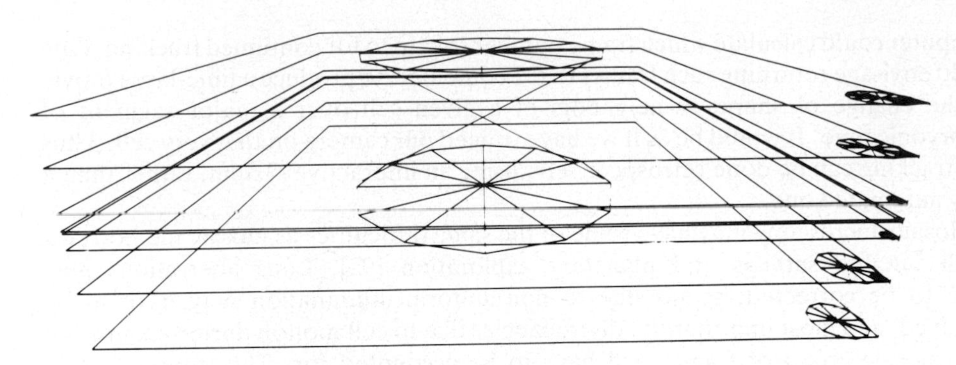

Fig. 28. Computer sketch of the double cone spread function of the light beam in a microscope. Note that two such cones with adjacent vertices will overlap considerably. The circles on the right show the sizes of the picture elements used for each plane. Computational speed is increased by using larger picture elements for planes further from the focal plane.

the material in a whole double conical volume of the object, rather than along a narrow strip. Moreover, this means that information from planes other than the plane of interest would be included. Thus I conclude that it is not desirable to imitate the geometry of computed tomography in high-resolution light microscopy.

X-ray microscopy [95] offers a possible means of carrying out computed tomography on embryos [94], especially those that are optically opaque. A number of instruments for single view X-ray microscopy has now been built [96], some of which acquire images directly in digital form. It would be a mere matter of rotating the specimen (or the microscope) to obtain the multiple views needed for computed tomography (cf. Sayre [97]). The major question to be resolved would be what could be seen. It has been shown, for instance, that chromosomes have a high contrast against cytoplasm using soft X-rays (Fig. 8c in Engström [95]). But what about nucleus versus cytoplasm, yolk granules, basement membranes, vacuoles, etc.? Can we develop vital X-ray stains, leaning on the experience of radiologists with what they call contrast agents? We have here a whole portion of the electromagnetic spectrum that has yet to be explored for its usefulness in embryology. If, for insance, we had sufficient contrast to see nuclei, we would have, for the first time, the prospect of following every cell as it moved inside a developing embryo.

5.6. *Computer sharpening of the focal plane*

Let us return to light microscopy and consider another way of approaching the problem of three-dimensional reconstruction. We obtain our highest spatial resolution in light microscopy by using the largest possible numerical aperture. Each point on a photograph or a digital image acquired through a microscope actually represents an integral over a double cone whose vertex is in the focal plane (Fig. 28). Thus when we think that we are looking at the focal plane, we are actually seeing the focal plane and all the planes above and below. All these planes except the focal plane are, however, blurred out. They contribute at best a general haze through which we must observe the focal plane. At worst, they can totally obscure it. Can we achieve an actual removal of the blurred planes from the image?

The answer is a qualified yes. We can note that two adjacent points in the focal plane have overlapping double cones (Fig. 28). If we were to set up a three-dimensional grid and assign an (as yet unknown) optical absorption ρ_{ijk} to each point (i,j,k), then each integral over the double cones would correspond to the sum

$$R_{mn} = \sum_{ijk} \gamma_{ijkmn} \rho_{ijk} \tag{15}$$

where R_{mn} is the reading or measurement off the corresponding point (m,n) on the film or pixel in the acquired picture. The summation is taken only over those points (i,j,k) that fall within the double cone whose apex is at the three-dimensional point$(m,n,0)$. (The focal plane is assigned z-coordinate 0.) Since the cone spread function is likely to be constant over the whole volume, the weight assigned to each point, γ_{ijkmn}, will depend on $i-m$ and $j-n$. The major factor in calculating the weight, at least for bright field illumination, will be the divergence of the cone relative to the volume elements represented by the (i,j,k). For other forms of microscopy, such as differential interference contrast, where the theoretical derivations of the three dimensional radiation field may not be accurate, S. Inoue (personal communication) has suggested that the weights be derived empirically.

These equations are now in the standard form of a set of simultaneous linear equations that may be solved by iterative computed tomography algorithms, such as ART (Algebraic Reconstruction Technique) [98, 99]. It does not matter that the geometry is not the same as in ordinary computed tomography. What we have here points to a rather unique development in the history of light microscopy. The image seen through the microscope is regarded merely as input data to a computer calculation, which will produce an image that is sharper and more reliable than that from the microscope itself. In the future we may cease looking through microscopes, but rather turn to the processed images for our visual information. (Cf. new forms of video microscopy described by Miller [100] Inoue [101] Allen et al. [102] Allen et al. [103] and Walter and Berns [104].)

We do not have to confine ourselves to the focal plane. One may think of the deconvolution of the double cone beam spread function as a sort of three-dimensional scrub bush, which cleans up the image wherever we move the vertex. Thus with focus motor control, and with algorithms that tell us when we are in focus, we could move our cones over any surface within an embryo. This should work because, in practice, reconstruction from projections can be treated as a nearly local operation [105]. Thus we may apply it only to the surface or volume we select. This can greatly decrease the computing time.

5.7. *Three-dimensional autoradiography*

Autoradiography is usually practiced by sectioning a (dead) specimen and making a contact print from the emitted radiation. In order that the tracks in the nuclear emulsion properly localize the sources of the radiation, the emulsion is kept thin, and the emittors, such as β rays (low-energy electrons), are deliberately chosen to have a short range. The exposure times vary from weeks to a good fraction of a year [106]. We can sometimes label cells and introduce them into an embryo [107], but we then do not know where they have gone unless we kill, fix and section the embryo. We of course then get only a single snapshot in time.

Suppose that we were to take the viewpoint of nuclear medicine. In this discipline one introduces radioactive emittors into a live patient in various ways, and then forms images by using lead collimators. The emitted particles are generally high energy gamma rays. They travel in straight lines and usually exit the body. By taking multiple views, we can even carry out computations just like those in computed tomography. (The weights in Eq. (15) must include factors for the inverse square effect and some absorption of γ rays by the patient.) Thus we could obtain a three-dimensional reconstruction of the sources of the radiation. If these were single labeled cells, our task would be even easier. This gives us a realistic prospect of following individual labeled cells within living embryos [108]. The visual opacity of the embryo need be no obstacle.

5.8. *Measurement of mechanical properties*

A complete understanding of the mechanics of embryos requires not only a knowledge of where cells go, but also the magnitude of the forces driving and resisting that motion. It is difficult enough to perform tensile tests on small, dead samples [109, 110]. In order to carry out similar measurements on microscopic living tissues, we enter the insufficiently developed realm of micromanipulation. The tools that have been developed include fine glass rods that may be bent [111] the force of small magnets on steel balls [112–114], and a host of shape distortion methods for measuring surface tension [115, 116]. The method of slitting embryos [33, 4] which, especially in neural plate, produces anisotropic gapes, has yet to be developed into a quantitative technique for measuring the elastic and viscous properties of tissues. Holographic methods, which promise to measure strains [117] nevertheless do not permit us to follow the trajectories of cells. The acoustic microscope [118] may be useful in embryology, but it can only measure elastic and viscous parameters to a very small depth. Harris et al. [119] have developed a method by which the forces exerted by parts of single cells in tissue culture can be measured from the wrinkling they cause on a very thin rubber membrane. Could a harmless vapor deposition method be developed which would polymerize such a rubber membrane around the whole outer surface of an embryo? If so, we might be able to directly observe the stress/strain relations between the cells in the intact, nearly undisturbed embryo. How do we grab onto an embryonic tissue to test its stress/strain relationship? Perhaps protein glues used in human surgery could be adapted to the necessary microtechniques. The whole technology of physical measurement of the properties of cells and intercellular connections is waiting to be developed. It is clear that making more than casual measurements here and there on an embryo will require computer-aided micromanipulation and analysis.

6. CONCLUSION

The problems we are trying to solve by computational embryology have a considerable history. Wilhelm His [120] made a concerted effort to explain the mechanics of embryos using the only two-dimensional analog 'computer' available at the time: an elastic sheet of rubber. He posed the major problems of neurulation we are still trying to solve. Thus His had the main ideas with which we still struggle. A proper development of computational embryology, with its testing of models and new techniques for in vivo observation, may bring these problems to solution. I can but end with the words of His [121] (cf. Coleman [122] and Picken [123]) which are unfortunately as true today as they were then: 'Embryology and morphology cannot proceed independently of all reference to the general laws of matter, – to the laws of physics and mechanics. This proposition would, perhaps, seem indisputable to every natural philosopher; but in morphological schools, there are very few who are disposed to adopt it with all its consequences.'

ACKNOWLEDGEMENTS

The work presented above that is in progress has been carried out with the help of Yuen Cheng (Section 4.3), Gillian King (Section 4.5), Aaron Stein (Section 4.2.6) and Allison Thurlbeck (Section 4.3). The work described in Section 5.6 was carried out in the laboratory of Lewis Lipkin. I would like to thank the following people for their critical reading of the manuscript: Leslie J. Biberman, Drummond Bowden, Yuen Cheng, Victor Chernick, Antone Jacobson, Gillian King, P. D. Nieuwkoop, and Aaron Stein. This work was supported by grants from the Winnipeg Children's Hospital Research Foundation, the Sellers Research Foundation, the Control Data Corporation, and the Canadian Medical Research Council.

REFERENCES

1. Kirkland, W. (1969) The Lure of the Pond. Henry Regnery Co., Chicago.
2. Robinson, A.L. (1981) Science 211, 1150–1151.
3. Gordon, R. (1982) A review of the theories of neurulation. In preparation.
4. Jacobson, A.G. and Gordon, R. (1976). J. Exp. Zool. 197, 191–246.
5. Gordon, R. and Jacobson, A.G. (1978) Sci. Am. 238 (6), 106–113.
6. Moore, K.L. (1977) The Developing Human, Second Edition, Clinically Oriented Embryology. W.B. Saunders Co., Philadelphia, London, Toronto.
7. Balinsky, B.I. (1975) An Introduction to Embryology, 4th ed., W. B. Saunders Co., Philadelphia.
8. Ballard, W. (1976) Biosci. 26, 36–39.
9. Keller, R.E. (1975) Dev. Biol. 42, 222–241.
10. Keller, R.E. (1976) Dev. Biol. 51, 118–137.
11. Malacinski, G.M. and Yoon, B.W. (1981). Dev. Biol. 88, 352–357.
12. Jacobson, A.G. (1980) Amer. Zool. 20, 669–677.
13. Bunch, W.H., Cass, A.S., Bensman, A.S. and Long, D.M. (1972) Modern Management of Myelomeningocele. Warren H. Green, Inc., St. Louis.
14. Bancroft, M. and Bellairs, R. (1975) Anat. Embryol. 147, 309–335.
15. Wilson, D.B. and Finta, L.A. (1980) J. Embryol. Exp. Morph. 55, 279–290.

16. Hoerstadius, S. (1950) The Neural Crest, Its Properties and Derivatives in the Light of Experimental Research. Hafner Publishing Co., New York.
17. Noden, D.M. (1973) In: J.F. Bosma (Ed.) Fourth Symposium on Oral Sensation and Perception. U.S. National Institutes of Health, Bethesda.
18. Hamburger, V. and Hamilton H.L. (1951) J. Morphol. 88, 49–92.
19. Rugh, R. (1962) Experimental Embryology. Burgess Publishing Co., Minneapolis, Minnesota.
20. Gordon, R. (1966) Proc. Natl. Acad. Sci. USA 56, 1497–1504.
21. Kastelic, J. and Baer, E. (1980) Symp. Soc. Exp. Biol. 34, 397–435.
22. Callen, H.B. (1960) Thermodynamics, An Introduction to the Physical Theories of Equilibrium Thermostatics and Irreversible Thermodynamics. John Wiley, New York.
23. Happel, J and Brenner, H. (1965). Low Reynolds Number Hydrodynamics, Prentice-Hall, Englewood Cliffs.
24. Gordon, R., Goel, N.S., Steinberg, M.S. and Wiseman L.L. (1972) J Theor. Biol. 37, 43–73.
25. Gordon, R., Goel, N.S. Steinberg, M.S. and Wiseman L.L. (1975) In: G.D. Mostow, (Eds.) Mathematical Models for Cell Rearrangement. Yale University Press, New Haven. pp. 196–230.
26. Odell, G., Oster, G., Burnside, B. and Alberch, P. (1980) J. Math. Biol. 9, 291–295.
27. Jacobson, A.G. (1958) J. Exp. Zool. 139, 525–557.
28. Glaser, O.C. (1914) Anat. Rec. 8, 525–551.
29. Glaser, O.C. (1916) Science 44, 505–509.
30. Rhumbler, L. (1902) Arch. Entwicklungsmech. Organismen 14, 401–476.
31. Moore, A.R. (1941) J. Exp. Zool. 87, 101–111.
32. Moore, A.R. (1945) The Individual in Simpler Forms. University of Oregon Press, Eugene.
33. Lewis, W.H. (1947) Anat. Rec. 97, 139–156.
34. Gierer, A. (1977) Quart. Rev. Biophys. 10, 529–593.
35. Odell, G.M., Oster, G., Alberch, P. and Burnside, B. (1981) Dev. Biol. 85, 446–462.
36. Waddington, C.H. and Perry, M.M. (1962) Proc. Roy. Soc. London B 156, 459–482.
37. Baker, P.C. and Schroeder, T.E. (1967) Dev. Biol. 15, 432–450.
38. Schroeder, T.E. (1970) J. Embryol. Exp. Morph. 23, 427–462.
39. Burnside, B. and Jacobson A.G. (1968) Develop. Biol. 18, 537–552.
40. Ziman, J.M. (1979) Models of Disorder. The Theoretical Physics of Homogeneously Disordered Systems. Cambridge University Press, Cambridge.
41. Berge, C. (1962) The Thoery of Graphs and Its Applications. John Wiley, New York.
42. Gordon, R. and King, G.M. (1982) Dev. Biol. Submitted.
43. Roux, W. (1888/1964) In B.H. Willier and J.M. Oppenheimer (Eds.) Foundations of Experimental Embryology. Prentice-Hall, Englewood Cliffs. pp. 2–37.
44. Huntley, H.E. (1952) Dimensional Analysis. Dover Publ. New York.
45. Massey, B.S. (1971) Units, Dimensional Analysis and Physical Similarity. Van Nostrand Reinhold, London.
46. Carnahan, B., Luther, H.A. and Wilkes J.O. (1969) Applied Numerical Methods. John Wiley New York.
47. Heilbrunn, L.V. (1958) The Viscosity of Protoplasm. Springer, Wein.
48. Gaunt, P.N. and Gaunt, W.A. (1978) Three Dimensional Reconstruction in Biology. University Park Press, Baltimore.
49. Spaeth, H. (1974) Spline Algorithms for Curves and Surfaces. Utilitas Mathetica, Winnipeg.
50. Newman, W.M. and Sproull, R.F. (1973) Principles of Interactive Computer Graphics. McGraw-Hill, New York.
51. Christiansen, H. and Stephenson, M. (1980) Graphics Utah Style – 80. Brigham Young University, Salt Lake City.
52. Carlbom, I. and Paciorek, J. (1978) Comput. Surv. 10, 465–502.
53. Schneider, G.M., Weingart S.W. & Periman, D.M. (1978) An Introduction to Programming and Problem Solving with Pascal. John Wiley, New York.
54. Steinberg, M.S. (1975) In G.D. Mostow (Ed.) Mathematical Models for Cell Rearrangement. Yale University Press, New Haven, pp. 82–99.
55. Adamson, A.W. (1967) Physical Chemistry of Surfaces, 2nd ed. John Wiley, New York.
56. Goel, N., Campbell, R.D., Gordon, R., Rosen, R., Martinez, H. and Ycas, M. (1970) Self-sorting of isotropic cells. J. Theor. Biol. 28, 423–468.
57. Goel, N. Campbell, R.D., Gordon, R., Rosen, R., Martinez, H. and Ycas, M. (1975) In: G.D. Mostow (Ed.) Mathematical Models for Cell Rearrangement. Yale University Press, New Haven, pp. 100–144.
58. Dormer, K.J. (1980) Fundamental Tissue Geometry for Biologists, Cambridge University Press, Cambridge.

59. Abbott, L.A. and Lindenmeyer, A. (1981) J. Theor. Biol. 90, 495–514.
60. Stein, M. and Gordon, R. (1982) J. Theor. Biol. In press.
61. Thompson, D.W. (1942) On Growth and Form. 2nd ed. Cambridge University Press, England.
62. Graham, N. (1980) Introduction to Pascal. West Publishing Co., St. Paul.
63. Hill, T.L. (1956) Statistical Mechanics. Principles and Selected Applications. McGraw-Hill, New York.
64. Wilf, H.S. (1962) Mathematics for the Physical Sciences. John Wiley, New York.
65. Sinnott, E.W. (1960) Plant Morphogenesis. McGraw-Hill, New York.
66. Jaffe, L.F. (1968) Adv. Morphog. 7, 295–328.
67. Larter, R. and Ortoleva, P. (1981) J. Theor. Biol. 88, 599–630.
68. Poo, M. (1981) Ann. Rev. Biophys. Bioeng. 10, 245–276.
69. Decker, R.S. and Friend, D.S. (1974) J. Cell Biol. 62, 32–47
70. Schoenwolf, G.C. and Kelley, R.O. (1980) Am. J. Anat. 158, 29–41.
71. Sheridan, J.D. (1968) J. Cell Biol. 37, 650–659.
72. Caveney, S. (1980) In: M. Locke and D.S. Smith (Eds.) Insect Biology in the Future, Academic Press, New York, pp. 565–582.
73. Sadourny, R. (1975) J. Atmos. Sci. 32, 2103–2110.
74. Jaffe, L.F. and Stern, C.D. (1979) Science 206, 569–571.
75. Spemann, H. (1938/1967) Embryonic Development and Induction. (Reprinted) Hafner Publishing Co., New York.
76. Jacobson, C.O. and Loefberg, J. (1969) Zool. Bidrag (Uppsala) 38, 233–239.
77. Campbell, R.D. (1976) J. Cell Sci. 21, 1–13.
78. Turing, A.M. (1952) Trans. R. Soc. London. 237 (B641) 37–72.
79. Meinhardt, H. (1978) J. Theor. Biol. 74, 307–321.
80. Kell, D.B., Clarke, D.J. and Morris, J.G. (1981) FEMS Microbiol. Lett. 11, 1–11.
81. Kell, D.B. and Morris, J.G. (1981) In: F. Palmieri et al. (Eds.) Vectorial Reactions in Electron and Ion Transport in Mitochondria and Bacteria, Elsevier, Amsterdam, pp. 339–347.
82. Gordon, R. (1980) In: G. Karreman (Ed.) Cooperative Phenomena in Biology. Pergamon Press, New York. pp. 189–241.
83. King, G.M. and Gordon, 09. (1981) Fed. Proc. 40, 556.
84. King, G.M., Gordon, R., Karmali, K. and Biberman, L.J. (1982) J. Exp. Zool. 220, 147–151.
85. Rosenfeld, A and Kak, A.C. (1976) Digital Picture Processing. Academic Press, New York.
86. Ferrie, F.P., Levine, M.D. and Zucker, S.W. (1980) Cell Tracking: A Modelling and Minimization Approach, Proc. 5th Int. Conf. on Pattern Recognition, pp. 396–402.
87. Harmuth, H.F. (1972) Transmission of Information by Orthogonal Functions 2nd ed., Springer, New York.
88. Okoshi, I. (1976) Three-Dimensional Imaging Techniques, Academic Press, New York.
89. Rhodes, M.L., Stover, H.S. and Glenn, W.V. Jr., (1982) True three-dimensional (3-D) display of computer data: medical applications. Proc. SPIE 318, 248–253.
90. Bradley-Moore, P.R. and Woloshuk, E.A. (1980) Stroboscopic Analysing Monitor: An Optical Instrument for Creating a Transparent Solid Three Dimensional Display, Proc. 10th Ann. Conf. on Sharing of Computer Programs and Technology. In press.
91. Stevens, J.K., Davis, T.L., Friedman, N. and Sterling, P. (1980) Brain Res. Rev. 2, 265–293.
92. Haralick, R.M. (1976) In: A. Rosenfeld (Ed.) Digital Picture Analysis. Springer-Verlag, Berlin, pp. 5–63.
93. Littleton, J.T., Durizch, M.L., Crosby, E.H., and Geary, J.C. (1976) Tomography: Physical Principles and Clinical Applications. Williams and Wilkins, Baltimore.
94. Gordon, R., Herman, G.T., and Johnson, S.A., (1975) Sci. Am. 233 (4), 56–58.
95. Engstroem, A. (1962) X-Ray Microanalysis in Biology and Medicine. Elsevier, Amsterdam.
96. Parsons, D.F. (1980) Ann. N.Y. Acad. Sci. 342.
97. Sayre, D. (1980) Ann. N.Y. Acad. Sci. 342, 387–391.
98. Gordon, R., Bender, R. and Herman, G.T. (1970) J. Theor. Biol. 29, 471–481.
99. Gordon, R. (1974) IEEE Trans. Nucl. Sci. NS-21, 78–93, 95.
100. Miller, J.A. (1981) Sci. News 119, 234–238.
101. Inoue, S. (1981) J. Cell Biol. 89, 346–356.
102. Allen, R.D., Travis, J.L. Allen, N.S. and Yilmaz, H. (1981) Cell Motility 1, 275–289.
103. Allen, R.D., Allen, N.S. and Travis, J.L. (1981) Cell Motility 1, 291–302.
104. Walter, R.J. and Berns, M.W. (1981) Proc. Natl. Acad. Sci. USA 78, 6927–6931.
105. Gordon, R. (1973) Artifacts in reconstructions made from a few projections. Proceedings of the First International Joint Conference on Pattern Recognition, Oct. 30 to Nov. 1, Washington, D.C.

IEEE Computer Society, Northridge, California, pp. 275–285.
106. Stumpf, W.E. (1981) J. Histochem. Cytochem. 29 (A1, Suppl.), 107–108.
107. Weston, J.A. (1963) Dev. Biol. 6, 279–310.
108. Lou, R.-S.Y. (1976) On the Fluid Mechanical Nature of Cell Rearrangement and the Modeling of Monolayer Cell Sorting, Ph.D. Thesis, State University of New York at Buffalo.
109. Ker, R.F. (1980) Symp. Soc. Exp. Biol. 34, 487–489.
110. Currey, J.D. (1980) Symp. Soc. Exp. Biol. 34, 75–97.
111. Harper, M.A. (1967) Br. Phycol. Bull 3, 195–207.
112. Waddington, C.H. (1939) Nature 144, 637.
113. Waddington, C.H. (1942) J. Exp. Biol. 19, 284–293.
114. Selman, G.G. (1958) J. Embryol. Expt. Morphol. 6, 448–465.
115. Harvey, E.N. and Frankhauser, G. (1933) J. Cell Comp. Physiol. 3, 463–475.
116. Phillips, H.M. (1978) Amer. Zool. 18, 81–93.
117. Vest, C.M. (1979) Holographic Interferometry. John Wiley, New York.
118. Quate, C.F. (1979) Sci. Am. 241 (4), 62–70.
119. Harris, A.K., Wild, P. and Stopak, D. (1980) Science 208, 177–179.
120. His, W. (1874) Our Body Form and the Physiological Problem of its Development. Letters to a Friendly Naturalist. Unsere Koerperform und das physiologische Problem ihrer Enstehung, Briefe an einen befreundeten Naturforscher. F.C.W. Vogel, Leipzig.
121. His, W. (1888) On the principles of animal morphology. Proc. Roy. Soc. Edinburgh 15, 287–298.
122. Coleman, W. (1967) The Interpretation of Animal Form. Johnson Reprint Corp, New York.
123. Picken, L. (1956) Nature 178, 1162–1165.

NOTE ADDED IN PROOF

The rotating microscope mentioned in Section 5.4 is described in: Gordon, R. (1982) SPIE Proc. 361, in press.

Manuscript received September 22, 1981.
Revised version received February 5, 1982.

Geisow & Barrett (eds.) Computiog in biological science

3

Computer analysis of bacterial cell walls and modelling of the surface growth process

A.N. BARRETT and I.D.J. BURDETT

1. INTRODUCTION

Complex biological problems, such as those concerning growth and development, are amenable to computer-assisted analysis. For structural studies, where it may be desired to attempt to reconstruct a particular developmental sequence, the initial analysis may be materially assisted by obtaining quantitative data using computer-directed tools such as a graphics tablet or digitiser. The objective is usually to attempt to identify critical features and to reduce the problem to a mathematically convenient form, without undue oversimplification, suitable for modelling or simulation studies. The modelling process may, in turn, suggest new variables which should be identified and measured. The examples we have chosen to illustrate the use of computers in biology concerns studies of a prokaryote, the bacterium *Bacillus subtilis*. Two aspects of our present studies will be described in this chapter: (i) attempts to reconstruct growth of the cell surface; (ii) the development of averaging techniques and the enhancement of the signal to noise ratio in electron micrographs of thin-sectioned cell walls.

Both these problems are dependent upon the availability of a small computer, a graphics display system and a graphics tablet. The tablet has proved invaluable for digital recording of features observed in the micrographs. Display facilities are necessary for selection and evaluation of the data both before and subsequent to analysis. Since the processing demands are relatively cheap by comparison with applications involving energy calculations of biomolecules or Fourier transforms of

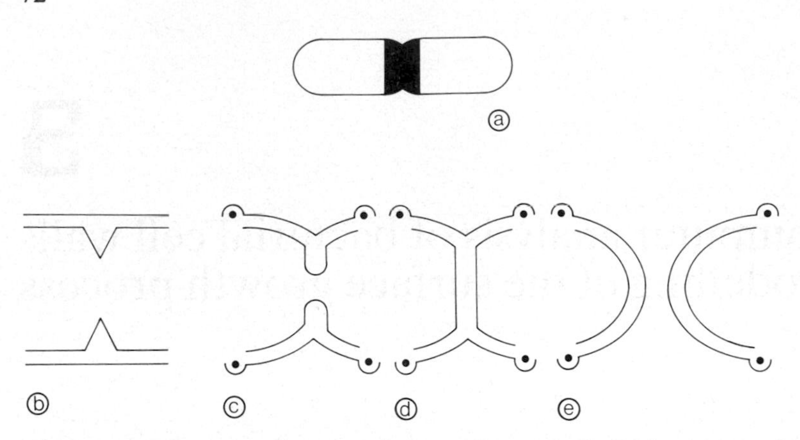

Fig. 1. Diagram showing sequence of cross-wall construction and pole formation in *B. subtilis*. (a) Location of division site and new polar surfaces at centre of cell; (b–c) stages of successive cross-wall in growth (b,c), septal closure (d), and cell separation (e), wall bands shown (●).

large arrays, a modest sized computer such as a PDP11/34 is quite sufficient for the recording, processing and analysis of the data.

2. CELL POLE FORMATION IN *BACILLUS SUBTILIS*

Bacillus subtilis is essentially rod-shaped and for convenience may be regarded as a cylinder with hemispherical caps. The organisms elongate only in length without measurable change of diameter. At division, the bacterium separates into two cells, forming new cell poles through the formation of a cross-wall. The cross-wall itself may be thought of as a perforated disc which gradually closes and splits into two curved layers of polar wall ([1]; see Figs. 1 and 2). Thus, the organism has a relatively simple shape and undergoes a series of morphologically defined events during the cell cycle. Bacteria are particularly favourable material for attempts to understand the basis for the generation of cell shape ([2] Koch and Burdett, in press). A revival of interest in this subject has arisen mainly from studies of the relationship of cell shape to possible physical forces acting upon the growing wall [2].

In cocci, such as *Streptococcus faecalis,* the major site of new surface extension is located at the cross-wall [3], and the organisms essentially consist of cell poles but no cylindrical surface. During division the 'old' cell poles are conserved and the new polar caps arise by septal ingrowth and cell separation. The junction of 'old' and 'new' cell surface is marked in *S. faecalis* by the presence of raised wall bands. The wall bands split during the ingrowth of the cross-wall and are bilaterally and symmetrically displaced away from the centre of the cell during pole construction [3]. A very similar process occurs in *B. subtilis* during septation but the amount of wall origination between the wall bands (see Fig. 1) is only 15% of the total wall volume. That is, the septal site is not usually the major site of surface extension. Only in certain temperature-sensitive *rod* mutants [4–6] does the septal site appear to

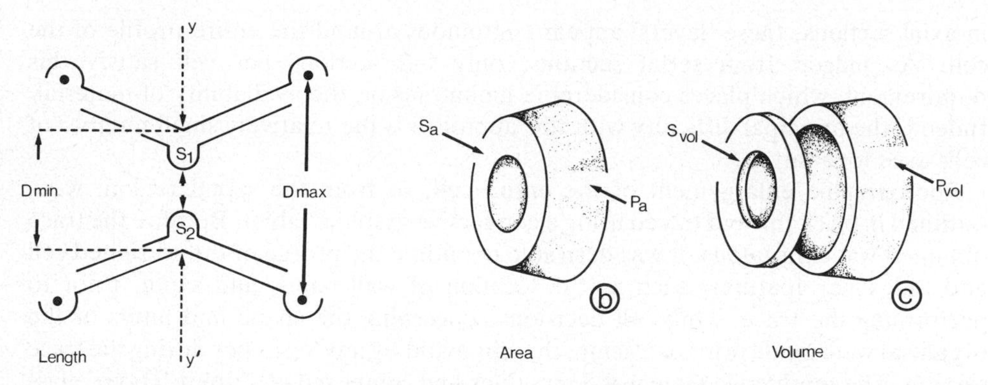

Fig. 2. Diagram of features used to describe geometry of developing poles in *B. subtilis,* (a) Linear measurement, and area (b) and volume (c) reconstructions obtained by mathematical rotation of one half of a dividing pole, as seen along YY'. Abbreviations: D_{min}, D_{max}, diameter at base of cross-wall . s_a, area of plane passing through cross-wall and at wall bands, respectively; S_a, area of plane passing through cross-wall at YY'; P_a, surface of nascent pole. S_{vol}, P_{vol}, volume of wall in cross-wall and pole, respectively.

generate the bulk of new cell surface [7, 8]. In this case, the *rod* organisms, at the restrictive temperature, are coccal shaped and have ceased to extend in length but grow only in diameter. Thus, at present, part of the cell cycle of *B. subtilis* can be analysed by electron microscopy but the mechanisms underlying extension of the cylindrical portion of thc cell, unlike that of the cross-wall, are still in debate (for review, see [9]).

The work to be described is the derivation of a model to describe the formation of new cell poles in *B. subtilis*. The essential features of septation may be summarised as follows: (i) ingrowth of the cross-wall by centripetal wall synthesis (forming an annulus); (ii) splitting of the cross-wall at its base to form two raised tears, analogous to the wall bands of streptococci. The raised tears mark the junction between septal and cylindrical wall; (iii) closure of the cross-wall; (iv) the gradual and continuous severance of the common cross-wall between dividing cells to form two new, curved polar surfaces. These stages are illustrated diagrammatically in Fig. 1.

The objective in attempting to derive a model for pole construction was to derive an equation which could be used to describe the shape of the nascent pole at all stages of growth. Our aim was to obtain this equation using features which appeared to play a key role in describing pole geometry. We then wished to explore the alterations in pole shape by modification of the variables using the framework of existing knowledge to place constraints on possible models of growth. For example, could the model provide evidence of the *precise* location of a growth site within the pole?

2.1. *Digitising and reconstruction techniques*

Reconstruction of cell poles was made from electron micrographs of glutaraldehyde-OsO_4-fixed and sectioned organisms. Only median, longitudinal sections were selected for analysis, and these came from the middle 15–20% of the cell [10, 1]. In thin section, the wall appears characteristically tribanded (dark-light-dark) and, in axial sections, these 'layers' appear continuous around the entire profile of the cell. As judged from serial sections, only 1–2 sections per cell satisfy this requirement, which places considerable limitations on the availability of material. Indeed, the principal difficulty with this approach is the relatively small number of cells used for analysis.

Photographic enlargement of the entire cell, or from the septal region, were outlined in black ink and traced using a conductive graphics tablet. Because the trace obtained was continuous, it was desirable to outline the profile of the sectioned cell and index key features, such as the location of wall bands and septa, prior to performing the trace. Thus, all decisions concerning the shape and limits of the organism were made prior to tracing, thereby avoiding any hesitancy during the trace session. The conductive plate was 3 mm thick and composed of Triplex 'Hyviz' glass coated with indium oxide; the available drawing area was 230 cm × 230 cm. The conductive plate was interfaced by Honeywell DDP-16 computer (more recently updated by a PDP11/34). Under command control the micrograph was traced and then displayed on a Tektronix 611 storage oscilloscope. Calculations of length were made by splitting the trace into strips, 0.6–2mm wide, and, for surface area and volume, rotated about their longitudinal axis [10, 1].

Analysis of the results was made by means of a companion program, which was used to plot graphs showing the individual data points and/or the mean ± standard deviation at some specified interval. A further subroutine fitted polynomials by least-squares regression, either to the original data values or to a suitable transformation of the variables.

Certain specific assumptions were made both in obtaining the data and in the derivation of a model for pole enlargement, namely: (i) that the cells were circular in cross-section; (ii) the wall bands extend around the circumference of the cell; (iii) the cross-wall is a perforated disc with a central, circular hole which, on septal closure, is converted into two curved surfaces (see [1]).

2.2. *Derivation of the model*

The first objective in developing a model for septation in *B. subtilis* was to relate three key linear measurements describing the size of the developing pole to increase of polar surface area (P_a). These key measurements are: (i) the diameter (D_{min}) between the bases of the cross-wall; (ii) the diameter (D_{max}) between wall bands on one side of the cell; (iii) the distance (h, Fig. 3) perpendicular to D_{min} and D_{max} (Fig. 2). These parameters were related to P_a by the following regression equations, obtained from the data by least-squares analysis:

$$D_{max} = 0.542 + 0.164P_a + 0.138P_a^2 \tag{1}$$

$$D_{min} = 0.538 - 0.197P_a - 2.54P_a^2 \tag{2}$$

$$h = 0.039 + 0.037P_a + 1.3P_a^2 \tag{3}$$

Thus, for a specified value of P_a the corresponding values of D_{max}, D_{min} and h could be readily determined. To satisfy the relationship between the increasing

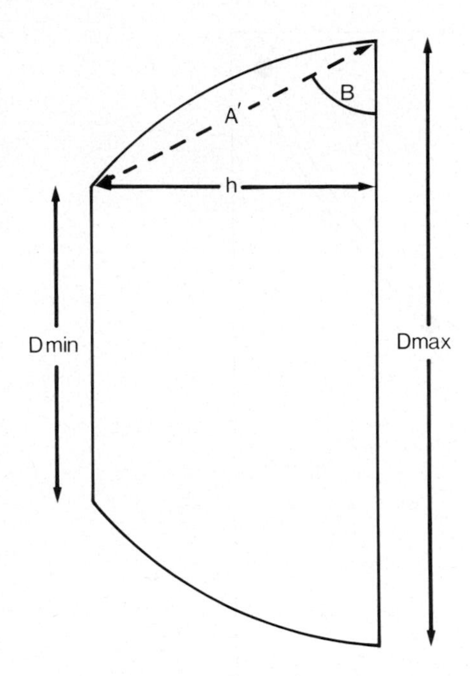

Fig. 3. Basic parameters used to define the model.

volume of the pole with increasing surface area, it was necessary to introduce another important feature characterising the growth process; namely that of curvature of the wall. It is this parameter which is possibly the most important in determining the overall shape of the pole. For a pole with specified surface area, the curvature of the wall (C) was defined as the ratio of the length of the wall (A) to the length of the chord determined by the distance between its end points (A') (Fig. 4). This relationship can be expressed as:

$$C = \frac{A - A'}{A} \tag{4}$$

The corresponding regression equation relating curvature to surface area was determined from a series of values obtained for C as a function of increasing surface area as:

$$C = 0.019 - 0.047P_a + 0.3P_a^2 \tag{5}$$

This relationship between C and P_a is shown in Fig. 4. Thus for a specified value of P_a the corresponding value for C could be easily obtained from equation (5). The value for A' is determined from simple trigonometry as:

$$A' = (h^2 + \tfrac{1}{4}[D_{max} - D_{min}]^2)^{1/2}$$

Substitution of the values for C and A' into equation (4) yields the corresponding value for A, the arc length of the wall. The remaining problem was to determine a curve of arc length A and chord A' consistent with the shape defined by the outer

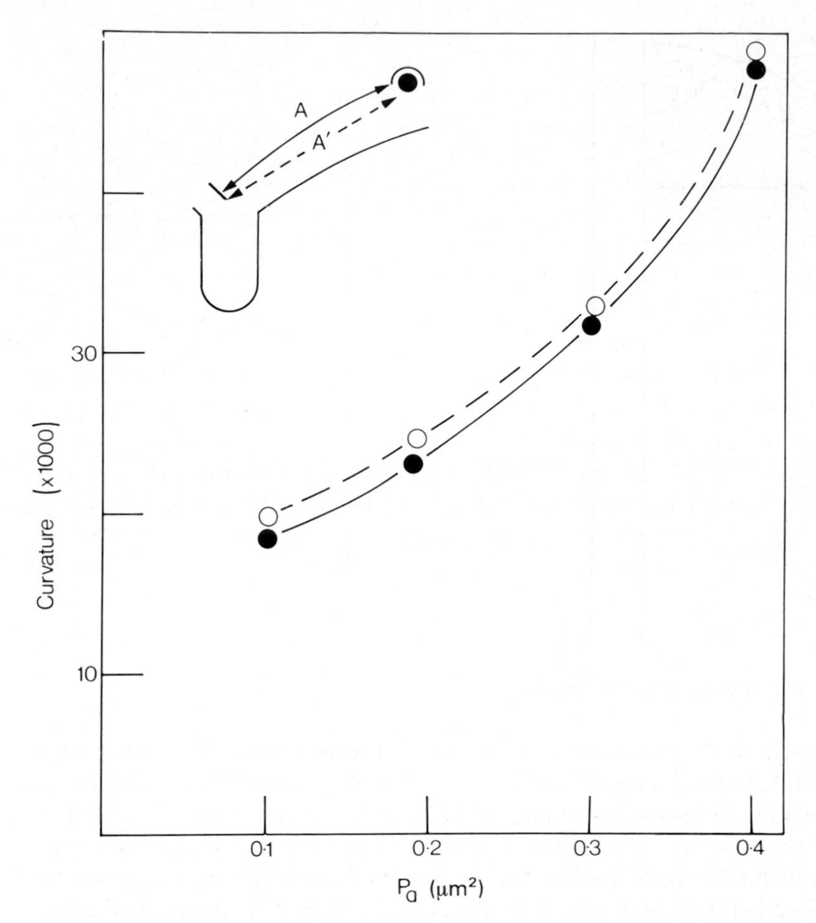

Fig. 4. Variation in curvature of the polar wall as a function of surface area for experimentally measured values (continuous line) and calculated values (broken line). The inset shows the parameters used to evaluate the curvature defined by equation (4).

wall. Since the wall profile is mostly regular in appearance, it was decided to explore the possibility of satisfying these criteria by fitting a suitable conic section or part of a section to the wall. The conic sections selected for analysis were the ellipse, circle, parabola and hyperbola. The equations for these sections are given below:

$$\frac{x^2}{a^2} + \frac{y^2}{b^2} = 1$$

$$x^2 + y^2 = a^2$$

$$y^2 = 4ax$$

$$\frac{x^2}{a^2} - \frac{y^2}{b^2} = 1$$

A computer programme was written to determine which of the above curves would best replicate the wall curvature for appropriate values of a and b. This approach involves finding those values for a and b giving a portion of curve whose arc length and corresponding chord length when substituted into equation (4) give a value for C consistent with that determined from equation (5).

As a result of the procedure it was found that values for C could be best satisfied by a linear function relating the focus (a) of the parabola to surface area, equation (5a). Figure 4 shows the agreement between the curvature obtained from a parabola with increasing focus and the experimentally determined values. This agreement was to within 5% for the values of P over which curvature was measured.

$$a = 0.253P_a + 0.023 \tag{5a}$$

Thus the shape of the outer wall could now be replicated for a specified surface area by constructing a parabola with focus determined from equation (5a).

To obtain a final reconstruction of the pole assembly, it is necessary to orientate the parabola into a position coincident with the upper and lower portions of the wall, i.e., the axis of the parabola is coincident with the bisectrix of the chord A' in Fig. 4. The operation of rotating a set of points (x,y) to a new position (x',y') is given by equations (6a) and (6b) below.

$$x' = x\cos\theta + y\sin\theta \tag{6a}$$
$$y' = y\cos\theta - x\sin\theta \tag{6b}$$

where

$$\sin\theta = h/A'$$
$$\cos\theta = (D_{max} - D_{min})/2A'$$

From equations (6a) and (6b), squaring and adding gives

$$x'^2 + y'^2 = x^2 + y^2$$

Substituting $y^2 = 4ax$ the equation for a parabola gives

$$x'^2 + y'^2 = x^2 + 4ax \tag{6c}$$

From equation (6a)

$$x' - x\cos\theta = y\sin\theta \tag{6d}$$

From equation (6b)

$$y' + x\sin\theta = y\cos\theta$$
$$\therefore y = (y' + x\sin\theta)/\cos\theta.$$

Therefore substitution in equation (6d) gives:

$$x' - x\cos\theta = y'\frac{\sin\theta}{\cos\theta} + x\frac{(\sin^2\theta)}{\cos\theta}$$

$$\therefore x\left[\frac{\sin^2\theta}{\cos\theta} + \cos\theta\right] = x' - y'\frac{\sin\theta}{\cos\theta}$$

Hence

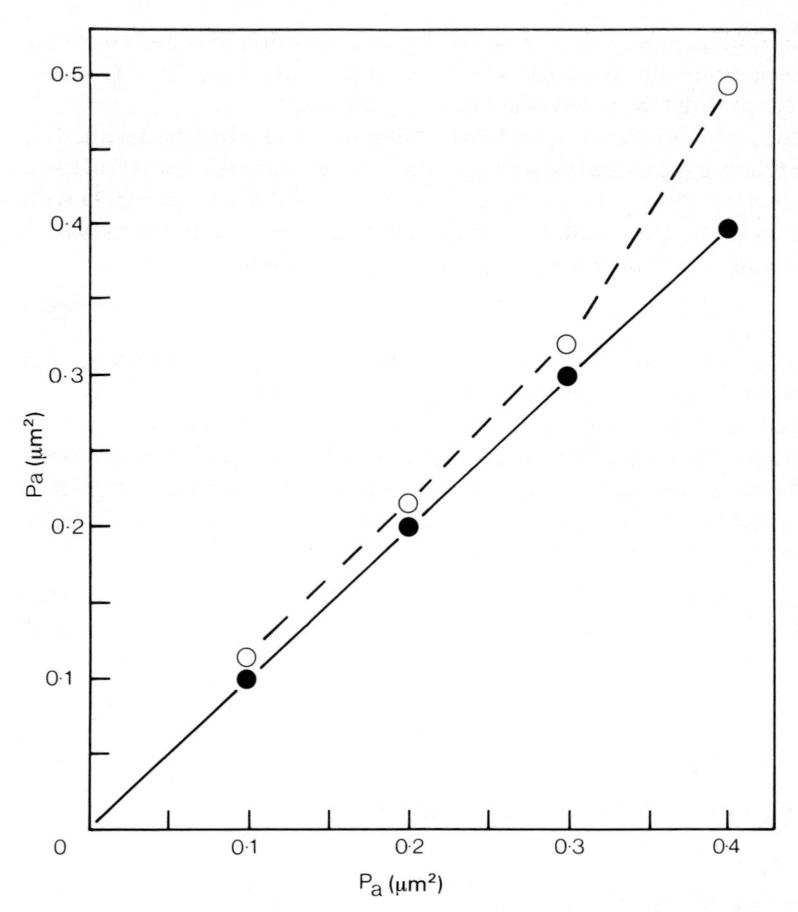

Fig. 5. Relationship between the experimental and calculated values for surface area values (--- calculated, ——— experimental).

$$x = x' \cos \theta - y' \sin \theta$$

Substituting for x into equation (6c) gives

$$y'^2 \cos^2 \theta + x'^2 \sin^2 \theta + x'y' \sin^2 \theta + 4ay' \sin \theta - 4ax' \cos \theta = 0 \qquad (6)$$

Equation (6) is the general form for a second order conic with variable coefficients and constant term zero. The three-dimensional equivalent surface is obtained by rotating the conic section about a central axis.

The agreement between the model defined by equation (6) and the experimentally determined results for surface area and volume of the pole assembly is shown in Figs. 5 and 6. The results show that for the range over which curvature exists, the agreement is to within 20% for both surface area and volume calculations. The profile corresponding to the inner wall was generated using equation (6) but with the parabola shifted by an amount corresponding to the measured thickness of the wall. The discrepancies between the model and experimentally determined results were

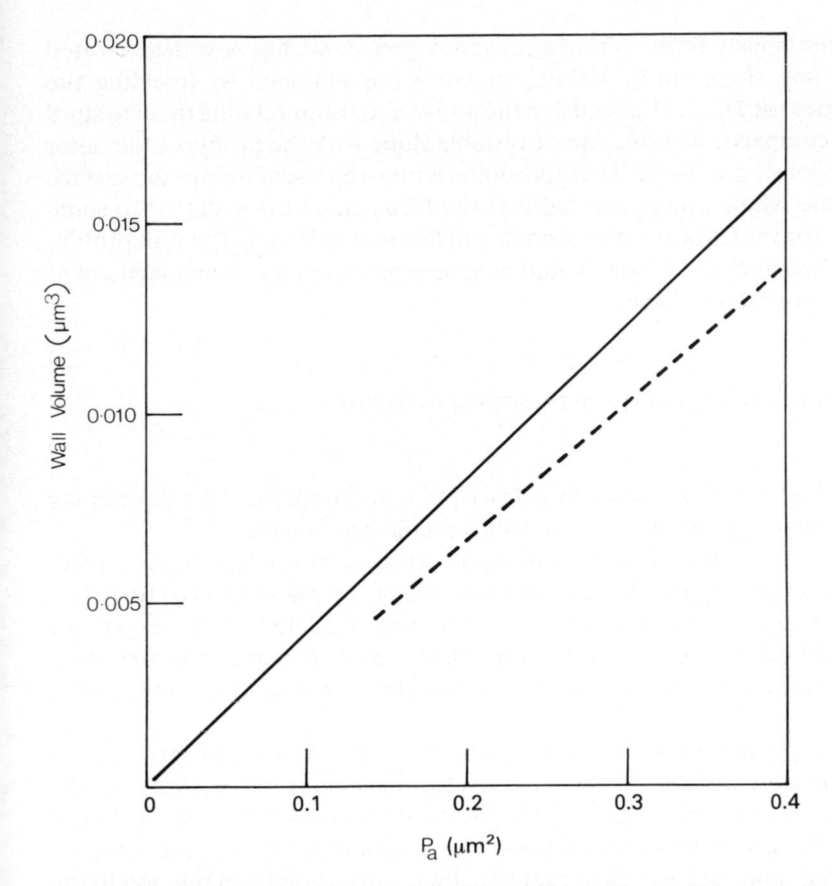

Fig. 6. Relationship between the experimental and calculated values for wall volume values (--- calculated, ——— experimental).

ascribed to errors incurred in deriving the initial parameters subsequently used to establish the model and to the differences between the experimental and theoretical values for the curvature.

For values of the surface area P_a of the nascent poles below around 0.07 μm observations show that curvature is non-existent and that the wall profile could be approximated by a straight line. The model in fact gives appreciable values for the curvature in this range and is therefore no longer representative of the experimental data. However, we now show that changing the value for the parabolic focus (a) for the smaller surface area values the model can be generalised to fit the observations even for the early growth stages.

Dividing equation (6) by (a) and letting (a) approach very large values gives:

$$y \sin \theta \simeq x \cos \theta$$

i.e.

$$y \simeq x \cot \theta$$

The model, previously representing a second degree conic has now degenerated into a straight line slope cot θ. Values for cot θ are obtained by inserting the appropriate values for D_{max}, D_{min} and h in the above equations relating them to sin θ and cos θ. The comparison of this line of variable slope with the profile of the outer wall shows very good agreement. Thus the model is now representative of the surface topology over the whole cycle provided that the focus value (a) is shifted to some very large value for values of P_a corresponding to less than 0.07 μm. The wall profile, being a straight line at $P < 0.07$ μm, would then be represented as a small segment of a parabola with a very large focus.

2.3. *The choice of the parabola for representing curvature*

The model outlined above is essentially a description of the shape of the developing pole, using a second degree conic surface with suitable coefficients.

Attempts were made initially to fit a parabola to sections through the complete pole at different stages of growth, i.e., with the axis of the parabola bisecting D_{min} and D_{max} and of length h. This model failed to represent the data however except for non-systematically changing values for the focal length (a) and was therefore regarded as unsatisfactory. Attempts at matching other conic sections to the entire pole also failed.

The choice of a parabola with its axis bisecting the chord A' not only gives a good fit to the wall curvature but has the advantage over other conic sections that only one parameter has to be determined namely the focal length (a) of the parabola. In our case (a) was found to vary systematically and was determined by a computer search procedure (search times being of the order of a few c.p.u. seconds) in relation to the curvature.

2.4. *Determination of absolute curvature*

The experimentally determined curvature gives a relative measure of the change in wall curvature over the growth cycle. One advantage in expressing the surface topology in an analytic form is that values for absolute maximum curvature can be obtained and subsequently related to physical forces, such as surface tension, which may in part determine the shape of the pole.

The expression for curvture $|K|$ of a curve $y = f(x)$ is given by

$$|K| = \frac{d^2y/dx^2}{\sqrt{1 + (dy/dx)^2}^{\,3}}$$

We are concerned with evaluation of the curvature across the wall at different stages of growth and for simplicity we determine the curvature for a parabola with varying focus passing through the origin since

$$y^2 = 4ax \text{ where } a = 0.253P_a + 0.023 \text{ (in our case)}$$
$$y = 2\sqrt{ax}$$
$$\frac{dy}{dx} = \sqrt{a/x}$$
$$\frac{d^2y}{dx^2} = \frac{-a^2}{2(ax)^{3/2}}$$

substituting into the expression for $|K|$ gives

$$|K| = \frac{\sqrt{a}}{2(a + x)^{3/2}}$$

The curvature for the parabola is always a maximum therefore where $x = 0$, i.e., for a parabola passing through the origin. Then $|K|$ is a maximum and varies as $\sqrt{a}/2a^{3/2} = 1/2a$.

This result may be simply interpreted as saying that when a is small, $|K|$ is large and when a is large $|K|$ is small which is to be expected. The maximised values of $|K|$ for the sequence of pole assembly are shown in Fig. 7.

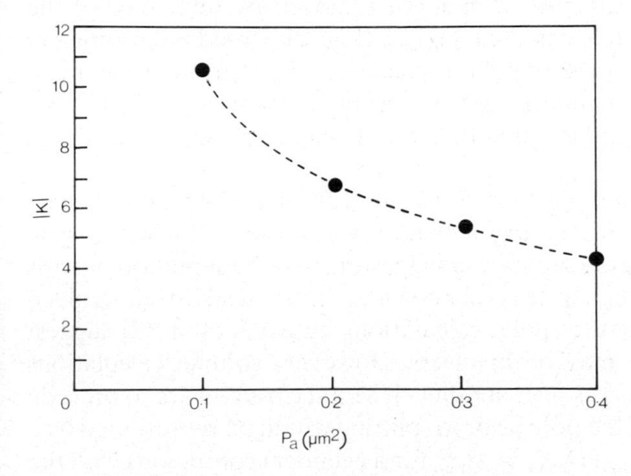

Fig. 7. Absolute curvature as a function of surface area of the nascent pole.

Assuming that surface tension is proportional to the curvature, the results obtained for the outer wall indicate that the maximum surface tension occurs as the wall begins to split and separate and fairly rapidly reaches a stable level where the tension is approximately half that of its previous value.

2.5. *Biological aspects of the model*

The model utilising a second degree conic surface with appropriate coefficients adequately describes the shape and size of developing cell poles in *B. subtilis*. A perturbation of the model is necessary during the earlier stages of growth in order to satisfy the experimental observations. Considerations of the curvature show that the perturbation occurs rapidly. That is to say that the absolute curvature goes from zero to a fairly significant value early in septation. The implication of this observation is that a substantial volume of wall is synthesised before splitting of the cross-wall occurs and when it does separate it does so very rapidly. In *Arthrobacter crystallopoietes,* a definite 'snapping' apart of dividing cells has been described [11]. This process also characterises the conversion of the flat disc or cross-wall into the curved layers of polar wall.

Under the particular growth conditions used, the thickness of the cross-wall was found to increase for about two-thirds of the septation sequence and then gradually decrease whilst the thickness of polar wall was maintained at a relatively constant thickness [1]. It is therefore tempting to assume that the separating layers of polar wall are maintained at a uniform thickness by material contributed by the ingrowing cross-wall. That is, the 'growth site' may be located in the cross-wall itself and that only very limited addition of wall material occurs at the polar surface. In rod B mutants of *B. subtilis* [7, 8] the thickness of the cross-wall is maintained at approximately twice the thickness of the separating layers of polar wall. The conversion of the cross-wall, essentially an annulus, into two inner layers of polar wall also involves an increase in area. Previous calculations for other gram-positive poles [3, 1] suggested that the amount of material removed by severance of the cross-wall, calculated as the decrease in area at D_{min} (Fig. 2), would be insufficient to account for the measured increase of polar surface area P_a. This increase might occur through the addition of wall material to the nascent polar surfaces. That is, the nascent poles may undergo secondary modification of shape and size by localised wall thickening.

In other strains of *B. subtilis*, and also in *S. faecalis,* gradients of wall thickness extend across the nascent poles, increasing in width towards the wall bands. Again septal closure may also precede pole separation suggesting that the septation process may be relatively flexible. Although little is yet known of the physical forces involved in moulding the shape of the mature pole, calculations by Koch et al. [2] suggest that surface tension-like forces may be involved. However, volume calculations indicate a preference for one process over another. If septal closure were to precede pole separation, the topology of the pole prior to splitting would be represented by a cylinder of diameter D_{max} or D_{min} ($D_{max} = D_{min}$ for a cylinder) continuous with the cylindrical portion of the cell. After splitting the pole adopts the shape given by the model and thus the cylinder has to transform from a cylinder to the rotated parabola shape. In undergoing this transformation the pole would have to externalise 40% of its surface structure since the volume for a cylinder at this stage of growth is 0.0848 μm^3 compared with the parabolic shape giving a volume of 0.0558 μm^3. This implies that considerably more work would have to be done in the case of septal closure preceding pole separation than when early separation occurs thus favouring the latter mechanism.

3. SIGNAL-TO-NOISE ENHANCEMENT IN STUDY OF CELL WALL STRUCTURE OF *BACILLUS SUBSTILIS* BY INTERACTIVE COMPUTATION

In the previous section we outlined methods for digitising electron micrographs of sectioned bacteria so as to obtain three-dimensional reconstruction of whole cells.

We now turn to another aspect of the organisation of bacterial cell surfaces, where we consider methods of measuring variations of density within the wall itself.

The problem which prompted the development of the technique to be described concerned the feasibility of locating Mg^{2+} bound to walls isolated from *B. subtilis*. Previous studies by Beveridge and Murray [12] showed that a variety of metals can be bound to cell walls and can be visualised by electron microscopy (for review, see [13]). The most obvious way in which measurements from wall preparations can be obtained is to produce single-line densitometer scans. However, in our experience, the variations in density presumed to correspond to sites of Mg^{2+} deposition, occurred not only in discrete patches but also within each patch. Some of these variations would depend upon the plane of sectioning. Other sources of variation might arise from sites having physiological significance, perhaps corresponding to domains of the wall which might represent sites of active biosynthesis and wall extension.

For such extended regions of density variation, we devised an averaging technique using multiple densitometer line scans of wall profiles. However, the problem is essentially one of establishing boundary conditions between the structure of interest and the background. That is, the limits to visualising boundaries are imposed by the ratio of signal to noise in the original electron micrograph. The use of the averaging technique effectively enhances the signal to noise ratio. The chapter by Shaw (this volume) shows the effectiveness of averaging several unit cells in a two-dimensional crystal lattice using the Fourier transform approach. Unfortunately, this technique is limited to the analysis of periodic arrays and is therefore inapplicable to the study of the *B. subtilis* wall, at least at the level of resolution obtainable with thin sections.

Technical details of wall preparation, electron microscopy and densitometry are given in [14]. In short, freeze-dried walls of *B. subtilis* were exposed to 10 mM $MgSO_4$, fixed in glutaraldehyde and embedded in Spurr's resin. Sections were examined without further poststaining or after brief exposure to uranyl acetate and lead citrate.

Figure 8 shows a field of view of isolated sectioned cell walls.

Several negatives of the stained sections were scanned on a modified Joyce-Loebl densitometer in the Computer Science Laboratory at the National Institute for Medical Research (NIMR) using a 40 μm scan spot size and a 256 × 256 scan area with 40 μm incremental steps corresponding to an image area of about 10mm × 10mm. The scanner records at a rate of approximately 200 density measurements per second and is controlled by a PDP11/34. The recorded values are stored on magnetic tape.

The enhancement of the image of the cell wall relies on the principle of time series analysis known as the superposition of multiple noisy copies [15, 16]. Suppose that it is required to measure a function s but that the only data available are a set of corrupted versions $f_1, f_2 \ldots, f_n$ given by

$$f_i(f) = s(t) + n_i(t)$$

in which the signal s is corrupted by noise n. For each t suppose that the values of $n_i(t)$ are independently distributed with zero mean. Let F denote the average of $f_1, f_2 \ldots, f_N$, i.e.

$$F(t) = s(t) + \sum_{i=1}^{N} n_i(t)/N$$

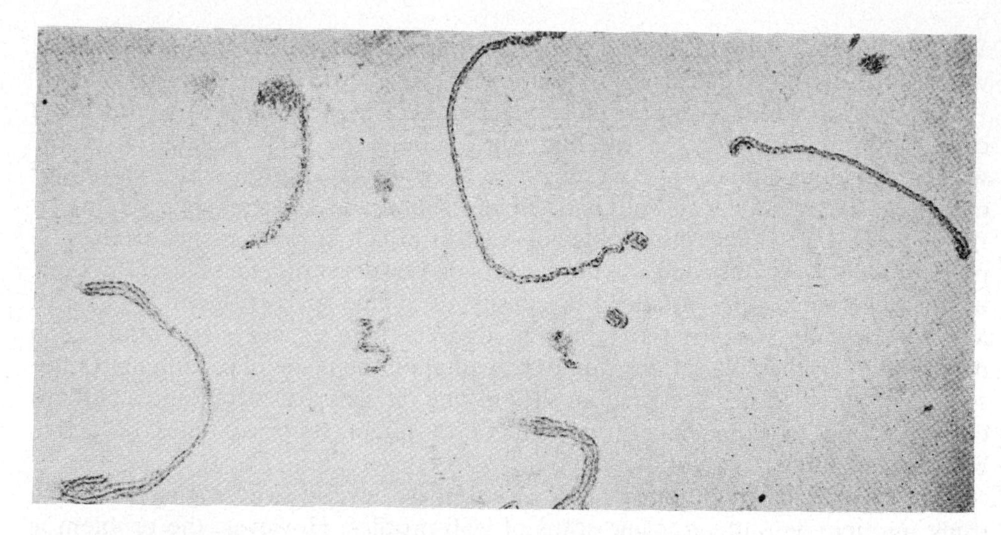

Fig. 8. Survey micrograph of *B. subtilis* strain 172 walls washed with 0.85% NaCl, exposed to 10mM $MgSO_4$ and fixed in glutaraldehyde. Section post-stained with uranyl acetate and lead citrate. Magnification 51,500.

Consider a single point t. The signal-to-noise ratio at t of an individual f_i is given by the root of

$$E(s^2(t))/E(n_i^2(t)), \text{ where } E \text{ denotes expected value}$$

The signal-to-noise ratio at t of F is given by the root of

$$E(s^2(t))/E(\sum_{i=1}^{N} n_i(t)/N)^2$$

Now

$$E((\sum_{i=1}^{N} n_i(t)/N)) = \frac{1}{N} E(n_i^2(t))$$

since $n_i(t)$ is independently distributed with zero mean. Thus the signal-to-noise ratio of F is $\sqrt{N}$ times as great as the signal-to-noise ratio of an individual f_i.

In the present case the signal s corresponds to the 'true structure of the cell wall' and the different corrupted versions available correspond to the density functions recorded in successive scan lines perpendicular to that of the cell wall. Thus the structure of the cell wall should be much more clearly visible in the function (marginal total) F given by

$$F(t) = \sum_{i=1}^{N} f_i(t)/N$$

than in any of the individual f_i.

A binary version of the resulting matrix of optical densities obtained from a scan of the film negative is displayed on a television monitor in such a way that each value lying between certain limits is represented by a bright dot at a position on the screen

Fig. 9. Binary version of cell wall. Points displayed are those for which optical density lies within the range 75–100.

corresponding to its position in the matrix. It is usually possible for the user to adjust the limits interactively in such a way that only a small fraction of the possible positions are illuminated, yet the dots which are shown indicate the general position and orientation of the cell wall. Figure 9 shows a representation of the cell wall displayed in this way. This binary version is now displayed on the visual display unit where a light pen can be used to define an appropriate rectangular window, i.e, one in which the cell wall is perpendicular to the scan direction, so that the density averaging will be effective.

In order for the technique to work at all it is necessary: (1) that the cell wall be more or less straight, and (2) that the summation is carried out in lines parallel to the cell wall.

Thus, as shown in Fig. 10, it may be necessary to choose a relatively short length (several hundred nanometres), in order to achieve (1) and to perform rather elaborate computation in order to achieve (2). Thus in order for the technique to be computationally efficient it is necessary for the cell wall to run more or less perpendicular to the direction of the scan-lines. This can be arranged by suitable orientation of the film in the densitometer and by the selection of an appropriate rectangular window within which to confine the evaluation of F. Such a window can

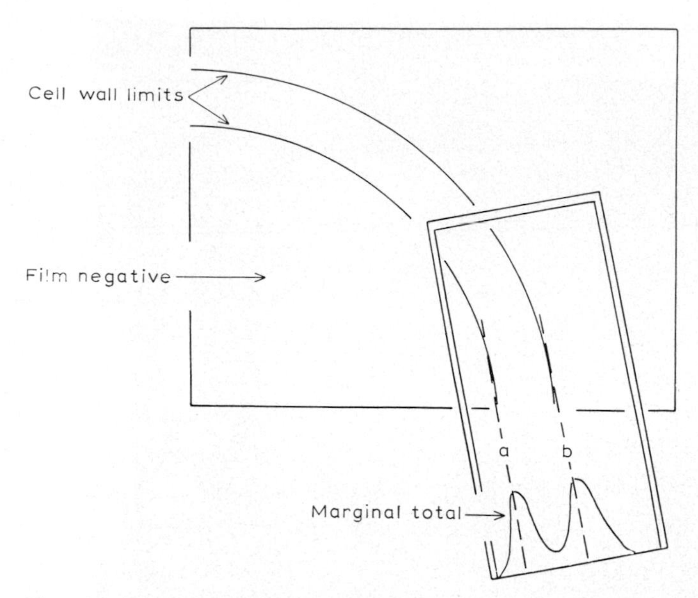

Fig. 10. Diagram showing orientation of film negative with respect to the direction of scan. The summation of optical densities is carried out in lines parallel to the limits a and b.

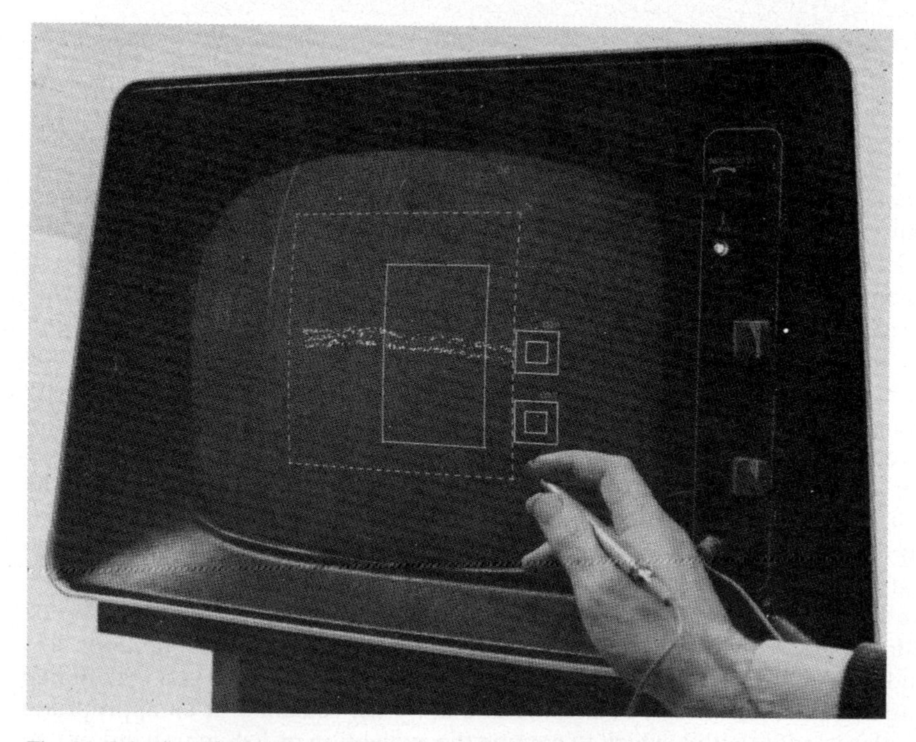

Fig. 11. Selection of relevant area using interactive graphics display.

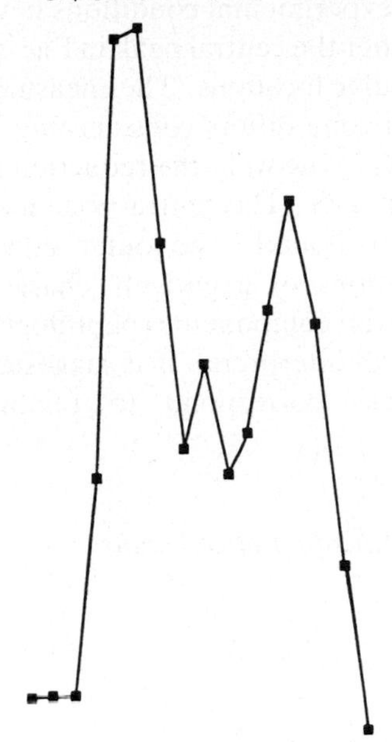

Fig. 12. Marginal total for the cell wall shown in Fig. 11. The region corresponding to the cell wall was found to lie between scan lines 65–78 (A). The magnitudes of the recorded optical densities corresponding to each scan line is given by (B).

be quickly selected using a light pen [17], as shown in Fig. 11. If we assume that the density matrix is d, and the selected window extends from i_1 to i_2 and from j_1 to j_2, then the marginal totals defined by

$$\sum d(i,j) \text{ and } \sum d(i,j) \text{ for } j = j_1,\ldots, j_2 \text{ and } i = i_1,\ldots,i_2$$

are calculated and displayed on the screen.

3.1. *Results and discussion*

Figure 12 shows the marginal totals for the cell wall shown in Fig. 11. The recorded range of density values was between 1 and 200. The density limits chosen to represent the cell wall region were defined as 75 and 100 nm and the dots shown on the display represent all densities lying within this range. The size of the density matrix was 256 × 256 and the selected window corresponded to almost half the total area scanned. The magnification recorded in the micrograph was approximately 27,000 and since the cell wall was defined within twelve scan lines with a 40 μm step between successive lines, the average wall thickness was calculated at 17.5

nm. The peak occurring between the two outer tracks of the wall shows the presence of localised subsidiary structure.

The biological significance of subsidiary structure will only become clear when enough isolated walls exposed to a range of different experimental conditions have been examined. For example, it is not yet certain whether the central peak in Fig. 12 occurs uniformly along the wall or is found at particular locations. The measurements available from Fig. 12 suggest that this central zone differs considerably in average optical density from the two outer dense tracks in the wall, the reduction in average optical density at the central peak being about 25%. The central zone may represent a site where Mg^{2+} ions are preferentially bound. The outer dense tracks in Fig. 5 may represent localised concentrations of negatively charged polymers (such as techoic acids), or regions where all wall components (peptidoglycan, teichoic acid) are packed more tightly [18]. The latter interpretation is suggested by measurements indicating that the outer tracks correspond to regions of maximal mass thickness [19].

Figure 13 shows a diagram of the whole process.

Automated quantification of cell walls

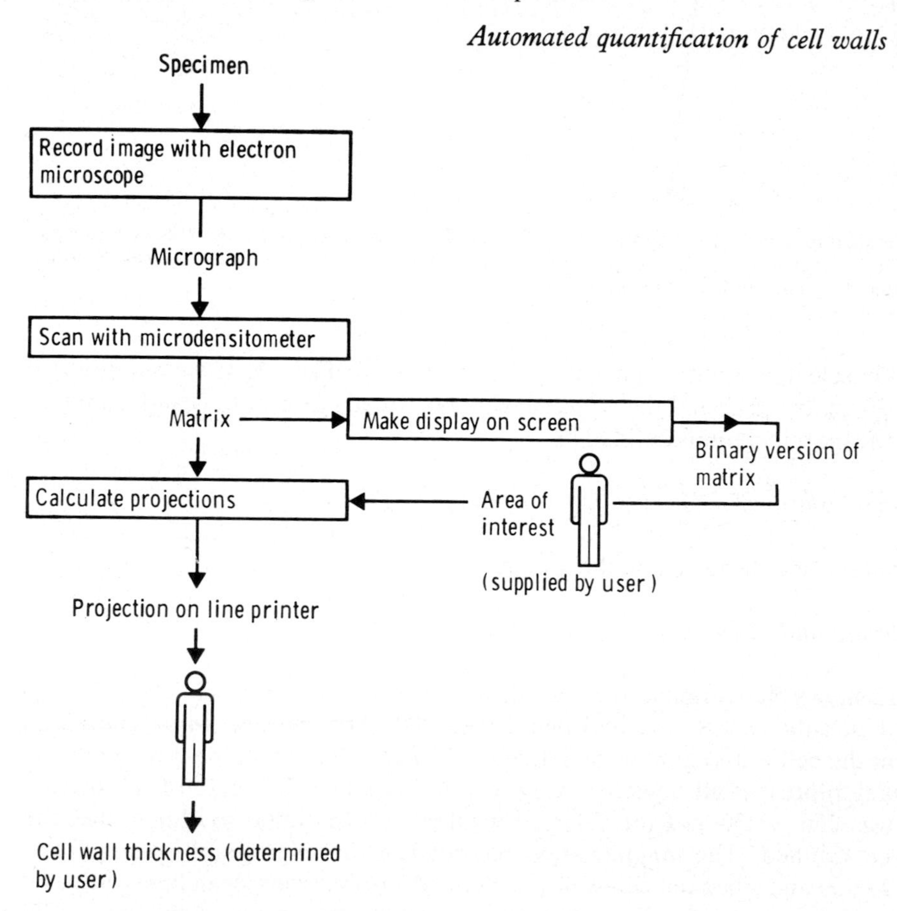

Fig. 13. A system to study cell wall structure.

3.2. *Conclusions*

The application of interactive computing and density averaging provides an objective way of analysing the structural details recorded in an electron micrograph. This approach has advantages over the alternative technique of microdensitometry using a long rectangular slit parallel to the cell wall. For example, the interactive facilities enable the user to easily select regions of varying size from which average totals can be readily computed. The density averaging process enhances the signal relative to noise and therefore allows the measurement of the cell wall thickness with much greater accuracy than is available from a single scan line.

ACKNOWLEDGEMENTS

We would like to thank: Academic Press for permission to reproduce diagrams, etc. from the J. Theor. Biol. (1981) 92, 127, and the Journal of Microscopy for permission to reproduce diagrams, etc., from J. Microsc. 113, 131.

REFERENCES

1. Burdett, I.D.J. and Higgins, M.L. (1978) J. bacteriol. 133, 959.
2. Koch, A.L., Higgins, M.L. and Doyle, R.J. (1981) J. Gen. Microbiol. 123, 151.
3. Higgins, M.L. and Shockman, G.D. (1976) J. Bacteriol. 127, 1346.
4. Rogers, H.J., McConnell, M.M. and Burdett, I.D.J. (1968) Nature, 219, 285.
5. Rogers, H.J., McConnell, M.M. and Burdett, I.D.J. (1970) J. Gen. Microbiol. 61, 155.
6. Rogers, H.J. and Thurman, P.F. (1981) FEBS Lett. 117, 99.
7. Burdett, I.D.J. (1979) J. Bacteriol. 137, 1395.
8. Burdett, I.D.J. (1980) J. Gen. Microbiol. 121, 93.
9. Rogers, H.J. (1979) Adv. Micro. Physiol. 19, 1.
10. Higgins, M.L. (1976) J. Bacteriol. 127, 1337.
11. Krulwich, T.A. and Pate, J.L. (1971) J. Bacteriol. 105, 408.
12. Beveridge, T.J. and Murray R.G.E. (1976) J. Bacteriol. 127, 1502.
13. Beveridge, T.J. (1981) Int. Rev. Cytology 72, 229.
14. Barrett, A.N., Burdett, I.D.J. and Paton, K.A. (1978) J. Microsc. 113, 131.
15. Rosenfeld, A. (1969) Picture Processing by Computer. Academic Press, New York.
16. Parzen, E. (1960) Modern Probability Theory and its Applications. Wiley, New York.
17. Newman, W.M. and Sproull, R.L. (1973) Interactive Computer Graphics. Addison Wesley, Reading, Mass.
18. Millward, G.R. and Reavelly, D.A. (1974) J. Ultrastruct. Res. 46, 309.
19. Misell, D.L. and Burdett, I.D.J. (1976) J. Microsc. 109, 171.

Geisow & Barrett (eds.) Computing in biological science

Calculation of biomolecular conformation

E. PLATT and B. ROBSON

1. INTRODUCTION

Ever since the pioneering studies of Pauling and Corey [1] in the 1950s, there has been considerable interest in calculating the conformations and conformational behaviour of biomolecules. In the earlier days, emphasis was placed on the use of theoretical reasoning to aid in the interpretation of ambiguous experimental data, though now efforts are being made to extend the scope of theoretical reasoning so that the minimum of experimental data may be utilized. In this work, the central question is: Can one predict the conformation of a biomolecule from its molecular formula and configuration around optically active atoms?

In the present review, we emphasize work undertaken in this laboratory and in other laboratories with which we have collaborated. Our approaches have involved both relatively 'exact' calculation (e.g., ab initio quantum mechanical methods, all-atom treatment of solvent-solute behaviour) and at the other extreme a search for valid, time-saving approximations. The main emphasis is on the study of the low molecular weight analogues of biological macromolecules, with some examples of the modelling of the larger systems using the results obtained from these low molecular weight analogues. No detailed account will be given of the procedures used when these are standard (e.g., ab initio quantum mechanical methods, Monte Carlo methods), since our use of these often differs primarily in the sense that the systems of interest to us are merely much more complex than those which chemists usually study. Rather, we wish to emphasize some of the uses to which these established techniques can be put, and some of the special problems which arise.

It is, however, worth noting that ab initio methods have only recently become practical standard procedures, and even now they are so time-consuming that their use in the calculation of the conformation of biomolecules cannot be described as 'routine'. We have employed the ATMOL/3 suite of programs on the CDC 7600 computer at the University of Manchester, with some earlier, comparative studies employing Gausian 70 and 100 programs. Here a (STO − 3G + 3d basis set) computation on one conformation of dimethyl phosphate ($CH_3.O.PO_2.O.CH_3$) would take about half an hour c.p.u. time, while (3s/2p basis set) computation on larger molecules such as 22 atom peptides may take an hour or so. The CDC 7600 is a very fast machine; the time required on an IBM360/67 is, in our experience, some three to four times larger. Naturally, the computation can be speeded considerably on a parallel processor such as the CRAY, though developments in ab initio calculations on parallel processors are still at an early stage of development.

The feasibility of an ab initio calculation is largely dictated by the time available, which depends not only on the machine used but the number of orbital functions to be employed. This number is determined in turn by the size of the molecule, and the precision with which its electronic structure is to be represented. In other words, it depends on the size of the basis set of orbital functions used. The choice of program is also affected by this aspect. Gausian 70 can handle up to 70 orbital functions, Gausian 100 up to 100 orbital functions, and ATMOL/3 up to 127 orbital functions. Our preference for the latter is obvious, since at least in pioneer studies on new molecules we wish to increase the number of orbital functions progressively until the results converge.

The computational expense of the ab initio method has been a major factor in our search for cheap, analytical methods which give comparable results. It is useful to note that such methods can be so fast as to be usable on a PET Commodore 3032 microcomputer in interpreted BASIC. Though our developments in this direction are still at an early stage, we can report that the least energy conformations of pentapeptides can be located on a PET in two to three hours. Analytical functions are thus an attractive possibility for a dedicated machine costing an initial, one-off outlay of £1,000 or less; and is a further justification for seeking analytical methods. Ab initio calculations are nonetheless necessary for the initial developing, testing, and refining of these analytical methods, as described in the present review. We discuss here the development of these alternative, very cheap methods using ab initio calculations, but emphasize from the outset that the resulting method obviously depends on the validity of the ab initio calculations. For this reason, our earlier studies on peptides also employed experimental data about conformational behaviour. Nonetheless, for many molecular species very little data of the appropriate type exists, and the ab initio results are thus of prime importance.

2. CALCULATION OF POTENTIAL SURFACES

The potential surface of a molecule is the representation of its energy as a function of conformation. From it, other properties including the corresponding free energy surface can be calculated. It represents the 'space state' of our problem, analagous to the set of all legal layouts of chess pieces on a chess board. The working solution, if a unique solution exists, is the minimum of least energy. More precisely we should say of least free energy, since a shallow minimum of higher energy can imply greater entropy, and possibly greater stability, than a deep narrow minimum. For the same reason, we should also speak of finding a state (or domain or zone) of least free energy on the potential surface, since every conformation implies a certain degree of thermal motion. These semantic difficulties disappear, of course, when the problem is properly addressed in the language of statistical mechanics. Nonetheless, location of a conformation of least energy is an important step and for a complex molecule like a globular protein or tRNA would be a considerable achievement at the present time.

Problems in calculating the conformation or conformations of least free energy include the need for accurate in vacuo potential surfaces (including the distortion of bond lengths and valence angles), the role of the solvent, the difficulty in calculating the entropic contribution, and the need to find the deepest energy minimum in a surface which is, for larger biomolecules, populated by many thousands of minima. However, the need for good in vacuo calculations is obviously the most important of these problems to be solved in that the testing of attempted solutions to the other problems presupposes satisfactory intramolecular potential surfaces.

Potential energy surfaces are calculated using empirical potential functions which represent the interatomic interactions as simple functional forms which have been parameterized against experimental data [2,3] semi-empirical quantum mechanical methods which seek solutions to the orbital approximation to the Schröedinger equation by the inclusion of some empirical parameters [4], or ab initio methods which use approximations to molecular orbitals (contained in a data base called the 'basis set') but otherwise do not use any empirical parameters.

Because of the sparsity of available experimental data ab initio calculations are currently of particular interest. Ab initio calculations are, nonetheless, very expensive in terms of computer time; it must be remembered that one needs to explore not only the potential surface as a function of rotation around single bonds, but ideally, the effects of 'flexible geometry', e.g., the contribution of variations in valence angles such as $N\text{–}\hat{C} = O$. As a result of this, ab initio calculations have been restricted to fragment approximations [5] and the use of minimal STO-3G basis sets for a sampling of limited areas of conformational space [6]. We have studied larger fragments with more extended basis sets where possible to test and extend previous calculations and make possible for the first time, an analysis of problems associated with intramolecular hydrogen bonding and flexible geometry in large fragments which have not been satisfactorily resolved by other methods.

For peptide systems we have explored *N*-formyl glycylamide ($HCO.NH.CH_2.CO.NH_2$), *N*-formyl glycyl *N'*-methylamide ($HCO.NH.CH_2.CO.NH.CH_3$), and *N*-formyl alanylamide ($HCO.NH.CH.(CH_3).CO.NH_2$) analogues of the dipeptide unit of polypeptides. Previous comparison [7] with the more widely employed *N*-acetyl glycyl *N'*-methylamide ($CH_3.CO.NH.CH_2.CO.NH.CH_3$) using empirical functions suggests that they are an excellent approximation to the latter more popular choice of analogue, and hence are suitable models for the 'short range' interactions dominating polypeptide behaviour [3].

In these calculations we used the extended basis (9s/5p) contracted to (3s/2p) of Gaussian-type functions derived by Dunning [8]. Some previously unpublished 4–31G calculations on *N*-formyl glycyl *N'*-methylamide glycyl analogue are also compared.

For nucleic acids, fragments such as stacked bases and the dimethyl phosphate anion (DMP^-) (as an analogue for the backbone of nucleic acids) are employed. The dimethyl phosphate analogue has been particularly extensively studied using (i) an STO-3G basis set (Basis 1A), (ii) STO-3G + a 3d function on phosphorus (Basis 1B) and (iii) extended basis set using the (10s/6p) orbitals for phosphorus and (7s/3p) [9] orbitals for carbon and oxygen, contracted to (4s/3p) and (3s/2p) respectively (Basis II).

For molecules of pharmaceutical importance, the fundamental 'core' of a drug series is of particular interest. In the case of the sulphonamide series, we used $PhSO_2NH_2$ with an STO - 3G + 3d function on sulphur basis set.

Finally, for small fragments we have also used ab initio results by other authors where possible: For example, $CH_2(OH)_2$, as a model for certain saccarhides. The 4–31G calculations on this molecule have been performed by the workers Jeffrey, Pople and Radom [10]. Studies on such small systems are fortunately fairly common in the literature, and provide for us a useful extension to the set of ab initio calculations against which the OFF approach (see below) can be developed. Having built up a 'dictionary' of potential surfaces of fragments in this way, these can be and have been used in the modelling of polypeptides, proteins, nucleotides and nucleic acid macromolecules.

3. PARAMETERISATION OF POTENTIAL FUNCTIONS

As simple functional representations of the energy of interactions between atoms, potential functions are economic and can be regarded as constituting the 'package' which the theoretician delivers to molecular biologists and pharmacologists for routine application. They thus deserve special consideration, in some detail.

Classically, workers have considered the conformational energy of a molecule, say E_{TOT}, to be the pairwise sum over all possible pairs of atomic centres each counted once

$$E_{TOT} = \sum_{i=1}^{n} \sum_{j=1}^{i-1} E_{ij} \tag{1}$$

where the E_{ij} are the potential functions typically of the form

$$E_{ij} = \sum_{K} \left(\frac{P_K}{r^{Q_K}} \right) \tag{2}$$

with the parameters P_K dependent on the types of atoms i, j and the powers Q_K of the distance r between atoms i and j.

The assignment of numerical values to the parameters is usually done such that the pairwise energies E_{ij}, or some properties derived from these energies, give the best possible agreement by a least squares method, with the energy or properties of a large number of molecules obtained from experiment or ab initio calculations. In practice, the powers Q_κ are parameters which should also be included in the fitting process, though these are usually fixed on the basis of prior studies. Actually, procedures are available for deriving the terms directly from an ab initio calculation, but the resulting functions are complicated and the technique has not yet found wide application.

There will technically be three or more such terms κ, representing van der Waals' repulsions, London dispersion forces, electrostatic interactions, and possibly hydrogen bonding contributions, and even hydrophobic interactions (strictly, a free energy contribution). These terms are thus each supposed to have a distinct physical significance, though individual terms with particular parameter values cannot necessarily be transferred to different contexts, i.e., the specific terms with specific parameters can only be safely used as first derived, in the context of the other specific terms with specific parameters. This is a question of consistency of the parameters. In a large part this is because potential functions are simply an analytical representation of the interatomic potential and errors in one term are allowed to compensate for errors in other terms without overall detriment. A typical error is the neglect of higher moments, e.g., of the dispersion term. These can be specified as further terms, of course, but there may not be adequate data for determining the parameters of these terms for all cases, and the potential functions become rather unwieldy. The principle of using the fewest number of parameters as possible consistent with fitting the experimental data holds good here. However, by reducing the flexibility of the mathematical description, it makes more difficult the problem of transferability of the functions to other chemical systems. It turns out that the van der Waals' repulsive, dispersion and electrostatic terms are sufficient to reproduce data to a first good approximation. The dispersion term is, however, in most usages, rather artificial, its main function being to add greater flexibility to the net description of the van der Waals' repulsive (steric overlap) component.

In addition to the above type of analytical term, potential functions are sometimes supplemented by an 'intrinsic rotational potential'. This is simply some trigonometric function of a dihedral angle, usually introduced to produce agreement between the experimental and calculated barrier heights, e.g., in ethane, when all else has failed. It should in our view, be seen only as an interim 'bootstrap' adjustment. Certainly there is no justification, in many cases, for identifying this term as a specific quantum mechanical phenomenum other than representable by atom-centred van der Waals' or electrostatic forces. On the other hand, there may be some justification in the case of systems rich in aliphatic hydrocarbons, in cases of a significant 'anomeric effect' (see below) and, of course, in the case of bonds of order greater than one (as in C_2H_2 and the peptide group).

An intrinsic rotational potential is, however, not very informative. Further, it may not be adequate in general. For example, the rotational barriers in hydrogen peroxide are likely to be due, at least in part, to interactions involving the lone pairs. An intrinsic rotational potential might provide an adequate description of rotational isomerism, but not of the interactions between hydrogen peroxide molecules, which will also depend on the lone pairs. Even when studies are confined to intramolecular interactions, one can readily think of examples (e.g., dimethyl phosphate, benzene sulphonamide, see below) where intrinsic rotational potentials could not be expressed as trigonometric functions of rotation round a single bond. For these reasons we have found it necessary to depart from the classical, atom-centred approach and to develop an 'Orbital Force Field' (OFF) description.

In essence, this is an empirical potential function approach modified to represent interactions between orbital centres rather than simply atomic centres. The principle point from a practical viewpoint is that it provides, in a natural way, good agreement with ab initio and thus a cheap and fairly reliable method of calculation which is readily refined as more ab initio and experimental data become available. In the simplest cases, it merely implies that lone pair orbitals or π orbitals are added to the computer representation, as 'dummy atoms'. Elements of this approach are contained in the EPEN method of the Scheraga School [11] and the MM1 method of Chung et al. [12]. However, it is by no means obvious where these orbitals are located, how one should best represent them as single points and what charges they carry, especially considering the variety of possible hybrid molecular orbital forms. The novel feature that we have introduced is an objective method of locating the position of such non-core centres using the ab initio wavefunction. The LCAO-m.o. SCF wavefunction is a single determinant of one-electron molecular orbitals which are a linear combination of atomic orbitals. These one-electron SCF m.o.'s are delocalised throughout the molecule. However, the energy and total wavefunction are invariant with respect to a unitary transformation of the active one electron m.o.'s. Thus a localised description of the m.o.'s may be obtained by the appropriate choice of unitary transformation. The Boys–Foster method [13] performs this localisation by maximising the sum of the squares of the distances of the dipole centroids of the active m.o.'s. The position of the dipole centroids of the l.m.o.'s taken as the centres of action of the molecule, and along with the atomic centres to give an Orbital Force Field for the molecule concerned.

In our calculations with potential functions, the α-6-1 functional form was used, so that equation (1) becomes

$$E_{\mathrm{ij}} = \frac{A_{ij}}{r^{\alpha}} - \frac{B_{ij}}{r^{6}} + \frac{C_i C_j}{r} \quad \alpha = 9 \text{ or } 12 \tag{3}$$

where E_{ij} is now the energy between two centres of action (atom centres, lone pair centres) i and j. Parameters A and B relate to the van der Waals' repulsion and attraction terms respectively. In order to reduce the number of parameters involved (and consequently to reduce the possibilities of an artifactual parameterisation) we adhere to the philosophy of Hagler et al. [14] by using the geometric mean law:

$$A_{ij} = (A_{ii}.A_{jj})^{0.5}| = A_i.A_j, \tag{4}$$

so that $A_{ii}^{0.5} = A_i$ is a parameter dependent on the type (sp^3 hybridized carbon, sp^2 hybridized oxygen, etc.) of the ith atom. The same geometric mean law was employed for B. Parameters C_i and C_j are the partial charges on atoms or orbital centroids. Parameters A and B yield E_{ij} in kcal mol^{-1}, with the coulombic parameters being reported in electron units (e.u.). No intrinsic rotational potential is introduced, and since the comparative ab initio calculations are on in vacuo molecules, no dielectric constants or other solvent dependent term is introduced into equation (3) unless stated otherwise.

The choice of values, A, B, C is the question to be answered. As a starting point we employed the (atom-centred function) parameter of the Hagler et al. [14], with the A_i and B_i parameter of dipole centroids being assigned the value of zero.

Note that the choice of the α power term in equation (1) is optional according to the original study of Hagler et al. [14] since both powers (with different parameters, A, B, C) give an equally good fit to the data base used in parameterisation. However, only the 9th power is consistent with experimental data from conformational behaviour in solution [15]. Further, the 9th power give a significant improvement over the 12th power term in a recent study of carboxylic acid crystals [16].

As partical charges represented by parameters C are less likely to be transferable between molecules, these charges are found by direct quantum mechanical methods. However, the charges of C, H given by Hagler et al. [14] were also investigated.

In calculations with only atomic-centred potential functions, the monopole charges were obtained from a Mulliken population analysis [17]. When utilising an orbital representation, these same net charges obtained for the atoms were distributed over the orbitals and the residual nuclear centres, monopoles being sited at the orbital centres and at the nuclear positions. The orbital centres were located as the dipole centroids by the Boys–Foster [13] localisation of molecular orbitals. The precise charges associated with the lone pairs were determined by parameterisation against the total ab initio energy, the whole conformational energy surface being taken into account.

4. THE EXAMPLE CASE OF PEPTIDE ANALOGUES

Although the glycyl dipeptide is of considerable interest in its own right as one of the naturally occurring dipeptide units of polypeptides, it also serves to help analyze the backbone interactions which dominate the peptide energy surface. Further, sidechain to backbone interactions characteristic of non-glycyl dipeptides contribute mainly in the vicinity of Φ (rotation angle C′–N–C_α–C′) = +120° ± 70° and to a much less extent in the vicinity of Ψ (rotation angle N–C_α–C′–N) = -−120° ± 70°. Many results for conformations well away from these regions of conformational space are thus, to a first approximation, of more general interest.

Calculations on *N*-formyl glycylamide (3s/2p) and *N*-formyl glycyl-*N*-methylamide (4–31G) are in excellent agreement, indicating the reasonable choice of the smaller analogue and the independence of our conclusion to the choice of extended basis set. They are also within 5 kcal of the energies calculated using the empirical 9-6-1 functions of Hagler et al. [14], except at Φ = −90, Ψ = 0, where there is a strong cis N---N overlap. One possibility is that the nitrogen-nitrogen or nitrogen-amide hydrogen repulsion may be too 'stiff' in crystal-derived potentials, and consequently further refinement might indicate the need to soften this interaction. Another is that an N---H–N hydrogen bond could be stabilizing the conformation at ($\Phi=-90$, $\Psi=0$). This may not be treated adequately in these empirical calculations (since lone pair and π orbitals are here not explicit) or possibly overemphasized in the ab initio calculations. Finally, the introduction of an empirically-parameterized valence force-field to allow for flexible geometry (variability of bond lengths, valence angles and all torsion angles) can account for the 'stiffness' of crystal potentials and 'soften' the energy surface in a manner consistent with the *experimental* data without needing to modify the empirical potentials or introduce an N---H–N hydrogen bond [15]. This presents an apparent dilemma, because the experimental data could be accounted for either by modifying the empirical calculations on the basis of the information from the ab initio calculations, or by the introducing of an empirical valence force field to represent the effects of flexible geometry.

Before considering this problem further, we considered it desirable to carry out calculations on another, more complex, dipeptide, in order to better appreciate the extent and nature of any disagreement between empirical and ab initio calculations when additional interactions between atoms are introducd. Adding a methyl sidechain to make the alanyl analogue serves this purpose and is of particular interest because alanyl dipeptides are the most generally used models for polypeptide units with sidechains [18]. The results did *not* suggest that a simple reduction in the size of the nitrogen or the 'stiffness' of its potential would resolve the dilemma (since it would enhance agreement in some regions and disagreement in others), which in any case would be inconsistent with the crystal data used by Hagler et al. to first derive the parameters. The major problem was therefore still to identify the origin of the discrepancy between the unmodified 6-9 potential and ab initio calculations.

To examine this in more detail, we plotted all ab initio energies reported here against the corresponding empirical 6-9 energies (Fig. 1). Several interesting points may be noted. If all conformations likely to contribute significant hydrogen bonds are considered separately, i.e., those within 30°, those within 30° of ($\Phi=180$, $\Psi=180$), ($\Phi=-90$, $\Psi=+60$), ($\Phi=+90$, $\Psi=-60$), ($\Psi=\pm90$, $\Psi=0$), then there is good agreement between the ab initio and empirical methods of about ±3 kcal, but may differ by as much as 5 kcal if they are considered together with the other points on the map. The relation for the non-hydrogen bonded configurations is linear with slope close to unity up to about 25 kcal, supporting the choice fo the 9th power functionality of the 6-9 potential. The departure from agreement above 25 kcal may be expected and there is certainly no justification for considering the higher energy points as departing from agreement because of differences other than the functionality of the repulsive term. The hydrogen bonded energies are 2–5 kcal lower than the non-hydrogen bonded configurations, and indicate that the ab initio calculations predict hydrogen bonds to be up to 5 kcal more stable than the empirical method. Further, and most important, the ($\Phi=\pm90$, $\Psi=0$) conformation for which an N...H–N hydrogen bond is possible clearly belongs to the hydrogen

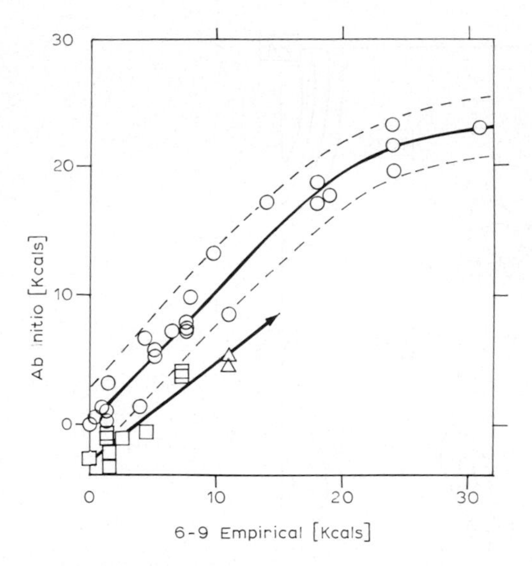

Fig. 1. Correlation between the conformational energies calculated by extended basis set ab initio methods with those calculated by the 9th power repulsion (9-6-1) potential functions of Hagler et al. [14]. The agreement confirms the 9th power functionality and is poor only for NH...O (□) and NH...N (△) hydrogen bonded conformations, since very expensive double zeta basis would be required in the ab initio approach.

bonded group, and *the discrepancy for this conformation can thus be accounted for solely in hydrogen bonding terms.*

This does not imply that the empirical potential is in error: the 9-6-1 potential fits both crystal [14], and solution data [15], (for both polar and non-polar solvents), without the need to introduce a hydrogen bonding contribution other than that already implied in the van der Waals' and electrostatic terms. It is our view that ab initio *calculations can provide a credible account of peptide conformational energies, except for hydrogen-bonded configurations.*

Such a result is in line with the observation that ab initio calculations using a double zeta basis, such as that used here, may overestimate hydrogen-bond energies by about 4 kcal. Polarization functions are needed in the basis to bring about a decrease in the hydrogen-bond energy [19]. Figure 2 represents our estimate, based on consistencies between recent almost wholly empirical and almost wholly non-empirical (ab initio) methods, of the topography of the alanyl dipeptide energy surface. Averaging and interpolation has been carried out by a least squares fit of Fourier curves. It predicts that the most stable conformation of the alanyl dipeptides is in the vicinity of ($\Phi=-90$, $\Psi=100$), representing a C_7 equatorial hydrogen-bonded ring configuration. The C_5 hydrogen bonded ring configuration ($\Phi=150$, $\Psi=-150$) is 1–2 kcal mol^{-1} higher in energy, while the C_7 axial configuration is 9–10 kcal mol^{-1} higher. The conformation in the right-hand α-helical regions ($\Phi=-60$, $\Psi=-50$) is 7–8 kcal higher than the C_7 equatorial conformation, while the left-hand α-helix region ($\Phi=+60$, $\Psi=+50$) is of the order of 1 kcal mol^{-1} higher than the right-hand α-helical region. The barrier to be crossed in going from

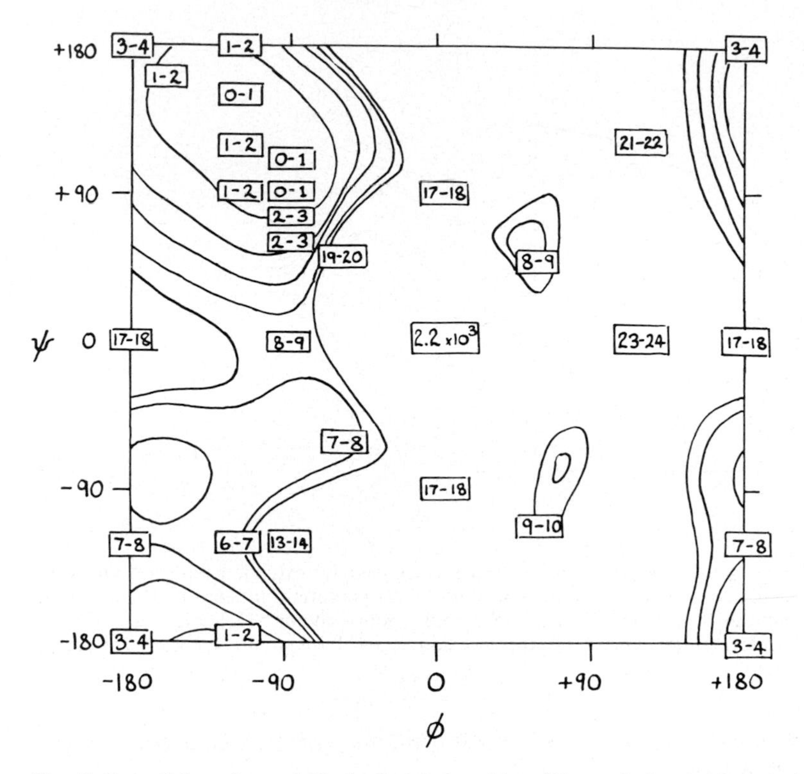

Fig. 2. Potential surface of blocked alanyl residue (*N*-acetyl alanyl *N'*-methylamide). The energy calculated by extended basis set ab initio methods is plotted as a function of rotation round the bond N–C_α (rotations angle ϕ) and the bond N–C′ (rotation angle ψ).

the C_7 equatorial to the right-hand α-helical regions in 8–9 kcal mol^{-1} in height, while that to be crossed in going from the C_7 equatorial to the left-hand α-helical region is 17–18 kcal mol^{-1} in height. This applied only to the artificial case of rigid geometry. Although flexible geometry had not been previously investigated by ab initio methods, and only recently by an empirical valence force field [15] relaxation of valence angles does appear to be important, particularly in the reduction of barrier heights. Considering the limited number of conformations for which we have, due to the time-consuming nature of the calculations, been able to study relaxation of geometry by ab initio calculations it would be premature to attempt to infer a map for which all geometry is minimized, at least on the basis of ab initio calculations. Figure 2 therefore only represents a new starting point for further calculations although the relaxation of geometry by ab initio calculations suggests that large energy changes are not to be expected, and this is particularly true of the low energy regions which are of greatest interest in the calculation of many properties.

Since a discrepancy in relation to hydrogen bonding was indicated and since this involves lone pairs, extension of the empirical potential approach to the status of an orbital-force field (OFF calculations, see above) is tempting. However, we are at this

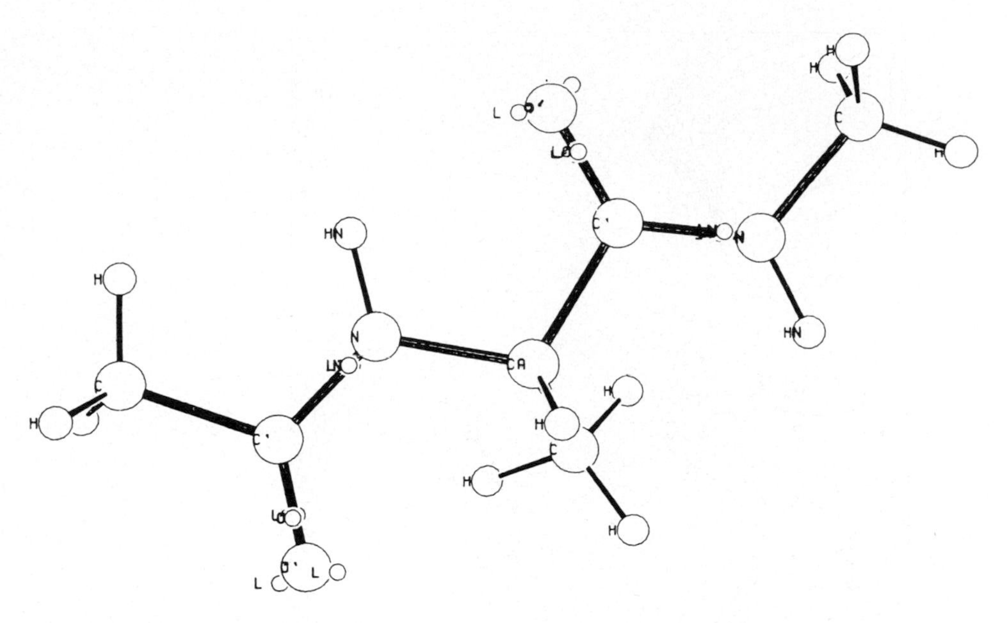

Fig. 3. The blocked alanyl residue *N*-acetyl alanyl *N'*-methylamide with non-core orbital dipole centroids localised by ab initio calculation.

stage working within the realms of the accuracy in empirical and ab initio techniques, and indeed experimental data is ambiguous at this level of detail. The potential functions of Hagler et al. [14] already agree tolerably well with experimental data [15] and the best result from an OFF calculation that can be hoped for is that it reproduces the original potential surface quite well. This is to say that if the OFF approach is in general the most valid approach, then peptide systems (or rather, the peptide backbone) must be shown to be a rather special case, in which, by the 'fortuitous' nature of the stereochemistry, inclusion of lone pairs, etc., makes little difference to the calculation. If this turns out to be the case, it is then presumably because peptides have been among the most widely studied biomolecules that the need for an OFF-type approach has not previously been seen as essential.

To explore this we proceeded as follows. The Boys–Foster localisation method [13] was employed on the SCF wavefunction of the dipeptide *N'*-acetyl alanyl *N'*-methylamide. A degree of localisation of greater than 0.97 was obtained. The molecule has 39 doubly occupied molecular orbitals, 10 of the localised molecular orbitals (l.m.o.'s) represent the 1s orbital of the C,N,O atoms, 17 of the remaining l.m.o.'s represent the σ-framework of the molecule, i.e., the C–H, N–H, N–C_α and C–C bonds. Four l.m.o.'s represent lone pair type orbitals which are located near to the oxygens, and are designated L in Fig. 3. The 8 remaining l.m.o.'s represent orbitals which are a combination of σ amd π type orbitals, and have centroids located above and below each C=O and each N–C (amide) partial double bonds (designated LO and LN respectively, in Fig. 3).

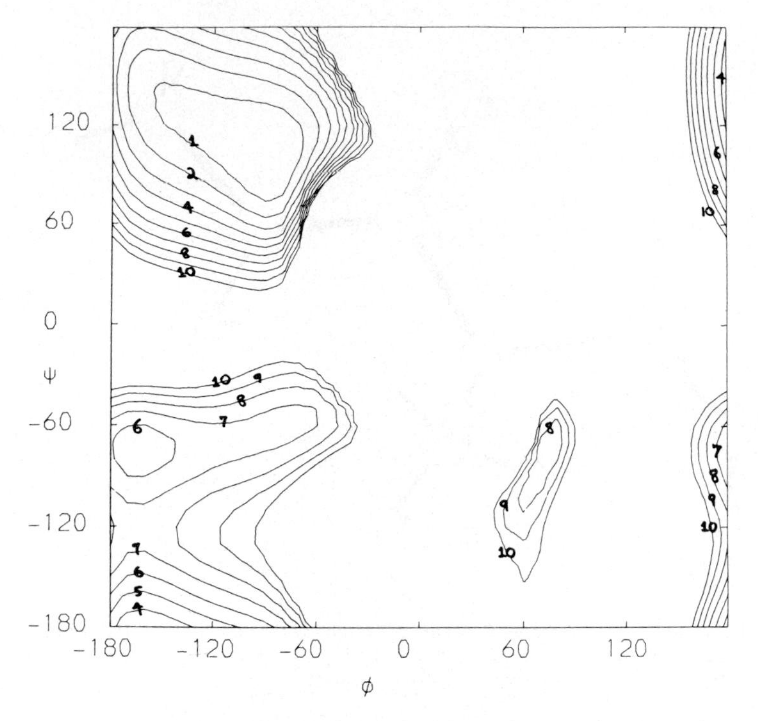

Fig. 4. The potential surface of the blocked alanyl residue *N*-acetyl alanyl *N'*-methylamide calculated with the 9-6-1 potential functions of the Hagler et al. [14], not including non-core interactions.

Figure 4 shows the (Φ, Ψ) conformational energy map for the dipeptide using the atom-centred 9-6-1 functions of Hagler et al [14]. Figure 5 shows the (Φ, Ψ) conformational map obtained using the OFF method on the dipeptide, with the charges L= −1, LO=LN=−.25 assigned to the dipole centroids. Comparison of these two maps, indicates that the essential qualitative features of Fig. 4 are maintained in Fig. 5. The quantitative differences between the two maps are small. Thus it may be concluded that the conformational energy map of a dipeptide may be reproduced accurately by using the atom-centred 9-6-1 potential functions alone and that it is not necessary to include centres of action located at positions near to the atom centres as in the OFF method. It must be recalled that this does not provide evidence against the OFF approach. Indeed, in conjunction with the studies in other classes of molecule (see below), it is a piece of negative evidence in its favour. By the same token, it is further evidence for the sufficiency of the potential functions of Hagler et al [14] for peptides, though as we shall see, these do indeed represent a special case.

5. THE EXAMPLE CASE OF THE NUCLEOTIDE BACKBONE

The relative sparsity of experimental data for nucleotides, makes ab initio calculations in their case even more important as a source of potential functions,

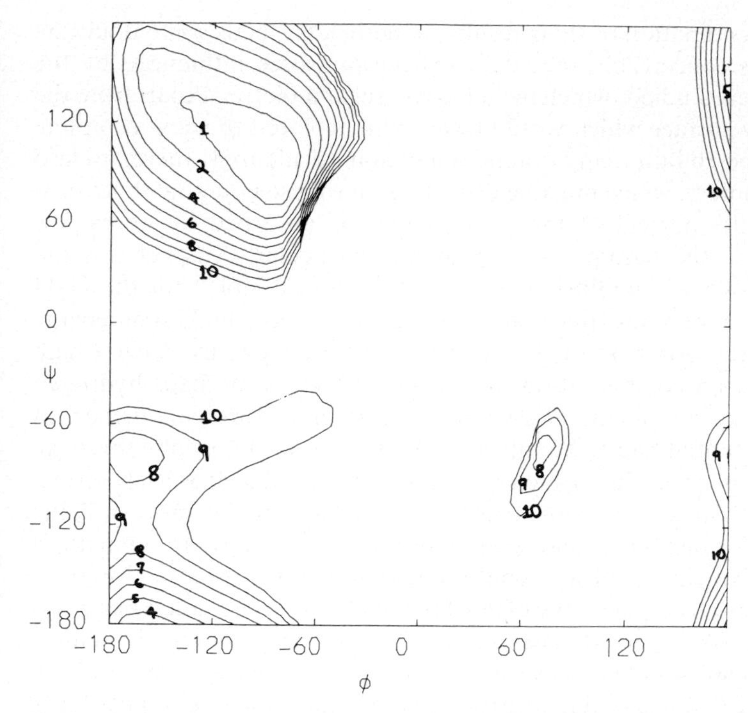

Fig. 5. The same potential surface as for Fig. 4, except that the OFF method (see text) is used to extend the potential functions of Hagler et al. [14] to include non-core orbital interactions. In this molecule, their inclusion has relatively little effect.

which may be parameterized so as to give agreement with the ab initio results for smaller analogues.

Recently, ab initio calculations have been used by Matsuoka et al. [20] to parameterise potential functions for nucleotides. Previously, such parameters as were available, had been derived on a somewhat ad hoc, piecemeal basis, without any extensive quantum mechanical or experimental justification. We were interested in determining whether these new potential functions placed conformational analysis of nucleic acids on a more similar footing with the peptide protein studies. We suspected that such a goal would be complicated particularly by a feature of the nucleic acid backbone which contrasts that of proteins: a priori, backbone behaviour might be strongly influenced by interactions involving lone pairs of the C–O–P(O_2)–O–C group (and perhaps of the oxygen of the adjacent sugar ring). It was this which first caused us to turn to an OFF calculation, because it could mean that one must depart from the classical approach of using potential functions of the distance between atom centres, and consider an orbital force field exploiting the centroids of certain molecular orbitals including the lone pairs. Matsuoka, et al [20], who were aware of this difficulty, used the potential functions of the atom-centred type and moreover they considered that specific inclusion of lone pair centroids was not required to obtain good agreement with their ab initio results.

A more cautious examination of their results, nonetheless, sheds some doubt on their conclusion. It is conceivable that their conclusions were influenced by the complexity of the system studied, which included two sugar moieties. Apart from the very extensive energy surface which would have to be explored to show complete agreement between the ab initio, and potential function calculations, there are also many potential parameters, which must be fitted to derive these potential functions. Errors arising from the neglect of lone pairs could be absorbed by unrealistic compensating errors in the parameter assignments. Particularly suspect was the value of the van der Waals' equilibrium energy of -2.5 kcal mol^{-1} for the O–H interaction given in their potential functions. The van der Waals' equilibrium energy is invariably only of the order of -0.1 kcal mol^{-1} in other systems. Such a high negative value suggests a compensation of uncounted factors, perhaps hydrogen bonding, or high-energy interactions elsewhere in the potential surface. Further, an intrinsic rotational potential had to be introduced, and in our philosophy this is an undesirable "boot strap" procedure which should only be introduced as a last resort.

Extended basis set calculations are generally considered to be more reliable because of their greater flexibility, and (computer time permitting) are the natural choice for an extensive survey of any conformational energy surface. However, charged molecules can pose special problems for the choice of extended basic set. Any practical choice of extended basis set might here be less reliable than a computationally cheaper minimal basis set because the relative inflexibility of the latter may resist unrealistic trends due to strong electrostatic effects. This possibility was supported by our first extended basis set calculation on the extended conformation, from which unexpectedly high charges on the phosphorus and oxygen atoms, particularly the backbone oxygen atoms, were obtained. This suggested that, in progressively extending a minimal basis set, results for DMP^- might get very much worse before beginning to improve.

We explored this possibility by further extension of this basis set and examined the resulting effect both on the total energy and the charge distribution.

It was, therefore, supplemented by: (1) an extra s-type gaussian orbital with an exponent of 0.8; (2) 3d orbitals represented by one gaussian with exponent 0.4 [21]; and (3) both extensions applied simultaneously. The extra orbitals were associated with the phosphorus atom. In all cases the charges on the phosphorus and oxygen atoms become somewhat less extreme and more importantly, the total energy fell. The two extensions applied simultaneously decreased the absolute values of these charges to the greatest extent. There was no significant effect on the carbon or hydrogen charges. As the basis sets are further extended, the energy would be expected to converge, and moreover, to do so as a function of changes in the resulting charge assignments. However, the energy is clearly far from convergence and the size of the basis set required to produce a tolerable degree of convergence would almost certainly be prohibitive.

Nonetheless, all basis sets gave results which were quantitatively similar in one important respect: we consistently found a large maximum at $\Phi=\Psi=180°$. (The energy difference between $\Phi=\Psi=180°$ and $\Phi=\Psi=60°$ using Basis II supplemented with an extra s and d functions on phosphorus is 9.85 kcal mol^{-1}.) The maximum at $\Phi=\Psi$ = 180° in the ab initio potential energy surface of DMP^- which seems to be independent of the choice of basis set, does not correspond to a steric clash. It arises instead, from electrostatic repulsion between the lone pairs (and their hybrid forms). This dominant feature of the potential surface cannot be reproduced in classical calculations with atom-centred potential functions. Thus, the use of OFF-type calculation is clearly indicated as the preferred approach. Which studies, however, should be carried out to derive the locations and interactions of the orbitals, since the charged species creates the difficulties described above?

First, an extended basis set used with a positive charge in the vicinity of the charged dimethylphosphate. Studies have been undertaken with a charged entity which has physical meaning, namely a counterion; this still leaves open the question of the proper location of the counterion, which is likely to depend on the molecular conformation and for any conformation might have considerable librational freedom. For this, Monte Carlo or Molecular Dynamics simulation of the dimethylphosphate and counterion in solution would be required in order to construct a solvent-modified potential energy surface of the dimethylphosphate (comparable studies are being undertaken by us for dipeptides, see [22].

Second, the uncharged, acid form of the dimethylphosphate could be studied. The results could then be used for conformations not likely to be grossly affected by a change in the ionisation state. Such studies have been carried out, but while complicating matters by introducing an asymmetry to the potential surface, they generally support the conclusions reached by the following, third approach.

The third (simplest and most obvious) approach would be to confine calculations to the more robust, if less flexible, partially extended IB basis set and this has been done in the following studies. This is because the features of importance are reproduced by the more extended basis sets, while on the other hand, the IB basis set satisfies the need to include d-orbital contributions in the molecular orbital description of the P–O bonds (see [23]).

Using the Boys–Foster localisation method [13] a 'degree of localisation' in excess of 0.97 was achieved for both basis sets. This is again a very satisfactory result which along with an analysis of the localised molecular orbitals obtained first suggested to us that orbitals might reasonably be represented as single centres of action for use with potential functions. DMP^- has 33 doubly occupied molecular orbitals, 11 of the localised molecular orbitals represent core orbitals (the 1s on carbon and oxygen, and the 1s, 2s, 2px, 2py, 2pz on phosphorus). In the case of Basis 1A, 12 of the localised orbitals provide the σ-framework of the molecule and thus the remainder must represent π or lone pair orbital lobes, or combinations of both. The latter are localised near the oxygens.

In the case of Basis set 1B, 10 of the localised orbitals provide the σ-framework of the molecule, with no σ-type molecular orbital localised along either the P1–O2 or P1–O3 bond. The remaining 12 localised orbitals, which have centres near to the oxygen atoms, represent lone pair orbitals or orbitals which are a combination of σ and π type orbitals.

The different localised description obtained from the two basis sets, with and without phosphorus d function, arises from the possibility of O2p→P3d donation

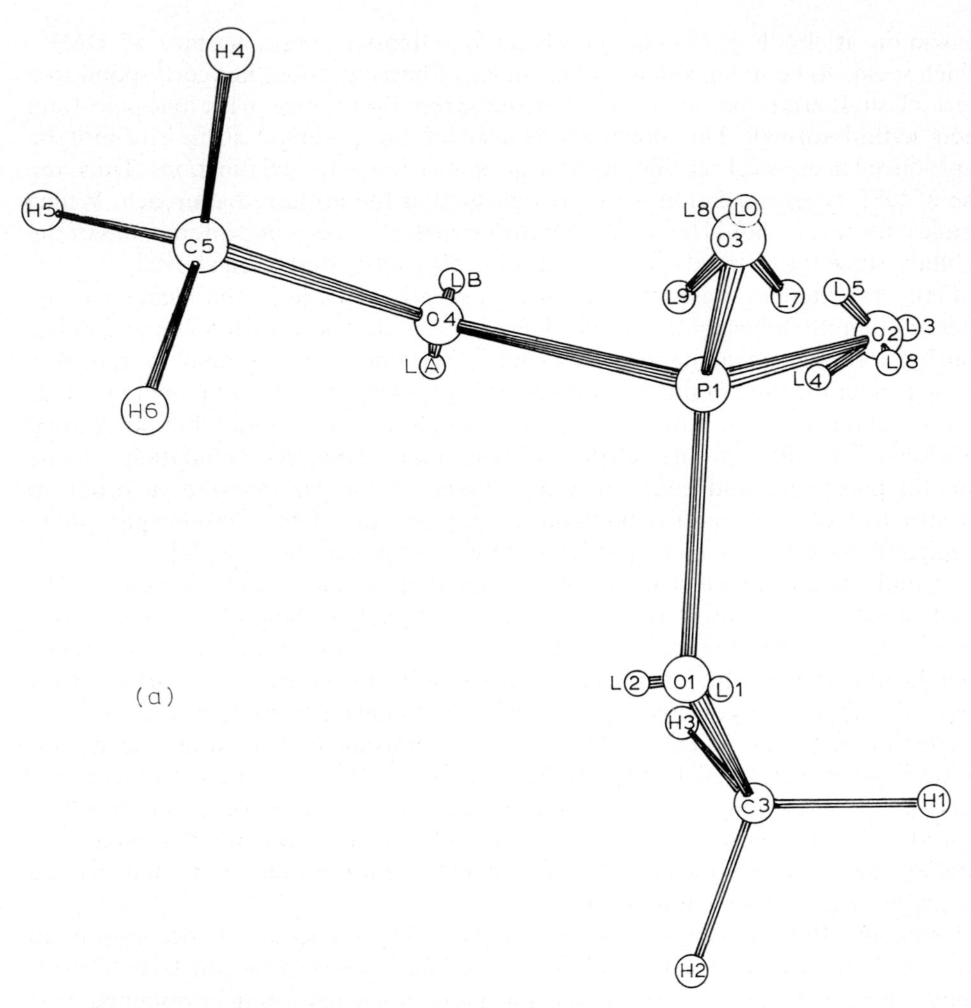

when P3d function are present and is fully discussed by Guest et al. [23]. Briefly and obviously, it is the 3d supplemented basis set which must be considered realistic.

The spatial positions of the orbital centroids (designated L) obtained using Basis IB, are shown in Fig. 6 for two conformations of DMP^-.

Since these localizations *might* depend on backbone geometry and resulting interactions, two conformations were studied. These may be expected to lead to the greatest possible difference in results due to different mutual approaches of orbitals L. These conformations are $\Phi = \Psi = 180°$ and $\Phi = \Psi = 60°$, with the dihedral angles H–C–O–P set at 60° in all cases. Certain valence angles of orbitals L on the pendent (not backbone) oxygen atoms undergo most variation, changing by some 18°. We find that, to a first approximation, a standard 'equilibrium' geometry can be used for all orbitals L irrespective of backbone conformation.

In contrast, a proper description does depend on the inclusion of phosphorus d orbitals. While the backbone oxygen orbitals L in the Basis set IB have the

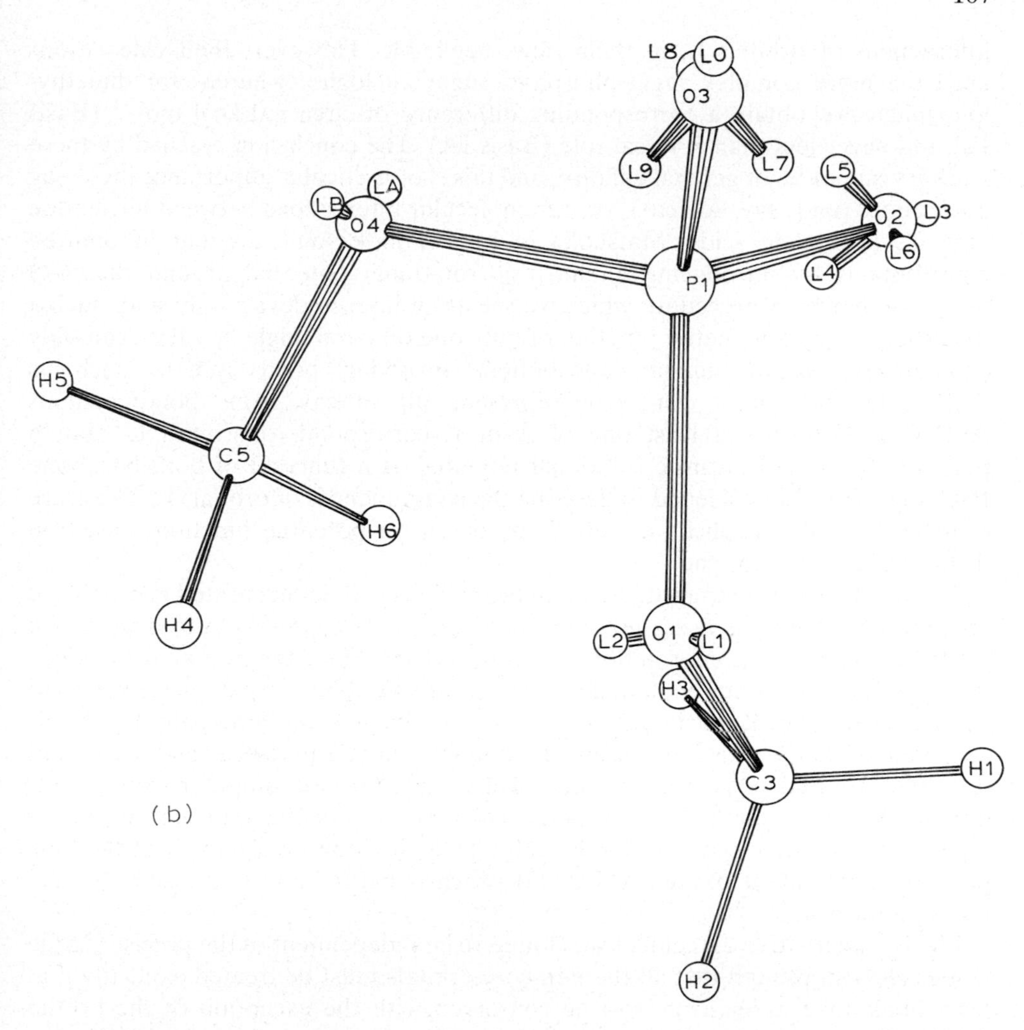

Fig. 6. Dimethyl phosphate (analogue of nucleic acid backbone fragment) with localised non-core orbital dipole centroids: (a) conformations I ($\phi = \psi = 90°$); (b) conformation II ($\phi = 90°$, $\psi = 340°$).

approximate expected geometry consistent with 2 lone pairs on each oxygen, the pendent oxygen atoms, each have 4 localised orbitals. Three of these four localised orbitals are deflected backward along the P–O, due to O2p→P3d interactions, the fourth localised orbital has the character of a lone pair on oxygen, but with the P–O2–L3 angle being nearly 180°. The bonding, thus decribed beween the phosphorus and pendent oxygens, is similar to that between sulphur and oxygen atoms in SO_2 [23].

In their parameterization from ab initio calculations, Matsuoka et al. [20] decided against including any centres of action other than the atomic centres with the justification that the energy difference between conformation $\Phi=\Psi=180°$ and $\Phi=\Psi=90°$ is only 0.61 kcal mol^{-1}. This difference arises largely from different interactions of orbitals L, in their view, negligible. However, their calculations used the more complex sugar-phosphate-sugar analogue, whereas for dimethylphosphate we obtain a corresponding difference of circa 8.0 kcal mol^{-1} (Basis IB) and never less than 5.4 kcal role (Basis IA). The conclusion reached by these workers cannot be of general validity, and this is of particular importance involving nucleotides (and, say, solvent), or intramolecular interactions between nucleotide units as in nucleic acids. Matsuoka et al [20] took some account of omitted contributions by introducing an intrinsic rotational potential around the P–O backbone bonds, a procedure which we see as an interim device only since such a potential as a trigonometric function of only one dihedral angle at a time can only compensate for uncounted contributions involving bond systems such as A–B–C–D. It cannot generally represent all effects from bond systems A–B–C–D–E where at least one of A or E correspond to orbitals L, and in particular a unified intrinsic rotational potential as a function of both backbone P–O angles would be needed to describe the consequences of orbitals L. These are thus best included explicity as centres of action for potential functions based on distances in Cartesian space.

The need for considering all the non-core orbitals was demonstrated in a series of computer experiments using potential functions, which also serve to analyse the contributions to the conformational energy surface. First, the potential functions used above were varied, but failed for any reasonable choice of parameter values to produce a significant maximum at $\Phi=\Psi=180°$. Second, the lone pairs L orbitals localized in the IB basis were assigned monopole charges in the range 0–2 e.u. and then the required peak was obtained. Third, in a further computer experiment, only the lone pairs on the backbone (O1, O4) were included, and this produced not a peak, but a saddle point at $\Phi=\Psi=180°$. Thus, it is the involvement of the lone pairs of the pendant oxygens (O2, O3) which is particularly responsible for the problematic peak.

Finally, qualitative agreement was found to be independent of the precise charge values chosen, though that all the non-core orbitals must be treated explicitly if at least qualitative agreement is to be conserved with the exception of the orbital disposed at circa 180° with respect to the P=O bond. Originally, we had intended to carry out a full optimization of the parameters by a least squares method. However, as noted above, the ab initio energy surfaces (Fig. 7) and the potential function surface (Fig. 8) are already in tolerable agreement of 1–2 kcal mol^{-1} which is better than the limit in the reliability of the ab initio method.

6. THE EXAMPLE CASE OF BENZENE SULPHONAMIDE, Ph SO_2NH_2

Benzene Sulphonamide, Ph SO_2NH_2, which is the core of a pharmaceutically important drug series, is of interest because the lengths of the covalent bonds to sulphur greatly reduce the strength of the atomic overlaps in the cis configuration of each atom pair across the bond, so that non-core orbital interactions dominate the conformational behaviour.

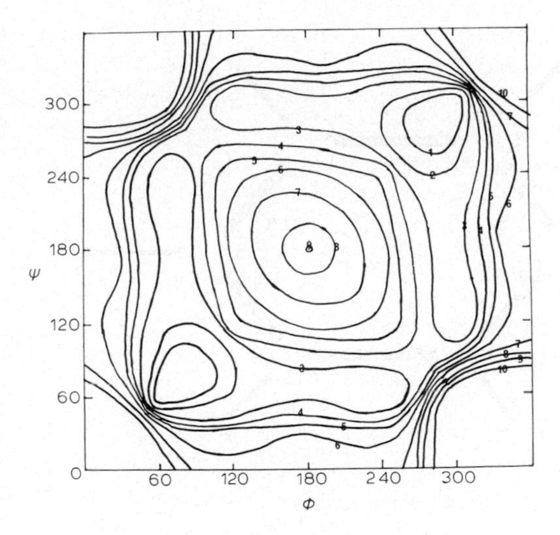

Fig. 7. The ab initio potential energy surface of dimethyl phosphate.

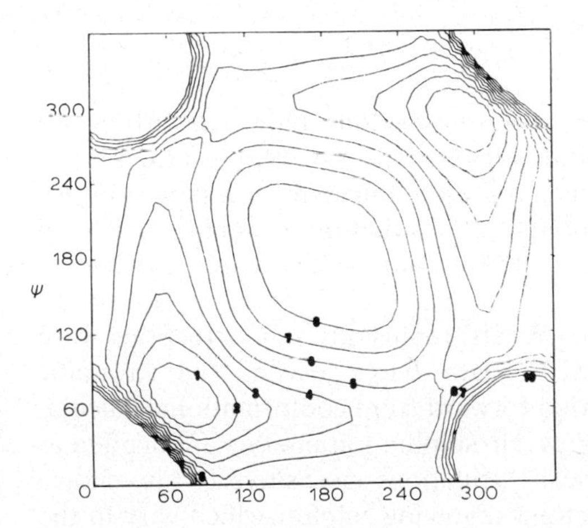

Fig. 8. The potential energy surface of dimethyl phosphate using the OFF method (see text).

After localisation, the 41 double occupied molecular orbitals of the molecule, may be classified as follows: (a) 14 l.m.o.'s represent the core orbitals, i.e., 1S, on C, N, O and 1s,2s, 2px, 2py, 2pz on S; (b) 9 l.m.o.'s represent the σ-orbitals of the 5 C–H, 2 N–H, 1 S–N and 1 S–C bonds; (c) 9 l.m.o.'s represent the σ or σ + π type orbitals of the phenyl ring. The dipole centroids can be located on every alternate bond lying above and below the ring plane adjacent to the midpoint of the bond. However, in OFF calculations the dipole centroids are placed above and below each bond of the phenyl ring, to take account of the bond equivalence of the phenyl ring. This assumption of equivalence may be unjustified for a substituted benzene. (d) 9 l.m.o.'s represent either lone pair type orbitals or lone pair + π type orbitals, associated with the oxygen and nitrogen atoms. These l.m.o.'s are designated LN, L1, L2, L3, and L4

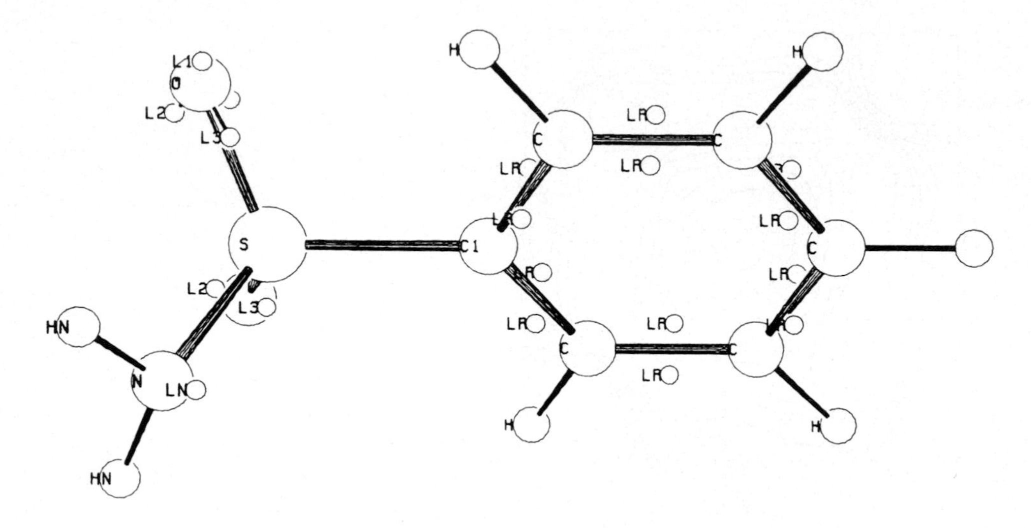

Fig. 9. Benzene sulphonamide with localised non-core orbital dipole centroids.

are shown in Fig. 9. One of these, LN represents a lone pair type orbital on nitrogen, two other, L1, lone pair orbitals attached to the oxygen atoms. The remaining 3 l.m.o.'s attached to each oxygen represent lone pair + π type orbitals, and are deflected back along the S–O bonds. This situation is similar to that in DMP^-, which is not surprising as $-PO_2^-$ and $-SO_2$ are isoelectronic.

Most of the parameters for the van der Waals' repulsions and attractions were carried forward from the peptide/DMP^- data bases except the aromatic carbon atom parameters which were carried forward from conformational calculations on molecules containing benzene rings. No sulphur parameters are needed as the chemical structure makes the empirical calculations insensitive to any choice (i.e., there are no van der Waals' interactions involving sulphur which vary in the empirical calculations on Ph SO_2NH_2).

Table 1 gives the results from the ab initio, 9-6-1 empirical atom-centred calculations (see fig. 10 also) and OFF calculations (see Fig. 11). the disagreement between the ab initio and OFF energy of any point on the energy surface is less than or equal to 1.3 kcal mol^{-1}. This is very acceptable. However, on neglecting the non-core orbitals and using the standard practice of parameterising only for interactions between atom centres, disagreements with the ab initio energy of 6–9 kcal mol^{-1} are encountered. Both the ab initio and OFF calculations predict that much of the conformational space is not significantly populated at biological temperatures, in stark contrast to the atom-centred calculations. Presumably, the classical procedures are grossly inadequte for conformational energy calculations on Ph SO_2NH_2 and it is reasonable to suppose that this is also the case in many other molecular systems of pharmaceutical interest.

TABLE 1
Comparison of the energy differences obtained by ab initio, atom-centred and OFF calculations

	Energy differences in kcal mol^{-1}		
Φ, Ψ	ab initio	atom-centred	OFF
0,0	1.266	0.609	0.875
0,45	5.423	1.73	5.591
0,90	12.25	3.457	10.99
0,135	8.185	3.725	7.437
0,180	2.08	3.464	3.254
45,0	0.725	0.141	0.755
45,45	4.857	1.370	5.688
45,90	9.150	2.380	9.803
45,135	4.992	2.374	5.377
45,180	1.094	2.668	1.330
90,0	0.0[1]	0.0	0.0
90,45	3.239	0.793	3.830
90,90	7.439	1.636	7.349
90,135	4.122	1.943	3.20
90,180	0.026	2.199	−0.658
135,45	3.75	.789	4.128
135,90	8.528	1.888	8.202
135,135	5.733	2.563	4.934

[1]The ab initio energy corresponding to the conformation $\Phi = 90$, $\Psi = 0$ is −822.719514 a.u.

7. OUTSTANDING PROBLEMS ENCOUNTERED IN ACCURATE CALCULATIONS OF CONFORMATIONAL BEHAVIOUR

When we are interested in carrying out calculations of conformation behaviour in solution, and in predicting average properties, such as can be derived from the potential energy surfaces by statistical mechanical methods, we find that there are still a number of outstanding difficulties to be overcome before accurate quantitative results can be obtained.

7.1 *Occasional need for an intrinsic rotational potential*

The need for an intrinsic rotational potential as an energy function of rotation around bonds is best regarded as a 'correction factor' to be added in when all else fails. In our view, the introduction of such a potential is an undesirable feature, which is fortunately circumnavigated in many cases by the OFF calculation. This usually produces results in tolerable agreement with ab initio calculations, without the obvious need for a correction factor. This *may,* however, not be true in some cases. Here we examine the test case of $CH_2(OH)_2$, which may be one such case.

As implemented so far, the OFF method may not be able to reproduce the ab initio energy surface as calculated by Jeffrey et al. [10], without the incorporation of an

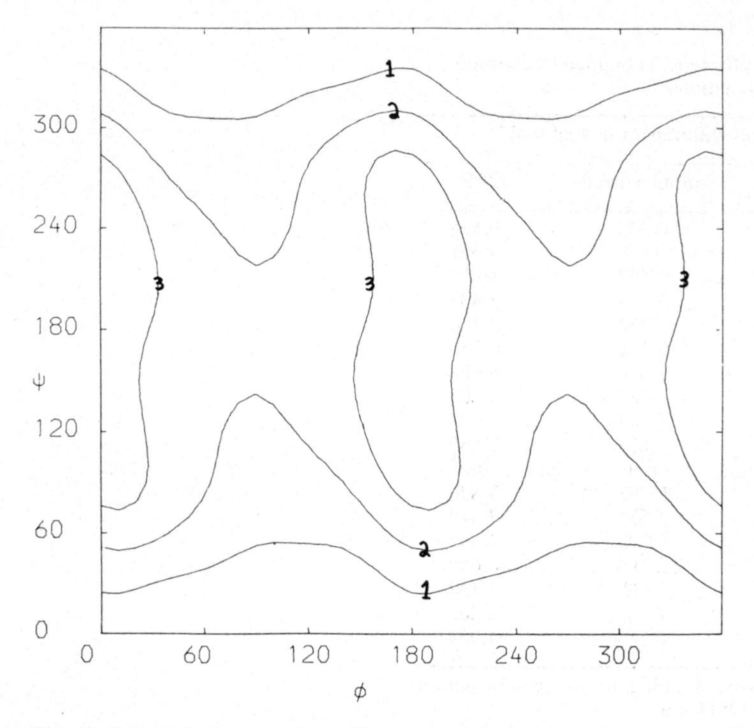

Fig. 10. Potential energy surface of benzene sulphonamide using the atom-centred potential functions of Hagler et al. [14].

intrinsic rotational potential, due to the so-called 'anomeric effect' operating in a molecule of the type O–X–O.

Jeffrey and Taylor [24] using the MM1 method which incorporates both lone pairs and an intrinsic rotational potential into the calculations, obtained agreement between the empirical and ab initio potential surfaces to within 2.3 kcal mol^{-1} for each of the 15 ab initio points calculated. The OFF method, without an intrinsic rotational potential, was able to match this agreement throughout the potential surface, except for conformations near either $\Phi=180$, $\Psi=0$ or $\Phi=0$, $\Psi=180$. Such conformations were 3.8 kcal mol^{-1} less in energy with respect to the minimum confrontation in the OFF calculation than in the ab initio calculation. On the introduction of an intrinsic rotational potential, the agreement between the empirical and ab initio calculations throughout the entire potential surface becomes as good as that obtained by the MM1 method.

Therefore, we are unable to deny significant contribution, of the anomeric effect requiring an intrinsic rotational potential. The discrepancies in the OFF calculation may result from ignoring the l.m.o.'s of the σ-framework. If charges were also to be located at the centroids of these l.m.o.'s in addition to those at the centroids of the lone-pair type or lone pair + π type l.m.o.'s located near the oxygen atoms then better agreement may be obtained with the ab initio surface without the incorporation of an intrinsic rotational potential. Similarly the EPEN method of

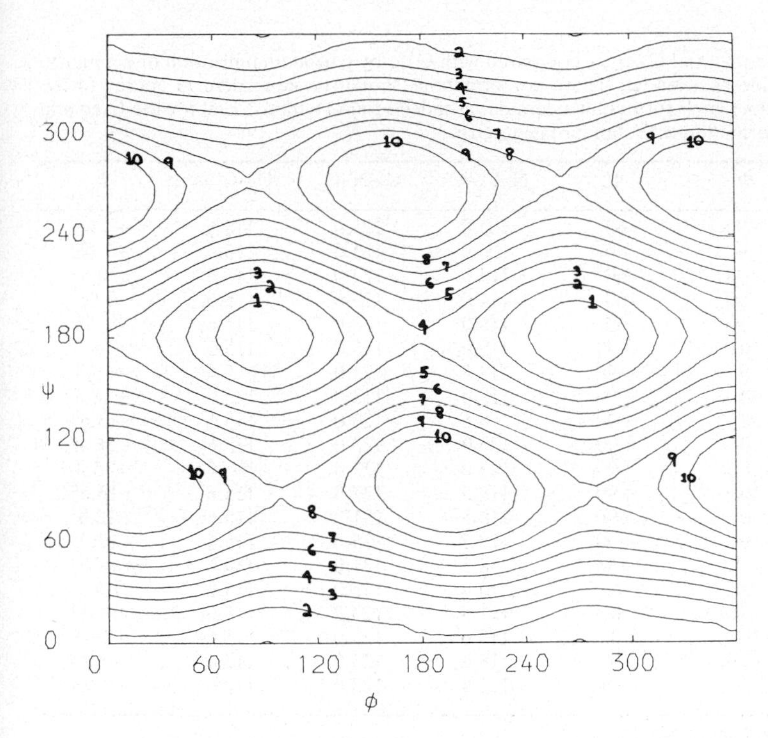

Fig. 11. Potential energy surface of benzene sulphonamide using the OFF method.

Scheraga, et al. [11] was able to obtain good agreement between many calculated and experimental properties of hydrocarbons by placing centres of charge on bonds, as well as on atomic centres.

7.2. *Flexible geometry*

The vast majority of calculations in the literature have employed the 'rigid geometry' approximation, which is to say that all bond lengths, all valence angles, and the dihedral torsion angles around bonds of order greater than 1 (e.g., double bonds, peptide links) are fixed at standard values. Robson et al [15], using the potential functions of Hagler et al [14] have shown this to give tolerable results (and in many cases accurate results) for the calculated NH–C^{α}H vicinal NMR coupling constant J_{α} of a variety of *N*-acetyl amino acyl *N'*-methylamides and dimethylamides, for the dipole moments of mixed L- and D-oligopeptides, and to some extent reproduce infra red data on intramolecular hydrogen bonding modes. However, they do not explain the high population of an cis N–C_{α}–C'–N barrier by residues in proteins, nor the experimental characteristic ratios (which relate to viscometric and light scattering properties) of long chain unperturbed random coil polypeptides. For example, while the coupling constant for *N*-acetyl alanyl *N'*-methylamides is in the experimental range of 7.8 ± 0.5 Hz, the calculated characteristic ratio for polyalanine is about 19, well in excess of the experimental

TABLE 2
Relaxation of barriers C_7eq-α_L and C_7eq-α_R compared with C_7eq by partial minimization of geometry, using *N*-formyl glycylamide as a model. These conformations should be equivalent in energy to the corresponding conformations of *N*-formyl alanylamide, as indeed found within 1 kcal for the C_7eq and C_7eq-α and C_7eq-α_L conformation with standard geometry.

Conformation	Φ	Ψ	N–C_α–C′	C′–N–C_α	C_α–C′–N	*E*
C_7eq	−90	60	102.0	123.0	115.6	+4.55
C_7eq	−90	60	106.0	123.0	115.6	+1.05
C_7eq	−90	60	110.0	123.0	115.6	−0.5
C_7eq	−90	60	114.0	123.0	115.6	−0.55[1]
C_7eq	−90	60	118.0	123.0	115.6	+1.55
C_7eq-α_L	0	+90	110.0	123.0	115.6	+18.7
C_7eq-α_L	0	+90	112.5	123.0	115.6	+16.4
C_7eq-α_L	0	+90	115.0	123.0	115.6	+14.6
C_7eq-α_L	0	+90	118.5	123.0	115.6	+13.65
C_7eq-α_L	0	+90	120.0	123.0	115.6	+14.45
C_7eq-α_L	0	+90	123.0	123.0	115.6	+15.4
C_7eq-α_L	0	+90	118.5	127.0	115.6	+13.65
C_7eq-α_L	0	+90	118.5	131.0	115.6	+12.3[1]
C_7eq-α_L	0	+90	118.5	135.0	115.6	+13.05
C_7eq-α_L	0	+90	118.5	131.0	119.6	+13.05
C_7eq-α_R	−90	0	110.0	123.0	115.6	+4.6
C_7eq-α_R	−90	0	112.5	123.0	115.6	+4.45[1]
C_7eq-α_R	−90	0	115.0	123.0	115.6	+4.95
C_7eq-α_R	−90	0	118.0	123.0	115.6	+6.1
C_7eq-α_R	−90	0	112.5	123.0	119.6	+3.95

[1]Minimum energy after relaxing geometry.

range of 5–11, for analogous homopolymers (e.g., of glutamate) which do not aggregate. Precise calculations on *N*-acetyl glutamyl *N*′-methylamide, and Monte Carlo calculations on long chain homopolymers, have failed to resolve the discrepancy.

However, when the rigid geometry assumption is discarded, and a valence force field introduced from parameterization of infrared data for vibration modes of *N*-acetyl *N*′-methylamide, all experimental data can be brought into tolerable agreement. Nonetheless, agreement is still only approximate. For example, the characteristic ratios are brought down through the experimental range to its lowest limits [15]. However, this demonstrates the importance of the effect, and a more detailed study of valence angle bending seems to be all that is required. The present IR force field appears to overestimate the freedom allowed to valence angle opening.

This has been investigated by ab initio methods. As the points of conformational space relevant to such an analysis are not strongly influenced by the presence of the sidechain, the *N*-acetyl glycylamide analogue was employed. We carried out flexible geometry ab initio (3s/2p) calculations for two important conformational barriers ($\phi=\pm90,\psi=0$) and ($\phi=0,\psi=\pm90$). A full minimization would be prohibitively time consuming, on the inspection of models and by use of empirical potentials, the valence angles N–$\hat{C}_\alpha$–C′ and C′–$\hat{N}$–C_α are likely to be of crucial importance for these conformations, as cis contacts are then relaxed. Table 2 shows that at ($\Phi=0$, $\Psi=\pm90$) a minimum in the space defined by these valence angles is found at (N–$\hat{C}_\alpha$–C′) = 118.5°, (C′–$\hat{N}$–C_α) = 131.0°, due to relaxation of the crucial C′---C′ cis interaction. The energy is brought down from 18.7 to 12.3 kcal mol^{-1}. Opening the $C_\alpha\hat{C}'$–N angle at this minimum does not significantly affect the energy. At ($\Phi=\pm90$, $\Psi=0$) the minimum energy with respect to the N–$\hat{C}_\alpha$–C′ angle so as to relax the N---N cis interaction was 4.45 kcal, a small decrease compared with the value of 4.6 kcal using fixed standard geometry. Opening the C_α–$\hat{C}'$–N angle to further relax the N---N in interaction was expected to be more effective than for the ($\Phi=0$, $\Psi=\pm90$) conformation, however, there was only a small decrease in energy to 3.95 kcal mol^{-1}.

The energies *after* relaxing geometry in the ab initio calculation agree to a first approximation with those predicted by empirical valence force field calculations at ($\Phi = \pm90$, $\Psi = 0$), respective energies of 3.95 and 4.7 kcal being obtained. At ($\Phi=0$, $\Psi=\pm90$) the resulting energies are 12.3 and 7.8 kcal respectively, in less satisfactory agreement. As the above minimization of the ab initio surface is confined only to a few crucial variables, it is probable that the latter agreement would be further enhanced in a full minimization.

Minimization of (N–$\hat{C}_\alpha$–C′) does not significantly reduce the energy for the C_7 eq conformation ($\Phi=-90$, $\Psi=60$), although causing the N–C_α–C′ angle to open from 110° to 114°. The global minimum on the standard geometry ab initio Φ–Ψ map liews very close to that on the flexible geometry ab initio Φ–Ψ map. The above quantum mechanical and empirical energies are thus referred to almost identical zero energy reference conformations.

An interesting (if expected) feature, is that at biological temperatures considerable fluctuations in the valence angles may occur (N–$\hat{C}_\alpha$–C′), can readily vary from 106° to 119°, (C′–$\hat{N}$–C_α) from 123° to 135°, and (C_α–$\hat{C}'$–N) probably from 115° to 125°, or more, the exact ranges depending on the rotation angles Φ and Ψ.

Rather more detail has recently been obtained, for the valence force field for *N*-acetyl *N*′-methylamide used to derive the original IR force field, and hence for the peptide link (unpublished work). The angles C_α–$\hat{C}'$=0 and N–$\hat{C}'$=0 are obviously inter-related, as are C′–$\hat{N}$–H and H–$\hat{N}$–C_α. The former pair, for example can be described by an elliptical energy well with quadratic form along the minor and major axes, the ellipse being centred at the energy minimum C^α–$\hat{C}'$=0 = 122° and C^α–C′–N = 116°; the energy rises 36 kcal mol^{-1} for a 15° variation in C^α–$\hat{C}'$=0 and a 10° variation in C^α–$\hat{C}'$–N (the ellipse lying with its major axis parallel to the C^α–$\hat{C}'$–N ordinate of the plotted energy surface). (S. Carr, I.H. Hillier and B. Robson, unpublished work).

Energy contributions of related angles of the type A–$\hat{B}$–C, A–$\hat{B}$–D can be expressed by

$$\Delta E_\theta = A_1(\theta_1-\theta_1^0)^2 + A_2(\theta_2-\theta_2^0) + B_{12}(\theta_1-\theta_1^0)(\theta_2-\theta_2^0)$$

where $A_1 = A_2 = 0.146$ and $B_{12} = 0.123$ for the above example system (C_α–$\hat{C}$–O and C_α–$\hat{C}'$–N).

For the system

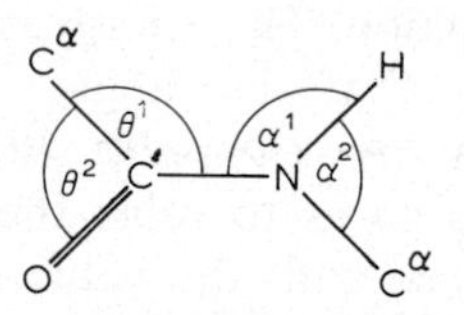

there are also significant cross terms representing the effect of interactions between the C^{α}–C'=0 and HNC^{α} groups. For small distortions of in-plane and out-of-plane bending of these groups, additional contributions such as

$$\Delta E_x = \kappa_x \cos(\omega - \omega^0)[(\theta_1 + \theta_2)^2 - (\theta_1^0 + \theta_2^0)^2]$$

(S. Carr, unpublished work) are indicated. Here $(\omega - \omega^0)$ is the distortion of the peptide group dihedral angle C^{α}–C'–N–C^{α} from the equilibrium value ω^0. This distortion also makes, of course, a strong independent contribution $\Delta E_{\omega} = \kappa_{\omega} \cos(\omega - \omega_0)$ where $\kappa_{\omega} = 23$ kcal mol^{-1} according to the ab initio results. The cross term ΔE_x (though not the contribution $\Delta E\theta$ and ΔE_{ω}) can be neglected when potential functions of distance C_{α}---C_{α}, C_{α}---H etc. are used to represent the cross interactions explicitly.

So far, the valence force field derived ab initio considered overall does imply higher energies for valence angle distortions than the preliminary IR-derived force fields. Although it is as yet incomplete and quantitative calculations of characteristic ratios etc. cannot be carried out; it is inevitable that the potential surfaces will be intermediate in detail between those obtained by the empirical force field and those calculated with rigid geometry, suggesting that the calculated properties will lie closer to the middles of the experimental ranges. In peptides internal rotation of the peptide group which is normally fixed in the planar configuration may make an important contribution, but no such consideration applies to the nucleotide backbone analogue dimethylphosphate (DMP^-). In the case of DMP^- the major changes likely to relax barrier heights are variations in the O1-$\hat{P}$-O4 angle and the two C–$\hat{O}$–P angles. These have been optimised independently at $\Phi=\Psi=90°$ (conformation I) and $\Phi=90°$, $\Psi=340°$ (conformation II), and results are given in Table 3. In reality, these angles are not entirely independent, so the optimisation is only partial. Also the results are preliminary because Basis IA (the minimal STO-3G basis set) was used in the calculations. Nevertheless, there is no reason to doubt that the following conclusions have qualitative validity. On minimisation, the O–$\hat{P}$–O angle fell from the standard value of 101.49° to 98°, and the C–O–P angle fell from the standard value of 118.78° to 114.78°. The studies on conformation II in which the standard value of the O–P–O angle was maintained while the C–O–P angle increased to 122.78° suggests that the fixed geometry used was reasonable.

Table 3 shows that (a) the partially optimised valence angles are varying up to about 4° above and below the standard values, (b) the energy of conformation I is expected to be reduced by about 2.2. kcal mol^{-1} to just over 2 kcal mol^{-1}, while that of conformation II is expected to be changed very little, when flexible geometry is considered.

TABLE 3
Flexible geometry calculations using Basis IA

O1-P̂1-O4	CÔP	$(E-E_1)$ in kcal mol^{-1} For $\Phi=\Psi=90$	$\Phi=90, \Psi=340$
94°	118.78°	10.133	–
98°	118.78°	–1.859	8.361
101.49°	118.78°	0	7.258
105°	118.78°	4.116	9.375
109°	118.78°	10.455	14.303
101.49°	110.78°	0.869	–
101.49°	114.78°	0.575	11.224
101.49°	118.78°	0.0	7.258
101.49°	122.78°	–	6.621
101.49°	126.78°	–	8.240

Strictly, the new barrier heights should be compared with the lowest conformational energy on the map, *after* the geometry of the latter has also been optimised. However, since this lowest energy conformation is by definition the least 'strained' as far as interactions between atoms and lone pair orbitals are concerned, this reference point is unlikely to vary significantly, as shown for the peptide case above. This is further suggested by the observation that even in the relatively strained conformation I and II, energy changes on optimisation arc fairly small. In general, it may be that optimisation of geometry is of less importance for DMP^- than it is for peptide systems. However, it will probably assume more importance for nucleic acids in the relaxation of clashes between groups. Thus, the observation in the present study that the energy may rise by roughly 7 kcal mol^{-1} as the valence angles are varied by approximately 8° provides a useful guideline to the magnitude of the changes in geometry which may be tolerated for nucleic acid backbones.

7.3. *Solvent effects*

It is well known that the solvent system which makes up the environment of peptides and other biological macromolecules plays a crucial role in their structure and function. Despite this, almost all calculations of conformationally dependent properties either neglect solvent effects altogether or treat them very approximately. However, Hagler et al [25] used the Monte Carlo method to stimulate the water structure and protein–solvent interactions in the triclinic lysozyme crystal. This technique has also been applied to the water in the crystal of the cyclic peptide cycle (–L–Ala–L–Pro–D Phe)$_2$, [26], which has been studied by X-ray crystallography [27]. There was a significant degree of agreement between the calculated and observed water molecule positions.

After establishing that the technique gave reasonable agreement with experimental data, we applied it to estimate the contribution of the solvent to the conformational behaviour of smaller peptide analogues. We carried out a Monte Carlo simulation of approximately 350 water molecules around *N*-acetyl alanyl

N'-methylamide [22]. We employed periodic boundary conditions in order to represent bulk solvent with minimal boundary effects.

Results suggest population enhancement of conformations with larger dipole moments in a way qualitatively resembling the results obtained using a continuum reaction field [28]. Large amounts of computer time (e.g., 20–40 hours/peptide conformation) are required because of the large number of water molecules explicitly included in the calculation and the need to sample a relatively large number of configurations (approximately 10^6) so that valid statistical averages can be obtained. One approximate method could involve the explicit inclusion of only a few hydration layers with the use of some continuum approximation to represent the bulk water.

7.4. Vibrational free energy contributions

The total free energy of a solute-solvent system is at present rather expensive to calculate, because of the time required to produce convergence of the entropic contribution. However, there are several possible simple procedures for estimating the vibrational entropy at a potential minimum, and some more general procedures which provide information about the general shape of the potential surface around a point (not necessarily a minimum) by calculating second derivatives. All of these are normally applied to in vacuo calculations, though in principle one could, of course, use the potential surface of the solute-solvent system.

As an example, we may cite the study of Hagler et al. [29] on alanine/methionine hexapeptides with host glycine residues inserted. The various minima in the potential surface were located and then their vibrational entropies calculated. Flexible geometry was taken into account, using the IR-derived force field discussed in Section 7.2. The important point here is that the free energy arising from the vibrational entropy was found to be of comparable magnitude to the potential energy contribution, and thus should always be included. The implications of this for the other calculations of properties described above have yet to be assessed, and this is the principal reason why quantitative results are only discussed above in terms of informative trends. Taken in conjunction with the solvent problem, it is clear that there is yet much to be done as regards accurate calculation of conformation-dependent properties in solution; though the theoretical apparatus seems to be available the computing time required is likely to be the major limiting factor.

8. PRACTICAL APPLICATIONS OF THE RESULTS TO MORE COMPLEX SYSTEMS

8.1. Modelling of nucleotide and nucleic acid structure

The results discussed in the preceeding sections suggest that for nucleotide systems 'Flexible geometry' might be neglected, to a *rough* first approximation. In contrast, however, the need for an OFF type force field is indicated, so that at least the

TABLE 4
Conformation angles and energies of local minima found for 5′-AMP

Method	ε[1]	conformation[2]	ω′	ω	Φ	Ψ	OH1	OH2	×	Energy
Expt.		gg	(−70)[3]	−69.04[4]	177.2	40.02	38.0	−82.87	−91.35	0.0[5]
OFF	1	tg	−93.6	−67.96	137.3	−51.88	−47.66	−18.65	−84.04	−18.84
	1	gg	No local minimum found							
	4	tg	−85.9	−73.59	160.0	−72.68	−45.9	−58.85	−102.7	−4.10
	4	gg	−75.5	−69.19	173.5	40.37	−57.02	−66.43	−106.7	−3.02
	*	tg	−89.8	−74.82	161.8	287.4	−43.25	−47.10	−108.8	−9.42
	*	gg	−77.6	−73.49	169.9	48.92	94.95	−275.9	−109.6	−5.39
Atom-	1	tg	−209.9	−73.99	138.8	283.7	−49.98	−20.96	−101.7	−11.16
Centred	1	gg	−122.9	−64.21	237.6	35.35	−65.53	−26.98	−95.16	−11.19
	4	tg	−70.5	−187.1	164.4	−76.68	306.4	−62.15	−104.6	−3.12
	4	gg	−84.5	−72.65	173.2	43.16	−55.47	−68.24	−110.9	−2.28

[1] ε is the dielectric constant
[2] as described by the notation of Sundralingham (1973)
[3] ω′ is not given by experimental X-ray data
[4] ω is a reasonable choice from three possible values
[5] Energy of X-ray structure is given the value of zero
* ε = 1, within PO_4^- group, but ε = 4 for other interactions

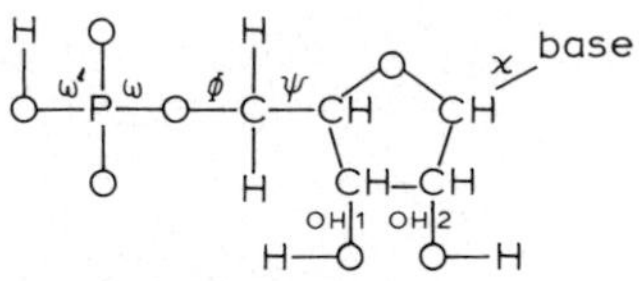

Fig. 12. Angle nomenclature describing bond rotation in 5′-AMP.

backbone should involve a representation of the non-core orbitals as obtained for dimethyl phosphate.

The biologically important nucleotide 5′-adenosine monophosphate (5′-AMP) contains seven rotatable bonds (see Fig. 12), but for discussion purposes its conformation can be classed as belonging to two major classes tg, gg in the notation of Sundralingham [30]. This approximately describes the conformation of one particular variant and important dihedral angle, the O–CH_2–CH–CH between the phosphate group and the sugar ring. The other dihedral angles affect only the orientations of peripheral groups, or, in the case of the sugar-base link, show only minor variation. All seven dihedral angles have been manipulated in a minimisation procedure combined with automatic and random selection of starting configuration in order to find all minima. The results are summarised in Table 4.

In the OFF calculations, the bond of the sugar link varies over a range of circa 25° for a variety of low energy conformers, even when the dielecric constant is varied. In contrast, both the tg and gg conformers are calculated to be possible, though the gg conformer is never significantly more stable than the tg conformer. In vacuo, only the tg conformer corresponds to a minimum. This is interesting because the gg conformer is the one found in the crystal [31]. In vacuo, the X-ray gg conformer is some 18.9 kcal mol^{-1} higher in energy than the minimum tg conformer.

Although compensating energies of this order can be provided by intermolecular interactions in the crystal of an ionic species, this rather high energy for an observed conformation still seems problematic. Further, when the non-core orbitals are removed (and a more classic type of calculation is carried out) the tg and gg conformers are of similar energy. This seems to provide evidence, albeit weak evidence, against the validity of the OFF calculation and thus the 5′-AMP crystal is another interesting test system.

This difficulty may however, only be an apparent one, since very different conformations would be expected for a polar or charged species in vacuo and in a high dielectric medium. Computer experiments were thus carried out with the following results.

The tg and gg conformers are comparable in energy in the OFF calculation, if a dielectric of 4 or more is introduced into the calculation. This is to be expected by comparison with the classical type of calculation, since a dielectric constant reduces the (purely electrostatic) effect of the non-core orbitals. The introduction of a dielectric constant, is however, only a purely artificial device, since a reaction field should be used in which the partial charges on atoms and non-core orbitals are never reduced from their effective in vacuo values. Nonetheless, if a dielectric constant is introduced only for interactions between the net charged phosphate group –PO_4^-– and the rest of the molecule, the tg and gg conformers still emerge as being dependent on the dielectric constant used. This is presumably the situation when the crystal nucleation proceeds in the solution phase, the smaller deficit of 4 kcal mol^{-1} being made up from crystal lattice forces.

For this reason, in calculations on nucleic acids, the charge on the phosphate groups is reduced without significantly affecting any partial charge in order to simulate counter-ion shielding. Calculations on DNA conformers are troublesome because of the number of conformations possible a priori and the time required to calculate each one. As a preliminary study the case of all-adenine all-thymine strands was considered. The following procedures were adhered to: (1) A periodicity constraint was applied so that each nucleotide pair has the same conformation. (2) In order to reduce computation time, the non-core orbitals of the bases were disregarded. The use of the atom-centred approach to the bases is suggested not to be a critical factor from calculations on the stacking of isolated uracil bases. (3) The starting conformation of each strand is that of B-form as given by Arnott ct al.

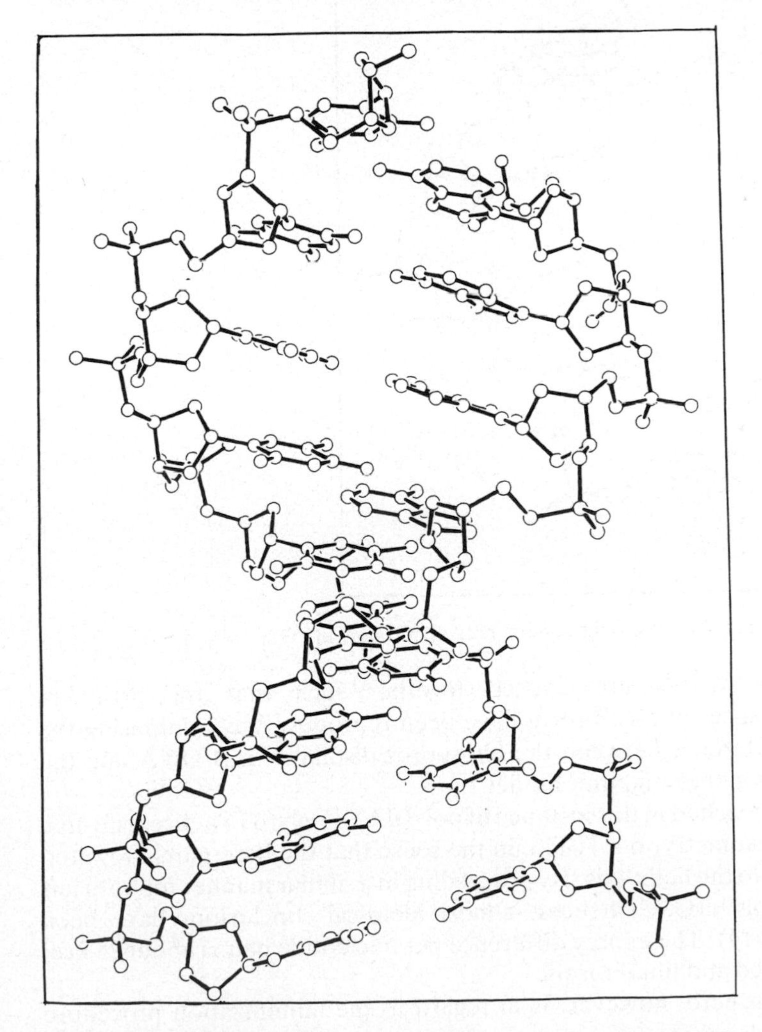

Fig. 13. Energy minimised structure of right-handed poly (dA).poly (dT).

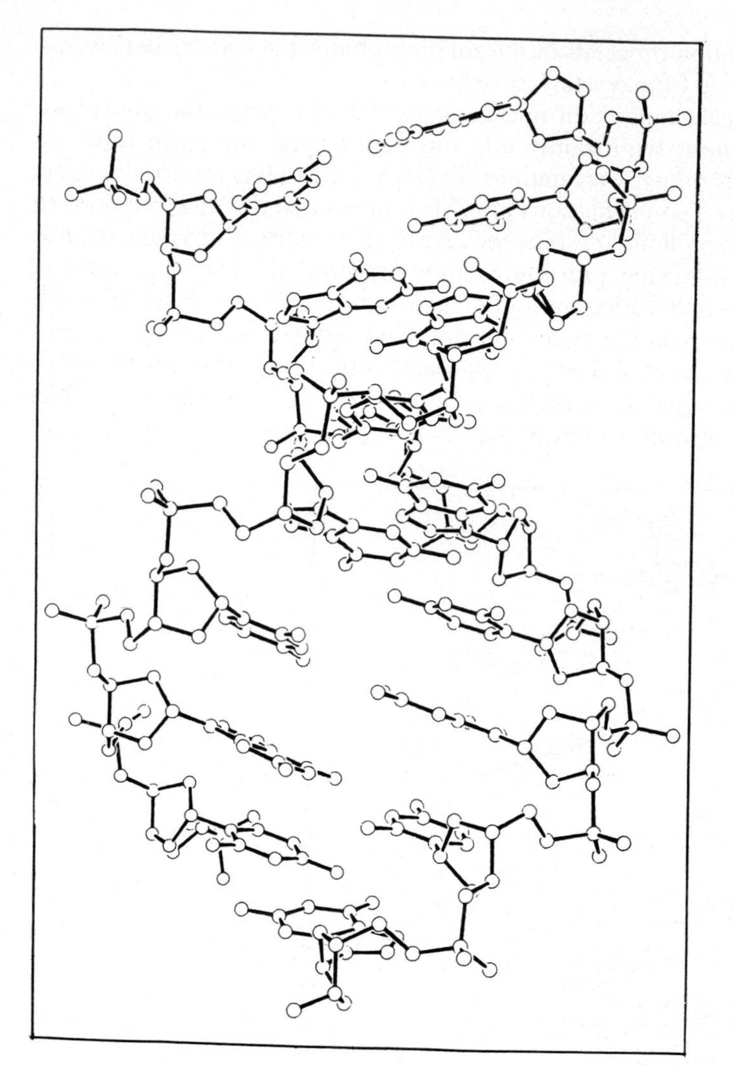

Fig. 14. Energy minimised structure of right-handed poly d(CG).poly d(CG).

[32]. (4) The two strands are docked together such that the array of interphosphorus distances of the B-form have been reproduced by minimising the root mean square difference between the phosporus distances in B-DNA and the phosphorus distances in the computer model.

Thc lowest energy reached in this instance of poly(dA)---poly(dT) helices is B-like in terms of internal geometry, but A-like in the sense that the base pairs are tilted *circa* 18° with respect to the helix axis. By proceeding in a similar manner for both left and right handed poly d(G–C) helices, almost identical conclusions have been reached (Figs. 13 and 14). The energy difference per nucleotide pair is about 25 kcal mol^{-1} between untilted and tilted forms.

An important point here, however, is in regard to the minimization procedure employed. A second derivative minimizer was used in conjunction with a SIMPLEX type. Whereas the first kind is highly efficient, it terminates on finding even trivial minima with barriers which may be negligible at biological temperatures. Indeed, using this procedure alone, the final 'minimized' structure would resemble very closely the starting configuration. Whereas the second type can escape or effectively ignore trivial minima, it is slower and can signal convergence at points which are not true minima, perhaps even broad saddle points. Used in conjunction, however, with convergence in one type of minimizer automatically involving the other minimizer, most difficulties are overcome. In the case of convergence without escape by both minimizers, a Monte Carlo sampling can then be carried out to verify that the deep minimum found is at least the deepest in the proximate area of conformational space, and this was also done here.

Calculations are commonly being carried out in a similar way for the topical problem of DNA helices of mixed handedness. So far all we can say is that such helices can indeed coexist in the same molecule, though with a significant boundary tension between regions of opposite handedness. This illustrates, however, that this kind of calculation, despite its limitations (neglect of solvent, etc.), can be highly informative in the treatment of important biological problems.

8.2. *Biologically active oligopeptides*

For calculations on peptide systems, previous discussion in this chapter has indicated that an OFF calculation need not be carried out, but that flexible geometry (especially variable valence angles) can be of paramount importance. Unfortunately, with *full flexible geometry* a comprehensive scan of a dipeptide energy surface as a function of even two rotation angles may take an hour, and this indicates that total energy minimization of a simple molecule can be extremely expensive. It turns out, however, that some important features of backbone flexibility can be reproduced simply by using the potential functions of Hagler et al. [14] but halving the van der Waals' repulsion contribution between vicinal atoms (as A and B in A–B–C–D), and this technique has been used in the following. An alternative technique, used for pancreatic trypsin inhibitor [33] is to pre-estimate 'relaxed' potential surfaces for the twenty types of residue, and possibly to include quantum mechanical and solvent effects here, and address these stored potential surfaces as required in the calculation. In neither case are accumulative effects of flexible geometry along the chain taken into account, in the sense that any one unit can be regarded as a flexible spring due to valence angle deformation, but at least a reasonable account is given of the effect of valence angle distortion on energy barriers.

For small charged and conformationally flexible oligopeptides, the solvent is also likely to play an important role. While specific inclusion of solvent molecules would be very time consuming, we know from the studies carried out so far (see above) that the reaction field makes a major contribution. For present purposes, we include this as

$$\Delta E_{RF} = -14.4 \, \frac{\mu^2}{R^3} \left(\frac{\varepsilon - 1}{2\varepsilon + 1} \right)$$

where μ^2 is the net dipole moment associated with a conformation, ε is the dielectric

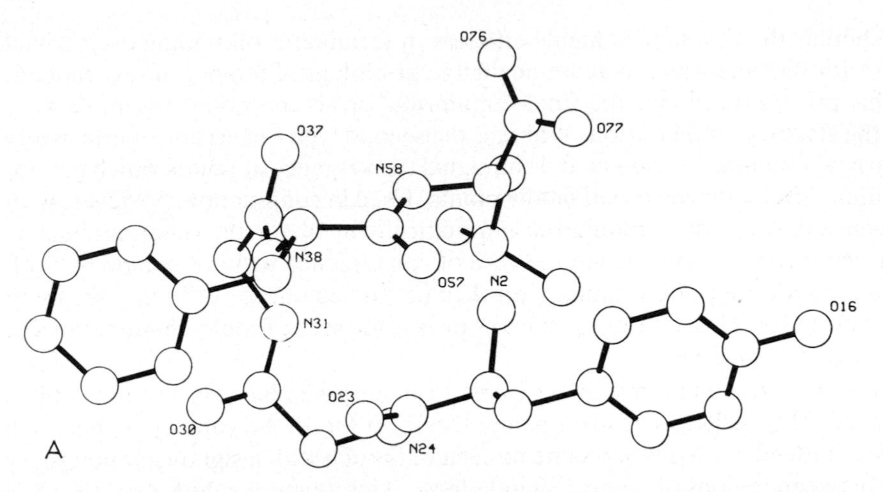

constant of the bulk solvent, and R is the radius of the spherical cavity in the solvent which the molecule occupies. This use of a spherical cavity, and particularly the choice of its radius R, is obviously problematic. After a number of investigations of systems of the type of interest here, we (G. Douglas and B. Robson, unpublished work) chose R to be the root mean squared distance between all n non-bonded atoms:

$$R = \left[\frac{\sum r_{ij}^2}{\frac{1}{2}n(n-1)} \right]^{1/2}$$

This generally lies between the Lagrange radius of gyration and maximum interatomic distance in the molecule, and is typically consistent with the spherical cavity estimated by eye, allowing for the 'raggy' surface of the molecule.

The effect of including the reaction field is complex but demonstrates its importance by revealing several possibly competing factors of comparable magnitude. These are: (1) The attractive interatomic forces which tend to compact the molecule. (2) The reaction field which tends to reduce the cavity radius R of the molecule. (3) The reaction field which tends to increase the net square dipole moment μ^2 which may, for certain configurations of charge or partial charge in the molecule, tend to *increase* the value of R and thus 'open up' the molecule.

The above third effect was expected for a simple oligopeptide charged only at the N- and C-termini, as exemplified by our current study as the conformations of the enkephalins and their derivatives (G. Douglas and B. Robson, work in progress). Minimization was carried again using combined second derivative and SIMPLEX minimization, and this applies to all studies considered below. From a variety of starting conformations, a small set of minima are again and again located on minimization. However, when the calculations are carried out with and without the reaction field, distorted β-bend type conformations are obtained as being of low energy. The two typical low energy structures, are shown in Fig. 15.

More complex effects can be 'programmed in to' the molecule by virtue of the way its charged groups are positioned. Bacterial cell wall mucopeptide is being extensively studied by us in this light (A. Metcalfe, G. Douglas, S. Thompson, and

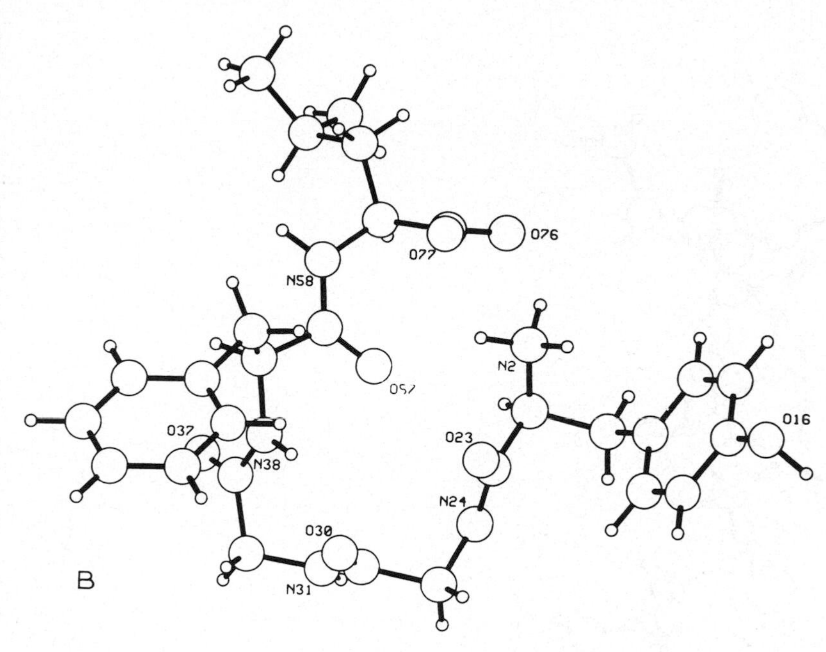

Fig. 15(a). Crystal structure of leu-enkephalin. (b) Energy minimised structure of leu-enkephalin, including the reaction field (see text).

B. Robson, work in progress), the system being of interest because of the marked expansion or contractions of the cell wall macromolecule (with mucopeptide units) according to solvent conditions incuding ionic strength. Briefly, charged groups are distributed along the chain in such a way that the reaction field would not be expected to extend the molecule. Indeed, there is, in the presence of the reaction field, a tendency to compact the structure by bringing two of the carboxyl sidechains together and thus create a large dipole moment with a positive (lysine) sidechain at the other end of the folded molecule.

A close approach of the carboxyl group is nonetheless prevented by the electrostatic repulsion between them. However, if sodium and chloride counterion are included, and their spatial positions included as variables, the combined effect of the ionic shielding and the reaction field allows a much closer approach and a more compact structure (see Fig. 16 and legend for further discussion). It is too early to say whether these *precise* effects are realistic, though one could obtain NMR and other experimental data on cell wall expansion and contraction under conditions of high and low ionic strength. What we *can* say, however, is that both the reaction field and the specific roles of ions can be very important, casting great doubt on the significance of any simple, in vacuo calculations.

As an example of a calculation of more direct pharmaceutical importance, we have studied (K. Woolley, G. Douglas, and B. Robson, work in progress) the preferred conformations of thyroid hormone releasing factor (TRF) of structure Pyroglu.His.Pro.NH_2, and two analogues Pyroglu.3-methyl-His.Pro.NH_2, Glu.His.Pro.NH_2. The latter two have respectively eight times the biological activity and effectively none of the biological activity of natural TRF. Briefly a

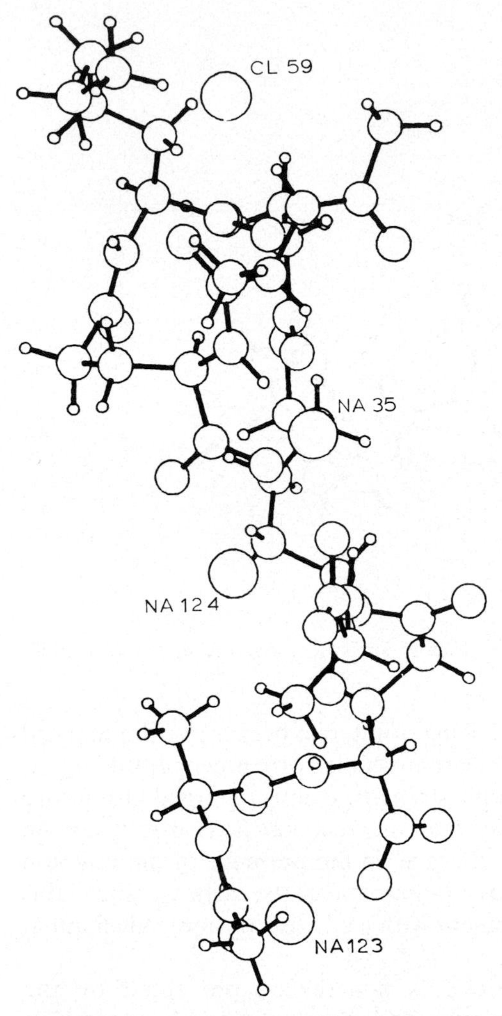

Fig. 16. Cross linking peptide of a bacterial cell wall peptidoglycan. The energy minimised conformation in the presence of freely moving counter ions and reaction fields (see text).

minimization from a variety of starting conformation, only three types converge as being stable, the 'propeller structure', the 'ring-stacked structure', and the 'cup structure' (Fig. 17). In natural TRF, the ring-stacked and cup structures are higher than the propeller by 2.8 and 6.8 kcal/mol respectively. In the 3-methylated histidine derivative, which is highly active, the ring-stacked and cup structures are 6.1 and 21.6 kcal mol^{-1} higher in energy respectively. In the inactive proline derivative the propeller structure is, however, *unstable*, by 2.3 kcal and 13.6 kcal mol^{-1} with respect to the ring-stacked and cup structures respectively. From this data so far, we may infer that the propeller structure is the active one: its conformational energy falls as the biological activity of the TRF rises with variation in chemical structure. In this case, the reaction field has a significant but less important effect, since the molecule has no net charges. Nonetheless, the effect of

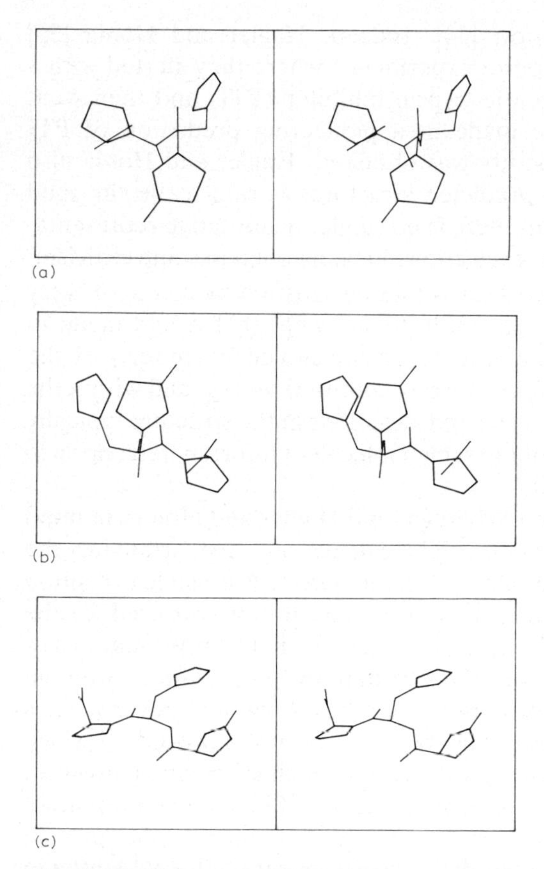

Fig. 17. Conformers of thyroid-stimulating hormone releasing factor (TRF); (a) propeller (P), (b) cup (C), and (c) ring stacked (Y).

the reaction field is in some sense 'programmed in' the TRF structure by its 'choices' of an N-terminal pyroglu and a C-terminal amidation.

Taken together, all these results underline the considerable importance of including solvent effects. What must be remembered is that certain solvent effects, e.g., the hydrophobic effect and water-peptide hydrogen bonding, have still been omitted. However it is our current opinion that the reaction field and the presence of the ions make a major solvent contribution for molecules of this type. Even the dipole moments of the peptide groups alone can be shown [22] to cause energy changes of about 5 kcal mol^{-1} between forms in which the peptide group dipole moments are roughly parallel or antiparallel in alignment.

8.3. *Globular proteins*

The challenging problem of predicting the three-dimensional structure of a globular protein from its amino acid sequence has long enticed workers in the field. For tackling this problem, the computer representation of the molecule is usually grossly simplified, and many assumptions made. Some authors have argued that this can make the results highly procedure dependent rather than sequence-dependent, and so lead to the result known as a priori [34]. Indeed, Hagler and Honig [35] demonstrated this possibility in a computer experiment, where they started with a *polyalanine* chain the length of pancreatic trypsin inhibitor (PTI) and then went through assumptions similar to those made in a pioneering prediction of PTI structure. A result of comparable quantity was obtained. Hagler and Honig also questioned the criteria by which the predicted structures is said to be 'in good agreement' with experiment. At present there is no single, quantitative representative method, but rather a set of criteria, each criterion being open to some criticism. What is clear, however, is that most workers would be satisfied by a case of 'very good agreement' without necessarily being able to say what that would mean in quantitative terms: probably, in such a case bench-top molecular models of the predicted and observed structures would be readily confused by eye, and within the limits of variation due to molecular libration and variations in the structure typically encountered in different crystal forms of the same molecule. Unfortunately, no such satisfactory result has yet been attained.

With these criticisms of Nemethy and Scheraga, and Hagler and Honig, in mind Robson and Osguthorpe [33] repeated the PTI calculation, first arranging the criticisms into a set of criteria which were met from the outset. A possible exception is that, as with the earlier studies, a 'united atom' representation was used for the sidechain, so that, for example, the tryptophan sidechain would be represented as a single 'giant atom' interacting with other atoms or dummy atoms with a potential function including a hydrophobic contribution. This has, however, never been a feature subject to major criticism, presumably because it is not a procedure which can obviously take account of, and pre-empt, the result which is known in advance. Rather, it is subject only to the more typical criticism that it is not a very accurate representation, and neglects some important effects. In contrast the protein backbone was represented in much greater detail. The pioneer study by Levitt and Warshal [36] had been strongly criticised because the simplification of the backbone used at that time did not allow for intramolecular hydrogen bonding, including that for formation of α-helices and β-pleated sheets which are fundamental and ubiquitous aspects of protein structure.

The cost of this increased detail is a larger member of conformational variables, a more detailed potential surface with more local minima, and hence greater cost in terms of computer time. Robson and Osguthorpe [33] avoided this problem to some extent by a new approach which does not depend on simplified representation of the protein. We have since found this technique to be very useful and it deserves some discussion. It is in essence a constraint method, which however, avoids introducing any discontinuities, reduces the number of variables used, and can exploit the natural interdependence between the values of the bond rotation angles in a protein.

In this treatment one considers certain selected bonds with rotation angles θ_1, θ_2, ... as parameters of a 'coupling function' F with value δ,

$$\delta = F(\theta_1, \theta_2, \ldots)$$

The function F describes some curve in the conformational space θ_1, θ_2, ... which can be considered as the most likely trajectory, or 'reaction path' for conformational transitions involving θ_1, θ_1, ... This function, however complex, is usually made periodic so that it has no boundary ends. The value δ is thus some measure of the distance round this path, though not necessarily a Euclidean distance in the (θ_1, θ_2, ...) space. However, δ must uniquely define a point in (θ_1, θ_2, ...) space.

In actuality, it is only necessary to write the *inverse* function F^{-1}. The minimizer changes directly only the values of δ_1, δ_2, ... (each for functions F_1, F_2, ... , each in turn coupling certain angles θ; so that all angles are *implicitly* represented). When the minimizer returns the *set* of values $\{\delta\}$, these are converted to the larger set of real angles $\{\theta\}$, and the structure is built up and its energy evaluated. As far as the minimizer is concerned, it is minimizing this energy only as a function of $\{\delta\}$. If MIN implies the generating of a new $\{\delta\}_{j+1}$ on the basis of inspecting certain previous $\{\delta\}_j$, $\{\delta\}_{j-1}$, ... with energies E_j, E_{j-1}, ... (the sample points), then we have the reiterative procedure

$$\{\delta\}_{j+1} \leftarrow \mathrm{MIN}[\{\delta\}_j, \{\delta\}_{j-1}, \dots E_j, E_{j-1}, \dots]$$
$$\{\theta\}_{j+1} \leftarrow F^{-1}\{\delta\}_{j+1}$$
$$E_{j+1} \leftarrow E\{\theta\}_{j+1}$$

where $E\{\theta\}_{j+1}$ implies the calculation of the energy of the conformation $\{\theta\}_{j+1}$ in the usual way. The minimizer used is of the combined second derivative-SIMPLEX type as described above in studies on DNA.

The function F will normally be chosen to define a trajectory: (i) which avoids high energy regions in the conformational space of the angles it couples (this is the major time saving feature), (ii) which connects all minima and passes close to all conformations of importance in proteins, and (iii) which takes the least energy pathway between these points. To select the function thus requires prior knowledge or calculation of the conformational space represented by (θ_1, θ_2, ...). Providing not too many angles θ are coupled per function, this is not a serious problem. An obvious choice is the angles ϕ (C'–N$\hat{-}$$C_\alpha$–$C'$) and ψ (N–$C_\alpha$$\hat{-}$$C'$–N) of each residue, the potential surface of which have been well studied for all 20 amino acid residues found in proteins. For Robson and Osguthorpe [33] the choice of pathway then a 45° tilted ellipse with its centre at $\phi = \psi = 0°$ and one major radius representing a vector pointing from $\phi = \psi = 0°$ to the β-pleated sheet region (circa $\phi = -160°$, $\psi = +160°$). Then the distance δ round the ellipse can be represented sufficiently by the angle γ between the vector pointing from $\phi = \psi = 0°$ to $\phi = --160°$, $\phi = +160°$, and the vector pointing from $\phi = \psi = 0°$ to any ϕ, ψ angles. In this way, any values of γ uniquely specifies values of ϕ, ψ which must however, lie on the ellipse, and which by choice of ellipse include the right and left hand α-helices, β-bend components, and major saddle points.

How well does this fare? We should ask this in three stages. First, since it is a constraint method, how close to the native observed structure can it get, irrespective of the calculation of energy? This is found by minimizing the overall root mean squared difference between distances between the non adjacent C_α atoms of the backbone in the computer model and the corresponding distances in the observed structure as a function of the conformation of the computer model, giving an answer of 1.1 Å for a variety of proteins. Second, how well does this describe the native structure when energy calculations is included? Here we begin from the computer model fitted to the native structure as described above, and minimize the energy as a function of conformation. The root mean square derivation as described above then drops to 2.5–3.0 Å (PTI, myoglobin, trypsin), which is not good, but well within the range generally obtained by folding proteins from an extended form, and of the magnitude found when comparing homologous proteins. This would suggest that the way we have calculated the energy is unsatisfactory, if promising. More precisely, it suggests that the potential functions may need refinement, though the crude way the solvent is included, treating the hydrophobic effect as a potential function, and the neglect of other entropic contributions, will have a role. Actually, the latter effects account for various difficulties. Computer experiments have been carried out in which the potential functions are refined at the observed structure of end protein in turn. Then the discrepancy falls to about 1.5 Å. This indicates that we can refine to almost any extent we like for any known structure. However, when the potential functions are transferred from a protein for which they were refined to one which they were not, the deviation falls to circa 3.0 Å again. Presumably, all we did in refinement was adsorb some uncounted factors.

Finally, how well does the method do for a real prediction, starting from an open unfolded form? The answer depends on how much we have unfolded from the native. For a small protein like PTI, we can now obtain a fit of 4.8 Å, not significantly better, but not significantly worse, than earlier studies. For myoglobin, the procedure is far too time consuming unless the helices are fixed in their observed helical conformation, and even here the result is a very poor fit of circa 9 Å. Nonetheless, it is interesting to note that the two halves of the molecule fold very well, to 3 Å fit with the observed: it is the coming together of the two lobes so formed which cannot be reproduced as yet without manual intervention. If we start from the predicted myoglobin structure of Sternberg and Cohen, derived by a helix-packing algorithm, we obtain about 3.5 Å deviation. In contrast, if trypsin is folded, starting with a model fitted to its homologue elastase, there is only a 2.4 Å deviation between observed and predicted trypsin structures.

This indicates that such procedures are useful if we have some information. Indeed, the above computer experiment with trypsin and elastase has been repeated with thrombin B and elastase, giving a useful preliminary proposal for the as yet unknown thrombin B chain conformation [37]. Regarding the equivalent treatment of trypsin as a control experiment, and noting that the expected relative positions of active site residues, highly conserved in this family of proteins, are reproduced without any prior constraint on those positions, there is some hope that this thrombin B structure is reasonable.

The above results also indicate, nonetheless, that much more powerful minimization procedures and means of seeking the global minimum are required.

To this end, the variable coupling approach has been extended so that any functional form F, not just an ellipse, can be approximated. Briefly, this is very efficient for small polypeptide hormones, and reproduces a real variable minimization very cheaply, but will not, of course, prevent entrapment in deep local minima. The combined second derivative-SIMPLEX minimizer can easily be supplemented by a Monte Carlo sampling procedure to seek deeper minima, but for large molecules it is simply not sufficiently informed to hit upon the global minima in reasonable time. To escape from this dilemma we have two (not necessarily mutually exclusive) possibilities. Either, we use faster computer machines and more rapid procedures, or we find ways of better 'informing' the search procedure. If the latter, then there is the question of whether to introduce information specific to the protein under investigation, which pre-empts the known result (but could have special, practical applications) or whether to devise procedures which still allow prediction from amino acid sequence alone, because they exploit *general* empirical observations on architectural principles in the reported protein structure. This aspect is taken up by Dr. Sternberg in the present volume, and a simple example is discussed in the appendix to this chapter.

REFERENCES

1. Pauling, L., Corey, R.B. and Bronson, H.R. (1951) Proc. Natl. Acad. Sci. USA 37, 205–211.
2. Scott, R.A. and Scheraga, H.A. (1965) J. Chem. Phys. 42, 2209–2215.
3. Brant, D.A., Miller, W.G. and Flory, P.J. (1967) J. Mol. Biol. 23, 47–65.
4. Maigret, B., Pullman, B. and Dreyfus, M. (1970) J. Theor. Biol. 26, 321–333.
5. Shipman, L.L. and Christofferen, R.E. (1973) J. Am. Chem. Soc. 95, 1408–1416.
6. Pullman, A. and Berthod, H. (1973) C.R. Acad. Sci. Ser. D. 277, 2077–2079.
7. Robson, B., Hillier, I.H. and Guest, M.F. (1978) J. Chem. Soc. (Far. Trans II) 74, 1311–1318.
8. Dunning, T.H. (1970) J. Chem. Phys. 53, 2823–2833.
9. Roos, B. and Siegbahn, P. (1970) Theor. Chim. Acta. 17, 209–215.
10. Jeffrey, G.A., Pople, J.A. and Radom, L. (1972) Carbohydr. Res. 25, 117–131.
11. Shipman, L.L., Burgess, A.W. and Scheraga, H.A. (1975) Proc. Natl. Aca. Sci. USA 72, 543–547.
12. Allinger, N.L. and Chung, D.Y. (1976) J. Am. Chem. Soc. 98, 6798–6803.
13. Boys, S.F. and Foster, J.M. (1960) Rev. Mod. Phys. 32, 300–302.
14. Hagler, A.T., Huler, E. and Lifson, S. (1974) J. Am. Chem. Soc. 96, 5319–5327.
15. Robson, B., Stern, P.S., Hillier, I.H., Osguthorpe, D.J. and Hagler, A. T. (1979) J. Chim. Phys. 76, 831–834.
16. Hagler, A.T., Lifson, S. and Dauber, P. (1979) J. Am. Chem. Soc. 101, 5111–5141.
17. Mulliken, R.S. (1955) J. Chem. Phys. 23, 1833–1840.
18. Ramachandran, G.N. and Sasisekharan, A. (1968) Adv. Protein Chem. 23, 283–438.
19. Kollman, P.A. (1977) In: (H. F. Schaefer, (ed.), Modern Theoretical Chemistry. Plenum Press, London and New York, Vol. 4, pp.
20. Matsuoka, O., Tosi, C. and Clementi, E. (1978) Biopolymers, 17, 33–49.
21. Perahia, D., Pullman, A. and Berthod, H. (1975) Theor. Chim. Acta. 40, 47–60.
22. Hagler, A.T., Osguthorpe, D.J. and Robson, B. (1980) Science 208, 599–601.
23. Guest, M.F., Hillier, I.H. and Saunders, V.R. (1972) Trans. Faraday Soc. 68, 114–120.
24. Jeffrey, G.A. and Taylor, R. (1980) J. Comp. Chem. 1, 99–109.
25. Hagler, A.T. and Moult, J. (1978) Nature 272, 222–226.
26. Hagler, A.T., Moult, J. and Osguthorpe, D. J. (1980) Biopolymers 19, 395–418.
27. Brown, J.N. and Teller, D.G. (1976) J. Am. Chem. Soc. 98, 7565–7569.
28. Renugopalakrichnar, V., Nu, S. and Rein, R. (1976) In: B. Pullman (Ed.) Environmental Effects of Molecular Structure and Properties. Reidel, Dordrecht, pp. 109–133.
29. Hagler, A.T., Stern, P.S., Sharon, R., Becker, J.M. and Nadier, F. (1979) J. Am. Chem. Soc. 101, 6842–6852.
30. Sundaralingham, M., (1969) Biopolymers, 7, 821–860.
31. Kraut, J. and Jensen, L.H. (1963) Acta. Cryst. 16, 79–88.
32. Arnott, S. and Hukins, D.W.L. (1972) Biochem. Biophys. Res. Commun. 47, 1504–1509.
33. Robson, B. and Osguthorpe, D.J. (1979) J. Mol. Biol. 132, 19–51.
34. Nemethy, G. and Scheraga, H.A. (1977) Quart. Rev. Biophys. 10, 239–352.
35. Hagler, A.T. and Honig, B. (1978) Proc. Natl. Acad. Sci. USA 75, 554–558.
36. Levitt, M. and Warshel, A. (1975) Nature 253, 694–698.
37. Robson, B. (1980) Trends. Biochem. Sci. 2, 240–245.
38. Robson, B. (1974) Biochem. J. 141, 853–867.
39. Robson, B. and Pain, R.H. (1971) J. Mol. Biol. 58, 237–259.
40. Garnier, J., Osguthorpe, D.J. and Robson, B. (1978) J. Mol. Biol. 90, 97–120.

Appendix

A simple method and BASIC program for secondary structure prediction on a microcomputer

B. ROBSON, G. M. DOUGLAS and J. GARNIER

1. INTRODUCTION

The information theory approach of Robson and Pain [38] and Robson [39] treats the amino acid sequence and sequence of corresponding residue conformations as two messages, related by an unknown code which must be broken. The basic idea is, as for other methods, to obtain information about the sequence-conformation relation from the small set of proteins of known sequence and secondary structure, and to apply this information to a protein of known amino acid sequence but supposedly unknown conformation to predict its secondary structure. This information is in the form of parameters which are, in fact, formally information measures. The formal justification for the procedure, which must take into account the finitude of the data and thus its limited significance in the statistical sense, is quite complex, which may have given the impression that the application of the parameters, once obtained, is also complex. In fact, it is one of the simplest and least ambiguous approaches.

In ten years of application we have seen no reason to change the methodology, though to be fair it should be said that the original studies were more in the way of describing a general formal procedure for secondary structure predictions, and subsequent papers were largely concerned with the effects of neglecting or including terms in the expansion of the basic information equation. The optical choice of terms, in our present view, is that described in the paper by Garnier et al. [40], and uses the so called 'directional information' parameters.

In this approach, the information in the sequence for the conformation S_j (say, α-helical) of the (j) residue R_j (say, alanine) is expanded as

$$I(S_j; \{R_{j-M}, \ldots R_{j+M}\}) = \sum_{m=-M}^{m=+M} I(S_j; R_{j+m})$$

where the $I(S_j; R_{j+m})$ are the parameters with values supplied. M is taken as 8, since for further separations along the sequence, the information tends to zero (i.e., the forces responsible for secondary structure are 'medium range' along the sequence, and other effects appear as more-or-less random 'noise').

This summation is repeated for each residue j in the sequence, and subsequently for each conformation S, which we wish to consider. In practice, we consider four, conformational studies $S_j = H$ (α-helical), $S_j = E$ (roughly extended, as in a β-pleated sheet), $S_j = T$ (one of two residues which together invert the direction of the chain by 180°: a 'β-bend' or 'reverse turn') and $S_j = C$ ('coil', which is effectively anything else not defined by the above). Thus there must be provided 1360 parameters $I(S_j; R_{j+m})$, for four states S_j, 20 types of amino acid residue R_j, and 17 values of $m(-8, -7, \dots 0, \dots +7, +8)$.

Having carried out the evaluation of $I(S_j; \{R_{j-M}, \dots R_{j+M}\})$ for all N residues in the sequence, and for all four possible states S_j per residue, each residue is predicted as H if $I(S_j = H; R_1, \dots R_j, \dots R_N)$ is the largest quantity, E if $I(S_j = E; R_{j-M}, \dots R_{j+M})$ is the largest quantity, and so on.

An advantage of this approach is that no additional rules are required for adjusting the result to make it appear physically meaningful. An isolated H state would in other methods be reassigned as 'coil' for example, since a very short α-helix is typically considered either as unstable or, from the point of view of definition, not an α-helix. The need to make any such reassignment is largely circumnavigated by the nature of the parameters, which automatically took into account the stability and definition aspect at the time of being generated from proteins of known sequence and conformation. Though problem predictions in the above sense can occur, this occurrence is a fairly rare event.

A disadvantage is that a large number of additions have to be made, so that a pencil and paper approach is only practical for very short sequences if predictions are to be performed routinely. On the other hand, the calculations barely justify the use of a large computer. It is thus ideal material for a mini computer, and a program is shown below in Commodore BASIC. The parameters are included as DATA statements. Note that the algorithm itself comprises only 23 lines (BASIC program lines 1740–1920); the rest of the program consists largely of input and sequence editing facilities found profitable for use in our laboratory and that of our French colleagues. These are, however, so advantageous for everyday use, especially when studying homologous proteins, that we have included the full listing.

2. PURPOSE OF PROGRAM

The purpose of the program is to predict each residue in a given sequence as α-helical (H), extended (E), turn (T) or coil (C), as described above. The results are tabulated with other data useful in studying protein conformation or in rational design of peptides. These further data are: (1) the strengths of the tendency to each of the four states, measured in centinats (the 'nat' is a standard unit of information and is discussed by Robson [38]); (2) the charge on the sidechain in electron units; (3) the hydrophobicity of the sidechain in kcal mol^{-1} for displacement of solvation shell water (see [33]); and (4) the *characteristic square length* of each residue, i.e., the value of $\langle r^2 \rangle_0 / n$ for a very long homopolymer of that residue when $\langle r^2 \rangle_0$ is the end-to-end distance of the unperturbed random coil chain and n is the number of residues. The magnitude of the characteristic length gives some useful indication of the role of the residue as a *flexible* hinge point in a sequence, this character *increasing* as the characteristic length *decreases*.

3. USE OF PROGRAM

The program requires commands and sequence data, which can either be entered interactively, by the keyboard, or automatically from BASIC DATA statements to be added by the user to the end of the program.

There are three levels of commands. The first level is active when the program is loaded and run, and is always interactive. Commands (without leading or embedded blanks) are:

Level 1 commands

.E for interactive mode with english prompts.
.F for interactive mode with french prompts.
.AUTO for automatic mode in which the user-supplied DATA statements are read.

Any other entry gives an interactive mode with minimal prompting and reports.

Level 2 commands

These are active once the above has been issued. Any input string not beginning with a point is considered as a further portion of the sequence, which must be in standard IUPAC one-letter code (G=Glycine, A-Alanine, V=Valine, L=Leucine, I-Isolenine, S=Serine, T=Threanine, D=Aspartate, E=Glutamate, N=Asparagine, Q=Glutamine, K=Lysine, H=Histidine, R=Arginine, F=Phenylalanine, Y=Tyrosine, W=Tryptophan, C=Cysteine, M=Methionine, P=Proline, X=Unknown).

The sequence may be introduced over several lines, any number of characters per line (no leading or embedded blanks) and possibly with commands interspaced between these lines. French equivalents are also available.

.ERASE	Begin a new sequence and reinitialize assays for a fresh prediction (not required for first prediction or after use of command '.' – see below).
.LIST	List sequence so far on the screen.
.PRINT	Send subsequent results to line printer (device 4) as well as the screen (already active in automatic mode by default).
.NONPRINT	Do not send subsequent results to the printer.
.N+	Add residues following to the N-terminus of the chain. e.g., .N+AVL.
.C+	Add residues following to the C-terminus of the chain. e.g., .C+APNE.
.N−	Delete specified number of residues from N-terminus. e.g., .N−6.
.C−	Delete specified numbers of residues from C-terminus. e.g., .C−1
.M=	Refer to, for example, the 10th residue viz: .M=10
.M+	Insert the following characters immediately *before* the residue specified by '.M=', e.g., .M+AEVG
.M−	Delete specified number of characters *starting* with the residue specified by '.M=', e.g., .M−3 For example,

.M=6
.M−3
.M+AEVG

Deletes three residues numbers 6,7, and 8, and replaces them with the four residues AEVG. The '.M+' or '.M−' are equivalent to '.N+' and '.N−' respectively, when preceded by '.M=1'.

.GO	Do prediction, but do not delete sequence or reset information in sequence to zero.
.	Equivalent to .GO followed by .ERASE. This is the most common command for normal use.
.PLUS	Erase sequence, but retain information for addition in the next production (used for adding together information for homologues and thus obtaining an 'averaged' prediction).
.+	Equivalent to .GO followed by .PLUS.
.AUTO	Read following commands and sequences from user DATA statements (i.e., involves automatic mode).
.INPUT	When encountered in user DATA statements, invokes the interactive mode.
.REM	Ignore following characters on line beginning .REM. Following '.PRINT', sends them to the line printer (device 4) as title, comment, etc.
N.B.	Any line entered which terminates with ! is deleted on entry (an input-error erase-line facility). However, .! erases the residue sequence, while printing it out for screen editing and re-entry if this facility is available on the microcomputer used.

Level 3 commands

These become active when an instruction to carry out a prediciton is encountered, *and* when an invalid symbol is found in the residue sequence string. The user is invited to correct the erroneous character. This is active even in automatic mode. The user may also enter '.INPUT' which forces the interactive mode at level 2, allowing editing, or '.!' which clears the residue sequence, first displaying it for screen editing, and forces the interactive mode at level 2.

4. PROGRAM LISTING

```
1 PRINT"]
2 PRINT"  ***********************************
3 PRINT"  * ┌───╲ ┌───╲ ┌──── ┌───╲ ┬  ╱───╲ *
4 PRINT"  * │   │ │   │ │     │   │ │  │     *
5 PRINT"  * └───╱ └───╱ └───  │   │ │  │     *
6 PRINT"  * │     │╲    │     │   │ │  │     *
7 PRINT"  * │     │ ╲   │     │   │ │  │     *
8 PRINT"  * │     │  ╲  └──── └───╱ ┴  ╲───╱ *
9 PRINT"  ***********************************
10 DATE$="28/05/81"
11 PRINT"■            DATE: ";DATE$
12 PRINT"
13 PRINT TAB(10);"TIME: ";LEFT$(TI$,2);":";MID$(TI$,3,2);":";RIGHT$(TI$,2)
14 PRINT"
15 OPEN 4,4
100 DIM IUPAC$(21):DIM HYDROF(21):DIM CHARGE(21)
120 PRINT"
150 DIM SEQ$(200):DIM P%(17,21,4)
160 DIM SEQ%(200):DIM SC(200):DIM SH(200):DIM CS(200)
170 DIM H%(200):DIM E%(200):DIM T%(200):DIM C%(200)
175 DIM C$(200):DIM CC(21):DIM PH(200):DIM PE(200):DIM PT(200):DIM PC(200)
180 PRINT "PROGRAM 'PREDICT' ------------ROBSON/PAIN/SUSUKI/GARNIER  1970-81■"
200 DATA "G","A","V","L","I","S","T","D","E","N"
210 DATA "Q","K","H","R","F","Y","W","C","M","P","X"
215 PRINT "        *****COMMAND LEVEL 1*****
220 PRINT "■WRITE/ECRIVEZ '.F' POUR LES COMMENTAIRES EN FRANCAIS, '.E'FOR ";
221 PRINT "ENGLISH COMMENTS, '.AUTO' FOR AUTOMATIC MODE (EN MODE AUTOMATIQUE)
230 INPUT S$
235 P$=".NONPRINT":IF S$=".AUTO" THEN P$=".PRINT"
260 IF S$=".E" THEN PRINT "■LEGAL CODE LETTERS ARE:-■"
270 IF S$=".F" THEN PRINT "■LES LETTRES PERMISES SONT:-■"
280 FOR I=1 TO 21
290 READ IUPAC$(I)
300 PRINT IUPAC$(I);
310 NEXT I
320 PRINT
330 PRINT
340 REM THE PARAMETERS (IN CENTINATS) FOR STATES H,E,T,C
350 PRINT
356 OPEN 1,4,1:OPEN 2,4,2
357 PRINT#2,"999 AAAAA S999  S999  S999  S999  SZ.9  SZ.9  99999.9"
360 DATA -5,-10,-15,-20,-30,-40,-50,-60,-86:REM G,H
370 DATA -60,-50,-40,-30,-20,-15,-10,-5:REM G,H
380 DATA 5,10,15,20,30,40,50,60,65:REM A,H
390 DATA 60,50,40,30,20,15,10,5:REM A,H
400 DATA 0,0,0,0,0,0,5,10,14,10,5,0,0,0,0,0,0:REM V,H
410 DATA 0,5,10,15,20,25,28,30,32,30,28,25,20,15,10,5,0:REM L,H
420 DATA 5,10,15,20,25,20,15,10,6,0,-10,-15,-20,-25,-20,-10,-5:REM I,H
430 DATA 0,-5,-10,-15,-20,-25,-30,-35,-39:REM S,H
440 DATA -35,-30,-25,-20,-15,-10,-5,0:REM S,H
450 DATA 0,0,0,-5,-10,-15,-20,-25,-26,-25,-20,-15,-10,-5,0,0,0:REM T,H
460 DATA 0,-5,-10,-15,-20,-15,-10,0,5,10,15,20,20,20,15,10,5:REM D,H
470 DATA 0,0,0,0,10,20,60,70,78,78,78,78,78,70,60,40,20:REM E,H
480 DATA 0,0,0,0,-10,-20,-30,-40,-51,-40,-30,-20,-10,0,0,0,0:REM N,H
490 DATA 0,0,0,0,5,10,20,20,10,-10,-20,-20,-10,-5,0,0,0:REM Q,H
500 DATA 20,40,50,55,60,60,50,30,23,10,5,0,0,0,0,0,0:REM K,H
510 DATA 10,20,30,40,50,50,50,30,12,-20,-10,0,0,0,0,0,0:REM H,H
520 DATA 0,0,A,0,0,0,0,0,-9,-15,-20,-30,-40,-50,-50,-30,-10:REM R,H
530 DATA 0,0,0,0,0,5,10,15,16,15,10,5,0,0,0,0,0:REM F,H
540 DATA -5,-10,-15,-20,-25,-30,-35,-40,-45:REM Y,H
550 DATA -40,-35,-30,-25,-20,-15,-10,-5:REM Y,H
560 DATA -10,-20,-40,-50,-50,-10,0,10,12:REM W,H
570 DATA 10,0,-10,-50,-50,-40,-20,-10:REM W,H
580 DATA 0,0,0,0,0,0,-5,-10,-13,-10,-5,0,0,0,0,0,0:REM C,H
590 DATA 10,20,25,30,35,40,45,50,53,50,45,40,35,30,25,20,10:REM M,H
600 DATA -10,-20,-40,-60,-80,-100,-120,-140,-77:REM P,H
610 DATA -60,-30,-20,-10,0,0,0,0:REM P,H
620 DATA 0,0,0,0,0,0,0,0,0,0,0,0,0,0,0,0,0:REM X,H
630 DATA 10,20,30,40,40,20,0,-20,-42,-20,0,20,40,40,30,20,10:REM G,E
640 DATA 0,0,0,0,-5,-10,-15,-20,-23,-20,-15,-10,-5,0,0,0,0:REM A,E
650 DATA 0,0,-10,-20,0,20,40,60,68,60,40,20,0,-20,-10,0,0:REM V,E
```

```
660 DATA 0,0,0,0,0,5,10,20,23,20,10,5,0,0,0,0,0:REM L,E
670 DATA 0,-10,-20,-10,0,20,40,60,67,60,40,20,0,-10,-20,-10,0:REM I,E
680 DATA 0,10,20,10,0,-5,-10,-15,-17,-15,-10,-5,0,10,20,10,0:REM S,E
690 DATA 5,10,15,20,15,15,10,10,13,10,10,15,15,20,15,10,5:REM T,E
700 DATA 0,5,10,15,20,0,-20,-30,-44,-30,-20,0,0,0,0,0,0:REM D,E
710 DATA -10,-15,-20,-25,-30,-35,-40,-45,-50:REM E,E
720 DATA -55,-60,-60,-50,-40,-30,-20,-10:REM E,E
730 DATA 10,30,50,30,20,0,-15,-30,-41,-30,-15,0,20,30,50,30,10:REM N,E
740 DATA 0,0,0,0,0,-5,-10,0,12,20,30,40,50,50,40,30,15:REM Q,E
750 DATA -5,-10,-15,-20,-30,-40,-50,-40,-33,-20,-10,0,10,10,0,0,0:REM K,E
760 DATA -10,-20,-40,-20,-10,0,-10,-20,-25:REM H,E
770 DATA -35,-30,-25,-20,-15,-10,-5,0:REM H,E
780 DATA 0,0,0,0,0,0,0,4,0,0,0,0,0,0,0:REM R,E
790 DATA 0,0,0,0,0,5,10,20,26,10,-10,-30,-60,-65,-60,-40,-20:REM F,E
800 DATA 0,5,10,15,20,25,30,35,40,35,30,25,20,15,10,5,0:REM Y,E
810 DATA 0,0,0,0,0,-10,-10,-10,-10,-10,-10,-15,-20,-25,-30,-20,-10:REM W,E
820 DATA 0,0,0,0,0,10,20,30,44,30,20,10,0,0,0,0,0:REM C,E
830 DATA -10,-20,-30,-40,-40,-30,0,10,23,10,0,-30,-40,-40,-30,-20,-10:REM M,E
840 DATA 10,20,30,30,20,10,0,-10,-18,-20,-10,10,30,40,30,20,10:REM P,E
850 DATA 0,0,0,0,0,0,0,0,0,0,0,0,0,0,0:REM X,E
860 DATA 0,0,0,0,10,30,55,55,57,40,0,0,0,0,0,0,0:REM G,T
870 DATA 0,0,0,-10,-20,-30,-40,-50,-50,-40,-30,-20,-10,0,0,0,0:REM A,T
880 DATA 0,0,0,0,-10,-20,-30,-40,-60,-40,-30,-20,-10,0,0,0,0:REM V,T
890 DATA 0,0,0,-10,-20,-30,-40,-50,-56,-20,-10,0,0,0,0,0,0:REM L,T
900 DATA 0,0,0,0,0,-10,-20,-30,-46,-40,-10,0,0,20,30,20,10:REM I,T
910 DATA 0,-10,-20,-20,10,15,20,25,26,25,20,15,10,0,0,0,0:REM S,T
920 DATA 0,10,20,20,20,15,10,5,3,5,10,15,20,20,20,10,0:REM T,T
930 DATA 0,0,0,0,0,0,5,10,31,10,5,0,0,0,0,0,0:REM D,T
940 DATA 0,-5,-10,-15,-20,-30,-40,-45,-47,-20,0,10,5,0,0,0,0:REM E,T
950 DATA 0,0,0,10,20,30,35,40,42,40,35,30,20,10,5,0,0:REM N,T
960 DATA 10,20,30,25,20,15,10,5,4,20,30,40,50,60,50,40,20:REM Q,T
970 DATA -10,-20,-30,-40,-25,-10,0,10,10,10,0,-20,-30,-20,-10,-5,0:REM K,T
980 DATA 0,0,0,0,0,0,0,0,-3,0,10,20,30,20,10,0,0:REM H,T
990 DATA 0,0,0,0,0,0,0,10,21,30,40,30,20,10,0,0,0:REM R,T
1000 DATA 0,0,0,0,0,-5,-10,-15,-18,-15,0,15,30,25,20,10,0:REM F,T
1010 DATA 0,0,0,5,15,15,20,25,29,25,20,15,15,5,0,0,0:REM Y,T
1020 DATA 0,0,0,10,20,30,40,80,36,-30,30,40,50,60,70,40,20:REM W,T
1030 DATA 20,40,50,60,60,55,50,45,44,40,35,30,25,20,15,10,5:REM C,T
1040 DATA -5,-15,-20,-25,-30,-35,-40,-45,-48:REM M,T
1050 DATA -45,-40,-35,-30,-25,-20,-15,-5:REM M,T
1060 DATA 10,20,30,40,50,70,10,-90,36,90,10,0,0,0,0,0,0:REM P,T
1070 DATA 0,0,0,0,0,0,0,0,0,0,0,0,0,0,0:REM X,T
1080 DATA 0,0,0,0,10,30,40,45,49,45,40,30,10,0,0,0,0:REM G,C
1090 DATA 0,0,0,0,-5,-10,-20,-25,-25,-25,-20,-15,-10,-5,0,0,0:REM A,C
1100 DATA 0,0,0,0,-10,-20,-25,-30,-35,-30,-25,-20,-10,0,0,0,0:REM V,C
1110 DATA 0,0,0,-10,-20,-30,-40,-30,-20,-20,-10,0,0,0,0,0,0:REM L,C
1120 DATA 0,0,0,0,0,-10,-20,-30,-33,-30,-10,0,10,20,30,20,0:REM I,C
1130 DATA 0,-10,-20,-20,10,15,20,25,50,25,20,15,10,0,0,0,0:REM S,C
1140 DATA 0,10,20,30,20,15,10,15,17,15,10,15,20,30,20,10,0:REM T,C
1150 DATA 0,0,0,0,0,0,0,0,0,0,0,0,0,0,0:REM D,C
1160 DATA 0,0,10,20,40,20,0,-10,-44,-40,-20,-10,0,0,0,0,0:REM E,C
1170 DATA 0,0,0,10,20,30,35,40,46,40,35,30,20,10,0,0,0:REM N,C
1180 DATA 10,20,30,25,20,15,10,0,-5,20,30,40,50,60,50,40,20:REM Q,C
1190 DATA -10,-20,-30,-40,-25,-20,-10,-8,-8,0,0,-20,-30,-20,-10,-5,0:REM K,C
1200 DATA 0,0,0,0,0,0,0,10,16,15,10,10,10,10,5,0,0:REM H,C
1210 DATA 0,0,0,0,0,0,0,0,-12,0,20,30,20,10,0,0,0:REM R,C
1220 DATA 0,0,0,0,0,-5,-10,-20,-41,-20,0,15,30,25,20,10,0:REM F,C
1230 DATA 0,0,0,0,0,0,0,0,-6,0,0,0,0,0,0,0,0:REM Y,C
1240 DATA 0,0,0,10,20,30,40,20,12,20,30,40,50,60,70,40,20:REM W,C
1250 DATA 0,0,0,0,0,0,-10,-30,-47,-30,-10,0,0,0,0,0,0:REM C,C
1260 DATA 0,-5,-10,-15,-20,-25,-30,-40,-41:REM M,C
1270 DATA -40,-30,-25,-20,-15,-10,-5,0:REM M,C
1280 DATA 0,0,10,20,30,40,50,55,58,50,10,0,0,0,0,0,0:REM P,C
1290 DATA 0,0,0,0,0,0,0,0,0,0,0,0,0,0,0:REM X,C
1292 DATA 36.8,135.,411.,136.,403.
1293 DATA 347.,395.,241.,144.,89.5
1294 DATA 120.,136.,308.,136.,308.
1295 DATA 308., 308., 346., 136., 1444., 136.:REM CC(I)
1296 FOR K=1 TO 4
1297 FOR J=1 TO 21
1300 FOR I=1 TO 17
1301 REM READ POSITIONAL INFORMATION FOR
1302 REM RESIDUE TYPE(J), CONFORMATION(K) AND POSITIONAL PARAMETER (I)
1310 READ PX(I,J,K)
1320 NEXT I
```

```
1330 NEXT J
1340 NEXT K
1350 REM READ (CHARACTERISTIC LENGTH)↑2 FOR RESIDUE 'I'
1351 REM (CHAR'C LENGTH)↑2=LIMIT(1/N(SIGMA R↑2)) AS N TENDS TO INFINITY
1352 REM R=END TO END DISTANCE OF HOMOPOLYMER,N=NUMBER OF RESIDUES
1360 FOR I=1 TO 21:READ CC(I):NEXT I
1361 DATA 0.0,-1.0,1.4,2.0,4.0,-1.2,-0.5,-1.2,-0.7,-0.7,-0.1,-0.9,1.1,0.3
1362 DATA 2.8,2.1,3.0,2.1,1.8,0.4,0.0:REM HYDROF
1363 DATA 0.0,0.0,0.0,0.0,0.0,0.0,0.0,-1.0,-1.0,0.0,0.0,1.0,0.3,1.0
1364 DATA 0.0,0.0,0.0,0.0,0.0,0.0,0.0:REM CHARGE
1370 FOR I=1 TO 21
1380 READ HYDROF(I):NEXT I
1390 FOR I=1 TO 21
1400 READ CHARGE(I):NEXT I
1401 REM INITIALISE INFORMATION ARRAYS
1402 FOR I=1 TO 200:H%(I)=0:E%(I)=0:T%(I)=0:C%(I)=0
1403 SH(I)=0.0:SC(I)=0.0:CS(I)=0.0:NEXT I
1420 SEQ$=""
1425 PRINT:PRINT "          *****COMMAND LEVEL 2*****"
1430 IF S$=".F" THEN PRINT "■TAPEZ LA SEQUENCE EN CODE MONOLETTRE"
1440 IF S$=".E" THEN PRINT "■ENTER SEQUENCE IN IUPAC ONE LETTER CODE"
1450 IF S$=".AUTO" THEN READ X$:PRINT#4,X$:IF X$=".INPUT" THEN S$=".E"
1451 IF S$<>".AUTO" THEN INPUT X$:REM READ SEQ OR COMMAND
1452 IF LEFT$(X$,1)<>"." GOTO 1498:REM IF NOT COMMAND, EXTEND THE SEQUENCE
1453 IF X$<>".X" GOTO 1455
1454 X$="XXXXXXXXXXXXXXXXXXXX":X$=X$+X$+X$+X$+X$:X$=X$+X$:GOTO 1498
1455 IF X$=".AUTO" THEN S$=".AUTO":GOTO 1450
1456 IF X$=".P" OR X$=".PRINT" THEN P$=".PRINT":GOTO 1450
1457 IF X$=".NP" OR X$=".NONPRINT" THEN P$=".NONPRINT":GOTO 1450
1459 IF X$=".PLUS" GOTO 1420:REM ADD IN INFO FROM PREVIOUS SEQ TO NEW SEQ
1460 IF X$=".GO" OR X$=".ALLEZ" OR X$="." OR X$=".+" GOTO 1500
1462 IF X$=".LIST" OR X$=".L" THEN PRINT SEQ$:GOTO 1450
1463 IF X$=".STOP" OR X$=".BYE" OR X$=".AU REVOIR" THEN CLOSE 4:END
1464 IF LEFT$(X$,4)<>".REM" GOTO 1467
1465 IF P$=".PRINT" THEN PRINT#4:PRINT#4,X$
1466 GOTO 1450
1467 IF X$=".ERASE" OR X$=".EFFACEZ" GOTO 1402:REM RESET FOR NEW SEQ
1468 IF X$=".!" THEN PRINT:PRINT SEQ$:SEQ$="":GOTO 1450:REM SCREEN EDITOR
1469 IF RIGHT$(X$,1)="!"GOTO 1450:REM IGNORE INPUT
1470 N$=LEFT$(X$,3):REM INTERNAL EDITOR
1471 IF N$=".N+" THEN SEQ$=RIGHT$(X$,LEN(X$)-3)+SEQ$:GOTO 1450
1472 IF N$=".C+" THEN SEQ$=SEQ$+RIGHT$(X$,LEN(X$)-3):GOTO 1450
1473 IF N$=".N-" THEN SEQ$=RIGHT$(SEQ$,LEN(SEQ$)-VAL(RIGHT$(X$,LEN(X$)-3)))
1474 IF N$=".N-" GOTO 1450
1475 IF N$=".C-" THEN SEQ$=LEFT$(SEQ$,LEN(SEQ$)-VAL(RIGHT$(X$,LEN(X$)-3)))
1476 IF N$=".C-" GOTO 1450
1477 IF LEFT$(X$,3)<>".M-" GOTO 1481
1478 X=VAL(RIGHT$(X$,LEN(X$)-3))
1479 SEQ$=LEFT$(SEQ$,Y-1)+RIGHT$(SEQ$,LEN(SEQ$)-Y-X+1)
1480 GOTO 1450
1481 IF LEFT$(X$,3)<>".M+" GOTO 1486
1482 SEQ$=LEFT$(SEQ$,Y-1)+RIGHT$(X$,LEN(X$)-3)+RIGHT$(SEQ$,LEN(SEQ$)-Y+1)
1483 GOTO 1450
1486 IF LEFT$(X$,3)=".M=" THEN Y=VAL(RIGHT$(X$,LEN(X$)-3)):GOTO 1450
1497 PRINT:PRINT "**ERROR/ERREUR**":GOTO 1450
1498 IF RIGHT$(X$,1)<>"!" THEN SEQ$=SEQ$+X$
1499 GOTO 1450
1500 IF LEN(SEQ$)>200 THEN SEQ$=LEFT$(SEQ$,200)
1501 IF S$=".E" THEN PRINT "■LENGTH OF SEQUENCE IS" LEN(SEQ$)
1502 IF S$=".F" THEN PRINT"■LE NOMBRE DES RESIDUS EST" LEN(SEQ$)
1503 IF S$=".E" THEN PRINT "■READ-IN HAS BEEN ACCEPTED"
1510 IF S$=".F" THEN PRINT "■ECRITURE ACCEPTEE"
1511 PRINT SEQ$
1512 PRINT"■          DATE: ";DATE$
1513 PRINT"
1514 PRINT TAB(10);"TIME: ";LEFT$(TI$,2);":";MID$(TI$,3,2);":";RIGHT$(TI$,2)
1515 PRINT"
1521 IF P$<>".PRINT" GOTO 1539
1522 PRINT#4:PRINT#4,"          DATE: ";DATE$
1523 PRINT#4,"
1524 PRINT#4, TAB(10);"TIME: ";LEFT$(TI$,2);":";MID$(TI$,3,2);":";RIGHT$(TI$,2
1525 PRINT#4,"
1538 PRINT#4,"SEQUENCE=  ";SEQ$:PRINT#4,"  "
1539 FOR II=1 TO LEN(SEQ$)
1540 Z$=MID$(SEQ$,II,1)
```

```
 1550 FOR JJ=1 TO 21:REM CHECK VALIDITY OF SEQUENCE
 1555 REM ASSIGN INTEGER CODE IN SEQ%
 1560 IF Z$=IUPAC$(JJ) THEN SEQ%(II)=JJ:GOTO 1630
 1570 NEXT JJ
 1571 PRINT:PRINT "         *****COMMAND LEVEL 3*****"
 1580 IF S$=".E" THEN PRINT "⊠INVALID SYMBOL '";Z$;"' AT ";II
 1581 IF S$=".F" THEN PRINT "⊠CARACTERE '";Z$ "' INCORRECT A LA POSITION";II
 1590 IF S$=".E" THEN PRINT "⊠         RE-ENTER IT NOW"
 1600 IF S$=".F" THEN PRINT "⊠         RETAPEZ VOTRE LETTRE"
 1610 INPUT "--->";Z$
 1611 IF Z$=".!" THEN PRINT:PRINT SEQ$:SEQ$="":GOTO 1450
 1620 GOTO 1550
 1630 NEXT II
 1670 PRINT
 1680 FOR I=1 TO LEN(SEQ$)
 1690 PRINT IUPAC$(SEQ%(I));
 1700 NEXT I
 1710 PRINT
 1720 PRINT
 1721 IF P$<>".PRINT" GOTO 1735
 1725 REM WRITE PRINTER HEADINGS
 1727 IF S$<>".E" GOTO 1730
 1728 PRINT#4,"H=HELIX,E=EXTENDED,T=TURN,C=COIL ";
 1729 PRINT#4,"HY=HYDROPHOBICITY,Q=CHARGE"
 1730 PRINT#4,"L↑2=(CHARACTERISTIC LENGTH)↑2":GOTO 1734
 1731 PRINT#4,"H=HELICE,E=FEUILLET BETA,T=TOURNANT,C=APERIODIQUE"
 1732 PRINT#4,"HY=HYDROPHOBICITE,Q=CHARGE"
 1733 PRINT#4,"L↑2=(DISTANCE CHARACTERISTIQUE)↑2"
 1734 PRINT#4,"PREDICTION   H       E       T       C      HY     Q       L↑2"
 1735 IF S$=".E" THEN PRINT " PREDICTION  H=HELIX E=EXTENDED T=TURN C=COIL"
 1736 IF S$=".F" THEN PRINT"PREDICTION H=HELICE ALPHA E=FEUILLET BETA";
 1737 IF S$=".F" THEN PRINT" T=TOURNANT C=APERIODIQUE"
 1740 FOR J=1 TO LEN(SEQ$)
 1741 SC(J)=SC(J)+CHARGE(SEQ%(J))
 1742 SH(J)=SH(J)+HYDROF(SEQ%(J))
 1745 CS(J)=CS(J)+CC(SEQ%(J))
 1750 FOR M=-8 TO +8
 1760 JM=J+M
 1770 IF JM<1 OR JM>LEN(SEQ$) GOTO 1840
 1780 R=SEQ%(JM)
 1790 MM=9-M
 1795 REM ADD IN INFO RES(M) CARRIES ABOUT RES(J)
 1800 H%(J)=H%(J)+P%(MM,R,1)
 1810 E%(J)=E%(J)+P%(MM,R,2)
 1820 T%(J)=T%(J)+P%(MM,R,3)
 1830 C%(J)=C%(J)+P%(MM,R,4)
 1840 NEXT M
 1850 H=H%(J):E=E%(J):T=T%(J):C=C%(J)
 1855 REM ASSIGN STATE OF RES(J)
 1860 Y$="C"
 1870 IF T>H AND T>E AND T>C THEN Y$="T"
 1880 IF E>H AND E>T AND E>C THEN Y$="E"
 1890 IF H>E AND H>T AND H>C THEN Y$="H"
 1899 IF P$<>".PRINT" GOTO 1910
 1900 PRINT#1,J;IUPAC$(SEQ%(J));"->";Y$;":",CHR$(29),H,E,T,C,SH(J),SC(J),CS(J)
 1910 PRINT J;IUPAC$(SEQ%(J));"->";Y$;": ";H;E;T;C;SH(J);SC(J);CS(J)
 1920 NEXT J
 1921 IF X$=".+" GOTO 1420
 1922 IF X$="." GOTO 1402
 1923 GOTO 1425
 5000 REM USERS REQUIRING AUTO MODE SHOULD PROVIDE INPUT IN DATA STATEMENTS
 5001 REM AFTER THIS LINE. SEE EXAMPLE FOLLOWING:-
 5005 DATA ".REM SCORPION VENOM V3",
 5010 DATA "KEGYLVKKSDGCKYGCLKLGENEGCDTE",
 5015 DATA "CKAKNQGGSYGYCYAFACWCEGLPESTPTYPLPNKSC",
 5016 DATA ".INPUT"
 5020 DATA ".+",
 6000 DATA ".INPUT"
 6005 DATA ".REM SCORPION TOXIN V3",
 6010 DATA "KEGYLVKKSDGCKYGCLKLGENEGCDTE",
 6015 DATA "CKAKNQGGSYGYCYAFACWCEGLPESTPTYPLPNKSC",
 6019 DATA ".INPUT"
 6020 DATA "."
READY.
```

5. NOTE ON DECISION CONSTANTS

The paper on Garnier et al. [40] also includes the values required for the 'decision constants' of the information theory approach. There are three such quantities, one for each of helix, extended chain, and turn, and they are to be simply subtracted from helix, extended, and turn information after adding in all the directional information parameters and prior to choosing which information measure, for helix, extended, turn, or coil, is the largest (thus making the 'prediction'). In that paper, decision constants are available for classes of protein which may be crudely, if adequately, assigned from estimates of helix and pleated sheet content by circular dichroism. If no such data are available, values for proteins containing medium amounts of helix and pleated sheet may be used, or preliminary predictions carried out to discover if the protein belongs to a helix rich (globin-type) class, and so on. Alternatively, the procedure of optimizing decision constants for maximum accuracy of secondary structure predictions for a large protein data base may be carried out, as used in our earlier studies.

The decision constants have not been implemented in the present BASIC program since it was not written to give optimal predictions as such, but to correlate conformational tendencies with the other properties, such as hydrophobicity, which the program prints out. Further, the predictions are also to be used to assign initial conformations for protein folding simulations based on energy calculation. Arguably, all decisions should be zero to yield secondary structures represented in the earliest stages of folding, since non-zero decision constants represent tertiary or supersecondary effects such as helix stacking, extended chain interactions to form pleated sheets, and so on [40]. If the user wishes to predict the secondary structure without the energy calculation, the decision constants are, of course, readily implemented. (This can, of course, be done manually, since information values are printed out.) The reader should note, however, that there is no excuse for adjusting decision constants to improve the prediction with reference to the answer known a priori, for the particular protein in which he is interested. This has never been a feature of our approach and it must be admitted that the omission of decision constant input from the present BASIC program was partly to emphasize this.

6. PREDICTION FOR SPERM WHALE MYOGLOBIN

```
6. Prediction for sperm wh ale my oglobin                                  (viii)

     DATE: 17/09/81

     TIME: 14:53:38

SEQUENCE=  VLSEGEWQLVLHVWAKVEADVAGHGQDILIRLFKSHPETLEKFDRFKHLKTEAEMKASEDLKKHGVTVL
TALGAIIKKKGHHEAELKPLAQSHATKHKIPIKYLEFISEAIIHVLHSRHPGNFGADAQGAMNKALELFRKDIAAKYKEL
GYQG

H=HELIX,E=EXTENDED,T=TURN,C=COIL HY=HYDROPHOBICITY,Q=CHARGE
L↑2=(CHARACTERISTIC LENGTH)↑2
PREDICTION   H     E     T      C     HY     Q        L↑2
  1 V->E: - 36  + 58  -105  + 25  +1.4  +0.0      411.0
  2 L->C: -  8  + 18  - 61  + 85  +2.0  +0.0      136.0
  3 S->C: -  4  - 37  +  1  +100  -1.2  +0.0      347.0
```

```
 4 E->H: + 96  -125  +  3  + 36  -0.7  -1.0    144.0
 5 G->H: +117  -202  + 37  + 44  +0.0  +0.0     36.8
 6 E->H: +181  -210  + 48  - 24  -0.7  -1.0    144.0
 7 W->H: +206  -135  - 39  - 68  +3.0  +0.0    308.0
 8 Q->H: +231  - 48  -141  - 70  -0.1  +0.0    120.0
 9 L->H: +228  + 53  - 96  - 60  +2.0  +0.0    136.0
10 V->H: +227  + 78  - 95  - 45  +1.4  +0.0    411.0
11 L->H: +210  + 43  - 81  +     +2.0  +0.0    136.0
12 H->H: +202  + 15  -  8  + 66  +1.1  +0.3    308.0
13 V->H: +162  - 12  + 45  + 70  +1.4  +0.0    411.0
14 W->H: +187  - 35  - 14  + 57  +3.0  +0.0    308.0
15 A->H: +175  - 43  -100  + 22  -1.0  +0.0    135.0
16 K->H: +208  - 38  - 80  - 33  -0.9  +1.0    136.0
17 V->H: +199  - 37  -135  - 55  +1.4  +0.0    411.0
18 E->H: +168  - 65  -122  - 74  -0.7  -1.0    144.0
19 A->H: +168  - 33  - 90  - 65  -1.0  +0.0    135.0
20 D->H: +173  - 24  -  9  - 20  -1.2  -1.0    241.0
21 V->H: +177  - 27  - 20  -  5  +1.4  +0.0    411.0
22 A->H: +173  - 88  + 15  + 35  -1.0  +0.0    135.0
23 G->H: +114  -102  + 42  + 49  +0.0  +0.0     36.8
24 H->H: +102  -120  + 42  + 61  +1.1  +0.3    308.0
25 G->C: + 34  -117  +  7  + 49  +0.0  +0.0     36.8
26 Q->H: + 70  - 43  -  6  + 10  -0.1  +0.0    120.0
27 D->H: + 63  + 61  - 39  - 35  -1.2  -1.0    241.0
28 I->E: + 76  +152  -106  - 73  +4.0  +0.0    403.0
29 L->E: + 82  +213  -141  -110  +2.0  +0.0    136.0
30 I->E: + 99  +192  - 76  - 98  +4.0  +0.0    403.0
31 R->E: + 79  +169  -  4  - 32  +0.3  +1.0    136.0
32 L->E: + 52  +103  + 54  + 55  +2.0  +0.0    136.0
33 F->T: -  9  + 46  +147  +101  +2.8  +0.0    308.0
34 K->C: - 29  - 63  +145  +177  -0.9  +1.0    136.0
35 S->C: - 79  -132  + 56  +190  -1.2  +0.0    347.0
36 H->C: - 88  -175  - 93  +136  +1.1  +0.3    308.0
37 P->C: - 37  -208  - 39  + 93  +0.4  +0.0   1444.0
38 E->H: + 31  -195  - 17  -  4  -0.7  -1.0    144.0
39 T->H: +152  -177  - 87  - 53  -0.5  +0.0    395.0
40 L->H: +205  -157  - 91  - 60  +2.0  +0.0    136.0
41 E->H: +241  -130  - 62  -107  -0.7  -1.0    144.0
42 K->H: +302  -138  - 50  -108  -0.9  +1.0    136.0
43 F->H: +344  -129  - 28  - 91  +2.8  +0.0    308.0
44 D->H: +363  -149  +  6  - 60  -1.2  -1.0    241.0
45 R->H: +339  -131  - 24  - 82  +0.3  +1.0    136.0
46 F->H: +291  -139  - 23  - 69  +2.8  +0.0    308.0
47 K->H: +236  -183  - 10  -  8  -0.9  +1.0    136.0
48 H->H: +197  -230  - 48  + 71  +1.1  +0.3    308.0
49 L->H: +152  -257  - 91  + 52  +2.0  +0.0    136.0
50 K->H: +183  -268  -110  + 12  -0.9  +1.0    136.0
51 T->H: +212  -252  -177  - 68  -0.5  +0.0    395.0
52 E->H: +288  -230  -197  -114  -0.7  -1.0    144.0
53 A->H: +318  -218  -205  -140  -1.0  +0.0    135.0
54 E->H: +351  -205  -187  -144  -0.7  -1.0    144.0
55 M->H: +379  -177  -153  -114  +1.8  +0.0    136.0
56 K->H: +414  -203  -160  -108  -0.9  +1.0    136.0
57 A->H: +448  -233  -170  -125  -1.0  +0.0    135.0
58 S->H: +444  -277  -169  -100  -1.2  +0.0    347.0
59 E->H: +426  -250  -167  -164  -0.7  -1.0    144.0
60 D->H: +383  -229  - 94  -130  -1.2  -1.0    241.0
61 L->H: +295  -172  - 41  - 53  +2.0  +0.0    136.0
62 K->H: +226  -138  + 40  -  6  -0.9  +1.0    136.0
63 K->H: +154  - 88  + 70  + 37  -0.9  +1.0    136.0
64 H->H: + 97  - 20  + 47  + 51  +1.1  +0.3    308.0
65 G->H: + 19  + 13  - 13  + 14  +0.0  +0.0     36.8
66 V->E: + 34  + 98  -105  - 50  +1.4  +0.0    411.0
67 T->E: + 42  +168  -162  - 83  -0.5  +0.0    395.0
68 V->E: + 44  +178  -175  - 85  +1.4  +0.0    411.0
69 L->E: + 77  +143  -176  - 90  +2.0  +0.0    136.0
70 T->H: +147  + 78  -147  - 93  -0.5  +0.0    395.0
71 A->H: +218  +  7  -140  - 80  -1.0  +0.0    135.0
72 L->H: +257  -  2  -161  - 90  +2.0  +0.0    136.0
73 G->H: +254  - 32  -133  -106  +0.0  +0.0     36.8
74 A->H: +293  - 18  -135  -110  -1.0  +0.0    135.0
75 I->H: +286  - 18  -141  - 83  +4.0  +0.0    403.0
76 I->H: +246  - 18  - 91  - 66  +4.0  +0.0    403.0
77 K->H: +193  - 23  - 20  - 21  -0.9  +1.0    136.0
78 K->H: +133  - 43  + 45  + 24  -0.9  +1.0    136.0
```

79	K->H:	+113	- 98	+ 25	+ 82	-0.9	+1.0	136.0
80	G->C:	+ 59	-157	- 8	+ 84	+0.0	+0.0	36.8
81	H->H:	+102	-160	- 73	+ 61	+1.1	+0.3	308.0
82	H->H:	+102	-155	-168	- 9	+1.1	+0.3	308.0
83	E->H:	+158	-140	-162	- 79	-0.7	-1.0	144.0
84	A->H:	+206	-138	-100	-100	-1.0	+0.0	135.0
85	E->H:	+221	-155	- 42	- 84	-0.7	-1.0	144.0
86	L->H:	+203	-152	- 76	- 63	+2.0	+0.0	136.0
87	K->H:	+197	-173	-155	- 43	-0.9	+1.0	136.0
88	P->H:	+279	-153	- 19	+ 8	+0.4	+0.0	1444.0
89	L->H:	+335	-137	+ 4	+ 25	+2.0	+0.0	136.0
90	A->H:	+365	-108	- 95	- 50	-1.0	+0.0	135.0
91	Q->H:	+358	-108	-121	- 95	-0.1	+0.0	120.0
92	S->H:	+286	- 72	- 74	- 20	-1.2	+0.0	347.0
93	H->H:	+247	- 70	- 48	- 29	+1.1	+0.3	308.0
94	A->H:	+185	- 68	- 5	+ 15	-1.0	+0.0	135.0
95	T->H:	+174	- 62	+ 58	+ 34	-0.5	+0.0	395.0
96	K->H:	+123	- 53	+ 85	+ 57	-0.9	+1.0	136.0
97	H->T:	+ 52	- 25	+ 97	+ 43	+1.1	+0.3	308.0
98	K->C:	- 22	+ 17	+ 20	+ 47	-0.9	+1.0	136.0
99	I->E:	- 59	+ 37	-121	+ 12	+4.0	+0.0	403.0
100	P->E:	- 2	+ 32	- 44	- 2	+0.4	+0.0	1444.0
101	I->E:	- 14	+ 37	+ 14	- 11	+4.0	+0.0	403.0
102	K->E:	+ 6	+ 47	-100	- 88	-0.9	+1.0	136.0
103	Y->E:	+ 35	+ 50	-111	- 71	+2.1	+0.0	308.0
104	L->H:	+ 57	+ 33	- 91	- 10	+2.0	+0.0	136.0
105	E->H:	+103	+ 5	- 92	- 34	-0.7	-1.0	144.0
106	F->H:	+142	- 14	- 83	- 76	+2.8	+0.0	308.0
107	I->H:	+214	- 63	- 71	- 48	+4.0	+0.0	403.0
108	S->H:	+259	- 92	- 84	- 20	-1.2	+0.0	347.0
109	E->H:	+286	-105	- 87	- 74	-0.7	-1.0	144.0
110	A->H:	+283	- 93	-100	- 95	-1.0	+0.0	135.0
111	I->H:	+279	- 53	-136	-118	+4.0	+0.0	403.0
112	I->H:	+234	- 23	-126	-108	+4.0	+0.0	403.0
113	H->H:	+178	+ 15	- 78	- 59	+1.1	+0.3	308.0
114	V->H:	+ 94	+ 83	- 65	- 25	+1.4	+0.0	411.0
115	L->E:	+ 42	+103	- 16	+ 15	+2.0	+0.0	136.0
116	H->E:	- 33	+ 85	+ 62	+ 66	+1.1	+0.3	308.0
117	S->T:	-131	+ 8	+176	+155	-1.2	+0.0	347.0
118	R->C:	-199	- 31	+176	+183	+0.3	+1.0	136.0
119	H->C:	-243	- 65	+102	+196	+1.1	+0.3	308.0
120	P->T:	-227	- 73	+226	+213	+0.4	+0.0	1444.0
121	G->T:	-241	- 87	+277	+234	+0.0	+0.0	36.8
122	N->C:	-191	- 46	+172	+176	-0.7	+0.0	89.5
123	F->C:	-144	- 19	+ 77	+104	+2.8	+0.0	308.0
124	G->C:	- 96	- 22	+ 27	+ 84	+0.0	+0.0	36.8
125	A->C:	+ 30	+ 2	- 15	+ 50	-1.0	+0.0	135.0
126	D->H:	+125	- 24	- 49	+ 25	-1.2	-1.0	241.0
127	A->H:	+175	- 63	- 50	+ 15	-1.0	+0.0	135.0
128	Q->H:	+180	- 58	- 61	- 30	-0.1	+0.0	120.0
129	G->H:	+154	- 87	- 43	+ 9	+0.0	+0.0	36.8
130	A->H:	+180	- 93	- 50	+ 10	-1.0	+0.0	135.0
131	M->H:	+178	- 77	- 83	+ 24	+1.8	+0.0	136.0
132	N->H:	+189	- 51	- 83	+ 23	-0.7	+0.0	89.5
133	K->H:	+231	- 38	-110	- 28	-0.9	+1.0	136.0
134	A->H:	+275	- 38	-150	- 65	-1.0	+0.0	135.0
135	L->H:	+285	- 37	-186	- 95	+2.0	+0.0	136.0
136	E->H:	+298	- 35	-187	-144	-0.7	-1.0	144.0
137	L->H:	+328	- 27	-181	-165	+2.0	+0.0	136.0
138	F->H:	+334	- 34	- 98	-146	+2.8	+0.0	308.0
139	R->H:	+312	- 76	- 74	-130	+0.3	+1.0	136.0
140	K->H:	+316	-113	00	-108	-0.9	+1.0	136.0
141	D->H:	+300	-144	- 59	- 80	-1.2	-1.0	241.0
142	I->H:	+276	-148	-111	- 68	+4.0	+0.0	403.0
143	A->H:	+225	-148	-135	- 70	-1.0	+0.0	135.0
144	A->H:	+165	-108	-100	- 48	-1.0	+0.0	135.0
145	K->H:	+ 98	- 68	- 75	- 53	-0.9	+1.0	136.0
146	Y->H:	+ 80	- 25	- 36	- 39	+2.1	+0.0	308.0
147	K->H:	+ 56	+ 2	+ 5	- 13	-0.9	+1.0	136.0
148	E->H:	+ 28	+ 20	+ 23	- 19	-0.7	-1.0	144.0
149	L->T:	+	+ 28	+ 29	- 5	+2.0	+0.0	136.0
150	G->T:	- 43	- 7	+ 87	+ 9	+0.0	+0.0	36.8
151	Y->T:	- 34	- 5	+ 94	+ 19	+2.1	+0.0	308.0
152	Q->T:	- 47	+ 2	+ 64	+ 55	-0.1	+0.0	120.0
153	G->C:	- 91	- 7	+ 87	+ 89	+0.0	+0.0	36.8

(x)

Geisow & Barrett (eds.) Computing in biological science
©Elsevier Biomedical Press, 1983

5

The analysis and prediction of protein structure

MICHAEL J.E. STERNBERG

1. INTRODUCTION

This chapter will consider the extent to which two questions about proteins can be answered by computer modelling.

(1) Can one predict the three-dimensional structure of a protein from its amino acid sequence?
(2) What additional information can be obtained from the crystal structure of a protein?

A detailed account of the current state of algorithms to answer these questions cannot be given in one chapter. I will therefore describe those methods that can easily be implemented by a non-specialist in this subject and give only a brief report or just references to more involved topics. For general reviews the reader is referred to refs. [1–6]. Given that a selection of subjects has to be made, I will emphasise the algorithms that I have used in the hope that I will be able to give some insight into the requirements and limitations of these methods. The cut-off date for this review is May 1981.

2. ASPECTS OF PROTEIN STRUCTURE

The achievement of protein crystallography over the last 25 years is that it has led to a description of the structure and function of proteins at the atomic level. This knowledge has encouraged other studies to further our understanding of this major class of biological macromolecules. In this section I will outline some experimental and theoretical studies on protein structure that are required in subsequent sections.

2.1. *Experimental studies on protein folding*

The classical experiments of Anfinsen [7] and his co-workers that showed that ribonuclease can be refolded, demonstrated that the three-dimensional structure of this protein is determined by its amino acid sequence and the environment. Subsequently, both proteins with and without disulphide bridges and both monomeric and multi-subunit proteins have been refolded (for reviews see [7–9]) showing that in principle it should be possible to predict theoretically the three-dimensional structure of a protein from its sequence. The notable exceptions to reversible unfolding with high yield are those proteins such as insulin and α-chymotrypsin that are derived by covalent modification of a precursor.

Although the native structure of a protein will be at a minimum of free energy, a protein will not fold by sampling every conformation as the time for this is excessive. The argument, as described by Levinthal [10] and by Wetlaufer [11], is that a protein with N residues, each of which has two rotatable bonds, can have 10^N main-chain conformations. If each bond rotation occurs in 10^{-13} s, the time to sample each conformation once is $(10^N) \div (2 \times 10^{13} \times N)$; for $N = 50$, the time is 10^{35} s. Therefore there must be one or more pathways along which the protein folds to reduce the time to acquire the active structure. Experimental evidence for the presence of a pathway has been obtained by Creighton [12] who showed that only some of the possible one- and two-disulphide intermediates are formed during the refolding of pancreatic typsin inhibitor, a protein with 58 residues and three disulphide bridges. Recently, it has been found that ribonuclease [13–15] has fast and slow refolding species. It is proposed that the slow refolding results from *cis/trans* isomerisation of prolines.

There have been several suggestions for the nature of the folding pathway. One common proposal (e.g. [7]) is that one or more regions of secondary structure, such as α-helices or a two-stranded antiparallel β-sheet, having marginal stability will act nucleation sites and direct the refolding. Several workers (Chantrenne [16], Dunhill [17], Phillips [18]) have suggested that in vivo the polypeptide chain folds as it is synthesised from its N-terminus on the ribosome. Evidence in support of N-terminal folding comes from experiments on β-galactosidase [19]. In Section 4.5, some computer simulations of the folding pathways of proteins will be described.

2.2. *Interatomic interactions, the hydrophobic effect and solvent accessibility*

A detailed understanding of the structure, mobility and function of proteins requires accurate descriptions of the interatomic interactions (for reviews see [1,20,21]). Several calculations of energy are described in subsequent Sections (3.3 and 4.1).

TABLE 1

Estimated contributions to the free energy of folding ribonuclease S, a tentative balance sheet.

Effects likely to stabilise folded conformation	Estimated magnitude of free energy change unfolded → folded protein kcals/mole.
Hydrophobic interactions	−130
Differences of the polar interactions between protein–protein and protein–water	−75
Entropy of release of water that was bound to polar atoms in the unfolded protein	−120
Effects likely to destabilize folded conformations	
Configurational entropy loss of chain	+70 to +190
Hydrogen bond distortion	+100 to +200
Unsaturated polar groups	+70
Total balance (experimental)	−10 to −20

Adapted from Finney et al. [21].

Experimental values [22] for the free energy change for the transition from an unfolded to a folded conformation for a protein in solution are for hen-egg white lysozyme and ribonuclease around −10 to −20 kcal/mol. Recently Finney et al. [21] have estimated the magnitudes of the different effects that contribute to this free energy change (Table 1). It is clear that no single effect dominates in the marginal stability of the native conformation. Protein/solvent interactions are clearly important and the hydrophobic effect is discussed further.

The hydrophobic effect [23] is primarily entropic resulting from the unfavourable ordering of water molecules that would occur around exposed non-polar atoms. Experimental values for this effect come from Tanford [24,25] and co-workers who determined the free energy change for the transfer of non-polar residues from water to organic mediums. Typical values for the changes for alanine and tryptophan side chains are −1.0 and −3.5 kcal/mol (1 kcal = 4.2 kJ). Various functions have been proposed to model the hydrophobic effect. I describe one possible approach based on the notion of the solvent accessibility of atoms as the application of this concept alone, and its relationship to hydrophobicity, will be used in subsequent Sections (3.6. and 4.2.).

In 1971, Lee and Richards [26] introduced solvent accessibility to quantify the extent of exposure of atoms to solvent. For a given atom in a structure this area is defined as the locus of the centre of a hypothetical probe (generally a water molecule) that rolls over the van der Waals' surface of that atom without penetrating the van der Waals spheres of the surrounding (non-water) atoms (see Fig. 1). The larger the accessible area, the more of the atom is exposed to solvent. A related measure is contact area [27,28] which is the area of the van der Waals' surface of the atom that is in contact with the probe. While the accessible surface is a continuous sheet that surrounds the protein, the contact area is formed by a series of disconnected patches. Re-entrant area is also a series of patches formed by the

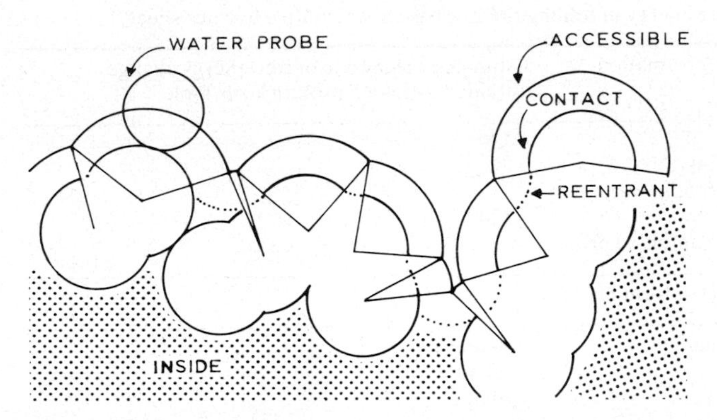

Fig. 1. The distinction between accessible, contact and re-entrant surface area. Adapted from [27]. See Section 2.2. for details.

internal facing part of the probe when it simultaneously contacts more than one atom. If the radii of an atom and the probe are R and r, then

$$\frac{\text{accessible area of the atom}}{\text{contact area of the atom}} = \frac{(R + r)^2}{R^2}$$

As the van der Waals' radii used to represent atoms or groups of atoms will generally range from 1.4 Å to 1.9 Å, there is no single multiplication factor to convert accurately accessible and contact area. However, a suitable approximation that will be used is 80 $Å^2$ of accessible area is equivalent to 23 $Å^2$ of contact area (see below).

One important application of accessible and contact areas is that they might well provide a convenient model for the hydrophobic effect. Chothia [29] has shown that the accessible surface area of a non-polar residue in an extended conformation is roughly proportional to the experimental transfer free energies obtained from Tanford and co-workers [24,25]. Suitable conversion factors are that each square Angstrom of accessible and contact area removed from solvent is equivalent to a free energy change of 23 and 80 cals/mol respectively. It must be pointed out that the validity of this conversion is controversial [30]. However, it might well provide a useful first approximation to model the hydrophobic effect. Indeed, the application of this relationship to predict the tertiary fold of proteins has, in my opinion, had some success.

The traditional method of calculating accessible and contact areas involves taking a planar section of the van der Waals surface of the protein, the evaluation of the length of the arcs, and a numerical integration over equally spaced sections to convert arc lengths to areas. (A computer program for this calculation, written by Dr. T. J. Richmond at Yale University, is available from me.) The calculation for all the atoms in myoglobin would take about 2 min of c.p.u. time on the IBM 360/195 at the Rutherford Laboratory, Didcot, Oxon, England. Other workers [31,32] have developed other numerical algorithms to quantify the exposure of atoms.

Recently, Wodak and Janin [33] have obtained an analytical approximation to the accessible area of protein atoms. The difficulty in obtaining an analytic expression is that the total surface buried on atom 1 is not the sum of the individual surfaces on atom 1 buried by atoms 2,3, ... as these individual surfaces generally overlap. Wodak and Janin [33] used a statistical approximation to estimate the extent of these overlaps. A trial on T4 lysozyme showed that the statistical approximation gave a total area 0.98 of that obtained by conventional integration and that the average fractional deviation between these calculations for each residue was 0.19. The statistical approximation was applied to a simplified model of protein structure. Following Levitt [34], each amino acid residue type is represented by a single sphere of specific van der Waals radius placed at the centroid of the residue position. The simplified model gave an accessible area 0.96 of that of the numerical value and the fractional deviation was 0.13. The algorithm is fast to compute, 0.1 s of c.p.u. on an IBM 370/168 for 58 residues. Furthermore, an important feature is that the function and its derivative with respect to position are analytic and can be used in minimisation procedures.

2.3. *The data bank of protein structures*

The achievement of protein crystallography is that over the last 25 years the three-dimensional structures of more than 80 proteins have been determined to near atomic resolution. Several workers have undertaken the collation and depiction of this vast body of information. Atomic co-ordinates for many of these proteins are available from the Brookhaven Data Bank [35]. For each protein, the co-ordinates are in a standard format and there are references to the papers describing the structures. The Data Bank is regularly updated and users are informed of these changes several times a year. Visual representations of these structures are most helpful. Richard Feldmann [36] has produced on microfilm a series of stereo diagrams of many proteins from different views with details such as the residues that surround any given residue in protein. Recently an additional aid has been developed by Feldmann, a set of computer-generated space-filling diagrams in colour of the structures of several proteins [37]. These diagrams, primarily designed as a teaching kit, succeed in giving an attractive alternative view of proteins that is different from the conventional chain tracings. Richardson [4] has produced a series of hand drawings of protein structures that highlight the relative positions of α-helices and β-strands using a ribbon and arrow representation.

As with any experimental technique, the results of protein crystallography must be carefully interpreted (see the book on the method by Blundell and Johnson [38]). The resolution of an electron density map is quoted in terms of the minimum interplanar spacing, a better resolved structure (i.e., to higher resolution) will have a lower value for the spacing. Generally a resolution of better than 2.8 Å is required for confidence in the positions of most of the side chains. However, flexible regions, particularly surface side chains, will be harder, or even impossible, to locate in the map. The error in an atomic co-ordinate from the intrepretation of a map will on average be at least 0.3 Å. There are several methods of improving the agreement of

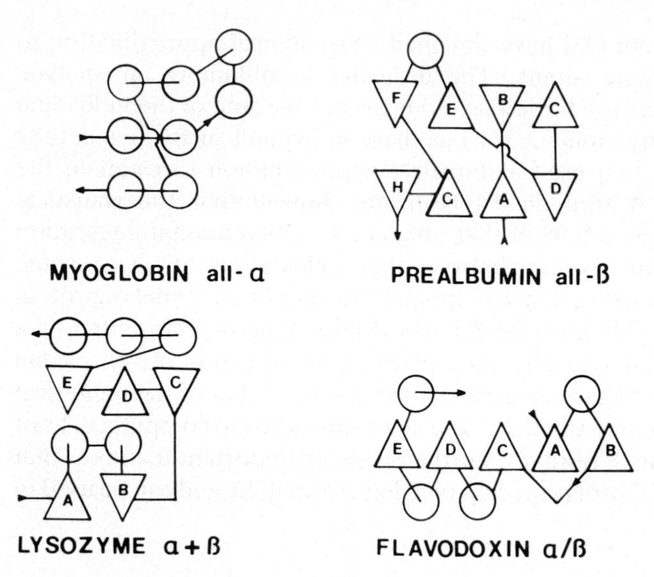

Fig. 2. The four classes of protein structure involving the packing of α-helices and β-strands. Reproduced with permission from [99]. The figure shows schematic diagrams of relative positions along the polypeptide chain of the α-helices and β-strands. Each α-helix is represented by a circle. The β-sheet is viewed along the direction of its component β-strands. A triangle whose apex points up (down) indicates a β-strand viewed from its N– (C–) end.

the model to the crystallographic observations and generally the structure will be regularised to have standard bond lengths and angles. The agreement between the model and the observations is expressed by an *R*-factor; unrefined structures having *R*-factors of up to 60% whilst for better refined structures at high resolution values of 18% are becoming common (e.g., human lysozyme [39]).

2.4. *The structure of globular proteins*

Examinations of the structures of the different proteins over the last ten years have begun to provide a set of principles governing their conformations (reviewed in [1–6]).

Protein structure results from the close packing [40] of atoms to create a hydrophobic core shielded from solvent [26,27,41] that consists of non-polar atoms and of polar atoms that either form hydrogen bonds or salt bridges. The surface of the protein is primarily formed from polar atoms but non-polar atoms are also exposed.

The tertiary structure adopted by the polypeptide chain consists of one or more distinct folding units or domains [8,42–44] each of which will be formed from forty to several hundred residues. Each domain can be classified into one of five classes [1,45] according to its secondary structure (Fig. 2). These classes are: all-α with just α-helices and no β-strands; all-β; α/β in which the chain alternates between α-helices and β-strands; α + β in which the α- and β-regions tend to segregate into separate regions; and coil in which there is little or no regular

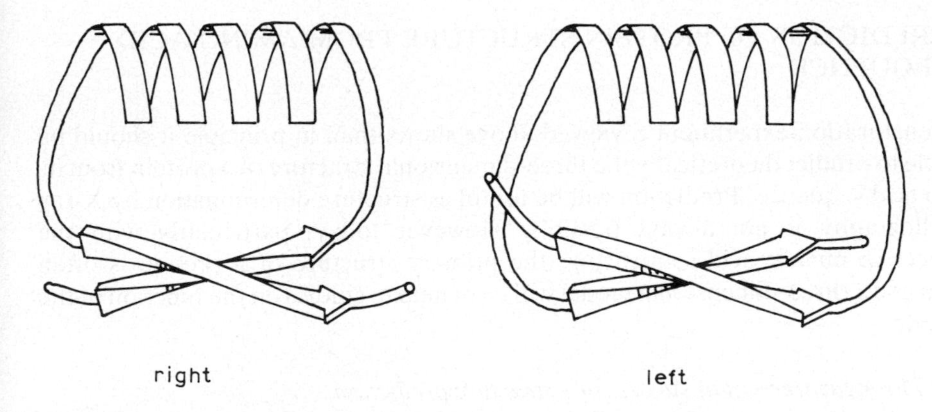

Fig. 3. Chirality in β-sheets. Two β-strands (shown as arrows) in a β-sheet with an α-helical connection between them. The usual left-handed rotation of around 20° between adjacent β-strands is indicated. If the α-helical connection lies above the β-sheet, then the relative positions of the two strands in the sheet defines the handedness of the connection. Nearly all the known connections are right-handed. Reproduced with permission from [99].

secondary structure (see [1] for details). The geometries of the α/α, β/β and α/β packings have been analysed and are described in Section 3.6.

The structure of the β-sheet is worth discussing further [46–55]. In some β-sheets [46,47] all the component β-strands make parallel-type hydrogen bonds with their neighbours whilst in other sheets all the hydrogen bonds are the antiparallel-type. However, in addition to these pure sheets, there are several mixed β-sheets involving both parallel- and antiparallel-type hydrogen bonds. Several workers have reported the occurrence of these types of sheets and analysed other features such as the tendency for β-strands sequential in primary structure to lie adjacent in a sheet. There is also a surprising feature of two types of chirality [48] associated with structures involving a β-sheet. Chothia [48] has commented that β-sheets are not the simple pleated structure suggested by Pauling and Corey [49] but are twisted, so that progressing along a β-strand, peptides i and $i + 2$ have a right-handed rotation of about +40° between them. The consequence of this is that for optimal hydrogen bonding between adjacent β-strands there must be a left-handed rotation of about 20° between them (i.e., when viewed normal to the strand direction). These chiralities have been observed in all β-sheets in globular proteins although the cause of this phenomenon is not clear [48,50,51]. The second structural chirality [52–55] occurs in the loop formed from two, parallel β-strands in a β-sheet and their connection, which can have an α-helical, β-strand or coil conformation. This chirality depends on the relative positions of the two β-strands given that the connection lies above the sheet (Fig. 3). Examination of all known structures has found only three left-handed connections while some hundred right-handed structures have been observed. One explanation for this chirality is that given the rotation [48] between adjacent β-strands, the right-handed unit allows a more direct path for the connection than the left-handed form.

3. PREDICTION OF PROTEIN STRUCTURE FROM AMINO ACID SEQUENCE

The renaturation experiment reviewed above shows than in principle it should be possible to predict theoretically the three-dimensional structure of a protein from its amino acid sequence. Prediction will be useful as structure determination by X-ray crystallography is not always possible. However today, particularly with the advances in nucleic acid sequencing, the primary structure of a protein is often known and a three-dimensional model will be of help to studies on the function of the molecule.

3.1. The measurement of success of a structure prediction

Some methods to assess the results of tertiary and secondary structure predictions will now be described.

Generally the root-mean-square deviation (r.m.s.d.) of the atomic (often just C^{α}) positions in a predicted structure from the crystallographic co-ordinates is reported to indicate the success of a folding simulation [56]. There are two procedures used to compute the r.m.s.d.. The first is the rotation method [56,57] that is used by crystallographers to compare structures. The r.m.s.d. is computed as:

$$\Delta r = \sqrt{\sum_{i=1}^{n} (X_i - y_i')^2/n}$$

with $\{x_i\}$ the coordinates of the crystal structure and $\{y_i\}$ the generated structure where the primed frame is the rotation that minimizes Δr. The second method is based on interatomic distances [34,57] is often quoted in folding studies. It is calculated as:

$$\Delta d = \sqrt{\sum_{i=1}^{n} \sum_{j=1}^{n} (d_{ij} - e_{ij})^2/n^2}$$

where $\{d_{ij}\}$ are the interatomic distances between atoms in the crystal structure and $\{e_{ij}\}$ the distances in the predicted structure.

The interatomic method suffers from the drawback [57] that it is unaffected by a change of hand ($x \rightarrow -x$). Furthermore any given comparison will report different r.m.s.d. according to which method is used as it can be shown empirically that, apart from changes of hand,

$$\Delta r \simeq 1.6\Delta d$$

To prevent confusion, I will only quote r.m.s.d. calculated by the rotation method.

The success of any folding simulation should be judged against the result one would expect from a random, compact structure that has no bad close contacts. Cohen and Sternberg [57] have generated a series of such random structures and found that the means ± standard deviations of their r.m.s.d. from the crystal structures are for: pancreatic trypsin inhibitor (58 residues) 11.2 ± 1.4 Å; myoglobin (153 residues) 16.2 ± 1.4 Å; and triose phosphate isomerase (247 residues) 20.6 ± 1.6 Å. This, and work by others [58–60], has established a benchmark to assess the results of tertiary structure prediction.

A variety of different measures of the success of a secondary structure prediction have been used and are reviewed in [1]. I will refer to an overall accuracy under $\%N$ defined as

$$\%N = \frac{100(N - Nx)}{N}$$

where N is the total number of residues in the protein and Nx is the total number of residues incorrectly predicted. A random prediction of κ different states would achieve an average $\%N = 100/\kappa$.

3.2. *Tertiary structure prediction from sequence homology*

An obvious approach to predict tertiary structures recognises that proteins with related sequences will have a similar tertiary fold. Accordingly there have been several structure predictions including those of α-lactalbumin [61,62] from the hen egg-white lysozyme structure and α-lytic protease [63,64] from elastase and α-chymotrypsin, and of an 8 α-helical structure for troponin C [65] from a 2 α-helical region in myogen. It is of course desirable to regularise a predicted structure by energy calculations as Lewis and Scheraga [62] did for α-lactalbumin. When faced with a more distant sequence relationship, it is important also to use the results of a secondary structure prediction (e.g., the work of Wootton [66] on glutamate dehydrogenase) or even secondary and tertiary algorithms (e.g., the prediction of thy-1 [67] see Section 3.6.6).

There are several computer methods to detect relationships between sequences. Dayhoff [68] has an extensive data bank of protein sequences (which is available in a book or on magnetic tape) and she also provides a service to compare any given sequence with those in her data bank.

3.3. *Simplified energy calculations*

At present it would be impossible to predict the three-dimensional structure of a protein by the minimisation of the free-energy of an all-atom representation. Apart from the problem of accurately modelling the atomic interactions, the number of degrees of freedom make the search for the native conformation impracticable. An attempt to simplify the energy surface was introduced by Levitt and Warshel [34,59]. They used a simplified geometry in which each side chain was modelled by a single sphere of suitable radius and placed at a fixed distance from the C^{α} atom (Fig. 4). Rough energy functions modelled the interactions between these side chains. The two main-chain dihedral angles (ϕ and ψ) were reduced to one variable C^{α}–C^{α} torsion angle. In the simulation they started with the sequence of pancreatic trypsin inhibitor in an extended conformation (sometimes the C-terminal α-helix was preset). A minimum energy conformation was found that resembled the native with the α-helix packing against a twisted β-sheet – the r.m.s. deviation between the predicted and crystal structure was 12.3 Å (10.4 Å with the preset α-helix). However, later work showed that the random r.m.s. for this prediction is

```
Sequence                        1 1 1 1 1 1 1 1 1 1 2 2 2 2 2 2 2 2 2 2
                1 2 3 4 5 6 7 8 9 0 1 2 3 4 5 6 7 8 9 0 1 2 3 4 5 6 7 8 9

                C D L P E T H S L D N R R T L M L L A Q M S R I S P S S C
Chou & Fasman

  α-helix       i I H B H i I i H I b[i i i H H H H H h H i i h i]B i i I

  B-strand      h B h B B h i b h B i i i[h h h h h]i[h h b i H]b B b b H

  turn              T           T T T                           T T T

  result        c c t t t t c c t t t α α α α α α α α α α α α α α t t t
Robson

  result        t c c α α α α c c t t α α α α α α α α α B B t t c t t t β
Lim

  type          H g H h G g G g H g g G G g H H H H H G H g G H g h g g H

  result        c c c c c c c c c c c α α α α α α α α α α α α α c c c c c
```

Fig. 5. An example of the prediction of secondary structure. The secondary structure of first 29 residues of leukocyte interferon is predicted by three methods – Chou and Fasman's, Robson and his co-workers, and Lim. See Sections 3.5.2 to 3.5.4 for details.

[75] have been described [76,77] and I will refer to this later work. From analysis of 29 proteins of known sequence and structure the observed propensity for a residue type to be in an α-helical, β-strand or turn conformation ($P\alpha$, $P\beta$, Pt) is obtained. The $P\alpha$ and $P\beta$ values are rank ordered and grouped so with decreasing $P\alpha$ or $P\beta$ there are strong α- or β-formers (denoted H), ordinary formers (h), weak formers (I), indifferent (i), weak breakers (b), and strong breakers (B). Clusters of formers and the absence of many breakers for the α- or β-structures are located and form the nucleus for the structure. Rules to delineate the boundary of potential structures are also given in terms of breakers and the average conformational propensities $\langle P\alpha\rangle$ and $\langle P\beta\rangle$, for α-or β-boundaries respectively.

Tetrapeptides that define turns are predicted both from the average turn propensity $\langle Pt\rangle$ and from a turn probability pt. Although these rules are easy to follow, and can be performed without a computer, there are several difficulties in applying their method. The exact boundary of a structure is difficult to predict as there are only guidelines and not definitions to decide whether to include in a structure a residue that is in a tetrapeptide breaker. The most serious problem occurs when a segment includes a potential α-helix and a potential β-strand that overlap. Although one is guided by the relative values of the $\langle P\alpha\rangle$ and $\langle P\beta\rangle$, one must also consider the length of the segments and the residue types that form the boundaries of the structures. No method is given to quantify the relative importance of these considerations. Despite these problems, the method is fairly easy to use and the authors quote an accuracy of $\%N = 77\%$ for the prediction of α-helices and β-strands.

Figure 5 illustrates the application of their method to the first 29 residues of leukocyte interferon [82,83]. In the prediction, the first step is to assign the turns. The first residue of each tetrapeptide than can form a turn is denoted by T. They have

$\langle Pt \rangle$ greater than $\langle P\alpha \rangle$ and $\langle P\beta \rangle$ and $pt > 1.0 \times 10^{-4}$. In two regions (the tetrapeptides starting at residues 9, 10, 11 and 25, 26, 27) there are overlaps of possible β-turns and only the tetrapeptide with the highest pt is taken. The central two residues of each β-turn can act as breakers for α-helices and β-strands whilst the first and last residue can be incorporated into an α-helix or a β-strand. The potential α-helical region is residues 12–25 with a $\langle P\alpha \rangle$ of 1.11 and a lower $\langle P\beta \rangle$ of 1.07. The inclusion of N_{11} into the turn delineates the N-terminal boundary and the C-terminal is defined, as P_{26} cannot be included at the C-end of an α-helix. Similar arguments delineate the two potential β-strand regions, residues 14–18 ($\langle P\alpha \rangle = 1.18$, $\langle P\beta \rangle = 1.23$) and residues 19–24 ($\langle P\alpha \rangle = 1.08$, $\langle P\beta \rangle = 1.09$). In the final prediction the α-helical segment is chosen, as for residues 12–25 $\langle P\alpha \rangle$ is greater than $\langle P\beta \rangle$.

3.5.3 *The method of Robson and co-workers*

Over the last 10 years Robson and his co-workers [78,79] have developed an empirical method of predicting secondary structure. In contrast to the method of Chou and Fasman [75–77], the principles behind their approach, being based on information theory, are hard to follow but the algorithm is very easy to use and can be programmed unambiguously. Robson's method is fully automatic and I strongly recommend its use.

The likelihood $L(\alpha,j)$ of a residue at position j in the sequence adopting an α-helical conformation is calculated from the location of residue types in a segment $j - 8$ to $j + 8$ by the formula

$$L(\alpha,j) = \sum_{m=-8}^{+8} I(\alpha,R_{j+m}).$$

The term $I(\alpha, R_{j+m})$ represents the effect of the residue type at position $j + m$ on the conformation at position j. The value of $I(\alpha, R_{j+m})$ does not depend on the nature of the residue type at j except when $M = 0$. For example Pro is very rarely found after the first 3 residues in an α-helix and this is modelled by $I(\alpha, \mathrm{Pro}_{j+3})$ having a very low value of -100. The values of I for α-helical, β-strand, turn and coil conformations are derived from analysis of proteins with known sequence and structure (see Tables 1 to 4 of reference [79]).

With reference to the leukocyte interferon sequence in Fig. 5, the value of $L(\alpha, 21)$ that express the likelihood that M_{21} is α-helical is obtained from the sum:

$$\begin{aligned} L(\alpha, 21) &= I(\alpha, R_{21-8}) + I(\alpha, T_{21-7}) + \ldots \\ &\quad + I(\alpha, M_{21-0}) + \ldots + I(\alpha, C_{21+8}) \\ &= -10 + 0 + 10 + 30 + 20 + 25 + 50 - 10 \\ &\quad + 53 - 35 + 0 + 20 - 20 - 60 - 10 - 5 + 0 \\ &= 58. \end{aligned}$$

In a similar way the values for the other 3 conformations were calculated to be:

$$\begin{aligned} L(\beta, 21) &= 68 \\ L(t, 21) &= -18 \\ L(c, 21) &= -31. \end{aligned}$$

The predicted conformation for M_{21} is the state that has the largest L value, ie., β.

With this approach the authors quote an accuracy of $\%N = 50\%$ for the prediction of α, β, c and t states in 26 proteins.

The accuracy of the prediction generally can be enhanced if a correct assumption is made about the α-helical and β-strand composition of the proteins. This assumption may stem from circular dichroism results and/or may be suggested by sequence homology (see the example of thy-1 below [67]). To incorporate this preference a value is added to each $L(\alpha,j)$ and $L(\beta,j)$. For example if leukocyte interferon was believed to have more than 50% α-helix and less than 20% β-strand, then the modified values of L (denoted as L') for M_{21} would be

$$
\begin{aligned}
L'(\alpha, 21) &= L(\alpha, 21) + 100 = 158 \\
L'(\beta, 21) &= L(\beta, 21) + \ \ 50 = 118 \\
L'(t, 21) &= L(t, 21) \qquad\quad\ = -18 \\
L'(c, 21) &= L(c, 21) \qquad\quad\ = -31.
\end{aligned}
$$

The state with the maximum L' is the predicted conformation, i.e., α. A further improvement in the results can be obtained if homologous sequences are used when the L or L' values can be averaged before the decision about the predicted state is made.

It is clear that this algorithm can be easily programmed. Alternatively, a computer program can be obtained from Dr. J. Garnier (see Appendix).

3.5.4. The method of Lim

Lim's method [80–81] is based on a series of stereochemical rules for patterns of residue types in α-helices and β-strands. The rules consider both the size of a residue and whether it is hydrophobic or hydrophilic. This approach is in contrast to Chou and Fasman's and to Robson's that are based on empirical parameters.

The reported accuracy for the prediction of α-helices, β-strands and coil regions is $\%N = 81\%$. Lim's algorithm is based on a rather formidable set of rules but once mastered it can be performed by hand considering 100 residues in about 1 h. A computer program that implements many of the rules, has been written by Lenstra [84] and is available from Professor Drenth.

In outline, residues are classed as large or small hydrophobics (H_L or H_S denoted by H and h in Fig. 5), large or small hydrophilics (G_L or G_S denoted by G and g), and Gly is considered independently. For an α-helix one requires a hydrophobic face formed from overlapping pairs or triplets of H residues at $i, i+4$ or $i, i+1, i+4$, or $i, i+3, i+4$. In Fig. 5 two such triplets are L_{15}, M_{16}, A_{19} and L_{17}, L_{18}, M_{21} which overlap between L_{17} and A_{19}. There are additional allowed patterns at the ends of α-helices including H–H pairs at i and $i+3$ (e.g., M_{21}, I_{24}) and H–G pairs at i, $i+1$ or $i, i+3$, or $i, i+4$ (e.g., R_{12}, L_{15}). The segment formed by the overlapping of these pairs and triplets is a potential α-helical region that is allowed if it obeys further rules. For example certain *H–H* pairs at i, $i+4$ (e.g., Tyr, Phe) are disallowed as the hydrophobic face cannot pack against the surface of the rest of the protein. In the example in Fig. 5 the segment R_{12} to I_{24} is predicted α-helical.

β-Strands are located only in regions not predicted as α-helical. Three types of β-strands are considered: (i) β-Strands that are buried inside a protein as typically found in the β-sheets of α/β proteins. The pattern involves 2 or more sequential *H* residues. (ii) β-Strands that have one face exposed to solvent and the other buried as typically found in β-sandwiches. The pattern is *HGHGH* where *G* is G_L or G_S. (iii) β-Strands that are termed 'semi-surface' and are formed from a run of Ala and Gly residues.

3.5.5. *Applications of secondary structure prediction*

During the development of both empirical and physico-chemical methods the authors use knowledge of the known structures to obtain the rules for their algorithms. It is important, therefore, to test the methods on a protein whose structure was not known prior to prediction. There have been two such 'blind' tests – on adenyl kinase [85] and on T4 lysozyme [86]. For both proteins more than two thirds of the residues were assigned to the correct secondary structure, although for T4 lysozyme the prediction of β-strand regions was poor. In general there are errors in the delineation of the exact termini of a secondary structure.

In the test on adenyl kinase an unweighted average of all the individual predictions was made and this gave better agreement with the native structure than any individual prediction. An alternative approach to obtain a better prediction is less rigorous but may prove more effective. This is to interpret the results of a series of predictions in terms of the merits and limitations of the different schemes. For example, suppose two empirical schemes suggest a region to be a β-structure although an α-helical conformation was a possibility. If Lim's stereochemical rules strongly favoured an α-helical structure then I would suggest that this region should be predicted as α.

There have been many applications of secondary structure predictions on unknown structures, e.g., see [1,77]. In general the methods can predict the structural class of a protein. Prediction is particularly useful in suggesting that a protein belongs to a family of related proteins, e.g., the analysis of thy-1 [67] (Section 3.3.3) and Wootton's [66] work on glutamate dehydrogenase. However, all the methods were developed from a consideration of water soluble proteins and the application of these algorithms to membrane proteins might well yield totally incorrect results.

3.6. *The docking of α-helices and β-strands*

3.6.1. *The combinational approach*

One approach to predict protein structure from sequence is: (i) predict the location of the α-helices and β-strands; (ii) dock the secondary structures into a native-like tertiary fold; and (iii) refine by energy calculations the rough fold into a three-dimensional structure. In this section step (ii) will be considered.

During the development of the algorithms it is generally assumed that step i has been successful and the docking starts from the correct assignment of α-helices and β-strands.

There are several approaches to the docking. Simplified energy calculations (see Section 3.3) can be used starting with preset secondary structures. Warshel and Levitt [88] have simulated the folding of myogen by docking of the 6 α-helices. They obtained a final structure whose r.m.s. deviation from the native was ~ 12 Å, compared to the random result of 13 Å. Alternatively the α-helical and β-strand regions can be used as constraints in the approach based on distance geometry (Section 3.4). Here the combinational approach is described.

In this method all combinations of packed α-helices and/or β-strands are first generated by docking the structures according to stereochemical rules. One member of this set of structures must be close to the native fold but the total number of possible structures still must be a restricted set of all possible combinations. Then one removes from the list those structures that are disallowed by stereochemical or topological constraints. The ultimate aim of this approach is to remove all but the native-like structure, but at present one is content to obtain a small set of possible structures that one considers further. Often the rules for the initial docking and the subsequent filtering are obtained by examination of known structures.

The most suitable candidates for this approach are those proteins or domains whose fold involve the packing of α-helices and β-strands according to one of three well-defined motifs [45,89]: (i) the association of packed pairs or α-helices as found in the α/α protein myoglobin; (ii) the stacking of two predominantly antiparallel β-sheets to form a β-sandwich as found in the β/β structure of immunoglobulin domains; and (iii) the packing of α-helices against a predominantly parallel β-sheet as observed in the α/β fold of flavodoxin and the glycolytic enzymes.

3.6.2. The packing of α-helices

Probably the first application of a combinational-type approach was by Ptitsyn and Rashin [90] who predicted a tertiary fold for myoglobin. They started with the crystallographic assignment of α-helices and docked the α-helices to maximally bury hydrophobic residues. The search was performed manually by use of rigid cylinders to represent α-helices and fliexible coils to model the interhelical connections. One of the two final structures resembled the crystallographic fold of myoglobin.

Recently a computer algorithm [91,92] to predict α/α structures has been developed based on rules derived from analysis of observed α-helix packing [28,89,93]. Richmond and Richard's examination [28] of myoglobin was based on the change in solvent accessible area that occurred when pairs of isolated α-helices pack into their native association. The value of this area change provided an assessment of the hydrophobic contribution to the free energy of α-helix association and accordingly they were able to select the principal dockings. For each α/α association, there was a central residue on each α-helix that points towards the other α-helix and that is in the middle of the patch of residues that have an accessible area change. The α/α pairings were classified into three types [28,91] (Fig. 6) each with a particular interhelical distance and dihedral angle (d,θ); pattern of hydrophobic residues that mediate the interaction; and chemical type of central residue (Fig. 6). Type I involves the close packing (d = 7.5 Å) of two glycine central residues at a typical acute dihedral angle θ of −80°. Type II is represented by d = 8.5 Å and θ acute = −60°. The packing in this interaction has been described by Chothia et al. [89,93] in terms ±4 of ridges and grooves. The surface of the α-helix is considered as ±4 ridges formed by a row of residues at positions i, i + 4,

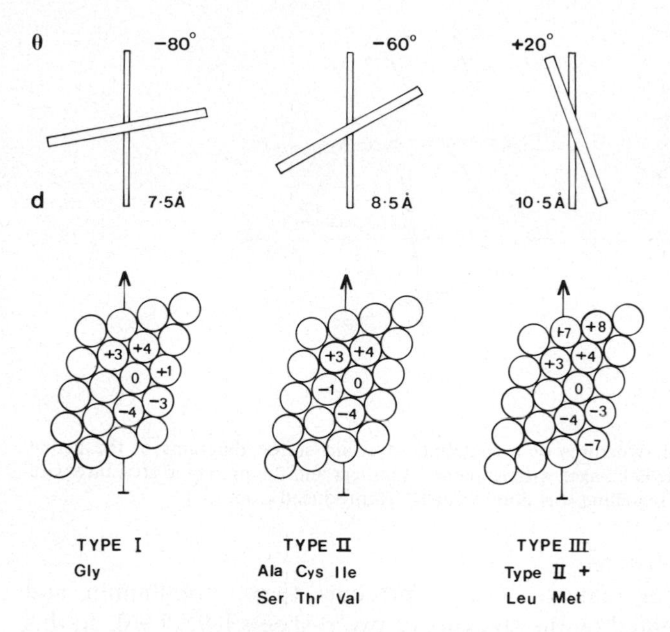

Fig. 6. The three classes of α-helix/α-helix packing. Above the dihedral angle (θ) and the separation (d) between a pair of α-helix axes are given. Below the residues on the surface of an α-helix are represented by circles and a typical pattern of residues that mediate the α/α pairing is shown for each class. The allowed central residues for each interaction are described.

$i + 8$ ... and an adjacent row at $i + 1, i + 5, i + 9$. Between these rows lies the ±4 groove. The ±4 ridges of one α-helix packs into the ±4 grooves of the other. Type III involves an approximately parallel (θ acute = +20°) alignment of two α-helices with a somewhat larger interhelical distance (d = 10.5 Å) than the other two types.

The first step to predict the tertiary fold from the sequence and the assignment of α-helices is to locate potential central residues based on the presence of a cluster of non-polar residues surrounding a residue of suitable chemical type. Richmond and Richard's algorithm [28] located all the central residues that are used in the principal dockings but also some that are not used. Next Cohen et al. [91] uses this list of central residues to generate $\sim 3 \times 10^8$ possible folds for the association of every combination of paired α-helices. Two constraints were imposed on the generated structures: (1) there had to be sufficient residues in a connection to span the distance; (2) there were to be few close contacts.

These constraints reduced the number of possible structure to 20. Finally Cohen and Sternberg [92] used two constraints on allowed Cα–Cα distances to model the binding of the haem group. These constraints were typical of those that would be obtained from solution studies. Two possible structures were thereby obtained, both resembled the myoglobin fold and one had the correct α-helix pairings (Figure 7) (r.m.s. deviation = 4.5 Å, random r.m.s. = 16 Å). Subsequently [94], this approach was applied to the docking of the 4 α-helices in the protein disc of tobacco mosaic virus protein. A reduced list of 62 structure was obtained one of which resembled the native with an r.m.s. deviation of 4.5 Å.

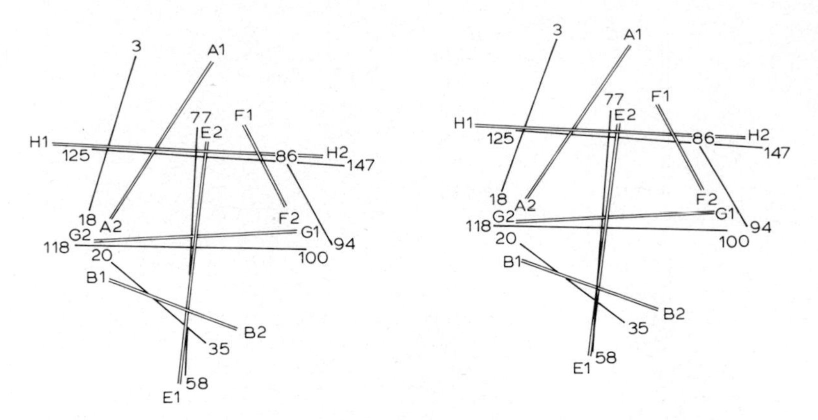

Fig. 7. The predicted and crystal structures of myoglobin. Overlaid stereo diagrams of the actual myoglobin structure identified as α-helix axes with sequence numbers and the predicted structure with the α-helix axes with letters. The matching was done visually. Reproduced from [91].

3.6.3. The stacking of two β-sheets

A major tertiary structural feature of all-β proteins (e.g., prealbumin and immunoglobulin) is represented by the stacking of two β-sheets [89,95,96]. In this fold, the polypeptide chain migrates between the two β-sheets that shield the hydrophobic core from solvent.

Cohen et al. [96,96] have analysed the packing in 10 such β-sandwiches. A standard geometry is observed: each β-sheet is formed from β-strands separated by ~ 4.5 Å and with an interstrand dihedral angle of −20°; the two sheets are separated by ~10 Å and there is an (anticlockwise) intersheet dihedral angle of ~ −30°. Details of which residues mediate the sheet/sheet interaction were obtained from the changes in solvent accessible area between the two isolated sheets and the pair stacked in its native conformation. As shown in Fig. 8 many, but not all of the in-pointing residues have an area change. There is a distinct pattern for this area change: in the top sheet it runs from the top left to bottom right corner whilst in the lower sheet it has an anticomplementary direction from the top right to bottom left corner. This anticomplementary pattern results from two effects of the twisted nature of β-sheets. First each β-sheet has two raised (U) and two lowered (D) corners, so the raised corners of the upper sheet and the lowered corners of the lower sheet do not interact (see Fig. 8). Second the positions of the side chains rotate clockwise by ~20° per residue as one progresses along the strand direction. The side chains that point out towards the solvent and away from the other sheet do not mediate the β/β interaction.

These notions formed the basis of a computer algorithm [95] to predict the tertiary structure of β-sandwiches by a combinational approach that starts from the sequence and the crystallographic assignment of β-strands. In outline, first all possible diagrams for the sheet structure are generated by permitting the segregation of strands to sheets, the position and direction of strands within a sheet, and the relative phasings of the β-strands (i.e., the hydrogen-bonding pattern). Then a reduced list of possible structures is obtained from a set of rules including

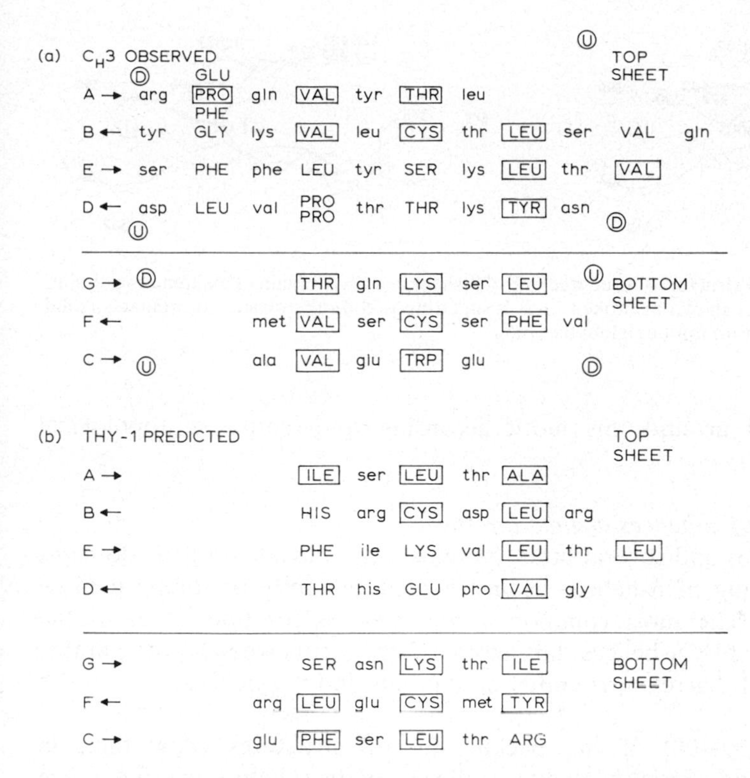

Fig. 8. The observed and predicted hydrophobic interactions that occur when two β-sheets stack. (a) C_{H3} domain of human Fc fragment [145]. The β-strands are lettered sequentially A to G and run horizontally to form two β-sheets ABED and GFC. In strands A, B and D Glu–Pro, Phe–Gly and Pro–Pro form β-bulges [146]. Residues vertically below one another in the same sheet can form interaction hydrogen bonds. Residues in capitals are in-facing (i.e., point towards the other sheet). Boxed residues are observed to have a change in non-polar contact area of > 5Å^2 on sheet stacking [95]. To model the stacking of the sheets, the top sheet diagram is slid down over the lower sheet diagram. U and D denote the raised and lowered corners of the twisted β-sheet. Note that the raised corners of the upper sheet do not interact with the lower sheet. (b) A predicted sheet diagram for thy-1 [67]. The non-polar residues that are suggested to mediate the sheet/sheet interaction are boxed – they have same anticomplementary direction as the boxed residues in C_{H3}. Figures a and b are adapted from [67].

requirements on: topology (such as the handedness of the connection between parallel β-strands); the location of non-polar residues with the correct anticomplementary pattern; and, if applicable, a sufficiently small distance between the Cys residues that form a disulphide bridge. In trials of 9 known β-sandwiches some 10^8 sheets were generated and a native-like structure was always high (top 6 to top 3,300) in the list of allowed structures rank ordered on the number of hydrogen bonds. The native-like structure was close to the crystallographic conformation with an r.m.s. deviation of 1.4–5.1 Å compared to a random result of ~12 Å (see Fig. 9).

A related combinational approach has been developed by Ptitsyn and co-workers [97,98] to predict a set of allowed topologies (i.e., strand position but not residue phasing) for β-sandwiches. Their algorithm starts with a motif [47], 'the Greek-key', that consists of 4 β-strands and is found in all β-sandwiches. Additional

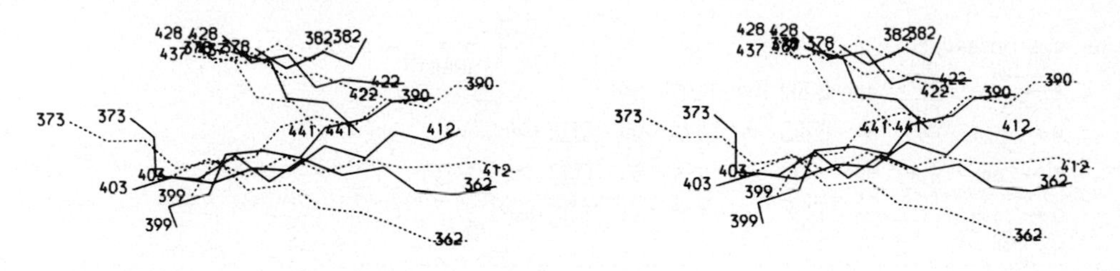

Fig. 9. Predicted and crystal structures of the stacking of β-sheet in a C_{H3} domain of an immunoglobulin. The predicted structure [95] shown in broken lines is superimposed on the crystal co-ordinates (solid lines) of the C_{H3} domain of an immunoglobulin [145].

β-strands are placed around this motif according to a series of topological restrictions.

3.6.4. *The packing of α-helices against a β-sheet*

The glycolytic enzymes and several other proteins, e.g., flavodoxin have domains that reflect the packing of α-helices against a predominantly or totally parallel β-sheet [89,99–102]. The most common arrangement is for both sides of the sheet to be surrounded by α-helices, but another form occurs when a pure parallel sheet forms a closed β-barrel surrounded on the outside by α-helices, e.g., triose phosphate isomerase.

Examinations [89,99,100] of the packing in α/β structures show that, in general, a standard geometry is observed with the axis of the α-helix lying 10 Å above and parallel (0° or 180°) to the β-strand direction (Fig. 10). The residues in the α-helix that mediate the interaction are predominantly non-polar and tend to lie along two adjacent $i \pm 4$ rows at positions $i, i + 1, i + 4, i + 5, i + 8, i + 9$. Of these residues 4 are particularly important and those lie at positions $i + 1, i + 4, i + 5$ and $i + 8$ forming a ‘diamond shape’ on the α-helix net (Fig. 10). The residues on the β-sheet that interact with the α-helix can lie on between 1 to 5 β-strands. These residues trace a distinct pattern on the sheet surface that results from the twisted nature of α-sheets. If the α-helix lies above the β-sheet, the lowered corners of the top surface of the β-sheet will not interact with the α-helix. Furthermore the side chains of these lowered corners will point towards the solvent and away from the α-helix and thereby will not interact. The concerted effect of several α-helices packing against the β-sheet is that there is a section of non-polar side chains on one or both sides of the β-sheet that is an extension of the pattern observed for the interaction with one α-helix. At a more detailed level it was found that the 4 α-helical residues that form the diamond shape surround one residue on the β-sheet that often is either Ile, Leu or Val. The steric clash that would be expected from this model is alleviated either by shortening the length of one of the interacting secondary structures, or by the location of a residue with a small volume (e.g., Gly or Ala) on either the α-helix or the β-sheet at the position of the possible clash, or by the α-helix lying with a large vertical angle to the β-strand.

In addition to these packing details, there are several [47,52–54,95,99,100]

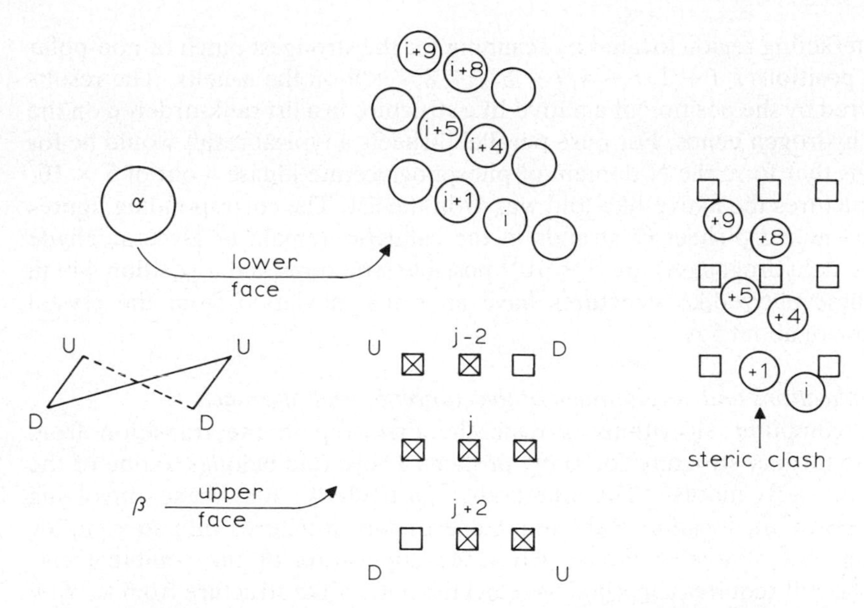

Fig. 10. A model for the packing of an α-helix against a β-sheet. (a) The lower surface of an α-helix packing against the upper face of a twisted β-sheet. U and D denote the raised and lowered corners of the sheet. (b) The position of the α-helical residues indicating the six residues (i, i+1, i+4, i+5, i+8, i+9) that generally interact with the β-sheet. The up-packing residues on 3 β-strands (running vertically) are shown. Those with a cross tend to interact with the α-helix. (c) The packing of the 6 α-helical residues on to the β-sheet with the position of a potential steric clash between one α- and one β-residues shown.

topological restrictions on α/β structures, particularly for those with pure parallel β-sheets:

(i) In all β-sheets, the connection between 2 parallel β-strands nearly always is right-handed.
(ii) In pure parallel β-sheets that do not form β-barrels (i.e., 'planar' sheets), there is only one chain reversal for the strand order. Thus with sequential strands lettered alphabetically, EDCAB is an allowed topology but DECAB is not.
(iii) The location of α-helices about a pure parallel β-sheet either shields both sides of a 'planar' sheet from solvent or lies on the outside of a β-barrel.

The application of these rules markedly restricts the number of possible topologies for the β-strands and their connections. For example there are 1×10^4 possible topologies for the 6-stranded sheet in the N-terminal domain of phosphoglycerate kinase but only 10 are allowed by these rules.

These topological restrictions provide the first set of filters in the combinational approach to predict the fold of α/β proteins with pure parallel β-sheets. A rather less restrictive set of topological constraints can be applied to mixed β-sheets. For each allowed topology one next generates all possible strand alignments and selects only those that give a pattern of non-polar residues compatible with the docking of α-helices. Finally α-helices are placed on the sheet with the standard geometry and with the interacting region located by scanning for the strongest patch of non-polar residues at positions i, $i + 1$, $i + 4$, $i + 5$, $i + 8$, $i + 9$ on the α-helix. The results are considered by the position of a native-like structure in a list rank-ordered on the number of hydrogen bends. For pure parallel β-sheets a typical result would be for the 6 strands that form the N-domain of phosphoglycerate kinase – out of 5×10^7 possible structures the native-like fold was 47 in the list. The corresponding figures for a typical mixed β-sheet (7 strands in the catalytic domain of glyceraldehyde 3-phosphate dehydrogenase) are 5×10^{11} possible structures and a position 649 in the list. These native-like structures have an r.m.s. deviation from the crystal conformation of about 5 Å.

3.6.5. Applications and development of the combinatorial approach

The above computer algorithms provide the first step in the transition from secondary to tertiary structure for many proteins whose fold belongs to one of the α/α, β/β or α/β motifs. The methods, particularly for those involving β-strands, rely on the location of the regular secondary structures with an accuracy greater than that attainable today. After the application of the combinatorial approach one still requires a method to select the native-like structure from the few other alternatives and a possible approach would be some adaptation of simplified energy calculations. Finally one envisages refinement of the predicted fold initially by simplified energy calculations [34,69,70] and then by all-atom calculations (see Section 4.1).

It is clear, therefore, that much work is needed before this approach can be applied in general to predict unknown structures. Indeed, although the algorithms are available from Dr. F. E. Cohen and myself, the programs were written to test the approach rather than to predict unknown structures and we expect that their use by others will, at present, be difficult. All the programs were written in FORTRAN and have been implemented on the ICL 2980 at Oxford University. Because the method involves the generation of every combination of docked secondary structures, the c.p.u. time increases rapidly as more units are docked. Typical c.p.u. times on an IBM 360/195 would range from 3 min to 3 h for structures such as myoglobin, an immunoglobulin domain, and flavodoxin. However, with a complete understanding of the combinational algorithm for certain proteins subjective judgments can be made to try and circumvent the problems of inaccurate secondary structure prediction and of the selection of the native-like structure from the list of alternatives. This method was used to suggest tertiary folds for the unknown structures of a cell surface antigen thy-1 [67], β_2-microglobulin [95], a fragment of the HLA-B7 histocompatibility [95] antigen, and interferon [83].

3.6.6. The prediction of a tertiary fold for thy-1

Rat thy-1 [67, 103] antigen, a polypeptide of 111 amino acid residues, is a major membrane molecule of thymocytes and of brain with no known function. The CD spectrum of thy-1 suggests there is extensive β-strand structure [103]. Furthermore, there are regions of thy-1 that have sequence resemblances with immunoglubulin (Ig) domains. In particular one of the two disulphide bridges in thy-1 seems homologous to the conserved disulphide in all Ig domains. We have, therefore, investigated whether thy-1 might have an Ig like fold – a stacked pair of β-sheets.

First, the method of Robson and co-workers [79] (see Section 3.5.3) was used to locate possible β-strand regions. One major advantage of this method was that one can bias the prediction towards an all-β protein. In particular the cut offs used were chosen on their accurate location of the β-strands in Ig domains. The result of the prediction was a series of β-strand regions suggestive of an Ig fold. Next, by use of both sequence resemblances between thy-1 and Ig domains and the stereochemical notations incorporated into the β/β algorithm, a possible hydrogen-bonding diagram for the β/β fold of thy-1 was established (see Fig. 8). In this predicted structure for thy-1 the in-pointing non-polar residues of the stacked β-sheets trace a similar anti-complementary pattern to that observed in Ig domains. Finally, the validity of this proposal for thy-1 was tested by the combinational algorithm and the predicted structure was 8 in a list rank ordered on the number of hydrogen bonds. A similar approach was used to predict Ig-like folds for β_2-microglobulin [95] and a fragment of the HLA-B7 histocompatibility antigen [95].

3.6.7. The prediction of a tertiary fold for interferon

Interferon has recently attracted much attention because of its possible medical role as an anti-viral and anti-tumor agent, e.g. [104]. Functional studies would be helped by a knowledge of the three-dimensional structure of this molecule. However, at present no crystal structure has been reported and Dr. Cohen and I have used the algorithms we have developed to try and predict a teritary model from the known amino acid sequences.

The sequences from fibroblast [105], leukocyte [82,106] and lymphoblastoid [107] interferons were considered. There is sufficient homology between these sequences to infer that they will adopt a common fold. First the secondary structure was predicted by the methods of Chou and Fasman [77], Robson and co-workers[79], and Lim [81]. There was considerable variation of the results both between the different sequences and also between the different methods on the same sequences. However, one expects that α-helices that will be important in the tertiary fold will have non-polar residues at positions i, $i + 4$ that are conserved between the different sequences. In addition one has ideas about the merits and limitations of the different predictive schemes. Together these ideas enabled us to interpret the secondary prediction and to suggest that there will be four α-helices (A,B,C,D) and possibly two short β-strands.

A probable structure for the α-helices A,B,C,D is a four-fold bundle (Fig. 11) that has been observed in several globular proteins [108,109]. Whether such a structure was feasible was investigated by the α/α docking algorithm (see Section 3.6.2). A total of 716 possible dockings for the four-fold bundle was generated, but many of the structures were similar. This list was then reduced to 2 by the requirements to form a compact structure in which polar side-chain atoms were not buried. These requirements were examined by placing a sphere of radius 2.5 Å centred 3 Å from the helix axis and then calculating the solvent accessible area (see

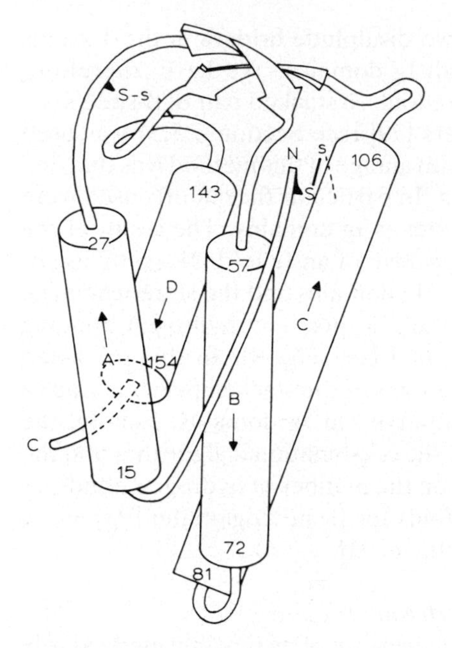

Fig. 11. A predicted structure for interferon. The pack of the four α-helices (A,B,C,D) in the postulated [83] structure of interferon. The numbering is from fibroblast [105–107] interferon and the disulphide linkages [101] are for leukocyte interferon. Reproduced from [83] with permission.

Section 2.2). Finally, a consideration of the recently determined [110] disulphide bridges provided constraints on intramolecular Cα–Cα distances that suggested one of the two structures is more likely.

This proposed structure is shown in Fig. 11. The sequence numbering is of fibroblast interferon but the disulphide bridges are those in leukocyte interferon. Certain residues are suggested as being functionally important based on sequence homology and spatial proximity in the tertiary fold.

In this work, several subjective judgements were made. This model was proposed *not* as a test of the algorithms but as a possible tertiary structure that can provide a basis for experimental work on the conformation and function of this molecule.

4. STUDIES BASED ON A KNOWN PROTEIN STRUCTURE

An important reason for the determination of a protein structure is that a wide range of studies are made possible once the atomic co-ordinates of the protein have been established. In this Section some calculations on the function, evolution, dynamics and folding pathways of proteins will be described.

4.1. Protein function

One of the major motivations for the determination of the three-dimensional structure of a protein is that it can lead to an understanding at the atomic level of the function of the molecule. Accordingly there have been a wide range of computer studies to model aspects of protein function.

This field is well suited to the calculation of conformational energies as many of the inadequacies in the description of interatomic interactions might well not markedly affect the results. Scheraga and his co-workers [111] have developed two packages for energy calculations that are available from the Quantum Chemistry Program Exchange. In one version all atoms including hydrogens are treated explicitly (ECEPP) whereas in the other [112] aliphatic and aromatic CH_3, CH_2 and CH groups are represented by united atoms (UNICEPP) thereby making the calculations faster. Using these packages, Scheraga and co-workers have studied various aspects of the binding of saccharide molecules to hen egg-white lysozyme. The earlier studies in which the conformation of the enzyme was held rigid have been extended to allow changes in the side-chain conformations of the protein. Warshel and Levitt [116] have also studied this enzyme–substrate complex using a combination of conformational and quantum-mechanical energy parameters. They found that electrostatic stabilization is an important factor in increasing the rate of the reaction step that leads to the formation of the carbonium ion intermediate. Steric factors, such as the strain of the substrate on binding to lysozyme, do not seem to contribute significantly in their calculations.

North and his co-workers [117] have investigated the attack by oxygen at each of the four methylene bridges in the haem of globin molecules. The energy calculation indicated the probability of finding one oxygen atom (O_B) of the bound molecule (O_A–O_B) at any orientation about the haem normal, i.e., the rotation about the Fe–O_A bond. The probabilities agree well with the experimentally observed proportions of the four isomers.

Other workers such as Hermans and Ferro [118], Levitt [119] have developed packages for energy calculations that can be applied to similar types of calculations. Gurd [120,121] and his group are pursuing a more restricted type of calculation in which the electrostatic energy is evaluated to model the pH dependence of the ionization of various residues in the globins.

The results of some of these studies, for example that of North and his co-workers [117], yield good agreement with experiment. This suggests that when energy calculations are applied to a well-defined system that has only a few degrees of freedom the results can provide reliable information about probable structure/function relationships.

4.2. *Solvent accessibility*

The residues of a globular protein that will interact with another large molecule must, of course, lie on the surface and can therefore be readily identified from a calculation of solvent accessibility (see Section 2.2). This obvious observation has recently been used [122] to propose a site on the surface of the C_{H2} domain of immunoglobulin G (IgG) that binds to a component of serum complement (C1q). The requirement for the receptor site is that it should consist of exposed residues that are strongly conserved in all C_{H2} domains but not strongly conserved in other constant domains.

The solvent accessibility was calculated from the crystal co-ordinates of the Fc fragment (C_{H2} and C_{H3}) domains of human IgG. Nine amino acid sequences for C_{H2} domains were used. One possible site would consist of a patch of hydrophobic residues but it was found that none of the observed patches were both accessible to C1q and formed from residues conserved in all C_{H2} domains. Instead a suitable patch consisting of several charged side chains was located on the surface of the last two β-strands of the C_{H2} domain and this was proposed as the C1q binding site. This conclusion was consistent with experimental work on chemical modifications.

There should be a variety of other biochemical problems such as the identification of antigenic sites that could be tackled by this simple approach.

4.3. Evolutionary relationships by the comparisons of tertiary structures

Ever since the early crystallographic studies on the globin [123] family and on the serine proteinases [124,125], it has been recognised that the three-dimensional structure of a protein evolves far more slowly than its amino acid sequence. The corollary of this is that distant evolutionary relationships resulting from divergence, both between different proteins and within one protein, might be detected from similarities in the tertiary structures even if undetectable from the amino acid sequences alone. Of course the alternative explanation of conformational similarities is that the proteins or the substructures within a protein (which must be larger than an individual α-helix or β-strand) have converged to a stereochemically and/or functionally favourable fold.

The problem is to assess the extent of the structural similarities. This can be expressed as the r.m.s. deviation between equivalence positions, generally C_α atoms, when one structure is optimally superimposed on the other (see Section 3.1). One wishes to equivalence substantial parts of the two structures whilst preserving a low r.m.s. deviation. Rossmann and Argos [126,127] have developed a computer program that performs an exhaustive search for the best set of equivalences. Other programs have been written by McLachlan [128] and by Remmington and Matthews [129].

Given a number of equivalences and the r.m.s. deviation between superimposed C_α atoms, the problem remains as to whether these values represent a significant structural similarity. Rossmann and Argos [126,127] have tackled this problem by evaluating the number of standard deviations that the best r.m.s. deviation is from the mean background of all r.m.s. deviations. Schulz [59] used an analogous approach to that employed to evaluate the significance of obtaining a given r.m.s. deviation in a structure prediction (see Section 3.1). He generated a series of random structures that were compact and self-avoiding structures and thereby obtained a distribution for the r.m.s. deviations of these from an actual protein structure. Both these approaches suffer from the drawback that there are major restrictions on the possible chain folds of protein that result in part from the packing of secondary structures (see Sections 2.4 and 3.6). Thus no protein structure is random and seemingly significant similarities in conformation may well result from convergence to a stable fold. In the long-term the best solution to the problem of assessing structural similarities will probably come from a detailed knowledge of the relationship between protein sequence and conformation.

4.4. *Protein mobility*

A variety of experimental techniques such as hydrogen isotope exchange and fluorescence quenching (for a review see [130]) has shown that proteins are not rigid structures but have considerable mobility. Recently two approaches have been used to provide a description of protein mobility at the atomic level – the interpretation of the thermal parameters [38,132–134] from crystal structure analysis and molecular dynamics calculations [135–137].

The spread of the electron density of an atom about its mean position is modelled in X-ray diffraction by a temperature factor or *B*-value [38,131]. *B* is related to the mean square amplitude of harmonic displacement along a given direction $\overline{U^2}$ by

$$B = 8\pi^2\overline{U^2}.$$

If isotropic displacement is assumed, the *B*-value has a single value; however, anisotropic displacement is modelled by decomposing *B* into 6 parameters whose values represent the axial lengths and orientation of an ellipsoid of displacement.

Over the last few years, the techniques of crystallographic refinement have been applied to protein molecules to yield *B*-values, generally isotropic, for individual or groups of atoms [39,132,133]. There are several problems in converting these *B*-values into true atomic motion. X-Ray measurements average both over time and crystal space and any motion could arise from intermolecular or intramolecular mobility. Furthermore the displacement of an atom from its mean position could equally well result from static disorder caused by different molecules in the crystal adopting different conformations or occupying different positions and orientations in the various unit cells. Finally it is well known that experimental errors such as the adsorption of X-rays by the crystal often lead to large variation in the *B*-values rather than in the mean atomic position.

Recently several studies have shown that to some extent information about protein mobility can be derived from *B*-values. Artymiuk et al. [39] have examined the *B*-values for two different crystal structures of lysozyme (hen egg-white and human-lysozymes). The two structures have quite different crystal packing and the data collection and processing were by different methods. A plot of the average main-chain *B*-value along the sequence found a broadly similar pattern for the two lysozymes. Thus these *B*-values probably reflect to a large extent intramolecular variations, both static and dynamic, from the mean position rather than the effects of crystal packing or experimental errors. Examination of the *B*-value plot showed that segments of regular secondary structure, both α-helices and β-strands, correspond quite closely to the relative smooth minima in the $\overline{U^2}$ distribution. The most mobile parts of the main chain occur over the lips of the active site suggesting that protein mobility might well be important in the catalytic mechanism of lysozyme. When the B-values for the side-chain atoms were examined, the sensible result was obtained that the exposed side chains had far more conformational variability than the side chains that were buried.

This study was performed at room temperature and it was not possible to distinguish between true motion and static variability. Frauenfelder et al. [133] have tackled this problem by obtaining B-values for metmyoglobin at a range of temperatures from 220°K to 300°K. The main conclusion was that the atomic displacements primarily result from the fluctuation of atoms between different conformational substates. The potential well $V(x)$ in which these substates lie is for most of the residues non-harmonic, i.e., does not obey the relationship

$$V(x) \propto x^2.$$

The well is either steep or has a broad square shape. However such examinations of the shape of the well are complicated by the fact that harmonic motion is assumed in the calculation of B-values from the diffraction data.

I envisage that these crystallographic results on probable motion can be used to interpret the finding of a variety of solution studies. As the refinement of protein structures to sufficient accuracy to obtain meaningful B-values has been possible only over the last few years, anyone interested in the B-values for a protein should write directly to the laboratory where the structure was refined.

A theoretical approach to model protein mobility has been developed by Karplus and his co-workers [135–137] using the technique of molecular dynamics calculations. In this approach the force on each atom is evaluated from potential functions and the atomic co-ordinates. Newton's Second Law is applied to obtain the resultant change in atomic position over a time step δt. The co-ordinates are altered and the procedure repeated so as to obtain the trajectory of each atom. The time step δt is of the order of 10^{-15} s and the calculations typically extend over $\sim 10^{-11}$ s. Earlier studies were on pancreatic trypsin inhibitor [135] but the results of a recent 16 ps simulation on ferrocytochrome C [136–137] have been compared to thermal parameters obtained crystallographically. The results of the simulation give the variation a position vector $\mathbf{r}$ of an atom from its mean position. If an isotropic B-value is calculated from the crystallographic data, then

$$B = 8\pi^2 \frac{\overline{\mathbf{r}^2}}{3}$$

which should be compared to the relationship between B and $\overline{U^2}$:

$$B = 8\pi^2 \overline{U^2}.$$

The difference [131] of a factor of three arises as $\overline{U^2}$ is the mean square displacement along a given direction, i.e., the mean square displacement of the projection of the position vector of the atom $\mathbf{r}$ on to this direction. Karplus and his co-workers found that about half of the observed B-factor value is accounted for by the mobility indicated from the molecular-dynamics simulation. Plots of motion along the main chain give similar results for the crystallographic and theoretical

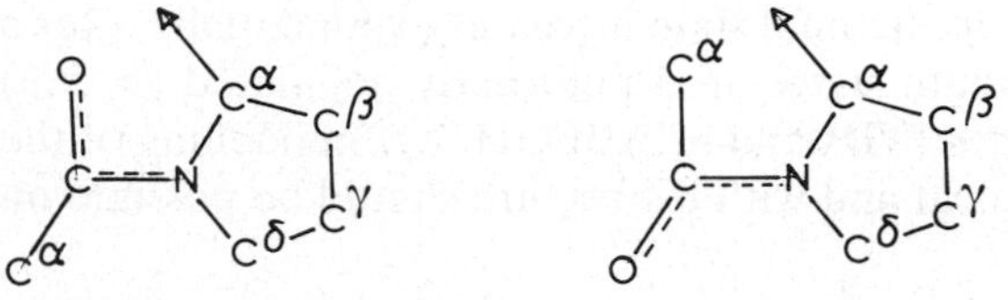

trans proline cis proline

Fig. 12. The *trans* and *cis* conformation of a proline residue.

studies. At present simulations by molecular dynamics for proteins are not routine calculations and I consider that it is a full-time effort to undertake such a study.

4.5. The kinetics of protein folding

There are two aspects to the modelling of the folding of a protein – the thermodynamic issue of predicting the final conformation and the kinetic question of characterising the pathway along which a protein folds. Various calculations that start with the sequence, and often the three-dimensional structure, of a protein have been applied to gain insight into the kinetic problem.

Recent experimental work [13–15] has shown that denaturated proteins can have slow and fast refolding species. The slow step has been attributed to the isomerisation of proline residues between the *cis* and the *trans* conformation (Fig. 12). However it is considered that only some of the prolines must acquire the correct conformer before the protein can fold into a native-like structure. The other prolines can initially be accommodated in the native-like structure with the wrong conformer and then later undergo isomerisation. To examine this suggestion, Levitt [138] has calculated the equilibrium energy of four conformations of bovine pancreatic trypsin inhibitor. In each conformation the molecule had roughly the conformation of the folded protein with the exception that one of the four prolines was set into the cis conformation. It was found that only one of the prolines (Pro 8) so destabilized the native conformation (by 35 kcal/mol) when in the cis isomer that it should block the rapid refolding of the molecule.

The probable rate-limiting step during fast refolding has also been considered. Karplus and Weaver [189] have performed rough calculations that suggest that the formation of a local structure by a random search is unlikely to act as a nucleus for subsequent refolding. Instead intermediate control is likely to be important in which the relative stabilities of the intermediates determine the pathway(s) and rate of folding. In their diffusion–collision–adhesion model local structures (e.g., α-helices) of transient stability diffuse together, collide and then pair. For the interaction of two α-helices or cluster(s) of α-helices, the time (τ) required for the interaction depends on the geometry of the interaction, the fraction of time that both segments are α-helical when they collide, and a diffusion coefficient. Recently Cohen et al. [140] have used the details of the packing of α-helices in the crystal structure of myoglobin obtained by Richmond and Richards [28] to calculate values for τ for all possible folding paths. Thus the rate and the pathway of folding can be calculated.

It was found that the concentration of the final state increases exponentially after a small (~0.1 ms) lag. Two intermediate states are significantly populated (> 5%) involving the associations of α-helices FGH and of ABFGH. The modelling of the folding of other proteins both α-helical and with β-structure should be possible on these lines.

Matheson and Scheraga [141] have developed a method that searches the amino acid sequence of a protein to locate probable nucleation sites on the basis of the formation of hydrophobic contacts. For bovine ribonuclease A the predicted site is a 13-residue pocket from Ile 106 to Val 118 which is consistent with experimental evidence such as the immunological identification of the site near the C-terminus (residue 124).

5. FUTURE DEVELOPMENTS

The central problem for many of the calculations on proteins is that our understanding of interatomic interactions is very limited. However, recently there has been advances in the refinement of the X-ray structures of proteins that is beginning to provide an accurate description of protein conformation at the level of hydrogen bond geometries, electrostatic interactions and the arrangement of the solvent structure around the protein (C.C.F. Blake, personal communication). It is to be hoped that this knowledge will lead to an improvement in the models of protein atomic interactions.

Ultimately any prediction of protein structure from sequence will involve the use of energy calculations (Section 4.1). However, in my opinion, the stepwise approach of first predicting the secondary structure (Section 3.5), then packing the α-helices and β-strands into a tertiary fold (Section 3.6), and finally refining the fold by energy calculations, is likely to be more successful than attempting to minimise the energy starting from an open-chain conformation. Central to this approach is the need to predict accurately the secondary structure from sequence. Many of the available algorithms were developed in the early 1970s and we now have a better understanding of the packing of α-helices and β-sheets. This stereochemical information on tertiary structure should be incorporated into secondary structure prediction. Indeed some suggestions on these lines have begun to be studied [142–144].

Despite the above problems, it is encouraging that in the review I have been able to cite a wide range of topics that are of interest to the general biochemical community for which answers have been provided by computer calculations. I hope that this chapter will help both the specialist and the non-specialist to select suitable algorithms to increase understanding of the diverse and complex properties of proteins.

ACKNOWLEDGEMENTS

I thank Sir David Phillips for his encouragement and advice during my research on protein structure. Many of the ideas expressed in this chapter arose from discussions with my colleagues – Dr. Janet Thornton, Dr. Fred Cohen and Mr. William Taylor. I am grateful to Professor Scheraga, Dr. Shoshana Wodak, Dr. J. Lenstra and Dr. Michael Geisow for information about computer programs. A Stothert Fellowship from the Royal Society provided financial support.

REFERENCES

1. Schulz, G.E. and Schirmer, R.H. (1979) Principles of Protein Structure, Springer-Verlag, New York.
2. Sternberg, M.J.E. and Thornton, J.M. (1978) Nature 271, 15–20.
3. Némethy, G.N. and Scheraga, H.A. (1978) Quat. Rev. Biophys. 10, 239–352.
4. Richardson, J.S. (1981) Adv. Protein. Chem. 34, 167–339.
5. Cohen, F.E. and Sternberg, M.J.E. (1981) In: R.C. Sheppard (Ed.) Amino-acids, Peptides and Proteins, Vol. 11, Chemical Society, London, pp. 233–240.
6. Cohen, F.E. and Sternberg, M.J.E. (1981). In R.C. Sheppard (Ed.) Amino-Acids, Peptides and Proteins, vol. 12, Chemical Society, London, pp. 255–262.
7. Anfinsen, C.B. (1973) Science, 181, 223–230.
8. Wetlaufer, D.B. and Ristow, S. (1973) Ann. Rev. Biochem. 42, 135–158.
9. Creighton, T.E. (1978) Prog. Biophys. Molec. Biol. 33, 231–297.
10. Levinthal, C. (1968) J. Chim. Phys. 65, 44–45.
11. Wetlaufer, D.B. (1973) Proc. Natl. Acad. Sci. USA 70, 697–701.
12. Creighton, T.E. (1979) J. Mol. Biol. 129, 235–264.
13. Brandts, J.F., Brennan, M. and Lin, L-N. (1977) Proc. Natl. Acad. Sci. USA 74, 4178–4181.
14. Baldwin, R.L. (1978) Trends Biochem. Sci. 3, 66–68.
15. Cook. K.H., Schmid, F.X. and Baldwin, R.L. (1979) Proc. Natl. Acad. Sci. USA 76, 6157–6161.
16. Chantrenne, H. (1961) Biosynthesis of Proteins, Pergamon, New York, p. 122.
17. Dunhill, P. (1965) Sci. Prog. 53, 609–619
18. Phillips, D.C. (1967) Proc. Natl. Acad. Sci. USA 57, 484–495.
19. Hamlin, J. and Zabin, I. (1972) Proc. Natl. Acad. Sci. USA, 69, 412–416.
20. Hopfinger, A.J. (1973) Conformational Properties of Macromolecules, Academic Press, New York.
21. Finney, J.L., Gellatly, B.J., Golton, I.C. and Goodfellow, J. (1980) Biophys. J. 32, 17–33.
22. Pain, R.H. (1978) In: F. Franks (Ed.) Characterisation of Protein Conformation and Function, Symposium Press, London, pp. 19-36.
23. Kauzmann, W. (1959) Adv. Prot. Chem. 14, 1–63.
24. Tanford, C. (1962) J. Am. Chem. Soc. 84, 4240–4247.
25. Tanford, C. (1962) J. Am. Chem. Soc. 84, 4240–4247.
25. Nozaki, Y. and Tanford, C. (1971) J. Biol. Chem. 246, 2211–2217.
26. Lee, B.K. and Richards, F.M. (1971) J. Mol. Biol. 55, 379–400.
27. Richards, F.M. (1977) Ann. Rev. Biophys. Bioeng. 6, 151–176.
28. Richards, T.J. and Richards, F. M. (1978) J. Mol. Biol. 119, 537–555.
29. Chothia, C. (1974) Nature 248, 338–339.
30. Karplus, M. (1980) Biophys. J. 32, 45–47.
31. Shrake, A. and Rupley, J.A. (1973) J. Mol. Biol. 79, 351–371.
32. Finney, J.L. (1978) J. Mol. Biol. 119, 415–441.
33. Wodak, S.J. and Janin, J. (1980) Proc. Natl. Acad. Sci. USA, 77, 1736–1740.
34. Levitt, M. (1976) J. Mol. Biol. 104, 59–107.
35. Bernstein, F.C., Koetzle, T., William, G.J.B., Meyer, E. Jr., Brice, M.D., Rodgers, J.R., Kennard, O., Shimanouchi, T. and Tasumi, M. (1977) J. Mol. Biol. 112, 535–542.
36. Feldmann, R.J. (1976) AMSOM – Atlas of Macromolecular Structure on Microfiche, Tracor Jitco Inc., Maryland.
37. Feldmann, R.J. and Bing, D.H. TAMS – Teaching Aids for Macromolecular Structure, Taylor-Merchant Co., New York.
38. Blundell, T.L. and Johnson, L.N. (1976) Protein Crystallography, Academic Press, New York.
39. Artmiuk, P.J., Blake, C.C.F., Grace, D.E.P., Oatley, S.J., Phillips, D.C. and Sternberg, M.J.E. (1979) Nature, 280, 563–568.
40. Richards, F.M. (1974) J. Mol. Biol. 82, 1–14.

41. Chothia, C. (1975) Nature 254, 304–308.
42. Rashin, A.A. (1981) Nature 291, 85–87.
43. Crippen, G.M. (1978) J. Mol. Biol. 126, 315–332.
44. Rose, G.D. (1979) J. Mol. Biol. 134, 447–470.
45. Levitt, M. and Chothia, C. (1976) Nature 261, 552–558.
46. Sternberg, M.J.E. and Thornton, J.M. (1977) J. Mol. Biol. 110, 285–296.
47. Richardson, J.S. (1977) Nature, 265, 495–500.
48. Chothia, C. (1973) J. Mol. Biol. 75, 295–302.
49. Pauling, L. and Corey, R.B. (1951) Proc. Natl. Acad. Sci. USA, 37, 729–740.
50. Weatherford, D.W. and Salamme, F.R. (1979) Proc. Natl. Acad. Sci. USA, 76, 12–23.
51. Nishikawa, K. and Scheraga, H.A. (1976) Macromolecules, 9, 395–407.
52. Sternberg, M.J.E. and Thornton J.M. (1976) J. Mol. Biol. 105, 367–483.
53. Sternberg, M.J.E. and Thornton, J.M. (1977) J. Mol. Biol. 110, 269–283.
54. Richardson, J.S. (1976) Proc. Natl. Acad. Sci. USA 73, 2619–2623.
55. Nagano, K. (1977) J. Mol. Biol. 109, 235–250 and 251–274.
56. Nyberg, S.C. (1974) Acta. Cryst. B30, 251–253.
57. Cohen, F.E. and Sternberg, M.J.E. (1980) J. Mol. Biol. 138, 321–333.
58. Hagler, A.T. and Honig, B. (1978) Proc. Natl. Acad. Sci. USA 75, 554–558.
59. Schulz, G.E. (1980) J. Mol. Biol. 138, 335–347.
60. Havel, T.F., Crippen, G.M. and Kuntz, I.D. (1979) Biopolymers 18, 73–81.
61. Browne, W.J., North, A.C.T., Phillips, D.C., Brew, K., Vanaman, T.C. and Hill, R.L. (1969) J. Mol. Biol. 42, 65–86.
62. Lewis, P.N. and Scheraga, H.A. (1971) Arch. Biochem. Biophys. 144, 576–583.
63. McLachlan, A.D. and Shotton, D.M. (1971) Nature 229, 202–205.
64. Delbaere, L.T.J., Brayer, G.D. and James, M.N.G. (1979) Nature 279, 165–168.
65. Kretsinger, R.H. and Barry C.D. (1975). Biochim. Biophys. Acta. 405, 40–52.
66. Wootton, J.C. (1974) Nature 252, 542–546.
67. Cohen, F.E., Novotny, J., Sternberg, M.J.E., Campbell, D.G. and Williams, A.F. (1981) Biochem. J. 195, 31–40.
68. Dayhoff, M.O. (1978) In: Atlas of Protein Sequence and Structure, Vol. 5, Suppl. 3, National Biomedical Research Foundation, Washington, D.C.
69. Levitt, M. and Warshel, A. (1975) Nature 253, 694–698.
70. Robson, B. and Osguthorpe, D.J. (1979) J. Mol. Biol. 132, 19–51.
71. Crippen, G.M. and Havel, T.F. (1978) Acta Cryst. A34, 282–284.
72. Crippen, G.M. (1979) Int. J. Pept. Protein Res. 13, 320–326.
73. Kuntz, I.D., Crippen, G.M. and Kollman, P.A. (1979) Biopolymers 18, 939–957.
74. Blout, E.R., de Loze, C., Bloom, S.M. and Fasman, G.D. (1960) J. Am. Chem. Soc. 82, 3787–3789.
75. Chou, P.Y. and Fasman, G.D. (1974) Biochemistry 13, 211–222.
76. Chou, P.Y. and Fasman, G.D. (1977) J. Mol. Biol. 115, 135–175.
77. Chou, P.Y. and Fasman, G.D. (1978) In: A. Meister (Ed.) Advances in Enzymology, John Wiley, New York, Vol. 47, pp. 45–148.
78. Robson, B. and Suzuki, E. (1976) J. Mol. Biol. 107, 327–356.
76. Garnier, J. Osguthorpe, D.J. and Robson, B. (1978) J. Mol. Biol. 120, 97–120.
80. Lim, V.I. (1974) J. Mol. Biol. 88, 857–872.
81. Lim, V.I. (1974) J. Mol. Biol. 88, 873–894.
82. Taniguchi, T., Mantei, N., Schwarzstein, M., Nagata, S., Maramatsu, M. and Weissmann, C. (1980) Nature 285, 547–549.
83. Sternberg, M.J.E. and Cohen, F.E. (1982) Int. J. Biolog. Macromolecules. 4, 137–144.
84. Lenstra, J.A., Hofsteenge, J. and Beintema, J.J. (1977) J. Mol. Biol. 109, 185–193.
85. Schulz, G.E., Barry, C.D., Friedman, J., Chou, P.Y., Fasman, G.D., Finkelstein, A.V., Lim, V.I., Ptitsyn, O.B., Kabat, E.A., Wu, T.T., Levitt, M., Robson, B. and Nagano, K. (1974) Nature 250, 140–142.
86. Matthews, B.W. (1975) Biochim. Biophys. Acta. 405, 442–451.
87. Green, N.M. and Flagan, M.T. (1976) Biochem. J. 153, 729–732.
88. Warshel, A. and Levitt, M. (1976) J. Mol. Biol. 106, 421–437.
89. Chothia, C., Levitt, M. and Richardson, D. (1977) Proc. Natl. Acad. Sci. USA 74, 4130–4134.
90. Ptitsyn, O.B. and Rashin, A.A. (1975) Biophys. Chem. 3, 1–20.
91. Cohen, F.E., Richmond, T.J. and Richards, F.M. (1979) J. Mol. Biol. 132, 275–288.
92. Cohen, F.E. and Sternberg, M.J.E. (1980) J. Mol. Biol. 137, 9–22.
93. Chothia, C., Levitt, M. and Richardson, D. (1981) J. Mol. Biol. 145, 215–250.

94. Cohen, F.E. (1980) D. Phil. University of Oxford.
95. Cohen, F.E., Sternberg, M.J.E. and Taylor, W.R. (1980) Nature 285, 378–382.
96. Cohen, F.E., Sternberg, M.J.E. and Taylor, W.R. (1981) J. Mol. Biol. 148, 253–272.
97. Ptitsyn, O.B., Finkelstein, A.V. and Falk-Bendzko, P. (1979) FEBS Lett. 101, 1–5.
98. Ptitsyn, O.B. and Finkelstein, A.V. (1980) Quart. Rev. Biophys. 13, 339–386.
99. Sternberg, M.J.E., Cohen, F.E., Taylor, W.R. and Feldmann, R.J. (1981) Philos. Trans. R. Soc. London B293, 177–189.
100. Cohen, F.E., Sternberg, M.J.E. and Taylor, W.R. (1982) J. Mol. Biol. 156, 821–862.
101. Cohen, F.E., Sternberg, M.J.E. and Taylor, W. R. (1981) J. Mol. Biol. submitted.
102. Janin, J. and Chothia, C. (1980) J. Mol. Biol. 143, 95–128.
103. Campbell, D.G., Williams, A.F., Bayley, P.M. and Reid, K.B.M. (1979) Nature, 282, 341–342.
104. Stewart, W.E. (1979) The Interferon System, Springer, Berlin.
105. Derynck, R., Content, J., DeClercq, E., Volckaert, G., Tavernier, J., Devos, R. and Fiers, W. (1980) Nature 285, 542–547.
106. Streuli, M., Nagata, S. and Weissman, C. (1980) Science 209, 1343–1347.
107. Zoon, K.C., Smith, M.E., Bridgen, P.J., Anfinsen, C.B., Hunkapiller, M.W. and Hood, L.E. (1980) Science 207, 527–528.
108. Argos, P., Rossman, M. and Johnson, J.E. (1977) Biochem. Biophys. Res. Commun. 75, 83–86.
109. Weber, P.C. and Salemme, F.R. (1980) Nature 287, 82–84.
110. Wetzel, R. (1980) Nature 289, 606–607.
111. Momany, F.A., McGuire, R.F., Burgess, A.W. and Scheraga, H.A. (1975) J. Phys. Chem. 79, 2361–2381.
112. Dunfield, L.G., Burgess, A.W. and Scheraga H.A. (1978) J. Phys. Chem. 82, 2609–2616.
113. Pincus, M.R., Zimmerman, S.S. and Scheraga, H.A. (1976) Proc. Natl. Acad. Sci. USA, 73, 4261–4265.
114. Pincus, M.R., Zimmerman, S.S. and Scheraga, H.A. (1977) Proc. Natl. Acad. Sci. USA 74, 2629–2633.
115. Pincus, M.R. and Scheraga, H.A. (1979) Macromolecules 12, 633–644.
116. Warshel, A. and Levitt, M. (1976) J. Mol. Biol. 103, 227–249.
117. Brown, S.B., Chabot, A.A., Enderby, E.A. and North, A.C.T. (1981) 289, 93–95.
118. Hermans, J. and Ferro, D. (1971) Biopolymers 10, 1121–1138.
119. Levitt, M. (1974) J. Mol Biol. 82, 393–420.
120. Matthews, J.B., Hanania, G.I.H. and Gurd, F.R.N. (1979) Biochemistry 18, 1919–1928.
121. Matthews, J.B., Hanania, G.I.H. and Gurd, F.R.N. (1979) Biochemistry 18, 1928–1936.
122. Burton, D.R., Boyd, J., Brampton, A.D., Easterbrook-Smith, S.B., Emanuel, E.J., Novotny, J., Rademacher, T.W., van Schravendijk, M.R., Sternberg, M.J.E. and Dwek, R.A. (1980) Nature 288, 338–344.
123. Perutz, M.F., Muirhead, H., Cox, J.M. and Goaman, L.C.G. (1968) Nature 219, 131–139.
124. Stroud, R.M., Kay, L.M. and Dickerson, R.E. (1971) Cold Spring Harbor Symp. Quant. Biol. 36, 125–140.
125. Shotton, D.M. and Hartley, B.S. (1970) Nature 225, 802–806.
126. Rossmann, M.G. and Argos, P. (1976) J. Mol. Biol. 105, 75–95.
127. Rossmann, M.G. and Argos, P. (1977) J. Mol. Biol. 109, 99–129.
128. McLachlan, A.D. (1979) J. Mol. Biol. 128, 49–79.
129. Remington, S.J. and Matthews, B.W. (1978) Proc. Natl. Acad. Sci. USA 75, 2180–2184.
130. Gurd, F.R.N. and Rothgeb, T.M. (1979) Adv. Prot. Chem. 33, 73–165.
131. Woolfson, M.M. (1970) In: An Introduction to X-ray Crystallography, Cambridge University Press, Cambridge, pp. 189–194.
132. Sternberg, M.J.E., Grace, D.E.P. and Phillips, D.C. (1979) J. Mol. Biol. 130, 231–253.
133. Frauenfelder, H., Petsko, G.A. and Tsernoglou, D. (1979) Nature 280, 558–563.
134. Huber, R. (1979) Nature 280, 538–539.
135. McCammon, J.A., Gelin, B.R. and Karplus, M. (1977) Nature 267, 585–590.
136. Northrup, S.H., Pear, M.R., McCammon, J.A. and Karplus, M. (1980) Nature 286,304–305.
137. Northrup, S.H., Pear, M.R., McCammon, J.A., Karplus, K. and Takano, T. (1980) Nature 287, 659–660.
138. Levitt, M. (1981) J. Mol. Biol. 145, 251–263.
139. Karplus, M. and Weaver, D.C. (1976) Nature 260, 404–406.
140. Cohen, F.E., Sternberg, M.J.E., Phillips, D.C., Kuntz, I.D. and Kollman, P.A. (1980) Nature 286, 632–634.
141. Matheson, R.R. and Scheraga, H.A. (1978) Macromolecules 11, 819–829.

142. Richards, F.M. and Richmond, T.J. (1977) In: Molecular Interactions and Activity in Proteins CIBA Foundation Symposium 60 (New Series) Excerpta Medica, Amsterdam, pp. 23–45.
143. Lifson, S. and Sander, C. (1979) Nature 282, 109–111.
144. Geisow, M.J. and Roberts, R.D.B. (1980) Int. J. Biol. Macromol. 2, 387–389.
145. Richardson, J.S., Getzoff, E.D. and Richardson, D.G. (1978) Proc. Natl. Acad. Sci. USA 75, 2574–2578.
146. Huber, R. Densenhofer, J., Colman, P.M., Matsushima, M. and Palm, W. (1976) Nature 264, 415–420.

APPENDIX – ADDRESSES

Computer programs and data banks are available from the following workers.

Solvent accessibility calculations

Professor F.M. Richards, Department of Molecular Biophysics and Biochemistry, Yale University, P.O. Box 6666, New Haven, CT 06511, USA.

Dr. S. Wodak, Université Libre de Bruxelles, Laboratoire de Chimie Biologique, Rue des Chevaux, 67, 1640 Rhode-St-Genèse, Bruxelles, Belgium.

also from
Dr. M.J.E. Sternberg.

Brookhaven Data Bank of Protein Co-ordinates

Dr. F.C. Bernstein, Protein Data Bank, Chemistry Department, Brookhaven National Laboratory, Upton, NY 11973, USA.

Data Bank of Protein Sequences

Dr. M.O. Dayhoff, National Biomedical Research Foundation, Georgetown University Medical Center, 3900 Reservoir Road, N.W., Washington, DC 20007, USA.

Distance Geometry

Professor I.D. Kuntz, Department of Pharmaceutical Chemistry, University of California, San Francisco, CA 94143, USA.

Secondary structure prediction by the method of Robson, Garnier and Osguthorpe

Dr. J. Garnier, Laboratoire de Biochimie physique, INRA, Bat. 433, Université de Paris Sud, 91405 Orsay, France.

Secondary structure prediction by the method of Lim

Professor J. Drenth, Department of Biochemistry, University of Nijmegen, P.O. Box 9101, 6500 HB Nijmegen, The Netherlands.

Several secondary structure prediction programs

Professor A.C.T. North, The Astbury Department of Biophysics, The University of Leeds, Leeds, LS2 9JT, UK.

The combination approach to docking α–helices and β-strands

Dr. M.J.E. Sternberg, Laboratory of Molecular Biophysics, Department of Zoology, South Parks Road, Oxford, OX1 3PS, UK.

Dr. F.E. Cohen, Department of Pharmaceutical Chemistry, University of California, San Francisco, CA 94143, USA.

Energy calculations developed by Scheraga and co-workers

Quantum Chemistry Program Exchange, Chemistry Department, Room 204, Indiana University, Bloomington, IN 47401, USA.

ECEPP is program No. 286
UNICEPP is program No. 361.

Other energy calculations

Dr. M. Levitt, Department of Chemical Physics, The Weizman Institute of Science, Rehovat, Israel.

Comparison of protein structures

Professor M.G. Rossmann, Department of Biological Sciences, Purdue University, West Lafayett, IN 47907, USA.

Professor B.W. Matthews, Institute of Molecular Biology, University of Oregon, Eugene, OR 97403, USA.

Dr. A.D. McLachlan, MRC Laboratory of Molecular Biology, Cambridge CB2 2QH, UK.

Image processing and analysis

Geisow & Barrett (eds.) Computing in biological science

Database techniques for two-dimensional electrophoretic gel analysis

PETER F. LEMKIN and LEWIS E. LIPKIN

1. INTRODUCTION

This article discusses GELLAB [1–6], a system for computer aided analyses of two-dimensional electrophoretic patterns. The 2D polyacrylamide gel electrophoresis (PAGE) technique [7] has been a rapidly developing biochemical tool, applicable to a wide variety of problems in molecular biology, basic biochemistry, genetics and clinical research [8]. The 2D PAGE technique can be used to separate hundreds to several thousand polypeptide components as a matrix of spots because the variables which determine electrophoretic mobility in each of the two dimensions are effectively orthogonal to each other. Isoelectric focusing over a pH gradient determines extent of movement in the first dimension. In the second dimension the sodium dodecyl sulfate (SDS) interaction with polypeptides results in a mobility which is a function of molecular weight.

The greatly-increased number of 'spots' detectable in 2D PAGE [7] is a major reason for efforts at computerized analysis [4,8–12]. As a result of these automation attempts, there has been an increase in complexity of intermediate analysis results that it would strain the limits of unaided human analytical ability. The need for analytic assistance is further increased by the added complication of non-linear spatial warping of corresponding moieties in comparable gels.

A 2D PAGE electrophoretic gel is a complex of distinct polypeptides, each one of which is characterized by position relative to other polypeptides ('spots') and density. However, unlike a geographic map, proximity of polypeptides on a gel is no particular indication of related genesis or biological function. Nonetheless the large number of discrete spots in a gel and the similarity that is preserved among gels from a similar source allows one to track many proteins through the effects of experimental variables. Hence, comparing biological specimens by comparing their corresponding gels for quantitative or qualitative differences has become an important means of determining protein manifested metabolic differences.

We are progressively more concerned with the generation of data structures, strategies and tactics for their employment in the analyses of *sets* of gels. Such comparisons, both qualitative and quantitative, among multiple gels might reflect, for example, successive values of a dose or time variable in an experiment or the clinical course of a patient.

In dealing with this material, human factor considerations place a practical limit on the number of spots for which manual density information can be obtained. Manual techniques such as optical flicker or dual-color comparison between two local regions on separate gels are useful for local alignment, especially in cases with obvious spot differences [9]. However, this method forces the user to deal with the gels sequentially in that only pairwise gel comparison can be made. This makes the process time consuming and difficult for the observer to visualize a pattern directly over a set of gels. Such manual comparison methods are probably capable of supporting a complete search for all major polypeptide differences, but the bookkeeping needed to identify the same spot in several gels makes computer aid attractive. Beyond some relatively small number of spots, some computer aid in matching, 'remembering' and retrieving images of preserved spot correspondences is seen as indispensable in thoroughly analogous sets of gels. An added benefit of this is that after the spots have been isolated, located and tagged, the machine can use this information to produce a variety of representations. These include pictorial, diagrammatic, numerical, etc., that aid the user to see patterns difficult to grasp when attention is focused on small regions. Final output of GELLAB includes labeled gel image maps and mosaics where statistically interesting spots have been marked as well as numeric spot data lists to support these findings.

Figure 1 illustrates the major steps performed in the GELLAB 2D gel analysis procedure. Gels are first accessioned (assigned a unique number) and related experiment information is entered at this time into the system. The gel images are then acquired (digitized and stored). Then spots are segmented in each gel followed by spots being paired between each gel and a standard gel using a small set of manually defined landmarks. Finally a multiple spot data base (DB) is constructed and analyzed. Final output of such a system takes several forms. This includes labeled gel image maps (superimposed on the original gel images) where statistically interesting spots have been marked as well as numeric spot data to support these findings. Various 1-, 2-, and 3D functions of spot data base features may be plotted both on interactive displays and later, on paper.

1.1. Gel characteristics

For the most part, a spot's resultant geometric position in a gel bears no relation to function or the origin of the protein it represents. Closely related polypeptides may be separated by considerable distances while functionally unrelated materials could

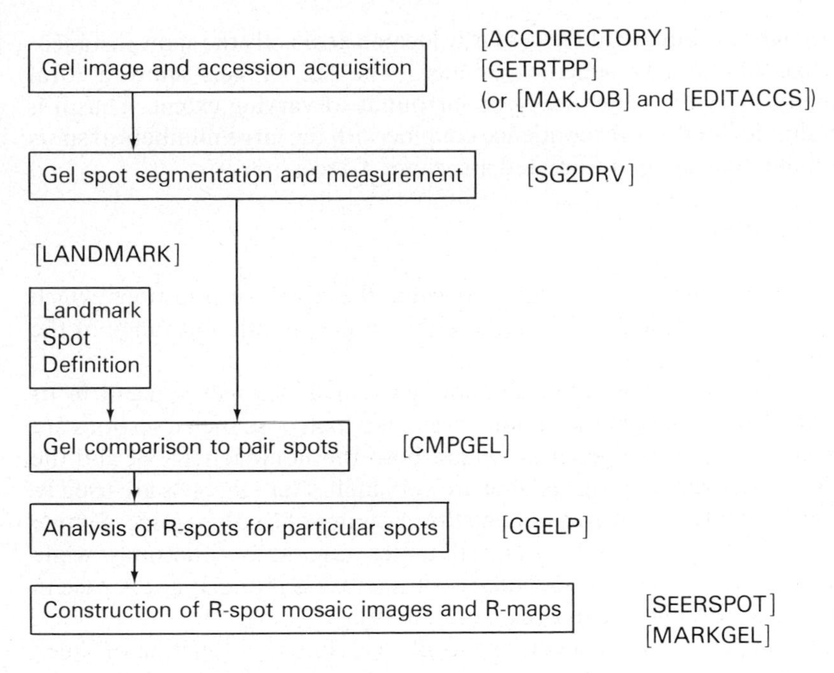

Fig. 1. Block diagram of the 2D-gel analysis GELLAB system. Programs associated with major steps of GELLAB are indicated in '[...]'. Gel images are acquired by scanning with a vidicon TV camera interfaced to a picture memory and saved on computer disk files. Accession information about the experiments which produced the gels is also used to update an accession file. The gel images are then segmented and measurements made of the detected spots. Landmark spots common to the *R*-gel and all other gels are then interactively selected. The landmark spots are aligned for all of the gels with a representative gel (*R*-gel) using gel image flicker alignment. This landmark spot information and the spot segmentation data is then used to pair congener spots in the remaining gels with the *R*-gel. The set of gel pairings with the same *R*-gel may be merged together to form a list of sets of equivalent *R*-spots called the congener gel data base (CGL) and subjected to further analysis.

be distributed in close proximity. In contrast to more conventional images such as microscopic fields or X-rays, the image 'structure' (i.e., local adjacencies, inclusions, etc.) in gels provides little information to facilitate the analysis. Individual spots unless contaminated by artifacts or overlapped by other spots are much simpler images than, say, the image of a cell in a blood smear. Thus once a spot has been isolated, its analysis and characterization as an *individual entity* at least in a single gel, is relatively simple.

Non-congruence is a more serious difficulty (i.e., the lack of point to point reproducibility of gels). It occurs in gels derived from the same sample and from a single run on the same apparatus. This is due to a large number of preparative factors including local temperature variations, local heterogeneities in polyacrylamide texture and/or local concentration, heterogeneities of ampholine concentration, etc. All of these variables and perhaps others less understood, combine to reduce the reproducibility of mobility of polypeptide fragments in one or another dimension. This net result is a set of gels which are *not congruent* but which are related by an affine transformation. In other words, comparable spots within a set of gels have corresponding neighbors but are not necessarily located at exactly the same distances from these neighbors in any specific instance. The set of gels show a local superimposeability, which is maintained for surrounds of varying extent. Thus it is this absence of simple direct correspondence coupled with the large numbers of spots in a set of gels that makes some automated assistance a necessity.

1.2. Classes of problems

It is necessary when viewing to carefully consider the kinds of questions which biological problems pose which will determine the nature, depth and range of the analyses to be performed.

(1) Is only one or a very few different spots present in one gel and not in its experimental pair? The paradigmatic biologic systems which pose such questions are those in areas such as bacterial genetics, where both the homogeneity of and the specificity of the product generating cell line are very high. Here the gels are used as detectors and serve simply to confirm or deny the existence of a polypeptide. Simple flicker analysis may be all that is required under favorable conditions while densitometery for this situation is a secondary consideration if one at all. A case in point is a single gene difference in an *E.Coli* mutant [9].

(2) Are there changes in any of several spots in a cell line as a function of time? These variations are often in polypeptide quantity so that densitometry is required. Moreover, the complexity of the analysis is increased so that a number of gels rather than a pair of gels must often be compared. The answer to such questions as this requires spot data structures and data base management software that are both significantly larger and significantly more complex than those required for the answer to the first question.

(3) Are there changes in several spots resulting from an applied stimulus? The less known about the outcome, i.e., the more exploratory the search, the more extensive and complex must be the gel analysis. When the cell line is only apparently homogeneous (as was the case for PHA stimulated lymphocytes [4]) and where the effect of the stimulus is both complex and a function of time, the gel analyses become correspondingly more extensive and complex. In such situations, where many new spots may result, there seems no alternative to automatic spot pairing using a computer.

(4) Is there a 'finger print' of morphologically homogenous but biologically and functionally different cell groups as for example differences among various lymphoblastic tumors? Here, especially if stimuli are required to elicit differences, the number of gels grows to an m (number of classes) times n (number of temporal samples) times p (number of levels of stimuli) number of comparisons. This assumes minimum problems in the reproducibility of the gels whereas often multiple gels of the same sample are run. Particular interest must be focused not only on differences but on subgroup similarities as might seem to be indicated by the clustering of density ratios in the histogram tables (cf., Table 10).

(5) Are known polypeptides present in normal or abnormal quantities in a body fluid? Here are essential problems of clinical chemistry, but multiplied by the large

number of spots present in the gel. At first, the answer to questions of this type might seem simpler than those above. The comparison of a single gel's contents to some internal or external standard cetainly involves future developments in the area which has been called by Anderson 'molecular anatomy' [8]. Because of the need for extensive bookkeeping in multiple gel analysis, it is likely that some of the data structures we present here will be an aid in this development.

Gels may be thought of as complex objects similar to a geographic map with individual polypeptides appearing in distinct morphologic regions. They are not however in any way certain reference points where unlike the geographic map, adjacency of morphologic polypeptides in the gel is no particular indication of related genesis or biological function. However, characteristic patterns are obtained with carbamylation and other biochemical treatments. Comparing biological specimens by comparing their corresponding gel maps is one means of determining major protein differences although this is tedious when done manually. Given a number of gels, polypeptide concentration values may be modeled as a density distribution for each set of corresponding spots and the analysis performed on the distributions.

1.3. Hardware and software implementation

This has been described in [9,13–17]. Briefly the system consists of a DECSYSTEM-2020 computer controlling a one-of-a-kind Real Time Picture Processor (RTPP) constructed in our laboratory [14–17]. The RTPP uses a PDP8e as a display processor controller for the sixteen 8-bit gray values (256 gray values 0 = white, 255 = black) 256 × 256 picture element (pixel) frame buffer image memories. A Quantimet 720 image analyzer is used for video TV camera input and TV display output. These memories may be configured as four 512 × 512 pixel images for use with the GELLAB programs. The RTPP has a TV camera and image (A/D) acquisition hardware which enables a TV frame to be acquired in 1/10 second into the frame buffer memories. The time shared 2020 (under TOPS10 monitor system) processor has 512 K words of 36-bit memory and three 160 Mbyte disk pack drives as well as two magtape drives. The RTPP's picture memory, as well as the PDP8e control teletype, has been interfaced to the 2020 via a UNIBUS – thus implementing a distributed picture processor. The GELLAB system is one of several image processing projects using the above hardware.

In addition, gel images scanned elsewhere on an Optronics scanner have been used on our system with a RT-11 magtape intermediary. The procedure is to scan the gel onto a RT-11 PDP11 disk file and then later to transfer one or more of these files to magtape. A GELLAB SAIL program called MT-11 is able to read images from these magtapes onto the DECSYSTEM-10 (or -20) file system.

The set of GELLAB programs SG2DRV, CMPGEL, CGELP, MARKGEL, SEERSPOT, DWRMAP, LMSEDIT, etc., are written in the SAIL programming language [18]. This language is currently implemented *only* on DECSYSTEM-10 and DECSYSTEM-20 computers. It has distinct advantages in its ease of algorithm expression, macro expansion, string, list, set and associative processing and record structure operations. Since SAIL strongly encourages structured programming it is an ideal environment in which to implement a set of complex interacting algorithms.

A subset of the GELLAB system has been constructed to run on any DECSYSTEM-10)or -20) *independent* of our special purpose hardware. Of course, interface software must be written to read and write image files to the local image display or plotting hardware systems. This export version of GELLAB will be discussed in more detail in the Appendix.

1.4. *Image acquisition*

Data acquisition is accomplished by scanning backlighted gels or gel autoradiographs with the Vidicon camera [9]. The Vidicon camera has a Nikon-N auto 1:2 28 mm objective lens routinely set to *f*8 with the autoradiograph film mounted 69 to 42 cm away and backlighted on an Aristo type T-12 uniformly illuminated light box (Port Washington, NY). The effective resolution of the image ranges from 250 to 170 microns/pixel (picture element) although other lenses can and have been used. A type 1009 NBS Neutral Density (ND) wedge is mounted at the bottom of the illuminated area. For use with 120 mm film size negatives (for use with the silver stain), a 55 mm micro-Nikkor lens is fixed at *f*8 and set for a distance of 55.5 cm.

The scanned images are acquired using the GETRTPP program of GELLAB and saved directly on the DECSYSTEM-2020's disk. At this step of acquisition, gels are assigned accession numbers and experiment information about the gels is entered. This information is used in tracking the gels throughout the analysis. An image consists of an 512 × 512 pixel array. The upper left hand corner of the image is defined to be (0,0) while (511,511) represents the lower right hand corner. The actual dynamic range of the gray scale data is slightly greater than 7-bits, however of this probably only 6-bits of gray scale resolution is actually valid. The maximum dynamic range of a vidicon is about 0 to 2.0 OD. Gels can be checked against the ND wedge using analog video detector circuits of the TV system to determine whether any spots are greater than the TV's dynamic range. Since most spot information is less than 2.0 OD there is generally no problem of gray scale distortion for photographically non-saturating spots. Fifty images are easily scanned in about one hour or less.

Although the transfer function of a Vidicon TV camera is non-linear, it may still be used to perform densitometry under some conditions. These include: (1) the majority of the material to be measured is below the saturating end of the camera's dynamic range; (2) the non-linear gray scale to OD transfer function be well behaved over the range of data to be measured; and (3) all calculations involving integrated OD/spot are done in the density domain. Even then, appreciable errors on very dark spots can occur due to saturation, noise and to optical problems (e.g., glare and under-representation of dark high OD pixels).

1.5. *Calibration*

Using the GETRTPP or EDITACC programs on the RTPP (see Fig. 1) the user interactively defines the computing window (the region in the gel image where the spots are located) taking care to omit the ND wedge and gross artifacts in the gel. Using these programs, the ND wedge (in the gel image) is calibrated by computing a gray scale histogram of a 25 pixel wide window positioned by the user across the wedge. Peaks in the smoothed histogram are then matched automatically with the actual OD values of the wedge. The wedge gray value peak information and a gel computing window position (interactively defined) are automatically updated into the gel accession file along with the gel experiment information. A piecewise linear function can be generated for the ND wedge as a function of gray value. This permits image gray values to be mapped to density (OD). Table 1 shows part of a typical accession file. Figure 2 illustrates a gel with the ND wedge 2a, its manually defined computing window 2b and ND wedge sample window 2c, and the resultant ND wedge histogram calibration curve 2d. The piecewise linear OD calibration function is drawn (black) over the histogram.

1.6. Gel segmentation: nature of the image

Gel image accessioning and acquisition is only the first stage in making spot information available for automated processing. Provision must be made for separating out 'pictorial information of interest' (in this case the spots) from 'noise' and 'background' before spot positions and spot properties can be compared. Consequently, a spot extraction algorithm must be capable, under a wide variety of actual gel image conditions, of (1) detecting, (2) defining the extent of, and (3) measuring the density of a spot.

The segmentation problem is one of the more important and ubiquitous, in the general field of image processing. Almost all real images resist simple gray scale thresholding as a solution to pictorial partitioning or segmentation and despite the simplicity of spot morphology, 2D gels are no exception. The thresholding operation applied to an image retains all values higher than a certain gray value while all others are set to white.

1.7. Spot morphology

The vagueness of spot morphology and inhomogeneity of gel background complicate these images. Spots often touch each other resulting in overlapping spots. The 'tails' of spots may extend for a considerable distance into an overlapping region. Spots have no distinct boundary, but occur most often as an effectively continuous Gaussian-like distribution which Lutin asserts that this distribution tends to be symmetric with respect to isoelectric point (pI) but is skewed in molecular weight (MW) [10]. In practice, this holds only for ideal non-conglomerate spots. In addition, spots may appear round, oblong, or take on various continuous shapes particularly when there is excessive loading of material in the gel. In all cases, except in the case of extreme overloading, however, the center of a spot is its darkest part. Spots may sometimes be obscured in certain regions of the gel which are susceptible to streaking in both MW and isoelectric axes. Spots can occur within these streaks which are of interest.

TABLE 1

Example of part of a gel accession descriptor file

```
ACCESS. #/PATIENT/BIRTHDATE/RACE&SEX/EXP DATE/EXP #/CULTURE REAG/AMPH,GEL/
INTRVL BEFR LBLNG/LBLNG ISOTOPE/DURTN LABEL/DURTN OF EXPSR/STUDY/
FILE #/TAPE #/OPT. BACKUP TAPE #/ CAMERA,LENS,DISTANCE/EXPRMNTR*
ND:.05,.20,.35,.50,.66,.80,.95,1.10,1.25,1.41,1.56,1.72,1.87,2.02,2.17
ASBSPT.DA = ASBESTOS E-SPOT FILE
    .
    .
    .
0250.1/P388D1/-/-/8-21-80/#A95/FISCHER'S/3:10, 10%/
0 HRS/C14/8/1 DAY/ALUMINUM,T0,CONTROL,BOTTLE#1/
L00153/-NONE-/--NONE--/VIDICON-AUTO,28MM F8,69CM/LIPKIN*
 41 67 89 110 127 144 159 171 182 189 195 205 0 0 0 0 53 425 78 425
0250.2/P388D1/-/-/8-21-80/#A95/FISCHER'S/3:10, 10%/
0 HRS/C14/8/1 WEEK/ALUMINUM,T0,CONTROL,BOTTLE#1/
L00157/-NONE-/--NONE--/VIDICON-AUTO,28MM F8,69CM/LIPKIN*
 40 65 88 109 126 144 159 170 181 190 196 208 213 0 0 0 31 460 51 400
0251.1/P388D1/-/-/8-21-80/#A95/FISCHER'S/3:10, 10%/
0 HRS/C14/8/1 DAY/ALUMINUM,T0,CONTROL,BOTTLE#2/
L00161/-NONE-/--NONE--/VIDICON-AUTO,28MM F8,69CM/LIPKIN*
 29 56 67 79 101 119 137 153 166 177 186 193 200 208 0 0 2 478 49 410
0251.2/P388D1/-/-/8-21-80/#A95/FISCHER'S/3:10, 10%/
0 HRS/C14/8/1 WEEK/ALUMINUM,T0,CONTROL,BOTTLE#2/
L00165/-NONE-/--NONE--/VIDICON-AUTO,28MM F8,69CM/LIPKIN*
 40 67 88 110 126 144 159 171 182 191 197 208 0 0 0 0 2 446 45 390
    .
    .
    .
0258.2/P388D1/-/-/8-21-80/#A95/FISCHER'S/3:10, 10%/
0 HRS/C14/8/1 WEEK/ALUMINUM,T24,CONTROL,BOTTLE#9/
L00221/-NONE-/--NONE--/VIDICON-AUTO,28MM F8,69CM/LIPKIN*
 40 65 88 110 127 144 159 171 182 191 197 208 213 0 0 0 2 491 53 405
    .
    .
    .
0264.2/P388D1/-/-/8-21-80/#A95/FISCHER'S/3:10, 10%/
0 HRS/C14/8/1 WEEK/ALUMINUM,T24,AMOSITE,TOXIC,PHAGOCYTIC,BOTTLE#15/
L00269/-NONE-/--NONE--/VIDICON-AUTO,28MM F8,69CM/LIPKIN*
 41 67 89 110 127 144 159 171 182 190 197 208 213 0 0 0 2 459 72 405
0265.1/P388D1/-/-/8-21-80/#A95/FISCHER'S/3:10, 10%/
0 HRS/C14/8/1 DAY/ALUMINUM,T24,AMOSITE,TOXIC,PHAGOCYTIC,BOTTLE#16/
L00273/-NONE-/--NONE--/VIDICON-AUTO,28MM F8,69CM/LIPKIN*
 62 84 105 123 140 156 168 178 187 193 208 0 0 0 0 0 2 446 40 440
0265.2/P388D1/-/-/8-21-80/#A95/FISCHER'S/3:10, 10%/
0 HRS/C14/8/1 WEEK/ALUMINUM,T24,AMOSITE,TOXIC,PHAGOCYTIC,BOTTLE#16/
L00277/-NONE-/--NONE--/VIDICON-AUTO,28MM F8,69CM/LIPKIN*
 37 63 86 108 125 143 158 169 180 189 194 201 210 0 0 0 6 454 23 360
    .
    .
    .
```

Example of part of a typical gel accession descriptor file for the P388D1 data base. Each data record contains four lines. The first four lines of the file define the record field descriptors and ND wedge values. The descriptors are separated by '/' and terminated with a '*'. The fourth line of each record is the set of gray value peaks corresponding to the ND wedge calibration. The last four value of that line are the computing window for that gel [$x1{:}x2,y1{:}y2$].

Polypeptides in the gel are not visible by themselves and must therefore be visualized in order to perform an analysis. At least four methods are currently used which include: Coomassie blue staining, autoradiography (on radioactively labeled proteins produced by growing the tissue culture in radiolabeled amino acids), silver staining [19–20], and fluorescent dyes. These spot detection methods have

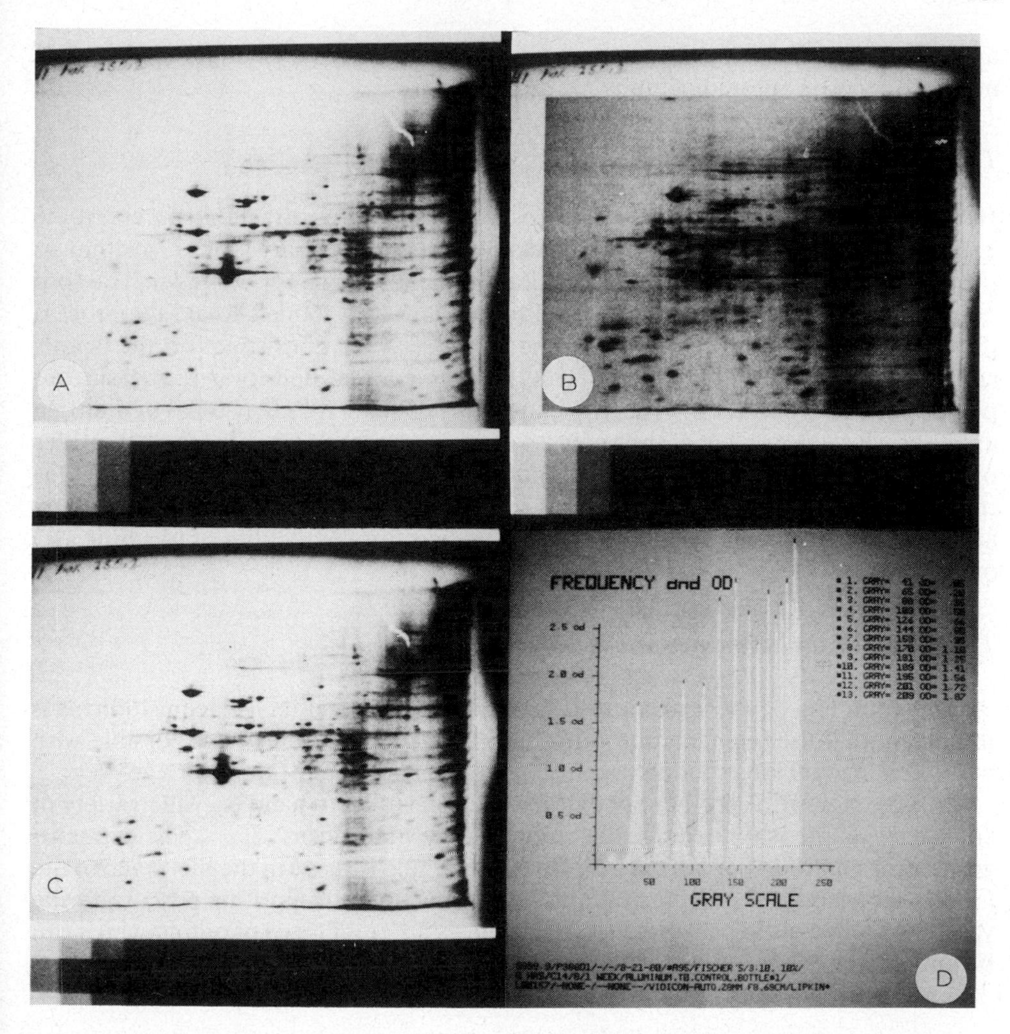

Fig. 2. Typical 2D gel with ND wedge for P388D1 macrophage like cells. (a) Original T0 gel image scanned at 250 microns/pixel; (b) image with computing window noted to define segmentation region; (c) ND wedge sample window defining wedge calibration region, and (d) smoothed ND wedge sample gray value histogram with computed piecewise linear calibration function superimposed. The ND value and gray value frequency are on the ordinate and gray value on the abscissa (0 to 255).

widely different dynamic ranges and stochiometry, as well as application for different types of biological material.

Care must be taken to ensure that autoradiographic film is used in the linear portion of the density versus log (exposure) curve otherwise saturation of some spots will occur. The dynamic range of spot detection may be covered using a series of increasingly long autographic exposures of the same gel. The Vidicon or other imaging detector is subject to similar saturation problems.

Because some spots will be recorded as saturated, it is useful to know which ones and furthermore to be able to track these spots throughout the entire analysis process. This is done in GELLAB. Spots saturating in one gel might not do so in another so that alternative measurements could be made as for example in the case of multiply exposed autoradiographs.

1.8. A segmentation model

In any locally determined (e.g., non-thresholding) feature extraction process, some explicit or implicit model of the pictorial objects is necessary. The algorithm as presented here embodies some of the ideas of the underlying spot model. The spot extraction methods previously reported use various spot models to aid the process [8,10–12]. A first order model is the triple (x,y,d) consisting of the spot's centroid (x,y) in cartesian space and its total integrated density d (a measure of polypeptide concentration). This triple appears adequate for many types of multiple analyses where the object of the analyses is to measure the amounts of polypeptides present. Segmentation is a method of spot extraction which results in obtaining this triple as well as other features. Our spot segmentation algorithm is based on a shape and density independent model and takes into account the realities of touching and overlapping spots.

1.9. Role of the segmenter in overall gel analysis

As shown in Fig. 1, the segmenter is applied following gel image acquisition. It is important that the segmentation procedure be made as automatic as possible with minimum manual intervention because of the large number of spots on a gel.

We present here a specific spot extractor which is able to handle a wide variety of spot shapes, density and cluster morphology. However, any spot extractor generating an ordered list of spot triples (x,y,d) could be used in the first stage of the GELLAB analysis. Parameterization of the segmentation algorithm permits a wide variety of gel stains to be handled, and produces different types of output which can be put to varied uses.

2. SEGMENTATION

2.1. The segmentation algorithm

The segmentation algorithm is a sequence of procedures applied to a locally averaged image. The first of these is the digital analog of the spatial second derivative; it is used to construct an image called the central core image consisting of the centers of spots. Second derivative (cf., Fig. 3) information delimits the extent of outward propagation resulting in an algorithmic limit on individual spot extent. Initial spot candidate generation is parameter independent. The decision function which later separates noise from valid spots is adjusted by user defined parameters. Auxillary information required by the segmenter, such as picture file name, and ND wedge calibration and computing window is obtained from the accession file. The segmentation algorithm, SG2DRV, is shown in flow chart form in Fig. 4.

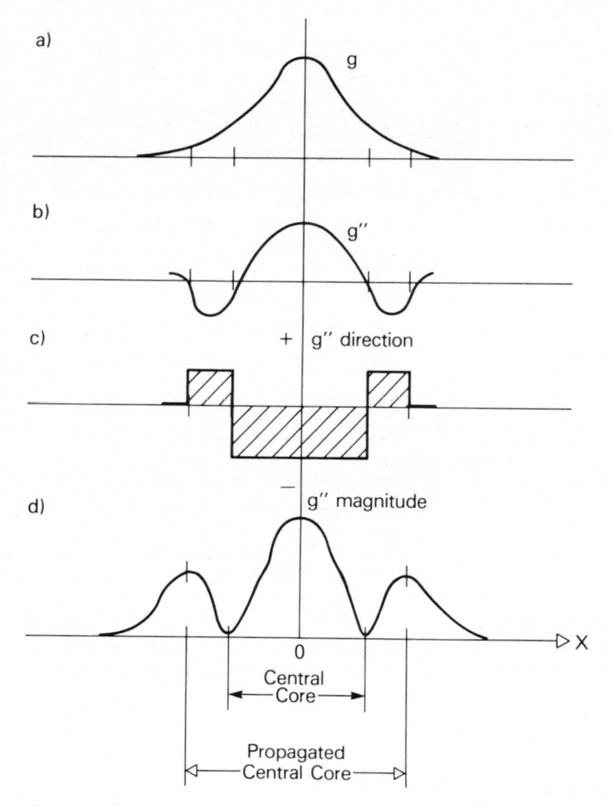

Fig. 3. 1D representation of a Gaussian like function g (a), its 2nd derivative g'' (b), its direction SIGN (g'') (c), and absolute magnitude (d) functions. In the central core region, the direction of g'' is less than 0 and changes sign in the propagated central core region. The outer extent of the propagated central core region is indicated by a second maxima in the g'' magnitude function.

2.1.1. *Principle*

Let g be a image gray scale point function whose mode, median and mean are all more or less central with respect to the extrema and let its second derivative be g''. The central region of a spot has a negative g'' direction and a g'' magnitude maximum. Beyond the mid-region where the direction of g'' changes sign, there is a second smaller peak in the magnitude of g''. Our segmentation procedure is based on finding these two maxima in 2-dimensions. The approximation to the boundary is operationally defined by the second maxima in the g'' magnitude function.

2.1.2. *Smoothing*

The original image is first smoothed to remove some of the high spatial frequency noise. This is illustrated using a 3 × 3 convolution filter [10]. Let matrix M_{ij} be defined as:

$$M_{ij} = \begin{matrix} 1 & 2 & 1 \\ 2 & 4 & 2 \\ 1 & 2 & 1. \end{matrix}$$

Then, for a central pixel (x,y), each smoothed pixel $f(x,y)$ is defined as:

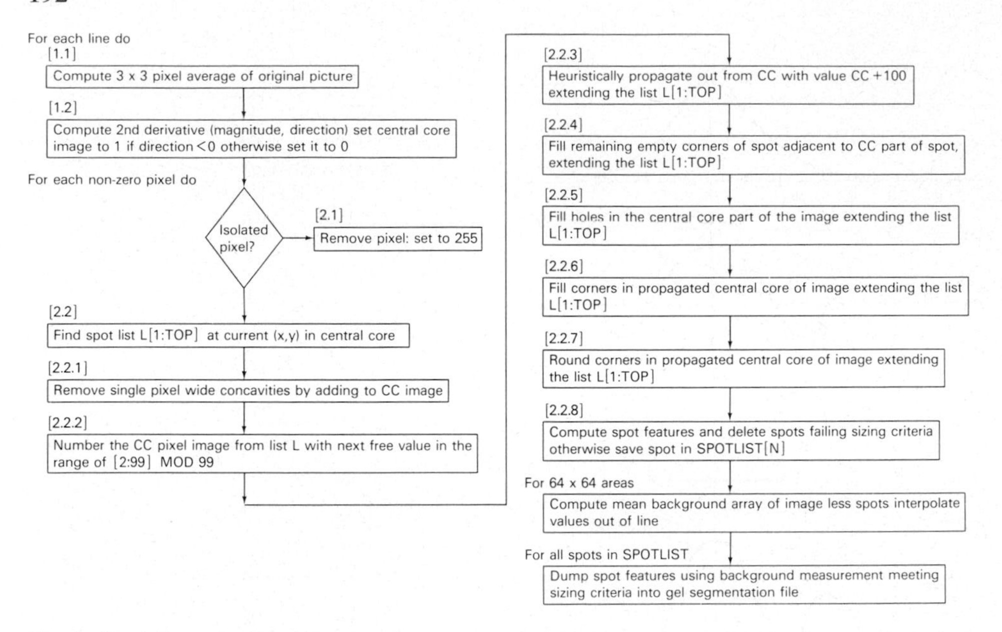

Fig. 4. Block diagram of the 2D-gel spot segmentation procedure performed during 2 passes through the image. Substeps k performed during passes 1 and 2 (through the image) are denoted by 1.k and 2.k respectively. The averaged and central core images are computed during pass 1 while pass 2 processes each spot to completion.

$$f(x,y) = (1/16) \sum_{i=1}^{3} \sum_{j=1}^{3} M_{ij} {}^* g(x + i - 2, y + j - 2).$$

It is applied over the entire picture, pixel by pixel in a top down left to right fashion. Each pixel in the 3 × 3 pixel neighborhood (defined by the center pixel) is multiplied by the corresponding 3 × 3 filter (pixel for pixel) and the total divided by 16. The result is saved in an *averaged* image. This filter removes enough of the high spatial frequency noise so the 2nd derivative analysis algorithm may be more successfully applied. Spot shapes are not distorted to any noticeable degree. The actual spot density measurements are made on the original image data. Larger filters, a 5 × 5, and a 7 × 7 [21], are also used.

$$\begin{matrix} 1 & 1 & 2 & 1 & 1 \\ 1 & 2 & 4 & 2 & 1 \\ 2 & 4 & 8 & 4 & 2 \\ 1 & 2 & 4 & 2 & 1 \\ 1 & 1 & 2 & 1 & 1 \end{matrix} \quad \text{divided by 52}$$

$$\begin{matrix} 4 & -6 & -12 & -14 & -12 & -6 & 4 \\ -6 & 9 & 18 & 21 & 18 & 9 & -6 \\ -12 & 18 & 36 & 42 & 36 & 18 & -12 \\ -14 & 21 & 42 & 49 & 42 & 21 & -14 \\ -12 & 18 & 36 & 42 & 36 & 18 & -12 \\ -6 & 9 & 18 & 21 & 18 & 9 & -6 \\ 4 & -6 & -12 & -14 & -12 & -6 & 4 \end{matrix} \quad \text{divided by 441}$$

2.1.3. Central core and magnitude 2nd derivative images

The 2nd derivative is computed as the vector (dx^2, dy^2) using the following difference formulae [22] (in a similar manner to the convolution filter of 2.1.2). These filters are applied to the averaged image just being computed.

$$dx^2 = \begin{matrix} 0 & 0 & 0 \\ 1 & -2 & 1 \\ 0 & 0 & 0 \end{matrix} \qquad dy^2 = \begin{matrix} 0 & 1 & 0 \\ 0 & -2 & 0 \\ 0 & 1 & 0 \end{matrix}$$

The *magnitude* image of g'' is approximated by the city block distance – the sum of the absolute values of dx^2 and dy^2. The direction image is not actually computed. Instead a *central core* image pixel is defined as having a '1' where ($dx^2 < 0$) and ($dy^2 < 0$), and being '0' everywhere else. Both the average and central core images are computed during the first raster scan through the image.

2.1.4. Extracting a spot

In a second raster pass through the image only those pixels coded as '1' are processed. Isolated pixels, defined as being 4-neighbor unconnected, are marked for deletion by setting them to a 255 code. Otherwise, each time a '1' code is encountered, a spot pixel list (SPL) is computed as in step [2.2]. A push down stack is used to keep track of all unexpanded pixels. An unexpanded pixel is defined as one found to be a neighbor of an expanded pixel but which has not been checked (i.e., is not on the SPL or push down stack). Each unexpanded pixel is expanded and checked to determine whether any of its 4-neighbor pixels have a '1' code and are not already in the SPL. Unexpanded pixels so identified are put into the push down stack while the pixel being investigated is saved in the SPL. The algorithm keeps processing the push down stack until it is empty. The spot will then be processed to completion using the SPL which will grow as the spot is propagated to the region approximated by the 2nd derivative magnitude function's second local maximum.

2.1.5. Removing concavities

Single pixel wide artifactual concavities which occasionally occur and are removed by checking each SPL pixel C for the following 4 neighborhood conditions. (By neighborhood we mean here the central pixel in question and its 8 adjacent neighboring pixels, see [23].) If a condition is found to be true, the '0' valued pixel in the central core image is changed to a '1'. The SPL is also updated. In each of the following cases, a '0' and '1' *must* occur and a '–' means 'don't care'.

$$\begin{matrix} 1 & 0 & 1 \\ 1 & C & 1 \\ - & - & - \end{matrix} \quad \text{or} \quad \begin{matrix} 1 & 1 & - \\ 0 & C & - \\ 1 & 1 & - \end{matrix} \quad \text{or} \quad \begin{matrix} - & - & - \\ 1 & C & 1 \\ 1 & 0 & 1 \end{matrix} \quad \text{or} \quad \begin{matrix} - & 1 & 1 \\ - & C & 0 \\ - & 1 & 1 \end{matrix}$$

2.1.6. Numbering the central core image

In the central core image, the spot is then assigned the next sequential number in the range of [2 : 99] modulo 100. All SPL pixels in the central core image for that spot get that number. It is very unlikely but in an extremely densely populated spot image, it is possible for two adjacent spots to have the same value for successive lines. This notation problem is easily solved by alternative coding schemes using larger numbers.

2.1.7. *Propagating the central core*

The numbered spot is then propagated with the value (C + 100) from the central core (C) value of the spot to the *propagated central core* region. This propagation from a central core edge point is performed in each of the 4-neighbor directions until it is terminated based on various constraints. (Whereas the 8-neighbor definition of a neighborhood included the corner pixels, the 4-neighbor definition does not [23].) The SPL is updated with the new pixels. The heuristic propagation termination conditions are:

(1) The 2nd derivative magnitude is increasing (starting 1 pixel out from the central core), outward from the central core indicating a second local maxima.
(2) The 2nd derivative magnitude outward from the central core has the same value twice in a row indicating a noisy edge.
(3) The propagation would impinge on another central core pixel.
(4) The propagation would extend beyond the computing window.
(5) The propagation would impinge on an isolated pixel.
(6) The gray value outward from the central core is increasing instead of decreasing indicating that the spot is overlapping a much larger spot.

2.1.8. *Corner filling*

This type of heuristic propagation sometimes forms small rectangular empty corner regions in the four corners of the spot. Such corners can be filled with propagated central core values. Both 0 and 255 (isolated pixel) corner values are candidates for filling (by changing to N, i.e., C + 100) if the central pixel is its central core. The four corner cases are expressed as neighborhood conditions as follows. In the corresponding positions of the neighborhood surrounding a pixel in question, C is the central core value, N is the propagated central core value, E is either 0 or 255, and '–' meaning 'don't care'.

$$\begin{matrix} E & N & - \\ N & C & - \\ - & - & - \end{matrix} \quad \text{or} \quad \begin{matrix} - & N & E \\ - & C & N \\ - & - & - \end{matrix} \quad \text{or} \quad \begin{matrix} - & - & - \\ - & C & N \\ - & N & E \end{matrix} \quad \text{or} \quad \begin{matrix} - & - & - \\ N & C & - \\ E & N & - \end{matrix}$$

2.1.9. *Hole filling the central core*

In very large saturating spots, the center of the spot may not be detected as such and thus not segmented. The spot will have a doughnut topology. This problem is repaired by filling any artifactual holes in the central core region. The leftmost and rightmost horizontal coordinates for each line of the central core are found and saved as run length codes [23].Then any 0's in the central core image between these points are changed to central core values.

2.1.10. *Concavity filling of propagated central core*

Occasionally, concavities may appear in the propagated central core image. These are filled by applying the same hole filling algorithm as in step [2.1.5] but for *all* spot pixels.

2.1.11. Round corners

The propagation algorithms applied above tend to leave the corners rather sharp. These are rounded out by applying the following neighborhood conditions (as in step [2.1.8]) to each central core pixel. If an exact match is made, then the 0 pixel on the diagonal is propagated and thus rounds out the corner (value N being $C + 100$).

0	N	–	–	N	0	–	–	–	–	–	–
N	C	–	–	C	N	N	C	–	–	C	N
–	–	–	–	–	–	0	N	–	–	N	0

2.1.12. Spot features and initial sizing

After the final SPL is computed, it consists of the pixels in the central core and propagated central core. Several features are computed using density values mapped from the average image. A preliminary spot sizing is performed to remove most of the background noise spots in step [2.1.8] where a 254 code is also placed in each deleted spot pixel in the propagated central core image.

2.1.13. Background correction

A background density correction is performed during a third pass through the image using a zonal notch filter algorithm similar to that described in [37]. A running average of the averaged image is computed (see [38] for algorithm description) for a $n \times n$ movable averaging window masked by the *complement* of the central core image. That is of background pixels which are not isolated pixels, deleted spots, central cores or propagated central cores. The result constitutes the background image and may be saved in the 2nd derivative magnitude scratch image since it is no longer needed. The mean background for a spot is then estimated by reading the background image at the centroid of the spot.

2.1.14. Computing corrected density and secondary sizing test

The features presently used to determine acceptance of a spot include: spot area (in square pixels), total integrated corrected spot density and range of pixel OD seen in the spot. The last is useful for eliminating small noise spots from the image. The corrected spot density D' is computed from D taking the background estimate into account. Those spots which meet the criteria are saved in the Gel Segmentation File (GSF). Spots failing the final density sizing criteria are deleted as before by having 254 placed in the central core image of each pixel. The gray value numeric data for the final set of spots may be optionally saved in an output image.

It is possible for the spot feature sizing parameter limits to be made more restrictive to eliminate some of the smaller noise objects. Table 2 shows part of a gel segmentation file for a typical gel.

2.2. Example of gel segmentation

We now show some image segmenter output which illustrates the wide range of effectiveness of the algorithm. Other results of segmentation are deferred to later. The segmentation algorithm appears to be applicable to a wide range of gel

TABLE 2

Example of Gel Segmentation File

```
SG2DRV :Version March 17, 1981 - 5:12AM
Today's date is 3/27/1981, 10:00:18 AM
User:[61,1]
Gel Segmentation File is: P20250.GSF
0250.2/P388D1/-/-/8-21-80/#A95/FISCHER'S/3:10, 10%/
0 HRS/C14/8/1 WEEK/ALUMINUM,T0,CONTROL,BOTTLE#1/
L00157/-NONE-/--NONE--/VIDICON-AUTO,28MM F8,69CM/LIPKIN*
 40 65 88 109 126 144 159 170 181 190 196 208 213 0 0 0 31 460 51 400
Switches: /7X7LOWPASS /ALLOWTOUCHINGEDGES
Window [31:460,51:400]
Area sizing limits (13.00: 500.00)
Density sizing limits (1.00: 500.00)
Density range sizing limits (.05: 2.70)
Saving output image in [61,2]Z00157.PIX
Mean background matrix in ND (std dev)
 .13(+-.03) .12(+-.03) .12(+-.03) .12(+-.03) .13(+-.04) .16(+-.05) .20(+-.06)
 .14(+-.03) .14(+-.03) .17(+-.04) .19(+-.04) .23(+-.04) .23(+-.05) .32(+-.12)
 .15(+-.03) .16(+-.03) .23(+-.05) .25(+-.05) .29(+-.06) .34(+-.05) .30(+-.05)
 .15(+-.03) .18(+-.04) .21(+-.04) .23(+-.04) .25(+-.05) .27(+-.05) .25(+-.05)
 .15(+-.03) .16(+-.04) .16(+-.03) .18(+-.04) .21(+-.04) .21(+-.04) .21(+-.04)
 CC#  1 M.E.R[ 436:444,  55: 59] D.R.=[  .13: .42] D/A= .259 MnB= .128
  1st MOM[ 440.26, 56.35] A=   22  D= 5.70 D'= 2.88 (D'/totalD')%=  .06%
  Sx= 1.68 Sy= 1.10 Sxy=  .79      V= 5.51
 CC#  2 M.E.R[ 448:453,  63: 69] D.R.=[  .28: .68] D/A= .527 MnB= .128
  1st MOM[ 451.63, 66.02] A=   25  D= 13.18 D'= 9.98 (D'/totalD')%=  .19%
  Sx= 1.38 Sy= 1.57 Sxy=  .99      V= 10.49
 CC#  3 M.E.R[ 436:439,  63: 69] D.R.=[  .21: .41] D/A= .327 MnB= .128
  1st MOM[ 438.01, 66.47] A=   20  D= 6.55 D'= 3.99 (D'/totalD')%=  .08%
  Sx=  .94 Sy= 1.63 Sxy=  .88      V= 4.47
       .
       .
       .
 CC# 685 M.E.R[ 144:150, 397:399] D.R.=[  .16: .34] D/A= .234 MnB= .128
  1st MOM[ 147.03, 398.31] A=   19  D= 4.44 D'= 2.01 (D'/totalD')%=  .04%
  Sx= 1.88 Sy=  .83 Sxy= 1.03      V= 3.72
Total of    686 accepted D spots accumulated density= 7872.56, area= 20625
Total of    685 accepted D' spots accumulated density= 5232.56, area= 20625
Total of   7265 omitted spots accumulated density= 8518.47, area= 42756
Omitted/Accepted density =  108%
```

Part of the gel segmentation file (GSF) output of the SG2DRV program for a ^{14}C-labeled P388D1 macrophage like cells autoradiograph gel *ACC#* 250.2. Segmenter parameters and some of the spot feature list data are presented. *CC* is the spot connected component number, *MnB* is mean background density, *A* is spot area, *D* is uncorrected total spot density and corrected density D' is computed as $D - (A)(MnB)$. D/A is the mean density and $(D'/\text{Total}D')\%$ is D' expressed as a percentage of total gel spot density. 1st MOM is the spot's centroid while *D.R.* is the density range of pixels seen in the spot. *Sx*, *Sy* and *Sxy* are the standard deviation and covariance of spot size with *V* being the Gaussian volume estimate of density for this region. Density values are in OD calibrated in terms of the associated ND wedge in the image.

magnifications and densities, i.e., autoradiographs of varying exposures and spot detection modalities, etc. It is also capable of resolving touching spots and other image complications over a wide range of conditions.

The segmenter has been applied to various types of gels (both autoradiographs and silver stain) of different types of material scanned at different magnifications with satisfactory results. At least 500 gels have been segmented using this program. Figure 5 is a composite photograph showing two P388D1 gel images before and after segmentation. As can be seen in Fig. 5b and 5d, the segmentations of spots were successful in a vast majority of cases. The number of spots segmented in these two gels were 547, and 672 respectively.

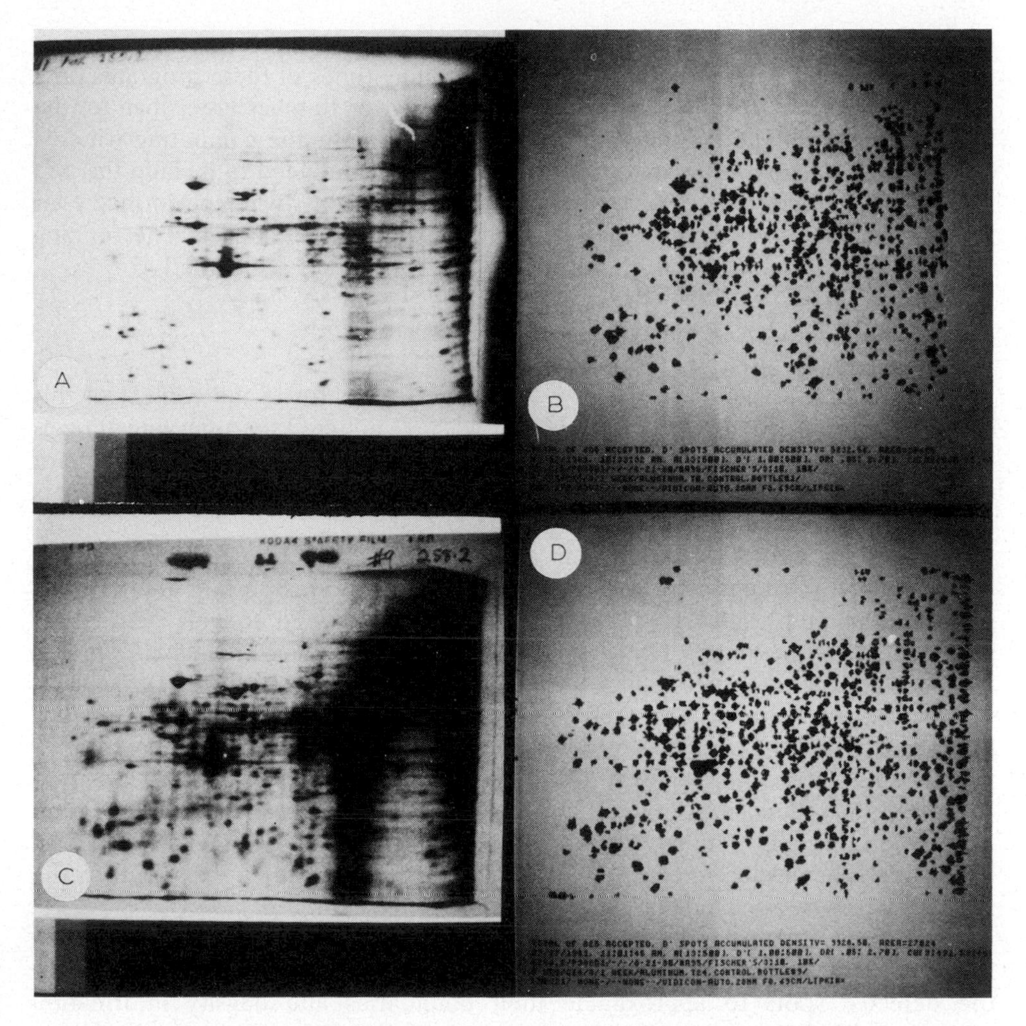

Fig. 5. Composite photograph showing two different P388D1 mouse macrophage gels before and after segmentation has been performed. (a) T0 gel 250.2; (b) segmentation of (a); (c) T24 control gel 258.2; and (d) segmentation of (c).

A set of sizing parameters for various classes of gels have been empirically determined. The parameters currently used for the 250 microns/pixel autoradiographs are: total spot area range [13–500] square pixels, total integrated spot density range [0.1–500] OD, spot pixel density range [0.03–2.7] OD. For the 170 microns/pixel autoradiographs, the total integrated density minimum limit is increased to at least 1.0 OD. For silver stained gels at 250 microns/pixel, the total integrated density minimum is 1.0 OD and the minimum spot pixel density range up to 0.15 OD.

Typical times for segmenting 250 microns/pixel resolution gels on the DECSYSTEM-2020 (KS-10 CPU) are on the order of 8 to 18 min/gel for gels on the order of 1000 spots depending on the segmentation options (such as low pass filter size, etc.) as well as number of spots. Running times of these programs on a KL-10 CPU TOPS-20 system seem to be of the order of 10 times faster than for the KS-10 processor. The 250 microns/pixel gel images have the computing window set to about 2/3 to 1/2 the area of the gel because of the need to include the ND wedge in the image. These computing times increase slightly when performed over the full 512 × 512 pixel image for the higher resolution gel images. The running times also increase somewhat with the number of spots being segmented.

2.3. Some other spot extraction algorithms

We presented the requirements for a gel image segmenter in the context of this particular problem domain. The segmenter we have adopted for analyzing 2D gels is only one of many applicable to this problem. We will show reasons for selection of our approach after consideration of some other segmenters and spot extractors which have been applied to gel images.

A large number of image processing techniques have been brought to bear on the detection and extraction of objects from images [22–23]. Some of these, using global techniques such as thresholding, fail when applied to gels because of inhomogenity of the image. Others using local algorithms which are globally applied have been much more successful.

A threshold segmentation technique [12,21] analyses the gel image density histogram to find the mean background value and then thresholds the image at 0.05 to 0.10 OD above this value to detect faint spots. Darker clusters of spots are segmented together using this technique into a single 'spot'. Each 'spot' is then tested to find maxima and minima regions within it to determine whether it should be iteratively split into subspots. Spots found in this way are then expanded to the region defined by a second derivative of the gray scale data.

Another segmentation technique [8,24] detects spots by first scanning in a raster direction and then orthogonally to find spot maxima. Ellipses are fitted over the detected spots to approximate their boundaries and density information measured. This group has also developed algorithms for spot extraction using convolution techniques [25].

Another method of spot detection is based on finding spots in the image starting at the darkest pixel in the image and then removing fitted spots at each step of the search [10]. The algorithm assumes that the spots are approximately Gaussian in x and skewed Gaussian in y. It fits parametric curves to the spots and then estimates the volume density from the parameters.

Spots may be detected by assembling detected line segments of spots from successive lines using a procedure called 'chain assembly' [11]. The edges of adjacent chains are smoothed and Gaussian curves fit to estimate the spot.

2.4. A distribution-free density independent spot segmenter

The algorithm we have presented uses the central core model of a spot. The central core was defined as the region of negative slope of the second derivative function of the image. Such a region occurs within the first minimum of g'' surrounding the spot's center. This model seems to work robustly on real gel data. It has failed only on those few spots of such a huge extent, saturated, or so noisy that no simple model exists for the spot. In addition, it is independent of assumptions as to exact spatial distributions of density as well as of orientation.

Because the shapes of gel spots corresponds to the physical diffusion process used in their generation, no explicit boundaries are present. Manual measurements made at what observers subjectively define as the boundary have resulted in up to 50% error in total integrated spot density in the case of fuzzy spots. We have chosen to algorithmically define a spot as its propagated central core which is found to be reproducible.

A common problem of many gels is artifactual streaking. One may attempt to remove streaks before processing or alternatively, the spots may be segmented in the context of the streaks. The central core algorithm finds spots regardless of whether the streak is present or not. Therefore streak removal by pre-processing is not necessary.

The central core algorithm finds a spot's peak if it is present and if there is sufficient resolution, in both the spatial and density domains of the image, will resolve overlapping peaks. The cases where it fails can be understood if one looks at the difference formula for g''. This discrete approximation to g'' can not resolve spatial position differences less than about 5 pixels. Thus gels scanned at higher resolution will show fewer unresolved touching spots after segmentation than gels scanned at lower resolution.

Two overlapping saturating spots will also sometimes be unresolved. Because the plateau effect occasioned by saturation obscures the second peak in g'' which is necessary for spot separation. If one wishes to keep track of and modify the segmentation of saturated spots it is necessary to employ a different morphologic analysis. In any case, saturated spots are tracked throughout the entire analysis process. Spots saturating in one gel might not do so in another so that substitute measurements could be made in the case of multiply exposed autoradiographs.

In practice, spots are frequently somewhat distorted so that an idealized Gaussian shaped spot may rarely be found. The segmenter works well for such spots because their extent is defined by the second maximum of the g'' magnitude over *all* of the spot's edge.

The sensitivity of the spot detector (e.g., stain) varies among gels so it is necessary to normalize spots in each gel in order to compare them. Normalization will be discussed later. However, the segmenter performs one type of normalization which is useful for well segmented gels. In addition to reporting each spot's total integrated density (D) and its background corrected density (D'). The segmenter also reports D' divided by the sum of D' for all spots accepted expressed as a percentage. These density features are illustrated in Table 2.

Scans of the same gel at two different resolutions provide an opportunity to investigate differences in segmenter behavior for the same set of spots. Use of PIXODT, image debugger [1], leads to the conclusion that the few instances where spots are incorrectly merged is due to either (a) lack of spatial resolution or (b) gray scale resolution (spots were close to or at saturation) or were very noisy. Overall, the correlation was good with most spots being correctly segmented.

3. SPOT PAIRING

We have treated the problem of spot extraction within a single gel using the SG2DRV program. Now we consider the first step in locating a particular spot in a set of gels, i.e., pairwise matching of the spot in two gels. This can be done by shifting one of the gels until the spot overlaps and recording the cartesian coordinates of the spot in each gel. This spot by spot pairing is a prerequisite to detecting whether individual polypeptides change with respect to experimental conditions. Furthermore, pairing of spots within a set of gels taken two at a time is the means whereby a multiple gel data base is gradually constructed.

Referring back to Fig. 1, gels are first acquired, then spots are segmented using the SG2DRV program. This resulted in a gel segmentation file (GSF) consisting of a list of spot (x,y, density) triples (as in Table 2). Note that in the table each spot has a spot index (which can be used to refer to the spot), a (x,y) centroid and a density measurement given in several formats. An algorithm for spot pairing between two gels using a small set of landmark spots to locally align subregions is presented which uses the GSF spot list files as data. The CMPGEL program implements this algorithm producing a gel comparison file (GCF).

3.1. Partitioned search in pairing spots

A major problem complicating spot localization is the local distortions in the gel such that neighboring spots in one gel will likely be neighbors in another gel while the intervening distances between them vary to some degree. Several semiautomated methods for aligning corresponding spots in two gels are discussed.

In one, a gel is transformed locally to the distortions of a second gel [12,21,25]. Sets of three evenly spaced corresponding landmark spots are manually defined for both gels covering regions to be transformed. A linear approximation to the affine transformation of this region is performed to translate, rotate and stretch the image locally. After the transformation, spots from the two gels are mapped to the same domain. Then a least squares fitting procedure matches those pairs less than a specified distance apart in the two gels.

An interactive mode using a color display allows comparison of spots between gels while geometric correction is done locally using a linear interpolation in localized regions of a pair of gels [8,24]. The corrected images are then used to produce protein maps for the local regions, can be investigated independently [24] and serve as a basis which to discuss a subset of spots.

3.2. A landmark driven spot pairing algorithm

We present an alternative view of the primary pairing algorithm. This involves in effect constructing a projected image composed from the two members of the gel pair. In the actual matching, computations are performed only on (x,y,d) data in a single plane, the representive gel plane. Central to the algorithm is the establishment of landmark spots that serve to 'anchor' the other spots in its vicinity. Essentially, landmark spots are manually aligned in the two gels at which point the computer automatically aligns all other spots with the corresponding spots in the other gel. The procedure is simple and is easily extended to align any number of gels.

Such partitioning by landmark region increases the efficiency of integral spot matching by providing an empirical basis for the partitioning of a gel image into tractable corresponding subregions.

A landmark spot should be selected according to particular criteria. It is a morphologically distinctive spot present in all gels such that neighboring spots and the landmark spot form a consistent morphologic structure. Moreover, this morphologic structure should be easily recognized across the set of gels. The landmark spot should not be a touching spot. The set of landmark spots are selected to fairly easily cover the regions of interest of the gel fairly. From 10 to 25 landmarks are generally selected depending on the quality of the gel with fewer required for better gels. This set of spots is called the landmark set. In practice, the operator aligns the landmark spots in the two gel images using the flicker algorithm [9], which permits one of the gels to move while keeping the other constant. Viewing time for each of the images may be independently set and varied until the user is satisfied that the two images of the same spot are locally 'superimposed'. The superimposed spot of interest is noted to the computer and the next landmark spot processed in the same fashion. The flicker procedure is described in [9] and landmark acquisition in [2].

GELLAB also offers alternative facilities for generating landmark spot data without using our special purpose RTPP interactive hardware. Program DWRMAP draws a labeled outline-plot of the segmented gel from the GSF file with the darkest spots (sorted by density) labeled in the plot with a table also given on the side of the plot. Spots are represented by an oval proportional to spot density. By manually comparing such R-map plots, corresponding lists of landmark spots can be defined using the CC#s of each gel. Program LMSEDIT then allows the manual entry of a landmark set as a list of pairs of CC#s from the two gels.

The landmark region surrounding a landmark spot is defined as a polygonal region having higher pairing certainty for spots closer to the landmark spot. The half-radius of certainty Ri for landmark i is a distance defined to be half the distance from the nearest landmark spot landmark i. Spots within the half-radius of a landmark set have a higher probability of being aligned (since the landmarks have 'perfect' integral alignment) than if the spot were outside of this radius. The landmark spots in each gel are compared with the two GSF spot lists and the best segmented spot's centroid is used rather than the coordinates manually produced. If no spot can be found for a landmark within specified error bounds (currently the dT2 distance; see below), then the manual landmark coordinates are used. The CGELP program to be discussed (cf. Section 4) has an operator VALIDLANDMARKS which computes a table and statistics of all valid landmarks for all gels in a multiple gel data base. It is

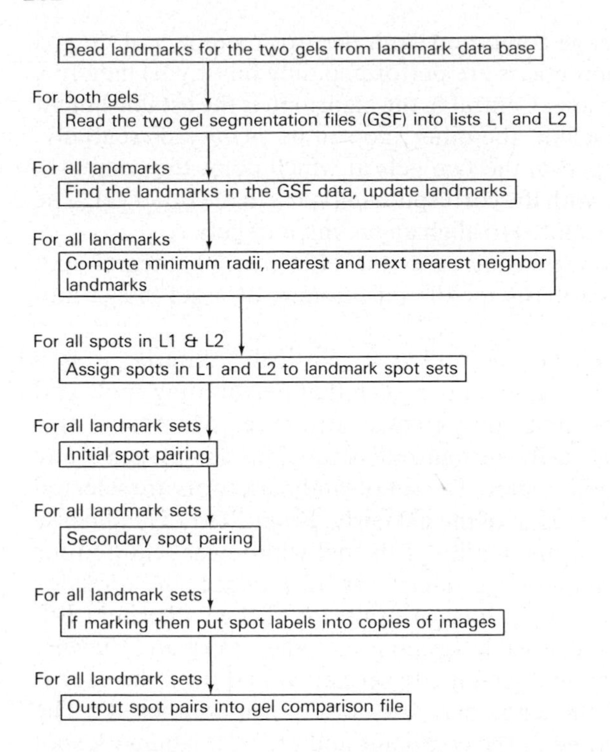

Fig. 6. Block diagram of the 2D-gel comparison procedure where spot segmentation lists are merged using a small set of landmarks to produce partitioned landmark sets of paired spots.

possible to ascertain how reliable the particular pairing actually was by backchecking paired spots in a set of pairing-labeled images to be discussed.

Partioned search has the added advantage that landmark regions contain an order of magnitude fewer spots than the total gel space. Therefore the combinatorics of performing the spot matching is greatly decreased as well.

3.3. Algorithm for landmark-oriented spot pairing between two gels

The spot pairing algorthim is illustrated in flow chart form in Fig. 6. It is implemented as the CMPGEL program. Pairing is performed in two passes through the landmark sets data using the primary and secondary pairing procedures. Each procedure operates on one landmark set at a time, in both gels.

In the primary pairing algorithm (Fig. 7), spots are first mapped to the Cartesian coordinate system defined by shifting the landmark spot to (0,0) relative to the origin in the two gels G1 and G2. Each spot in G1 is provisionally paired to the spot that is its nearest neighbor (by minimum Euclidean distance) in the projected image of G2. Because of possible asymmetry of the two landmark regions, the reverse comparison is also performed so that each spot in G2 is provisionally paired with its nearest neighbor spot in G1. This nearest neighbor distance is denoted dP (pair

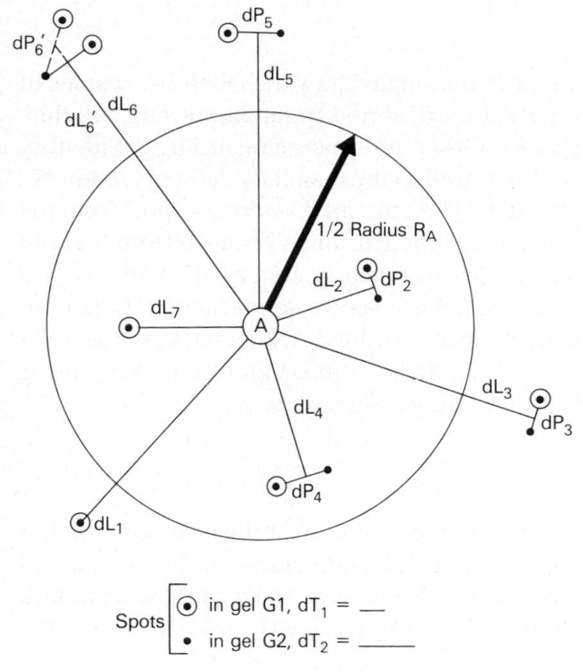

Fig, 7. During primary spot pairing labeling assignment, each potential nearest neighbor spot pair in a landmark set is assigned one of four labels: SP, sure pair; PP, possible pair; AP, ambiguious pair; US, unresolved spot. The labelings are defined by the following cases: [1] US – unresolved spot (no *dP*); [2] SP – $dL2 < Ra$ and $dP2 < dT1$; [3] PP – $dL3 > Ra$ and $dP3 < dT2$; [4] PP – $dL4 < Ra$ and $dP4 > dT1$ and $dP4 < dT2$; [5] PP – $dL5 > Ra$ and $dP5 < dT1$; [6] PP – $dL6 > Ra$ and $dP6 < dT2$. For the other spot AP′ – $dL6' > Ra$ and $dP6' < dT2$ and $dP6' > dP6$; [7] US – unresolved spot (no *dP*).

distance). The distance from the landmark spot to the mean locus of the two spots in the provisional pair is denoted dL. Two user specified parameter distances are empirically defined: dT1 and dT2. Spots closer than dT1 to the landmark spot are relatively well paired. Spots greater than dT2 are poorly paired and possibly should not be paired. The current values of dT1 and dT2 (5 and 10 pixels respectively) were determined empirically, by examination of the nearest neighbor values of many sets of paired gels under gel resolution range of 170 to 250 microns/pixel. Figure 7 shows various pairing which can occur. Four types of pairing labels can be defined. These are sure pair 'SP', possible pair 'PP', ambiguous pair 'AP' and unresolved spot 'US'. The primary spot pair labeling assignments are defined in Fig. 7.

The primary pairing algorithm is a simple first order model not taking some spots on the periphery of the landmark region into account. These spots may be misclassified as an AP or US whereas they would be a SP and PP classification in another adjacent landmark region. To correct these few misclassification errors, a secondary pairing algorithm is applied in order to possibly re-pair AP and US spots in the next-nearest landmark set using AP and US spots from those sets. The resultant re-paired spot pair (either a SP or PP if it meets the threshold criteria) is then placed in the landmark set with the smallest dL value.

3.4. CMPGEL output

Finally, after spot pairing, the program can optionally draw the labels into copies of the original images. The paired spot data, (x,y,d) sorted by landmark sets, are then output into the gel comparison file (GCF). Other information regarding the identity of the two gels and gel segmentation files as well as the manually defined landmarks is part of the permanent preface to the GCF. The estimated landmark spots from the GSF found in the GSFs are also reported as is the Euclidian distance from them to those manually defined by the user. If this distance is greater than dL from a landmark for either G1 or G2, then that landmark spot is so marked and the GSF spots are partitioned using the manually defined coordinates, landmark spot sets. At the end of the GCF is a statistics summary for both the primary and secondary pairing reguarding the number of each of the four pairing assignments.

3.5. Example of spot pairing

The spot pairing algorithm just described has been in use over the past year and has been applied to over 500 gels. Figure 8 shows the landmark spots for the pair of P388D1 gels with each of these spots marked with a small plus sign and the landmark name to its right. Table 3 illustrates a typical landmark set entry for the 250.2/258.2 pair of gels.

After the spot labels have been assigned, copies of the original images may be overwritten with the label names for all spots in the spot lists. SP, PP and AP labels appear in the marked image as 'S', 'P', and 'A' while US appears as a small '+' (because using the larger 'U' symbol would crowd the image where it is noisy). Figure 9 shows the labeled pairs for the P388D1 gel pairs where landmark spots in these images have a box around them. Table 4 shows part of the output of a typical GCF for gels 250.2 and 258.2. The first part of the table illustrates the landmark registration and parameters while the rest shows some representative paired spots as well as pairing statistics.

Secondary pairing for gels 250.2/258.2 increased the (SP+PP)/total spots percentage from 62.7% to 64.0%. In general we observe a 1.5% to 3% increase in (SP+PP) labeling which seems to indicate that most pairing is performed (as expected) during primary pairing. When two very different gels (with widely different number of total spots extracted) are compared as is expected a fairly high number of US and AP spot labels result. This does not mean that the spots that are SP or PP paired are not paired well. For similar gels, the (SP+PP) ratios are in the range of 65% to 85% depending on the artifactual noise of the gels.

3.6. Global gel matching

In general, the global matching of two gels would involve a D'arcy Thompson type transformation [26] of one of the gels to bring it into congruence with the other. The set of points on the grid are displaced by a continuously differentiable 2D distortion function in this transformation. One approach to gel analysis is to remove

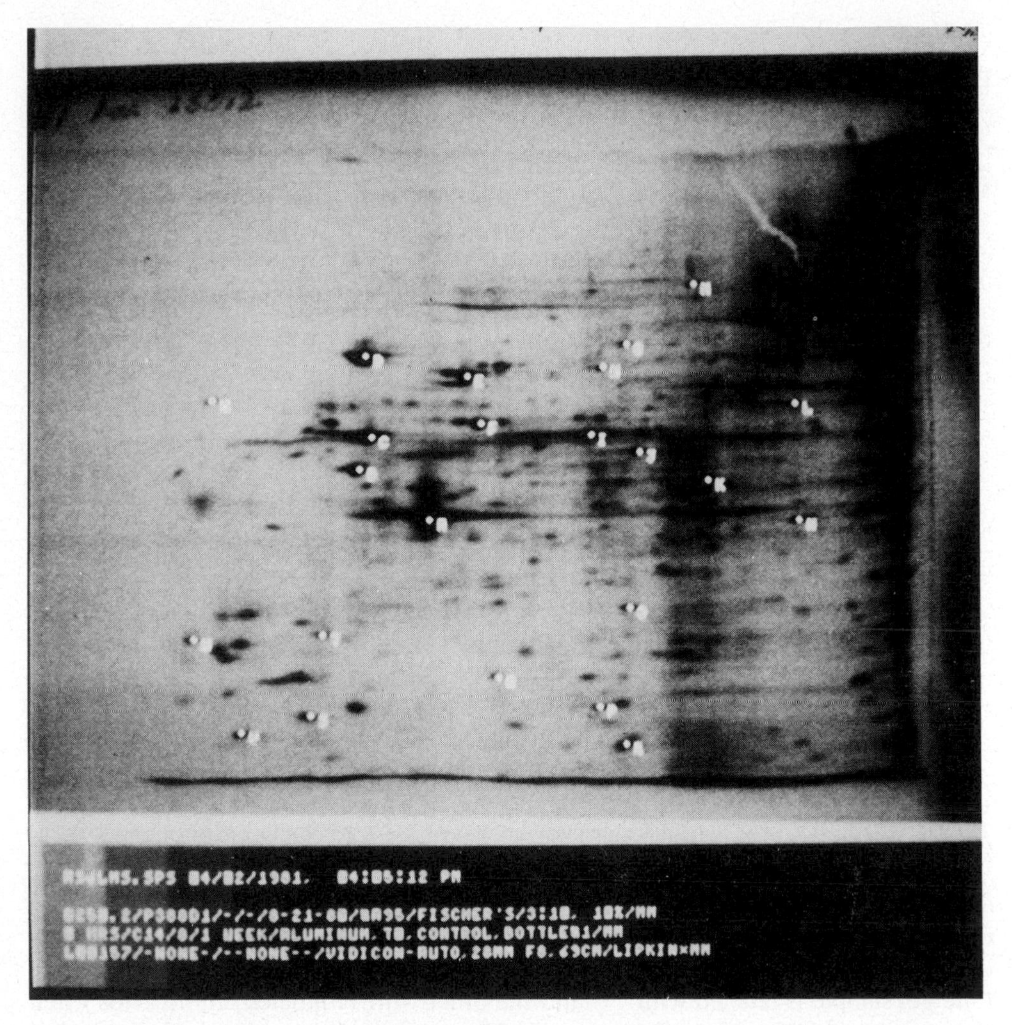

Fig. 8. Landmarks used in pairing the two P388D1 gels superimposed on R-gel 250.2. The landmark is defined at the small '+' sign with its landmark set name its right.

the distortion (as is currently done with satellite images) and then perform a point by point comparison [22]. This correction implies some knowledge of the complete inverse distortion function that is by and large lacking for 2D gels although estimates are computed using landmark triples [12,21,25].

The amount of computation required to perform the D'arcy Thompson image transformation for every pixel in the image is considerably greater than for simply pairing spots which may be thought of as sublandmark registration. Since the actual gel analysis requirement is to pair spots for comparison, it is more efficient to simply pair locally corresponding spots rather than to transform the gel images themselves and then compare all of the spots.

The landmark driven pairing algorithm has an additional advantage, i.e., it is also

TABLE 3

Example of a landmark set from the landmark data base file

```
/ CMPGEL: VER# 9/23/80 - 9:09AM
/ INTO SYS@:JUNK@@.DA FROM GSF FILES: P20250.GS AND P20258.GS
/ SURE!PAIR THRESHOLD= 5, POSSIBLE!PAIR THRESHOLD= 10
01/11/1981,    11:14:04 AM
  LANDMARK #A   G1[211, 262], G2[199, 265]
  LANDMARK #B   G1[173, 235], G2[160, 236]
  LANDMARK #C   G1[180, 219], G2[166, 220]
  LANDMARK #D   G1[177, 176], G2[162, 177]
  LANDMARK #E   G1[231, 185], G2[221, 186]
  LANDMARK #F   G1[239, 211], G2[229, 211]
  LANDMARK #G   G1[318, 168], G2[312, 168]
  LANDMARK #H   G1[305, 182], G2[297, 183]
  LANDMARK #I   G1[292, 218], G2[285, 219]
  LANDMARK #J   G1[325, 226], G2[317, 227]
  LANDMARK #K   G1[363, 241], G2[354, 244]
  LANDMARK #L   G1[410, 200], G2[400, 201]
  LANDMARK #M   G1[355, 138], G2[350, 137]
  LANDMARK #N   G1[413, 261], G2[399, 268]
  LANDMARK #O   G1[322, 308], G2[311, 311]
  LANDMARK #P   G1[307, 362], G2[293, 368]
  LANDMARK #Q   G1[321, 382], G2[308, 388]
  LANDMARK #R   G1[248, 346], G2[235, 349]
  LANDMARK #S   G1[149, 367], G2[134, 371]
  LANDMARK #T   G1[154, 321], G2[133, 324]
  LANDMARK #U   G1[ 89, 324], G2[ 65, 324]
  LANDMARK #V   G1[115, 375], G2[ 94, 381]
  LANDMARK #W   G1[ 99, 198], G2[ 71, 198]
```

An example of a typical landmark set entry in the landmark data base file. Entries are accessed by gel name pairs, (e.g., by the gel name pairs (250.2,258.2) or (258.2,250.2)). G1 (G2) corresponds to gel 250.2 (258.2) The G1 $[x,y]$ 2-tuples are the positions of the spots in gel i for the specified landmark and similarly for G2.

applicable to automatically defined landmarks, should they become available. One interesting technique that has been used to circumvent the gel distortion problem is double labelling [27]. Spot pairing is greatly simplified if two gels are congruent. One gel sample is labeled with ^{14}C and the other with ^{3}H-labeled amino acids when the samples are grown in deficient media. The two samples are then added together just prior to running the gel. Two-step autoradiography is performed for the ^{14}C- and for ^{3}H-labeled gel. Unfortunately, this technique can only be used with material which can be radioactively double labeled. An analogous role could be played by fluorescent labeling of known spots for stained gel images.

3.7. *Biases due to landmark selection*

The partition of the plane into variable sized polygonal landmark regions is based essentially on the local spread of landmarks. A priori one would think that pairing would be more likely to be correct in regions of high landmark concentration. However, a small radius of confidence may have undesirable effects on pairing, i.e., spots that would otherwise be matched as sure pairs might be entered into the probable pair category. It is possible for a spot to be paired to be found in the *next* to

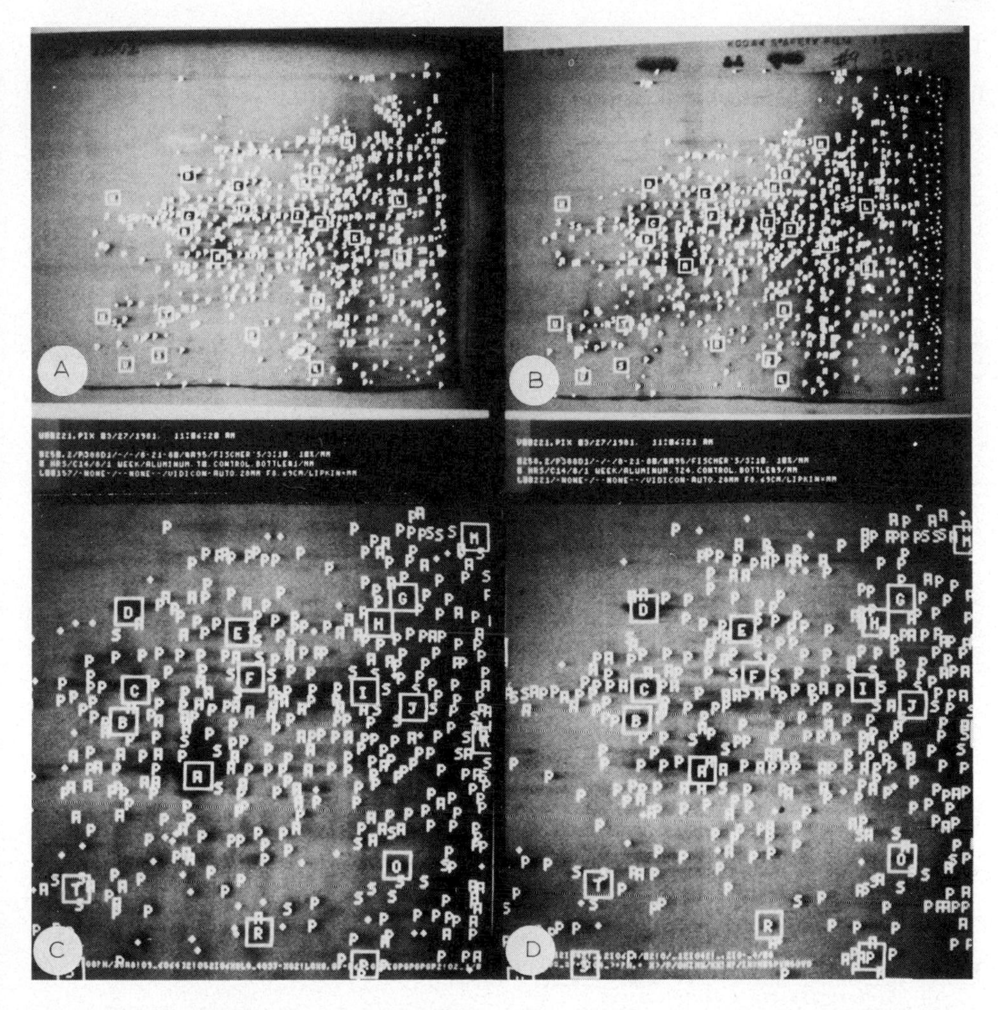

Fig. 9. CMPGEL labeled marked images with pair labels: S, sure pair; P, possible pair; A, ambiguous pair; '+', unresolved pair. Gels 250.2 and 258.2 are labeled in (a,b), a 2× magnified central region of these images is in (c,d).

next nearest neighbor landmark set rather than the landmark or next nearest landmark sets. In digital space, the problem of a possible shift of a spot from one landmark region to another, such that pairing would be affected, as a result of increasing the concentration of landmarks is obscure and does not seem easily treated.

The consideration of correctness and completeness of the primary pairing algorithm is not simple although the algorithm in itself is quite straightforward. Performance should not be gauged exclusively on the results when gels of widely different spot numbers are compared. On the other hand, comparisons of closely similar gels should yield good results.

TABLE 4

Example of Gel Comparison File

```
CMPGEL[50,32]: Version Jan 26, 1981 - 1:50PM
Today's date is 03/27/1981,  11:01:56 AM
User: [61,1]
Gel Comparison File is: C20258.GCF from P20250.GSF and P20258.GSF
0250.2/P388D1/-/-/8-21-80/#A95/FISCHER'S/3:10, 10%/
0 HRS/C14/8/1 WEEK/ALUMINUM,T0,CONTROL,BOTTLE#1/
L00157/-NONE-/--NONE--/VIDICON-AUTO,28MM F8,69CM/LIPKIN*
 40 65 88 109 126 144 159 170 181 190 196 208 213 0 0 0 31 460 51 400
0258.2/P388D1/-/-/8-21-80/#A95/FISCHER'S/3:10, 10%/
0 HRS/C14/8/1 WEEK/ALUMINUM,T24,CONTROL,BOTTLE#9/
L00221/-NONE-/--NONE--/VIDICON-AUTO,28MM F8,69CM/LIPKIN*
 40 65 88 110 127 144 159 171 182 191 197 208 213 0 0 0 2 491 53 405
Distance sizing limits (dP1 =  5.00, dP2 =  10.00):
Switches:  /MARK
LMSLL.LM from gel ACC#'s 0250.2 and 0258.2
The (Representitive)R-gel is: 0250.2
 LANDMARK #A  G1[211, 262], G2[199, 265]
 LANDMARK #B  G1[173, 235], G2[160, 236]
 LANDMARK #C  G1[180, 219], G2[166, 220]
        .
        .
        .
 LANDMARK #W  G1[ 99, 198], G2[ 71, 196]
G1[A, 437][ 208, 263],E.Diff= 3.2, G2[A, 502][ 198, 266],E.Diff= 1.4-OK
G1[B, 358][ 173, 236],E.Diff= 1.0, G2[B, 413][ 158, 237],E.Diff= 2.2-OK
G1[C, 287][ 177, 219],E.Diff= 3.0, G2[C, 326][ 164, 220],E.Diff= 2.0-OK
        .
        .
        .
G1[W, 228][  96, 199],E.Diff= 3.2, G2[W, 260][  68, 198],E.Diff= 3.0-OK
R[A]=  22 to nearest LMs[B,B], next nearest LMs[C,C]
R[B]=   9 to nearest LMs[C,C], next nearest LMs[A,A]
R[C]=   9 to nearest LMs[B,B], next nearest LMs[A,A]
        .
        .
        .
R[W]=  42 to nearest LMs[D,D], next nearest LMs[C,B]
Marked gel comparison files are: U00221.PIX and V00221.PIX on [61,2]
G1 HAS  685, G2 HAS  825 SPOTS
TOTAL DENSITY G1= 5232.56, G2= 9928.50
OMITTED TOTAL DENSITY G1= 8518.47, G2= 11116.49

LM[A] G1 HAS   46, G2 HAS   45 SPOTS
#A G1: 549[-22, 43]&G2: 619[-25, 34] PP,DP=9.5,DL=48,D1=1.4,D2=1.6od
        .25Maxd1 .29Maxd2  15A1  13A2 .19MinD1 .23MinD2
        1.40sX1 1.18sX2 .81sY1 1.05sY2
#A G1: 352[ 1,-29]&G2: 403[ -2,-32] PP,DP=4.2,DL=32,D1=5.1,D2=11.0od
        .44Maxd1 .56Maxd2  20A1  34A2 .34MinD1 .41MinD2
        1.60sX1 1.64sX2 1.06sY1 1.64sY2
#A G1: 369[ 0,-23]&G2: 429[ 6,-23] AP,DP=6.0,DL=24,D1=10.5,D2=17.5od
        .54Maxd1 .74Maxd2  36A1  43A2 .30MinD1 .36MinD2
        2.11sX1 1.53sX2 1.34sY1 2.25sY2
#A G1: 376[ 5,-20]&G2: 429[ 6,-23] PP,DP=3.2,DL=24,D1=9.0,D2=17.5od
        .57Maxd1 .74Maxd2  26A1  43A2 .32MinD1 .36MinD2
        1.06sX1 1.53sX2 2.01sY1 2.25sY2
#A G1: 394[-5,-17]&G2: 450[ -7,-19] SP,DP=2.8,DL=20,D1=23.7,D2=29.2od
        .73Maxd1 .97Maxd2  75A1  56A2 .25MinD1 .35MinD2
        2.96sX1 2.02sX2 1.87sY1 2.06sY2
#A G1: 514[-30, 24]&G2:   0[  0,  0] US,DL=39.0,DL=39,D1=1.4,D2= .0od
        .25Maxd1 .31Maxd2  16A1  19A2 .18MinD1 .24MinD2
        1.08sX1 1.14sX2 1.35sY1 1.38sY2
#A G1:   0[  0,  0]&G2: 616[-20, 34] US,DL=33.0,DL=33,D1= .0,D2=1.6od
        .23Maxd1 .28Maxd2  15A1  14A2 .18MinD1 .23MinD2
        1.15sX1 1.27sX2 1.31sY1 .92sY2
                .
                .
                .
LM[B] G1 HAS  19, G2 HAS  30 SPOTS
#B G1: 306[-23,-15]&G2: 355[-26,-14] PP,DP=3.2,DL=30,D1=5.4,D2=1.7od
        .41Maxd1 .30Maxd2  29A1  13A2 .20MinD1 .25MinD2
        1.83sX1 1.08sX2 1.52sY1 .91sY2
#B G1: 481[  4, 37]&G2: 549[  1, 38] PP,DP=3.2,DL=38,D1=1.7,D2=2.0od
        .28Maxd1 .30Maxd2  14A1  15A2 .23MinD1 .25MinD2
        1.16sX1 1.05sX2 1.15sY1 1.14sY2
                .
                .
                .
PAIRING STATISTICS
------------------
        After Initial pairing:
US 194
SP 178
PP 732
AP 348
(SP+PP)/(US+AP+SP+PP)= 62.7%
```

The current pairing algorithm defines a sure pair (SP) as being within the landmark radius *R/i* for a given landmark *i*. We have found that most of the possible pairs (PP) are actually paired and should be pooled with the sure pairs as well-matched spots. Large numbers of ambiguous pairs and unresolved spots result when comparing two widely different or noisy gels. Currently, nothing is done with these (AP and US) spots. Although they are tracked through the data base. A possible extension to GELLAB processing would be to incorporate additional procedures to further process the AP and US spots such as merging AP fragments with the spots they belong with. Conglomerates of spots sometimes appear as single spots and other times as several spots, e.g., actin complex, so that merging spots is an attractive idea under the right conditions.

We have found that highly populated spot regions should have somewhat more landmarks, however landmarks should not be 'on top of' each other. Other criteria in landmark selection include using fewer landmarks if the regions have little distortion and line up fairly well. A landmark spot should be well defined morphologically and non-touching being part of a locally consistent pattern in all of the gels to be compared.

Manually landmarking a pair of gels takes from 3 to 30 min depending on the comparability of the gels with an average time being about 4 to 7 min. These times depend on gel quality (the single most important factor) and the set of landmarks selected (which must be in all gels to be compared). CMPGEL processing times are on the order of 2 min.

The ability to pair most of the spots in a set of gels enables examination of larger gel data bases where subtle shifts and correlations in the spot data can be more easily detected. Figure 10 shows a log density – log density scatter plot of two normalized paired (SP and PP labels only) P388D1 gels. Most of the spot pairs are close to the 45 degree line. Some of the outliers are real and some are due to noise in the entire gel-image processing system. We will now consider techniques for further resolving noisy data using multiple gels and means for facilitating the checking of outliers.

```
        After secondary pairing:
US 186
SP 178
PP 751
AP 337
(SP+PP)/(US+AP+SP+PP)= 64.0%
```

This is an example of the first part of the gel comparison file (GCF) output of the CMPGEL program applied to gels 250.2 (T0) and 258.2 (T24) in the effect of time experiment. The manually selected landmark spots are listed followed by the number of spots in each gel. The best spot estimates of the landmark spots from the GSF data are then given. The Euclidian distance betwen the segmented and manually defined landmark spots indicates how well these spots fit the landmark estimation. Half-radii and next-nearest neighbor landmark spots are then listed followed by total spot densities listed for each gel for both included and noise (omitted) spots. The second part of the GCF contains labeled landmark sets GCF with pairing statistics for gels 250.2 (G1 is the *R*-gel) and 258.2 (G2). In each paired spot, the bracketed 2-tuple is the relative cartesian distance from the spot to the landmark. The segmentation spot index preceeds the '[' for both G1 and G2. The spot reliability labels are (SP,PP,AP,US). Distance DP is defined as that between spots in a pair and DL is the distance of a pair from the landmark. D1 (D2) is the *D'* (background corrected) density measurement of the spot in G1 (G2). *Od* is the maximum OD value seen for either spot in the pair. MaxD1 (MaxD2) and MinD1 (MinD2) are the maximum and minimum density values seen for any pixel in the spot.

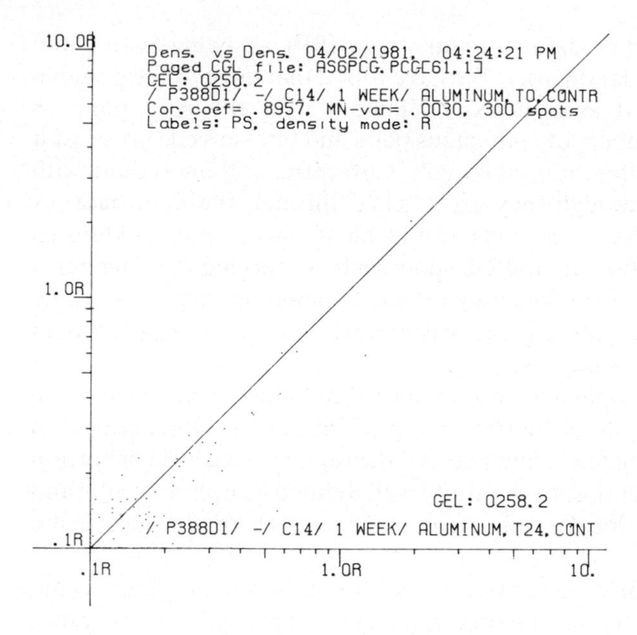

Fig. 10. Density vs. Density scatter plot (on a log–log scale) of P388D1 gels 250.2 (T0) and 258.2 (T24). The density values were first normalized using least squares (cf., Section 4) to gel 250.2.

3.8. *Role of pairing 2 gels in multiple gel data base analysis*

When comparing corresponding spots among a number of gels, the pairing performed by this algorithm is a prerequisite for the analysis of multiple gels where one treats the values of particular spots within a set of gels. Figure 11 shows the steps in the data reduction performed in the multiple gel analysis in GELLAB. Gel segmentation files (GSF) consisting of spot lists are merged using the gel pairing algorithm into gel comparison files (GCF). These in turn are merged into a gel data base (PCG). We next discuss this last procedure and its ramifications for 2D gel analysis.

4. MULTIPLE 2D GEL ANALYSIS

Earlier we have discussed the need for computer support of 2D gel electrophoresis analysis. Such support along largely data structural lines has been shown to be essential. We have treated the problems of spot extraction and pairwise spot comparison and in the process have indicated that experiments involving time or dose variables require comparisons of spots from multiple gels. We now deal with multiple gel comparisons, the most powerful and demanding mode of application of 2D electrophoresis to biological and clinical investigation and describe a computer

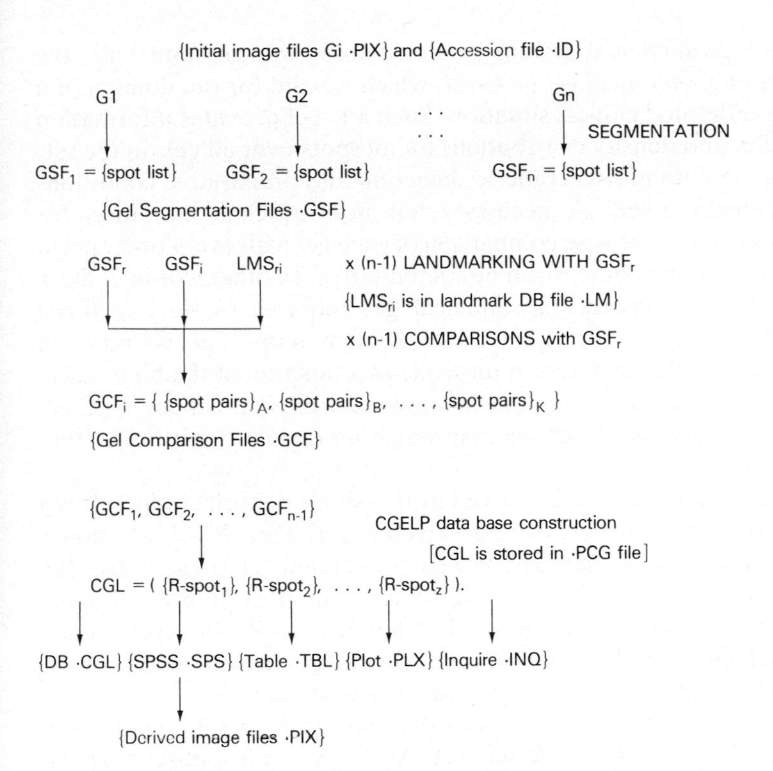

Fig. 11. Data file structures and corresponding file extensions used in the sequential steps of gel analysis. The GSFs (gel segmentation files) are produced by the spot segmentation of the gel images. The GCFs (gel comparison files) are result in comparing the GSFs using a set of landmark spots to pair spots between gels. The paged CGL data base is constructed by merging the set of GCFs. Note that the arrows indicate direction of data reduction.

program, CGELP for multiple gel analysis. An earlier version of CGELP called CGEL is discussed in [3].

Associated spots and their characteristics can be partitioned by one criterion and then repartitioned as one attempts to 'see around' the data from several perspectives. From our early efforts at gel analysis with the FLICKER system, it became evident that what was required was a system which could automatically find and measure all (or most) spots in a gel. Spots from two or more gels should be comparable which implies that the program needs to be able to partition and to cocatenate lists acquired at different times and from different gels. Without checking *all* or at least most of the spots in the set of two or more gels, no complete statement of the types of spot differences can otherwise be made. These constraints imply both a gel pairing program and a spot data management system.

In a given gel, the majority (if not all) spots, once isolated, can be characterized by (to the first approximation) a triple, comprising x and y position (centroid) and an adjusted integrated density value proportional to polypeptide concentration. Among gels, the idiosyncratic variations of these triples due to variation in gel and sample preparation, detection, etc., confound what are the 'real' variations produced in the biological/clinical system by time, dose, clinical state, etc. We propose the concept of a *canonical gel or C-gel*, which is valid for the domain of a given experiment or a defined clinical situation. Such a *C*-gel provides information characterizing position and density distributions for all spots over all gels in the set. Further, it excludes the data idiosyncratic to detection and preparative conditions unrelated to the biologic issue. A necessary but not sufficient condition for construction of a *C*-gel is the pair wise comparison of each gel with every other gel in the set, with the condition that comparison be commutative. In other words, if there are *n*-gels in the set, to construct a canonical gel requires $(n - 1)$ factorial comparisons times the number of spots. Since each element of the C-gel is a function expressing the variation of the spot descriptor triple as a function of the biomedical variable, it is not easily constructed. Though not easily realized in practice, the *C*-gel provides a model object against which we may weigh a pragmatic substitute, the *representative or R-gel.*

The *R*-gel, in contrast to the *C*-gel, is derived from a single pictorial object. It is a real gel chosen from a set of gels representing a given experiment. *R*-gel selection is detailed below, but it may be considered to be what it is named, a representative (by experimenter criteria) gel which is believed to contain most if not all spots encountered in any of the members of the set. It is not necessarily an experimental control gel, but its selection by the biologist certainly reflects his knowledge of the experiment and of the resulting individual gels that constitute the set.

The *R*-gel is used as the basis against which other gels in the set are compared. Each spot in the *R*-gel is the index to a *R*-spot set. An *R-spot set* is that set of spots, with at most one from each gel in the set of gels, which corresponds to a given spot in the *R*-gel. The *list* of *R*-spots under ideal conditions includes all spots in all gels. Until biochemistry can provide essentially noise free gels and extensions to the analysis including methods for handling missing or very noisy spots in the *R*-gel, such a complete and ideal accounting is simply not attainable.

4.1. *The general system of analysis*

The design philosophy underlying the part of the GELLAB system that deals with multiple gels is the interactive and flexible manipulation of spot data organized by congeneric association. Paired spots and their locations and densities are recorded in a congener-oriented database denoted the ‘CGL’, which can be searched in a variety of ways. Various representations, numeric, diagrammatic, pictorial, textual or tabular, of this data base or of its derivatives can be rapidly displayed in order that the researcher may quickly grasp patterns and implications. Hypothesis verification is performed by interactive partitioning and testing new representations of segments of the data.

Fundamental to our system of analysis of multiple gels is the concept of congeneric polypeptides which give rise to sets of corresponding spots across gels. A congeneric set of polypeptides is one in which each member arises from a common group of biologic processes. The quantitative expression of such production may be muted or exaggerated under varying experimental conditions. But in each gel where it is

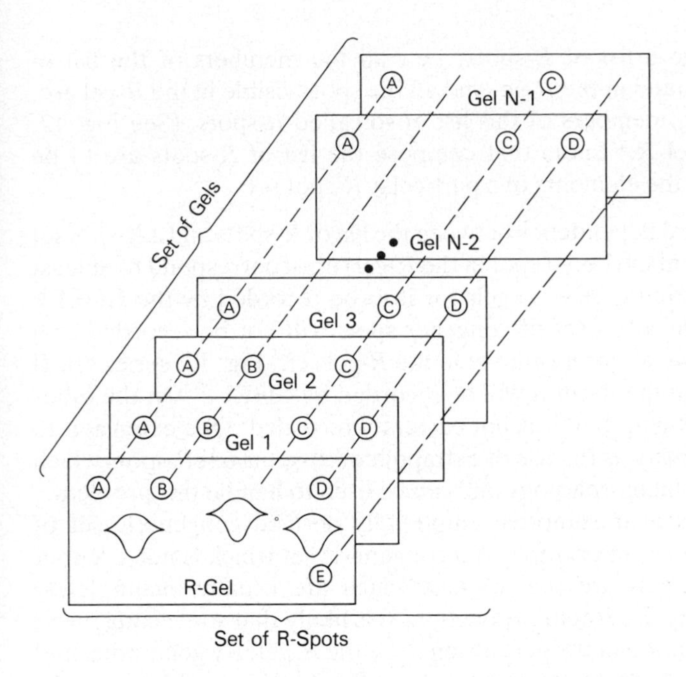

Fig. 12. Spots A, B, D and E in the R-gel are R-spots. R-spot set A consists of $\{Ar, A1, A2, \ldots, An - 1\}$ which has one member from each gel. Other R-spot sets such as B and D are not represented in all gels. The assumption is that spots which are members of an R-spot set arc congeneric, i.e., the formation of the polypeptides which they denote represents the action of a chain of biologic processes common to all of them. One possible form of systematic error could occur: Suppose that in gel 3 the spot perceived as B was in fact displaced by severe local preparative distortion, so that its expected position was occupied instead by another polypeptide. This kind of false positive event is largely dependent on the pairing algorithm. The set C, is not an R-spot set even though represented in all other gels since there is no member in the R-gel. The presumption is that the members of set C are also congeneric, but whether by chance or biologically determined, the absence of a congener in the C position of the R-gel keeps the entire set outside of the data base. This problem can be resolved two ways: (a) by peforming experiments involving more than one R-gel (constructing other CGL data bases); or (b) extrapolating a spot C into the R-gel. This latter approach is being investigated by our group.

detected, *the spot denoting the congeneric polypeptide occupies the same relative position in the local gel morphology*.

4.1.1. The R-spot set and the list of R-spots

Formally, we represent these two constructs, the list of R-spots and the R-spot set, as follows:

(1) *A R-spot set* – This is a set of congeneric spots of a particular polypeptide, having at most one member from each gel (but definitely including a particular member of the list of *R*-spots), corresponding to a given spot in the *R*-gel. (Cf. Fig. 12 The B series of spots). The *R*-spot set may be regarded as a vector, each element of which is taken from a single plane of the three-dimensional (3D) stack of gels.

(2) *The list of R-spots sets* – All the distinguishable R-spot sets in the *R*-gel, taken

together, constitute a list of *R*-spots; i.e., all the members of the list of *R*-spots are to be found in the *R*-gel and all the spots visible in the *R*-gel are, at least potentially, members of this list of so called *R*-spots. (See Fig. 12, A,B,C, etc., in Gel *R*.) Spots that compose the list of *R*-spots are to be distinguished from the elements of a particular *R*-spot set.

The linkage and reciprocal dependency between the list of *R*-spots and a *R*-spot set is this: (1) A *R*-spot set member (i.e., a spot in the *R*-gel) must correspond to at least one other spot in the remaining $(n - 1)$ gels for it to be recorded by the CGELP program as a spot pair; and (2) a set of congener spots will not be recorded as a R-spot set if it does not have a representative in the *R*-gel (cf., Fig. 12, series C). If in (1) the spot only exists in the *R*-gel it will be recorded by CGELP. On the other hand, if it exists in a non *R*-gel, then it is not currently recorded. One extension to GELLAB, currently underway, is the use of extrapolated *R*-spots or *eR*-spots which can be inferred from the local morphology and can be used to handle this problem.

A *R*-spot set represents a presumptive empirically derived congeneric set of polypeptides. Figure 12 shows an example of a congeneric set which is not a *R*-spot set (spot set *C*). Since *R*-gels are real objects which are usually incompletely representative of the totality of protein production, it is likely that some congeneric sets will not have representation in the gel chosen to be the *R*-gel. By generating and testing several data bases built on complementary *R*-gels, this problem can be handled at present (until the *eR*-spot GELLAB extension is completed).

4.1.2. Local morphology

We have found that for gel analysis a most effective strategy is to concentrate on sets of local morphologies (both within and across gels) rather than to treating one object at a time. Even if the task is defined as detection of the presence or absence of a single spot, some consideration of local morphology is necessary for any decision to be made by machine or requiring human confirmation.

Recognition and identification (as opposed to detection) is quite difficult because of the absence of fixed shape and size of the individual spot. In dealing with spot identification, we are actually concerned with problems of local morphology, in which we are aided by the machine to (1) establish the proper region of regard, (2) maintain a local coordinate system, and (3) perform pictoral and numeric comparisons.

4.1.3. The congener spot data base

We have discussed procedures that have been preparatory in that they deal with operations on individual spots or spot pairs. After constructing the set of *R*-spot sets, we are now in a position to use these data so as to construct a data base which can be ordered as a function of biological, clinical, experimental or temporal variables. The richness of the data base does not limit us to any one of these as the facilities which we now describe allow a multiplicity of orderings. A variety of representations may be chosen which may best be determined by the nature of the experiment. The biology demands that the analytic process be limited in its 'attention' to the set of congeneric spots, one from each gel, a process that transcends the constraints of the individual gel. The CGELP data management system permits this type of analysis to be applied successively to the majority of such *R*-spot sets.

The types of operations performed consists of many computational or representational operations on the list of *R*-spot sets or sublists of *R*-spot subsets. The latter subsetting may be automatically accomplished based on an experiment dependent characteristic of a gel (from the accession file), on a statistical property of spot or *R*-spot set features, etc. Alternatively, the user may construct at will a working set of gels taken from the entire set of gels. A wide variety of representations of the data, both image and numeric, is available with many modes of display including superimposition on the original image. Important data structures include:

(1) The set of working gels used to restrict the CGELP operations to a subset of the gels in the data base. Only gels in the working set are used in the computations.
(2) The gel subsets structure which is used to manipulate gel subsets in easily redefining working set or gel classes.
(3) The classification sets which contain the names of the gels in each of up to 9 classes. Thus, the user can, depending on the problem he is dealing with, classify gels by temperature, by disease, by metabolic condition, etc.
(4) A 'search results list' of *R*-spots set number names which were found by one or more of the various available search options (or explicitly defined) is available to many of the CGELP operators.

In dealing with real data, it is frequently necessary to create a working subset of gels taken from the original data base in response to different questions. The same data may be used to analyze different aspects of the same experiment by being partitioned in various ways. A related requirement is the facility to declare classes of gels and to create further subsets based on class membership. As an aid to manipulating subsets of gels, gels may be put into named subsets and treated as an entity.

4.1.4. Solution strategies

The properties that characterize spots, the principle of local morphology and the different objectives of different users require of an analysis system which is capable of varied analysis. Such a system offers the capability of designing a solution strategy or set of strategies rather than a direct and single solution. Among the important tools available for such strategies is the experimenter directed creation of multiple representations of the same data. Many of the system procedures are essentially procedures of presentation which allow the user to alternate between say numeric position or density data and synthetic images.

These tools include *R*-maps and mosaic images which facilitate the backchecking of any *R*-spot set in both a global (the *R*-map) and a local but multiple gel (mosaic) context. A mosaic is an image constructed by concatenating, in a 4×4 array, corresponding subregions from each gel, ordered in the array by spot density, surrounding the spot of interest. The mosaic provides a powerful tool whereby the user may be assured, on the basis of visual evidence, that a spot belongs to a given *R*-spot set. The *R*-map image provides the link between the global location of an individual spot as seen numeric *R*-spot set data or local mosaic image. The *R*-map is invaluable for rapid evaluation of the validity of spots found to be of interest by GELLAB statistical searches or manual examination of the data base. Mosaics are insufficient for establishing a spot's context because of their locality and thus the *R*-map fills this void. Numeric data, particularly functions of density presented in rank ordered tabular form is useful for evaluating magnitude differences between spots in an *R*-spot set. The gray scale numeric representation of each pixel comprising a spot in a small window of the image is occasionally useful in determining whether a spot is actually one or two or whether a spot was fragmented by the segmenter. The accession file information is always available for use with a data set or its derivatives. Any portion of it may be used as the associative key with which to regroup gels within the gel data base.

Tools such as the foregoing are invoked as needed at user discretion to establish and/or confirm membership in a biologically significant congeneric vector, i.e., a *R*-spot set. Moreover they can be used to quantitate substantive changes as a function of the biologic variable at issue.

In sum, the GELLAB set of programs represents a general method to organize and selectively compress the data of 2D gels so that the user may more efficiently perceive patterns out of the welter of individual spots. Once *R*-spot sets of interest have been found, it is a direct process to quantitate their individual components by merely printing their *R*-spot sets.

4.1.5. Analyzing multiple gels as a continuum

Each congener polypeptide visualized as a spot may be thought of as having a distribution of spot densities when sampled in a set of gels. It is expected that this distribution will cluster multimodally in the case of significant spot density differences according to the biological state of the sample. Therefore, it is important that biologically non-significant variances be controlled and minimized (false positives). Adequate numbers of gel samples must be obtained for the data base to aid in detecting these multimodal distributions.

We must assume that not all spots will be accounted for (false negatives). No automatic procedure can account for the almost infinite variety of image noise found in these gels. The semiautomation of the gel analysis may be sufficient to find spots for those biological problems where the changes are above the noise level and resolvable by the system.

4.2. CGELP spot data base analysis system

The role of the spot data base may be seen in the overview of the entire gel analysis procedure (cf., Fig. 1). The hardware environment mentioned in Section 1.3 is of some interest in understanding the processing and data structure manipulations. The GELLAB system is currently implemented using two hardware systems: the Image Processing Unit's Real Time Picture Processor (RTPP) and a Digital Equipment

Corporation DEC-2020. Figure 11 illustrates the data structures required and generated at the different stages of processing. Image acquisition and landmarking are currently performed using the interactive RTPP system with distributed processor software (2020/RTPP) – GETRTPP and LANDMARK. However, images have been acquired on an Optronics scanner and entered into GELLAB via magtape. Using the primary gel image files, accession file and landmark spot sets file, the secondary spot segmentation, gel spot pairing and CGL data base construction and analysis are performed on the DEC-2020. A cost accounting estimate of the various steps in the complete analysis of an average set of 20 gels found typical DEC-2020 times were about 25 min of CPU time/gel with average 150 K-word memory program sizes and less than 15 min of RTPP real time/gel. Using the manual landmarking approach with programs DWRMAP and LMSEDIT (cf. Section 3.2), much of the RTPP interactive time would be eliminated in cost of several more minutes of 2020 CPU time.

Such an environment imposes some practical limits on the capacity of GELLAB. Gel analysis, as illustrated in Fig. 11, is primarily a series of data reduction steps mapping gel image information into a set (of over 3000) *R*-spot density distributions for up to 128 gels. Images are reduced to spot lists, then spot lists are reduced to spot pair lists and finally, spot pair lists are reduced to a list of *R*-spot sets. Clearly, comparing hundreds to thousands of spots in over 100 gels would be a monumental task if done manually. Most of the computation is involved in the initial image data reduction phase whereas the amount of additional computation is dependent on the type of questions to be asked about a particular set of gels. For example, 60 searches on a 15 gel data base of over 600 *R*-spots with the 15 gels being repartitioned as 2 gel working sets took less than 1.5 h of CPU time.

Procedures used in this later phases of analysis are based on the analytic principles discussed above and carried out by the experimenter using an interactive program called CGELP. CGELP is an interpreter program and is used to construct a representative spot data base and then to analyze all or part of this data base in various ways.

4.2.1. Generation of the R-spot set data base

The first step in the construction of the data base is the generation of the list of *R*-spot sets. The set of $n - 1$ GCFs (Gel Comparison Files, cf., Fig. 11) are read one spot pair at a time for each gel pair where one of the spots is a *R*-gel spot. Each gel pair referenced by a 'key' for the *R*-gel spot, is formed for this pair (cf., Appendix in [3]) and the data base is tested to determine whether a *R*-spot set currently exists for that *R*-gel spot. If it does not, then a new *R*-spot set is created and both spots are put into that set. If it does, then the other spot in the pair is inserted into that set. In either case, the *R*-spot set is initially rank ordered by density, darkest first. Alternate *R*-spot set orderings may then be routinely performed as part of the analysis.

Parts of the CGELP system are illustrated here by examples of operations and results obtained on one of several projects to which it has been applied. We have chosen the work on some of the P388D1 gels for the majority of our illustrations. Table 5 lists the top level CGELP commands. A subset of CGELP commands are

TABLE 5

CGELP top level commands

```
CReate - create a CGL data base from a set of CMPGEL .GCF files.
DDplot - Draw (plot .PLX) LOG %density/%density spot plots from CGL DB.
EDit - Edit spots from the CGL data base.
EXIt - Exit CGELP to the monitor to save paged CGL data base for later use.
EXTrapolate - missing spots in Rsets from mean (dx,dy)+LM position.
Gels - Gels lists the names and total densities of the current gels.
HElp - Print this message.
HIstogram - compute histograms of functions of Rspot sets (.TBL).
Inquire - Interrogate the CGL data base for particular spots.
Limits - print the current statistical limits.
PLot - feature vs. feature plot of 2 (or 3) spot features over CGL DB.
PRotect - paged CGL data base for read-only (toggle).
REOrder - Rspot sets (after changing density mode).
RMap - plot the Rmap surrounding the Rspot using density estimates.
SAve CGL- Save the CGL data base in a (.CGL) file.
SET Accession file name - change the default GEL.ID name.
SET Classes - Define gel class partition.
SET DAta base file - setup the new or old CGL data paged data base.
SET DEnsity mode - report results in Abs (D'), Percent, Ratio or Volume units
SET Fields - Set the list of fields desired for gel labeling.
SET Gel subset - define a gel subset.
SET Label - Set the 'Label' code to (S, P, A, U, E) used in searching.
SET RAtio - list of Rspots for normalizing spot densities for Ratio mode.
SET RGel - Set the name of the R-gel used in searching.
SET Statistics limits - Set statistics limits for use in searching.
SET Working gels - Define working set of gels from CGL data base.
SPss - Dump an SPSS .SPS summary file of part of the CGL database.
TAbulate - Print set Rspot dens-rank order, ratio, Mn-variation (.TBL).
TImer - print the run and cpu times for commands (toggle - normally on).
Valid landmarks - list the valid landmarks for each gel in a table.
```

The top level CGELP program commands listed here are available to the user on an interactive basis. The first part of the command that is required for it to be unique is capitalized.

described in [3] while the complete detailed description will be published in the GELLAB user's manual [39].

Table 6 illustrates a simple sequence of CGELP commands which creates a CGL data base and in the process partitions the CGL data base into T0 (initial harvesting) and T24 (cells harvested 24 h later with no media change inbetween) classes. Table 7 illustrates some typical CGL R-spot set data base entries where spots are rank ordered by density. Table 8 illustrates a rank order table for some selected *R*-spots from the P388D1 data.

4.2.2. *CGELP commands*

The user employs the CGELP interpretive system to analyze a set of gels as determined by successive partition parameter selection. Particularly when dealing with a new data base, the user employs CGELP 'experimentally'. Procedures are invoked. Intermediate results are displayed and examined for confirmation or rejection of the tentative hypothesis, other procedures are then invoked, and so forth. Graphics displays are performed on the user's graphics terminal (if it is a Tektronics 40××-series or DEC GT40). A plotter file may be optionally generated which may be plotted later on a hardcopy plotter or redisplayed on the graphics terminal. The nature of the interaction is highly dependent on the nature of the scientific questions to be asked of the gel data base.

TABLE 6

Example of constructing a CGL data base

```
.RUN CGELP
*SET ACCESSION FILE
        *gelll.id
*SET DATA BASE FILE
        *as5pcg.pcg
*SET FIELDS - to be used in accession file information
        *2,3,10,12,13
*CREATE
        *c20251.gcf
        *c20252.gcf
        .
        .
        .
        *c20258.gcf
        .
        .
        .
        *c20265.gcf
        *
        *spau - (the expected spot labling SP+PP+AP+US)
*VALIDLANDMARKS - list valid landmark statistics
*SET CLASSES
        *auto - class naming based on accession file field information
        *yes - change class names
        *t0 - class 1
        *t24 - class 2
        *
*SET DENSITY MODE
        *ratio
*SET RATIO LIST
        *least squares
*REORDER
*SET GEL SUBSET - define T24 gels as the toxic subset
        *yes
        *toxic
        *250.2,258.2,259.2,260.2,261.2,262.2,263.2,264.2,265.2
*EXIT - save the current data base
.RUN CGELP
*SET DATA BASE - restore saved data base
        *as5pcg.pcg
*GELS - list the gels and numberspots/gel in the data base
*SAVE
        *as5all.cgl - dump all spots in the data base
*SET LABEL - restrict the data base to SP+PP
        *ps
*TABLE - compute and save the inter-gel correlation matrix
        *mn-variation
        *as5mnv.tbl
*INQUIRE - search for the landmarks in the data base
        *landmarks
*SPSS - save the list of landmark spots in an SPSS formated file.
        *as5lms.sps
        ** - '*' indicates the search results list from INQUIRE search
;       Perform a 5% confid level F-test search on classes 1,2
*SET STATISTICS
        *0,512 - relative distance from landmark
        *0,512 - DP (distance between pair of spots)
        *0,512 - DL (distance between LM spot and mean of spot pair)
        *0,500 - mean area of an R-spot set
        *0,1000 - mean density of an R-spot set
        *0,1000 - standard deviation of an R-spot set
        *0,10.0 - std/mean density
        *0,10.0 - spot texture
        *.95 - significance level
        *0  - any number of gels in the data base
*INQUIRE
        *f-test/file
        *as5f95.inq - save search results in file as well
        *1,2
*SPSS - Save search results list in an SPSS formated file
        *as5f95.sps
        ** - use search results list
*HISTOGRAM
        *Set mean Rspot set density
        *tty
        *as5mrd.tbl
*SET WORKING SET
        *define
        *toxic
*SET CLASSES - redefine classes to include toxic and controls
        *yes - set classes manually
        *yes - change class names
        *control
        *toxic
        *
        *0 - 250.2
        *1 - 258.2
        *1 - 259.2
        *2 - 260.2
        *2 - 261.2
        *2 - 262.2
```

```
        *2 - 263.2
        *1 - 264.2
        *1 - 265.2
*SAVE - save the toxic data base in print file
        *as5tox.cgl
*EXIT
```

This example illustrates a typical CGELP command sequence used to construct a normalized CGL data base file (AS6PCG.PCG). After its construction, a *F*-test search is performed to find statistically significant spots. The CGELP commands are given in capitals and the answers to the CGELP prompts are indented and in lower case. The '.' prefix indicates a TOPS-10 monitor command while the '*' indicates a CGELP command. Comments are preceeded by '–'.

TABLE 7

Examples of CGELP data base

```
Output file: AS6NEW.CGL 03/09/1981,  12:20:16 AM
Pairing labels: PSUA
Using least square normalization.
Relative distance limits are[ .00, 512.00]
DL limits are[ .00, 512.00]
DP limits are[ .00, 512.00]
MN area limits are[ .00, 500.00]
MN density limits are[ .00, 1000.00]
S.D. density limits are[ .00, 1000.00]
Coef. variation: S.D./Mean Rset density limits are[ .00, 10.00]
Spot texture limits are[ .00, 10.00]
Class difference t-Test, F-test Rank order significance limit is .90:
Check if # gels in R-spot set [0:1000]
There are 16 gels with 546 Rspot sets consisting of 8239 spots.
Spot free store has 384977 spots available.
   1179 sure pairs,
   4072 possible pairs,
   2873 ambiguous pairs,
   115 unresolved spots.
[1]Total density[0250.2]=4739, # spots=547, Mj= 1.000      bj= 0.000
   Study: /P388D1/C14/1 WEEK/ALUMINUM,T0,CONTROL,BOTTLE#1
[2]Total density[0251.2]=5638, # spots=569, Mj= .9056      bj= 1.481
   Study: /P388D1/C14/1 WEEK/ALUMINUM,T0,CONTROL,BOTTLE#2
[3]Total density[0252.2]=3961, # spots=464, Mj= .8851      bj= 5.562
   Study: /P388D1/C14/1 WEEK/ALUMINUM,T0,AL2O3-HC,BOTTLE#3
[4]Total density[0253.2]=3022, # spots=396, Mj= .8969      bj= 6.813
   Study: /P388D1/C14/1 WEEK/ALUMINUM,T0,AL2O3-HC,BOTTLE#4
[5]Total density[0254.2]=3816, # spots=502, Mj= 1.134      bj= 4.716
   Study: /P388D1/C14/1 WEEK/ALUMINUM,T0,AL2O3-18U,BOTTLE#5
[6]Total density[0255.2]=3408, # spots=504, Mj= 1.448      bj= 3.779
   Study: /P388D1/C14/1 WEEK/ALUMINUM,T0,AL2O3-18U,BOTTLE#6
[7]Total density[0256.2]=4687, # spots=528, Mj= 1.006      bj= 4.449
   Study: /P388D1/C14/1 WEEK/ALUMINUM,T0,AMOSITE,BOTTLE#7
[8]Total density[0257.2]=5336, # spots=607, Mj= .8750      bj= 5.672
   Study: /P388D1/C14/1 WEEK/ALUMINUM,T0,AMOSITE,BOTTLE#8
[9]Total density[0258.2]=8627, # spots=672, Mj= .5724      bj= 2.926
   Study: /P388D1/C14/1 WEEK/ALUMINUM,T24,CONTROL,BOTTLE#9
[10]Total density[0259.2]=8403, # spots=733, Mj= .5616      bj= 4.770
   Study: /P388D1/C14/1 WEEK/ALUMINUM,T24,CONTROL,BOTTLE#10
[11]Total density[0260.2]=6756, # spots=643, Mj= .8159      bj= 2.771
   Study: /P388D1/C14/1 WEEK/ALUMINUM,T24,AL2O3-HC,TOXIC,PHAGOCYTIC,BOTTLE#11
[12]Total density[0261.2]=7043, # spots=629, Mj= .8454      bj= .9399
   Study: /P388D1/C14/1 WEEK/ALUMINUM,T24,AL2O3-HC,TOXIC,PHAGOCYTIC,BOTTLE#12
[13]Total density[0262.2]=6550, # spots=692, Mj= .8376      bj= 1.733
   Study: /P388D1/C14/1 WEEK/ALUMINUM,T24,AL2O3-18U,PHAGOCYTIC,BOTTLE#13
[14]Total density[0263.2]=11268, # spots=896, Mj= .5539      bj= 3.019
   Study: /P388D1/C14/1 WEEK/ALUMINUM,T24,AL2O3-18U,PHAGOCYTIC,BOTTLE#14
[15]Total density[0264.2]=7001, # spots=674, Mj= .7301      bj= 4.131
   Study: /P388D1/C14/1 WEEK/ALUMINUM,T24,AMOSITE,TOXIC,PHAGOCYTIC,BOTTLE#15
[16]Total density[0265.2]=2387, # spots=355, Mj= 1.672      bj= 5.192
   Study: /P388D1/C14/1 WEEK/ALUMINUM,T24,AMOSITE,TOXIC,PHAGOCYTIC,BOTTLE#16

R-spot[ 34] [0250.2] XYabs=(142,228) MnD= 3.29 SD= 2.06 SD/MnD= .63 #gels=13
 ACC#[Index]C  RDensL area maxOD  D'   Lbl LM  DP  DL  Dx   Dy  Xabs Yabs txtr
-------------- ------- ---- -----  ---- --- -- ---- -- ---- ---- ---- ---- ----
0250.2[ 219]1    9.10R   62   .50   9.1  PP B   2.8 35 (-32, -8) (142,228)  .12
0252.2[ 215]1    5.49R   48   .36   6.2  PP B   2.8 35 (-34,-10) (125,244)  .04
0252.2[ 193]1    4.69R   18   .62   5.3  AP B   9.8 40 (-51,  1) (108,255)  .18
0263.2[ 374]2    3.88R   37   .39   7.0  PP B   4.1 33 (-31,-12) (147,191)  .04
0255.2[ 175]1    3.33R   16   .31   2.3  PP B   8.0 36 (-32,-16) (115,220)  .03
0258.2[ 273]2    2.98R   39   .34   5.2  PP B   5.1 35 (-33,-13) (126,225)  .02
0262.2[ 270]2    2.43R   28   .29   2.9  PP B   4.0 34 (-32,-12) (144,229)  .01
0256.2[ 182]1    2.11R   18   .28   2.1  PP B   7.1 34 (-31,-15) (148,241)  .01
0254.2[ 221]1    2.04R   17   .29   1.8  AP B   9.1 42 (-74, 10) ( 67,265)  .02
0254.2[ 224]1    1.82R   15   .26   1.6  PP B   6.1 39 (-38, -9) (103,246)  .01
0259.2[ 314]2    1.74R   31   .28   3.1  PP B   4.1 35 (-33,-12) ( 99,196)  .01
0264.2[ 256]2    1.68R   20   .31   2.3  PP B   4.5 36 (-34,-12) ( 96,219)  .01
0260.2[ 259]2    1.47R   19   .24   1.8  PP B   3.2 35 (-33,-11) (139,229)  .01

R-spot[ 65] [0250.2] XYabs=(177,176) MnD= 75.81 SD= 20.83 SD/MnD= .27 #gels=16
 ACC#[Index]C  RDensL area maxOD  D'   Lbl LM  DP  DL  Dx   Dy  Xabs Yabs txtr
-------------- ------- ---- -----  ---- --- -- ---- -- ---- ---- ---- ---- ----
0250.2[  99]1 125.90R  262  1.39 125.9  SP D*   .0  0 (  0,  0) (177,176) 1.51
0252.2[  95]1 108.07R  274  1.30 122.1  SP D*   .0  0 (  0,  0) (163,194) 1.39
0251.2[  95]1 102.15R  266  1.34 112.8  SP D*   .0  0 (  0,  0) ( 78,177) 1.51
0254.2[ 105]1  88.71R  240  1.21  78.2  SP D*   .0  0 (  0,  0) (145,194) 1.21
```

```
0255.2[  79]1  87.34R  200  1.13  60.3  SP D*   .0  0 (  0,   0) (149,176) 1.06
0253.2[  37]1  77.94R  204  1.20  86.7  SP D*   .0  0 (  0,   0) (186,166) 1.14
0257.2[ 104]1  72.89R  199  1.30  83.3  SP D*   .0  0 (  0,   0) (195,151) 1.35
0256.2[  93]1  71.83R  203  1.16  71.4  SP D*   .0  0 (  0,   0) (182,197) 1.10
0265.2[  59]2  69.73R  148  1.10  41.7  SP D*   .0  0 (  0,   0) (151,129)  .92
0260.2[ 146]2  66.58R  205  1.11  81.6  SP D*   .0  0 (  0,   0) (171,180)  .94
0263.2[ 233]2  60.76R  232  1.25 109.7  SP D*   .0  0 (  0,   0) (175,148) 1.21
0261.2[ 103]2  60.62R  184  1.18  71.7  SP D*   .0  0 (  0,   0) (190,157) 1.04
0262.2[ 144]2  59.47R  134  1.24  71.0  SP D*   .0  0 (  0,   0) (175,184) 1.12
0258.2[ 135]2  59.07R  212  1.17 103.2  SP D*   .0  0 (  0,   0) (161,177)  .96
0264.2[ 131]2  53.23R  187  1.17  72.9  SP D*   .0  0 (  0,   0) (129,176) 1.08
0259.2[ 184]2  48.75R  210  1.18  86.8  SP D*   .0  0 (  0,   0) (132,149) 1.08

R-spot[376] [0250.2] XYabs=(307,321) MnD= 1.79 SD= .54 SD/MnD= .30 #gels=13
 ACC#[Index]C  RDensL area maxOD  D'   Lbl LM  DP  DL  Dx   Dy  Xabs Yabs txtr
------------- ------- ---- -----  ---- --- -- ---- -- ---- ---- ---- ---- ----
0258.2[ 501]2   2.98R   36   .32   5.2  SP O   1.4 17 (-12, 12) (296,325)  .01
0263.2[ 616]2   2.66R   24   .35   4.8  SP O   1.4 17 (-14, 10) (298,286)  .01
0265.2[ 296]2   2.17R   18   .23   1.3  AP O   8.1 20 ( -9, 18) (276,284)  .00
0250.2[ 394]1   2.10R   19   .28   2.1  SP O   3.0 19 (-13, 11) (307,321)  .01
0260.2[ 464]2   1.96R   16   .28   2.4  AP O  10.0 18 ( -5, 17) (322,332)  .00
0261.2[ 448]2   1.69R   13   .32   2.0  SP O   3.6 17 (-15,  8) (323,299)  .00
0263.2[ 617]2   1.61R   15   .32   2.9  AP O   3.6 17 (-18,-25) (294,251)  .00
0262.2[ 514]2   1.59R   14   .29   1.9  SP O   1.0 18 (-13, 12) (299,332)  .00
0252.2[ 356]1   1.50R   17   .27   1.7  AP O   8.5 17 ( -5, 14) (307,345)  .01
0253.2[ 252]1   1.35R   13   .25   1.5  PP O   7.1 17 ( -6, 10) (326,316)  .00
0251.2[ 391]1   1.27R   14   .24   1.4  SP O   3.0 19 (-16, 11) (216,322)  .00
0257.2[ 436]1   1.23R   16   .23   1.4  SP O   3.0 19 (-16, 11) (322,296)  .00
0256.2[ 375]1   1.21R   16   .24   1.2  SP O   1.4 17 (-14, 10) (316,345)  .01
```

A CGL data base for P388D1 macrophages, with *R*-gel 250.2, contains 546 *R*-spots. The state of CGELP at the time the CGL DB is saved (with the SAVE command) is illustrated indicating the current partitions. Three examples of *R*-spot sets from this DB are shown: [34], [65] and [376]. Correspondences to *R*-spot [34] and [376] are missing some of the gels. A spot's *RDensL* is its normalized least squares density gel. *Dx* and *Dy* are the spot's position relative to its associated landmark. Metrics (*MnD*,SD) are the mean and standard deviation of the density measurement in the *R*-spot set. Table entry 'C' is the class partition name which in this case has the class partition of 1 = T0 and 2 = T24. *D'* is the background corrected absolute density of the spot while area and maxOD metrics are also recorded. The absolute position of the spot in the gel image is (*Xabs*,*Yabs*). *Lb*1 is the pairing label SP, PP, AP, US or EP. Note that the heuristic pairing values *DP* and *DL* are similar for most spots as are the (*Dx*,*Dy*) relative distances to the landmark spot. Because of this particular consistency, any spot in a *R*-spot set with a large deviation in one of these position features may be regarded as a possible outlier and so treated. *R*-spot [65] is a landmark spot (*D*) (denoted by the * in the *LM* field) with corresponding values of *DP*, *DL*, *Dx* and *Dy* being zero by definition. *R*-spot [65] shows significant class differences in the mean density.

TABLE 8

Rank order density table for selected spots

```
          File: RNKTB1.TBL 02/26/1981,   10:17:23 AM
          RANK-ORDER table: <ACC#>&<LMset>&<Class #>
          Paged CGL data base file: AS6PCG.PCG[61,1]
          Using least square normalization.
          User defined spot list

Density
 125.9 |             0250.2D1
 122.8 |
 119.7 |
 116.6 |
 113.5 |
 110.4 |
 107.2 |             0252.2D1
 104.1 |
 101.0 |             0251.2D1
  97.9 |
  94.8 |
  91.7 |
  88.6 |             0254.2D1
  85.5 |             0255.2D1
  82.4 |
  79.3 |
  76.2 |             0253.2D1
  73.0 |
  69.9 |             0257.2D1
       |             0256.2D1
  66.8 |             0265.2D2
  63.7 |             0260.2D2
  60.6 |             0263.2D2
       |             0261.2D2
```

```
  57.5 |           0262.2D2          0250.2F1
       |           0258.2D2
  54.4 |                             0251.2F1
  51.3 |           0264.2D2
  48.2 |           0259.2D2          0253.2F1
       |                             0254.2F1
       |                             0264.2F2
       |                             0262.2F2
       |                             0256.2F1
  45.1 |                             0255.2F1
  41.9 |                             0252.2F1
       |                             0265.2F2
       |                             0257.2F1
       |                             0263.2F2
  38.8 |                             0258.2F2
       |                             0261.2F2
  35.7 |                             0260.2F2
  32.6 |                             0259.2F2
  29.5 |
  26.4 |
  23.3 |                                      0263.2G2
       |                                      0261.2G2
  20.2 |                    0251.2F1          0260.2G2
       |                                      0262.2G2
  17.1 |  0261.2B2          0250.2F1
       |  0260.2B2
       |  0263.2B2
  14.0 |  0262.2B2          0252.2F1          0264.2G2
       |                                      0259.2G2
       |                                      0265.2G2
       |                                      0255.2G1
       |                                      0250.2G1
       |                                      0254.2G1
       |                                      0258.2G2
       |                                      0252.2G1
  10.9 |  0264.2B2          0256.2F1          0256.2G1
       |  0265.2B2          0257.2F1          0251.2G1
       |                    0262.2F2          0257.2G1
       |                    0261.2F2          0253.2G1
       |                    0255.2F1
   7.7 |  0259.2B2          0263.2F2
       |  0258.2B2          0258.2F2
       |  0254.2B1          0253.2F1
       |                    0264.2F2
       |                    0254.2F1
       |                    0265.2F2
       |                    0260.2F2
       |                    0259.2F2
   4.6 |  0250.2B1
       |  0256.2B1
       |  0251.2B1
       |  0257.2B1
   1.5 |  0252.2B1
       |  0255.2B1
-----------------------------------------------------
R-spot |         45       65       98       99      119
Class #  1=T0
Class #  2=T24
```

A rank order density table may be constructed for a small number of selected spots. Five spots were selected from the P388D1 gel data base. Each entry presents three kinds of information; the accession #, the landmark set associated with the spot and finally the class assigned to the spot by the SET CLASS operator.

4.2.3. Data base normalization

Integrel density variation makes some scheme for normalization necessary. The density data initially transmitted to CGELP is already normalized with respect to percent of total gel density. However, this is usually not satisfactory, hence other normalization modes are available. One may normalize the CGELP data base by a subset of well defined 'stable' spots common to all gels or selected for some particular reason. Other normalization methods are also available. Changes in the density

TABLE 9

Example of R-spot set searches

```
a. Landmark set constraint search (LIST LANDMARK subcommand)

        LM[ A ]=R-spot[10]
        LM[ B ]=R-spot[38]
        LM[ C ]=R-spot[58]
        LM[ D ]=R-spot[65]
        .
        .
        .
        LM[ W ]=R-spot[536]

b. T-test constraint search (T-TEST subcommand at .99 significance)

R-spot[ 45] [0250.2] XYabs=(130,266) MnD= 9.28 SD= 5.47 SD/MnD= .59  gels=16
 [45](ml,m2)= 4.72, 13.84, Liml[ 2.17: 7.27], Lim2[ 10.28: 17.39], m2/ml= 2.93

R-spot[119] [0250.2] XYabs=(318,170) MnD= 16.55 SD= 3.79 SD/MnD- .23  gels=16
 [119](ml,m2)= 13.87, 19.22, Liml[ 12.50: 15.25], Lim2[ 15.57: 22.87], m2/ml=
1.39
        .
        .
        .

c. F-test constraint search (F-TEST subcommand at .99 significance)

R-spot[ 45] [0250.2] XYabs=(130,266) MnD= 9.28 SD= 5.47 SD/MnD= .59  gels=16
 [45](ml,m2)= 4.72, 13.84, m2/ml= 2.93
 |m2-ml|= 9.12, t(1-a/2)SQRT(vl+v2)= 4.69, f=11

R-spot[ 51] [0250.2] XYabs=(170,202) MnD= 9.18 SD= 4.77 SD/MnD= .52  gels=15
 [51](ml,m2)= 6.02, 12.80, m2/ml= 2.13
 |m2-ml|= 6.79, t(1-a/2)SQRT(vl+v2)= 4.47, f=6
        .
        .
        .

d. Rank order constraint search (RANK ORDER subcommand at .99 significance)

R-spot[ 45] [0250.2] XYabs=(130,266) MnD= 9.28 SD= 5.47 SD/MnD= .59  gels=16
nl=     8 n2=     8 n=    16 R=   100 R'=    36 Ralpha=    43
R-spot[ 51] [0250.2] XYabs=(170,202) MnD= 9.18 SD= 4.77 SD/MnD= .52  gels=15
nl=     7 n2=     8 n=    15 R=    28 R'=    84 Ralpha=    34
        .
        .
        .
```

This table gives examples of search output with four different constraints used in the INQUIRE command linear search: landmark, *t*-test, *F*-test and rank-order test.

normalization method change the rank of the spots in relation to each other in a R-spot set. CGELP permits reordering the data base.

4.2.4. CGELP data base searching and investigation

One of several tests, statistical or otherwise, is performed as a governing condition during execution of a linear search through the CGL data base. The search results list (SRL) is a composite tabulation of *R*-spots selected by the current search meeting all conditions as to gel working set, statistical metrics (see below) and class statistics (see below) where applicable. Table 9 illustrates the results of several types of searches including finding the landmarks, *t*-, *F*-, Wilcoxson–Mann–Whitney rank order tests [28].

For the *t*-, *F*-, and *WMW* rank order-test searches, a histogram of the ratios of the two class mean densities (*m2*/*m*1) is computed at search completion of spots in the SRL. Table 10 illustrates this ratio histogram for a .90 significance level in the INQUIRE *F*-test search.

A conjunction of statistical limits of *R*-spot set metrics may be used to define the SRL. This can be useful for finding *R*-spot sets which: (1) consist of spots with high or low variance, (2) have primarily dark or primarily light spots, or (3) are complete in having all spots present, etc., *R*-spot set parameters tested include: Relative distance of spot from *R*-spot center, *DL*, *DP*, mean *R*-spot set area and density, standard deviation and coefficient of variation, spot texture and number of gels in the *R*-spot set.

4.2.5. Use of the search results list

As mentioned previously, the search results list is a sublist of *R*-spot sets selected by various CGELP commands and is available either for CGELP further processing or output as SPSS numeric files. SPSS numeric files can be read by the SPSS program [29], MLAB [30], or other statistical analysis packages. Other GELLAB programs (MARKGEL and SEERSPOT – see [3,39]) use the SPSS file as part of their input to generate *R*-maps and mosaic images respectively. Figure 13 shows a typical *R*-map image, produced by the MARKGEL program, of the *R*-gel in one set of P388D1 gels. The spots selected were the result of applying the *F*-test in the search with a .90 significance level. These spots also appear in the 2 class density ratio histogram Table 10. Figure 14a–d shows some typical mosaic images generated from the P388D1 gel data base.

4.3. Some results from a P388D1 data base

GELLAB has been applied to PHA stimulation of lymphocytes [31–32], the effect of asbestos on P388D1 macrophages [33–34], as well as other projects both inside and outside of NIH. We have presented some preliminary results of the effect of time on the P388D1 mouse macrophages in tissue culture here as an aid in understanding the types of questions which may be asked of a gel data base system.

5. DISCUSSION

The GELLAB system for multiple gel analysis has been defined to the point where we can now re-examine system tactics and system problems in the overall biological context (as discussed in Section 1.1).

5.1. System characteristics and limitations

Gel scanning, segmentation and pairing are all finite resolution digital processes and each introduce some error. The computer analysis of a continuous process (for all practical purposes in this case a continuous gel) is performed in a digital space at both finite spatial and finite density digitization.

When multiple gels of split samples are run there is additional variance beyond that due to gel scanning alone. Samples of the tissue cultures of the same material result in multiple gels with an additional source of variation. Sampling of a biological

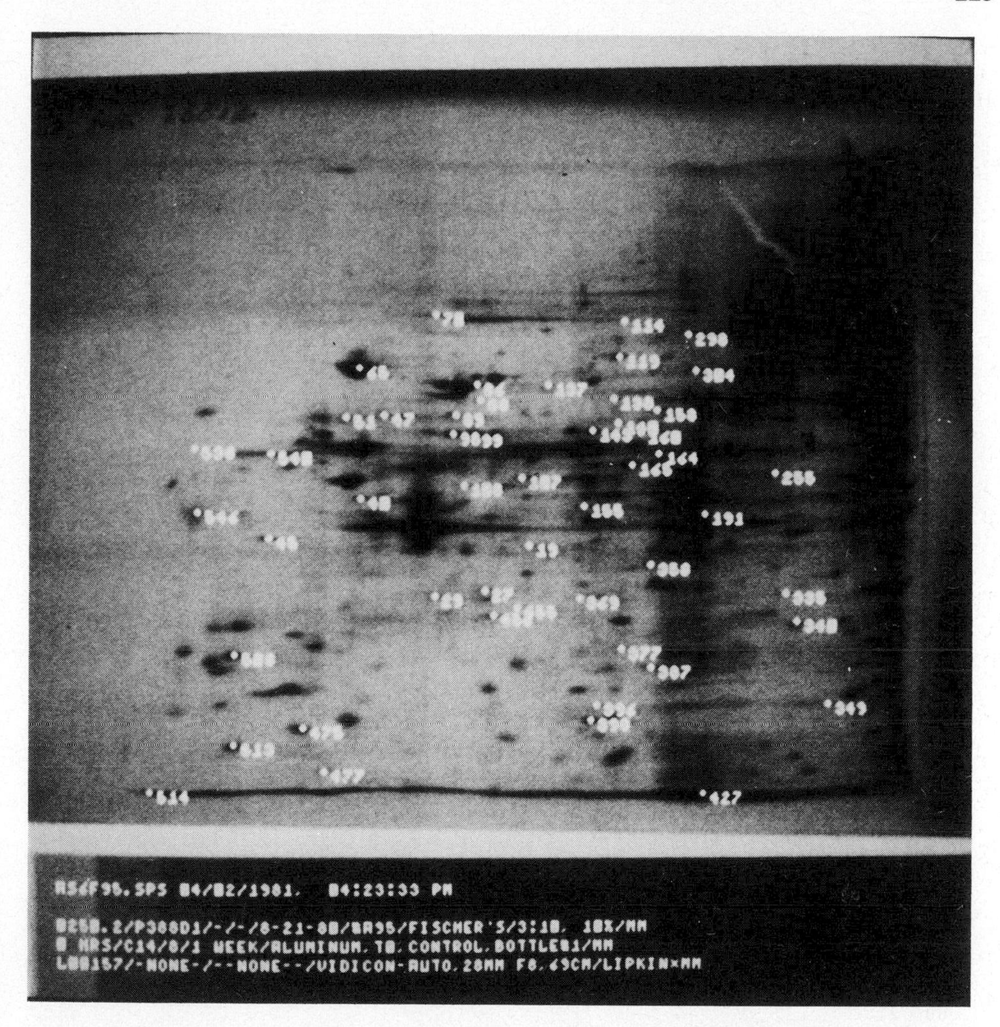

Fig. 13. An *R*-map image of a .95 significance *F*-test search on a least square normalized CGL data base for P388D1 gels was produced using the MARKGEL program. The *R*-map was performed on the *R*-gel (250.2) but could have been performed on any gel in the data base. There are three labeling options: NONUMBER = no label numbers (just the '+' at the centre of the spot); USELANDMARKS = if an *R*-spot is a landmark spot, use the landmark letter name rather than a number for it; the final and default option is to always label with a number.

process at various stages of its progression in synchronized or partially synchronized cultures is also another source of error.

Overall GELLAB system variance was explored using a set of duplicate scans of split sample gels of *E. Coli*. The reproducibility of repeated scans of the same gel at resolutions of 250 microns/pixel was the first test where correlation coefficients between gels ranged .98 to .99. In the case of duplicate gels, correlation coefficients were in the range of .97 to .99. Spots which differed markedly were checked by direct visual examination of the segmented gel image and in some cases, the central core

TABLE 10

Histogram of ratios of density means found with F-test

```
File: AS6F90.INQ 03/09/1981,   12:37:44 AM
Pairing labels: PS
Using least square normalization.
Class #  1(T0)=0250.2, 0251.2, 0252.2, 0253.2, 0254.2, 0255.2, 0256.2,
               0257.2,
Class #  2(T24)=0258.2, 0259.2, 0260.2, 0261.2, 0262.2, 0263.2, 0264.2,
               0265.2,
F-test class search at  .90 significance
Found   80 R-spots,  mean sd/mn= .46
-----------------------------------------------------------------
m2/ml    R-spot sets
  .35 104
  .40 517
  .45  22  71  87 427
  .50
  .55 415
  .60 255 335
  .65  65
  .70  98 191 349
  .75
  .80
  .85  38  99
  .90
       .
       .
       .
 1.15
 1.20 372
 1.25
 1.30  39  92
 1.35  40 107 165 315 398
 1.40  48 119 164 475
 1.45  47 163
 1.50 455 501 508
 1.55  86  88 121 143 175 358 376
 1.60 133 138 247
 1.65 120 122 229 456
 1.70
 1.75  29 166 510
 1.80 137
 1.85 387 476 499 513
 1.90 369
 1.95 158
 2.00  27
 2.05 108 116 160 538 546
 2.10  19 298 396
 2.15  51 304 540
 2.20
 2.25
 2.30 340 477
 2.35  89
 2.40 114
 2.45  41 155
 2.50 502
 2.55 140
 2.60
 2.65
 2.70 377
       .
       .
       .
 2.95  45
       .
       .
       .
 3.25 514
```

```
3.30
        .
        .
        .
4.85
4.90
4.95   70
```

A histogram of ratios of density means found during a F-test (2-class) search at .90 significance is given as an example of this histogram output. The histogram is computed after a 2-class search (for either the *t*-, *F*- or rank order-search). The ratio value is computed as (*m2*/*m1*) for classes 2 and 1. *R*-Spot set numbers are entered in the histogram to permit easy backchecking to the spots in the data base or *R*-map.

image was checked at the pixel level for spot definition using PIXODT [1]. Problem spots were found to be due to (a) very light small spots, (b) fuzzy spots, and (c) rarely, very light spots in the tail of a very large spot.

The use of higher spatial resolution scanning conditions although advantageous for spot resolution, etc., imposes some burden on the factor of field of view. Dynamic range in the density domain using a Vidicon camera (approximately 0–2.0 OD) is less than that for the class of photomultiplier scanners. For analyzing most spots in most autoradiograph gels, this is not a major problem as the average spot is usually less than about 1.5 OD peak density when care is used to avoid saturation of the autoradiographs. Silver stained gels [35–36] constitute more of a problem as more spots tend to saturate in the dynamic range of the Vidicon. By controlling the silver stain development process, the maximum OD of the gel can be controlled to within a workable range for most spots.

A representative gel (*R*-gel) is used as the approximation to the canonical gel (*C*-gel). As a consequence of this approximation some problems arise which include: mis-pairing a spot because it is poorly defined in the *R*-gel; missing spots which are in other gels but not in the *R*-gel; and noise in the *R*-gel masquerading as true *R*-spots. Spots in the GELLAB data base may be manually edited to correct errors discovered by the user.

False positives may appear in a statistical search of the CGL data base. These can result from incorrect inclusion of one or more noise points in a *R*-spot set, which nevertheless meets statistical criteria. Mosaic and *R*-map images are the major tool for handling such false positives by backchecking. Direct visual examination of the *R*-spot numeric data itself is useful in finding outliers. The false negative spot rate may be decreased by finding additional *R*-spot sets of interest by manually scanning the CGL data base for interesting *R*-spot sets with one or two outlier spots which caused problems with the current statistical tests. It is possible to automatically *ignore* or, alternatively, to *find* *R*-spot sets with outliers since they have a large *R*-spot set coefficient of variation as well as significant differences in other features.

5.2. *Future directions*

Interpreting the constellation of *R*-spots as multiclass distributions facilitates finding subtle shifts in spot quantitation. Viewed in this manner, the expected variance of

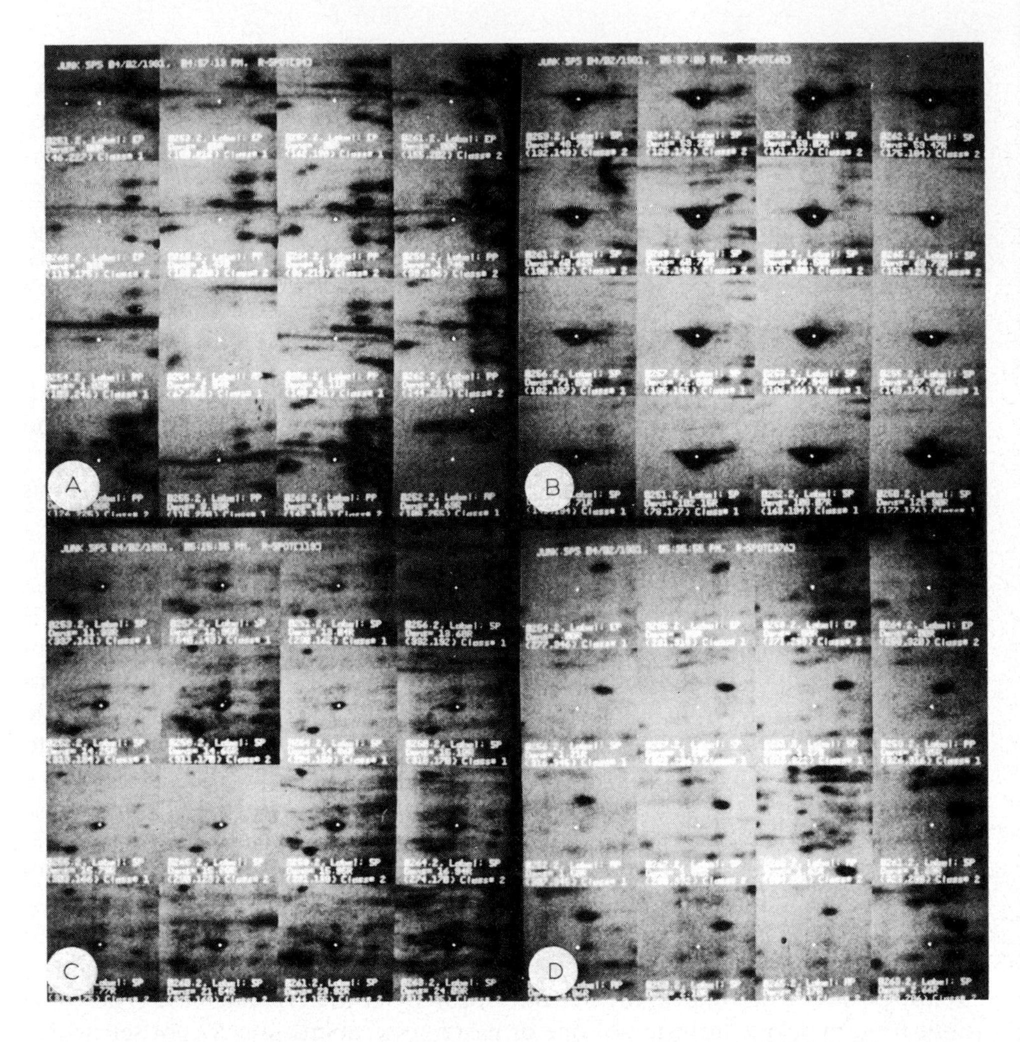

Fig. 14. A mosaic mapping of subregions of a set of gels is performed using the SEERSPOT program on a *R*-spot set of (P388D1 gels). This displays an ordered (lightest to darkest, right to left and top to bottom) illustrate congener spots for a particular *R*-spot. (a) *R*-spot [34] is a poorly defined spot which sometimes appears in a streak and has 5 EPs (extrapolated pair). (b) *R*-Spot [64] has a relative density decrease with time. (c) *R*-Spot [119] has a relative density increase with time. (d) *R*-Spot [376] has a very light spot relative with density increase with time and four EP spots. The *R*-spot region extracted from each corresponding image (magnified by a 2× zoom) is inserted into 128 × 128 pixel subregions of the mosaic image. At the bottom of each panel is an accession number label, gel class and density information. The name of the SPSS file and date is written at the top of the image. The *R*-spot in each panel has a 3 × 3 white '+' drawn in its center. Since the mean density of the images vary – it would be difficult to display and photograph such a mosaic, therefore, the normal mode of operation is to compute the mean background density of each subregion and then to adjust each panel to have the same mean background. The default 2× zoom parameter may be changed to 1×, 4× or 8× by specifying "ZOOM:*n*×".

particular *R*-spot sets can be easily measured and used as a basis for further gel analysis experiments. By having all of the *R*-spot distributions available simultaneously to the data management system, it is now possible to correlate *R*-spot set changes such that sets of *R*-spots changing in the same or in opposition as a function of independent experimental variable can now be determined.

Because of the variety of biological applications and problems in gel preparation, we do not anticipate a fully automated system. However, once a sequence of parameterized operations are identified as habitually used for a class of gels, they can be set up for automated running on a stripped-down system. Such automatic sequencing of operations is now routinely available in GELLAB using the MAKJOB program. It accepts a list of gel accession numbers which are to be analyzed as well as class information. Three batch processing jobs are created to interactively landmark, segment and pair gels, and build an initial CGL data base with several 2-class *F*-test searches. In the P388D1 set of gels, a number of spots were found automatically by the system showing significant statistical differences in the cell line with time. Some of these differences are seen in Figs. 13 and 14, Table 8.

6. CONCLUSION

A set of algorithms in the GELLAB system for the analysis of multiple 2D electrophoretic gels image spot lists using a spot segmentation, spot pairing and congener gel data base analysis has been presented. These algorithms have been successful in analyzing spots under a wide variety of gel conditions. A gel data management system such as CGELP opens the way for asking and answering questions about lists of spot density distributions. Such data reduction applied to a set of gel images has greatly reduced the amount of redundant information retained. Furthermore, by constructing the data base using the inverted file concept, it is possible to rapidly access and update the database. Treating the CGL data base as a set of density distributions leads to the application of various statistical tests for automaticaly determining spot significance which is crucial when investigating a large number of gels with potentially of the order of thousands of spots each.

Significant problems, statistical, operational and others, still remain which must be resolved before reliable reproducible multiple 2D PAGE gel analysis can be routinely performed. That is not to say that useful intermediate results can not be obtained. On the contrary, using backchecking with mosaic and *R*-map images much useful data can be resolved. We are optimistic that many of these problems can be handled by improvements at all levels of gel analysis including better gel preparation, spot extraction and pairing, and the use of better analytical and statistical techniques which take some of these problems into account.

ACKNOWLEDGEMENTS

The constant help afforded by Morton Schultz, Bruce Shapiro, and Earl Smith, our colleagues in the Image Processing Section has been invaluable. Our collaborators Carl Merril and David Goldman of NIMH and Eric Lester (formerly of NCI, now at University of Chicago Medical School) have provided stimulating ideas and critical evaluation of the methodology as it has developed. At Chicago Eric has done much in our effort to export our system at NCI to a TOPS20 system at another site where we are attempting to eliminate as many problems as possible before making it generally available. Bob Connors of NCI syggested the Kruska-Wallis rank order test for comparing more than two classes of gels simultaneously. Parts of this paper are derived from material which has appeared in part in [1–3,6].

REFERENCES

1. Lemkin, P. and Lipkin, L. (1981) Comp. Biomed. Res. 14, 272.
2. Lemkin, P. and Lipkin, L. (1981) Comp. Biomed. Res. 14, 355.
3. Lemkin, P. and Lipkin, L. (1981) Comp. Biomed. Res.
4. Lipkin, L.E. and Lemkin, P.F. (1980) Clin. Chem. 26, 1403.
5. Lester, E.P., Lemkin, P.F. and Lipkin, L.E. (1981) Anal. Chem. 53, 390A.
6. Lemkin, P.F. and Lipkin, L.E. (1981) In: R. Allen, Arnaud (Eds.) W. Electrophoresis '81, De Gruyter, New York.
7. O'Farrell, P.H. (1975) J. Biol. Chem. 250, 4007.
8. Anderson, N.G. and Anderson, N.L. (1979) Behring Inst. Symposium 1977, Mitt. 63, 169.
9. Lemkin, P., Merril, C., Lipkin, L., Van Keuren, M., Oertel, W., Shapiro, B., Wade, M., Schultz, M. and Smith, E. (1979) Comp. Biomed. Res. 12, 517.
10. Lutin, W.A., Kyle, C.F. and Freeman, J.A. (1978) In: Electrophoresis '78, N. Catsimpoolas (Ed.), Elsevier/North-Holland, Inc., pp. 93–106.
11. Garrels, J.I. (1979) J. Biol. Chem. 254, 7961.
12. Bossinger, J., Miller, M.J., Kiem-Phing, V., Geiduschek, P. and Xuong, N.H. (1979) J. Biol. Chem. 254, 7986.
13. Lemkin, P. (1978) NCI/IP Technical Report #21b, Nat. Tech. Info. Serv. PB278789 (listing PB278790).
14. Lemkin, P. and Lipkin, L. (1980) Comp. Prog. Biomed. 11, 21.
15. Carman, G., Lemkin, P., Lipkin, L., Shapiro, B., Schultz, M. and Kaiser, P. (1974) J. Histochem. Cytochem. 22, 732.
16. Lemkin, P., Carman, G., Lipkin, L., Shapiro, B., Schultz, M., Kaiser, P. (1974) J. Histochem. Cytochem. 22, 725.
17. Lemkin, P., Carman, G., Lipkin, L., Shapiro, B. and Schultz, M. (1977) NCI/IP Technical Report #7a, Nat. Tech. Info. Serv. PB269600/AS.
18. Reiser, J.F. (1976) SAIL, Stanford University Artificial Intelligence Laboratory memo AIM-289, August, 1976. Also available from U.S. Dept. Commerce. Nat. Tech. Inform. Serv. No. Ad-A045-102, Springfield, Va.
19. Merril, C., Switzer, R.C. and Van Keuren, M.L. (1979) Proc. Natl. Acad. Sci. USA, 76, 4335.
20. Merril, C.R., Goldman, D., Sedman, S.A. and Ebert, M.H., (1981) Science 211, 1437–1438.
21. Vo, K-P, Miller, M.J., Geiduschek, E.P., Nielsen, C. and Xuong, N.H. (1981) Anal. Biochem. 112, 258.
22. Rosenfeld, A. (1969) Picture Processing by Computer, Academic Press, New York.
23. Rosenfeld, A. and Kak, A. (1977) Digital Picture Processing, Academic Press, New York.
24. Anderson, N.G., Anderson, N.L. and Tollaksen, S.L. (1979) Clin. Chem. 25, 1199.
25. Taylor, J., Anderson, N.L., Coulter, B.P., Scandora, A.E. and Anderson, N.G. (1979) In: Proceedings of Electrophoresis '79, Radola, B.J. (Ed.),W. de Gruyter, New York.
26. Bookstein, F.L. (1978) The Measurement of Biological Shape and Shape Change, Springer-Verlag, New York.
27. McConkey, E.H. (1979) Anal. Biochem. 96, 39.
28. Natrella, M.G. (1966) Experimental Statistics, NBS Handbook 91, U.S. Govt. Printing Office, Wash., D.C.

29. Nie, N.H., Hull, C.H., Jenkins, J.G., Steinbrenner, K. and Bent, D.H. (1975) SPSS – statistical package for the Social Sciences, McGraw-Hill, New York.
30. Knott, G.D. (1979) Comp. Prog. Biomed. 10, 271.
31. Lester, E.P., Lemkin, P., Lipkin, L.E. and Cooper, H.L. (1981) J. Immunol. 126, 1428–1434.
32. Lester, E.P., Lemkin, P., Cooper, H.L. and Lipkin, L.E. (1980) Clin. Chem. 26, 1392.
33. Lemkin, P., Lipkin, L., Merril, C. and Shiffrin, S. (1980) Envir. Health. Perspect. 34, 75.
34. Lipkin, L. (1980) Envir. Health. Perspect. 34, 91.
35. Merril, C., Switzer, R.C. and Van Keuren, M.L. (1979) Proc. Natl. Acad. Sci. USA, 76, 4335.
36. Goldman, D., Merril, C.R. and Ebert, M.H. (1980) Clin. Chem. 26, 1317–1322.
37. Schwartz, A.A. and Soha, J.M. (1977) Appl. Opt. 16, 1779–1781.
38. Lipkin, L., Lemkin, P., Shapiro, B. and Sklansky, J. (1979) Comp. Biomed. Res. 12, 279–289.
39. Lemkin, P., GELLAB User Manual, in preparation.

APPENDIX

GELLAB A user's guide to this system will be available from our laboratory. Copies of the software may be available on application to the authors, but this will only be exportable to groups having access to SAIL (on a 'DECSYSTEM-10 or -20).

Geisow & Barrett (eds.) Computing in biological science

7

Nucleic acid morphology: analysis and synthesis

BRUCE A. SHAPIRO and LEWIS E. LIPKIN

1. INTRODUCTION

Within the last few years nucleic acid biochemists have begun to understand the morphologic significance of nucleic acid (DNA and RNA) secondary structure. The observation of various features, e.g., 'hairpin' turns, 'bulge' loops, 'internal' loops and 'multibranch' loops which may be seen in relatively constant position is indicative of specific structures of given lengths within the molecules. These structures are believed to have functional significance for the expression of genetic information. Codons, groups of three consecutive bases which select specific amino acids for assembly into proteins, that are enclosed within double stranded portions of these structures are believed to be less accessible for genetic activity than those in single stranded looped sections. Thus, for example, a more accessible codon may more readily partake in the translation process of protein production than one that is less accessible. It is apparent, then that the ability to measure and/or determine as objectively as possible time locations of these structures is exceedingly important.

To understand nucleic acid morphology it is necessary to comprehend the fundamentals that are involved in the formation of the molecular configurations. Nucleic acid molecules consist of a sequence of bases linked together to form strands. The bases that comprise these strands consist of guanine (G), adenine (A), cytosine (C), and thymine (T) in DNA. In RNA molecules uracil (U) appears in place of thymine. DNA and RNA generally consist of double stranded structures, although single stranded structures may also exist. Each strand twists around the other to form a double helical structure. The double helix is formed by each strand binding to the other by hydrogen bonds between complementary pairs, an A to a T (or U) and a G

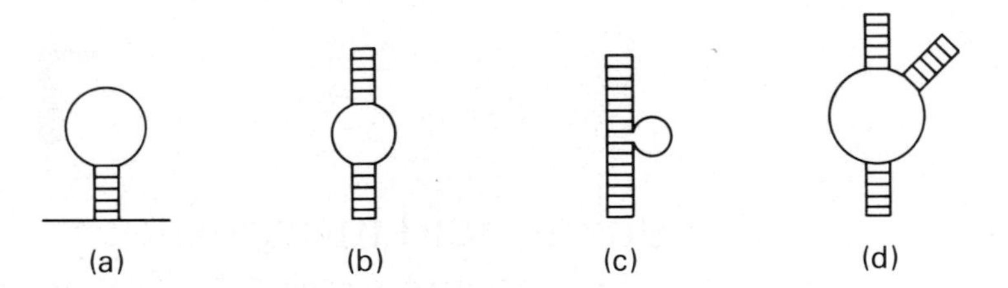

Fig. 1. Examples of morphologic structures present in nucleic acid secondary structures: (a) hairpin loop; (b) internal loop; (c) bulge loop; and (d) multibranch loop.

to a C. These bonds may be broken by a process called denaturation, which usually results in structures that are less compact than undenatured molecules thereby making them more amenable to analysis with an electron microscope. Denaturation may be accomplished by high temperature, by an alkaline pH, or by using substances such as formamide that compete with the bases for hydrogen bonding sites. Regions richest in AT base pairs (lower stabilization energies) will break apart first followed by regions rich in GC base pairs (higher stabilization energies; see Section 3.2) as conditions for denaturation improve. Thus, molecules which have been denatured tend to have more single stranded regions (unpaired bases).

Double stranded regions may form when some portions of a single strand base pair to other portions of the same strand. If a double stranded region has an intervening single strand, the loop that is formed is known as a 'hairpin'. Generally speaking, 'hairpin' turns (single stranded portion) range in size from 3 to 20 bases. 'Internal' loops are defined by two double stranded regions separated by two intervening single stranded regions where each intervening single strand is greater than or equal to one base in length. A 'bulge' loop is an 'internal' loop where the size of one of the intervening single strands is zero. Finally, a 'multibranch' loop consists of three or more double stranded regions emanating from a loop(see Fig. 1). It then follows, that given a single stranded molecule, it will fold up into a configuration that potentially contains all of the morphologic substructures defined above. In general it will seek out that configuration which tends to minimize the molecule's free energy. In its native state, the molecular configuration is almost certain to be quite compact. However, if denaturation techniques are applied, a degree of simplification occurs. Some biochemical techniques [1–7] have also been emerging which attempt to fix and/or mark specific regions of the molecule with respect to its double or single stranded regions thereby adding more evidence for the existence of specific structures.

This chapter discusses two complementary aspects of nucleic acid morphology. One is the computer analysis of electron micrographs of nucleic acid molecules with the intention of producing secondary structure maps [8], i.e., a representation which shows the location and length of various structures seen in the electron micrographs and second, the synthesis by computer of the most energetically stable nucleic acid configuration given sequence information. The eventual combination of the analytic and synthetic elements of this problem will permit one to reinforce the other, thus enabling a further understanding of nucleic acid morphology and how it relates to gene expression [9].

2. COMPUTER GENERATION OF NUCLEIC ACID SECONDARY STRUCTURE MAPS

Secondary structure maps attempt to capture the essence of the structure of a molecule and to make it relatively easy to visualize the structure. When several secondary structure maps from a homogenous group of molecules are merged together, occupancy histograms may be generated. These aid in determining the stability of particular structures. Each of the secondary structure maps are normalized to the length of the molecule. The maps may be manually generated by either projecting an electron micrographic negative onto paper and then tracing the outline of the molecule with a writing implement or by using a digitizing tablet. In general an attempt is made to trace the molecule starting from its 5′-end. Sometimes an end may be visually determined by splicing onto one end a specific marker nucleic acid with a known morphology [10]. During the tracing process the positions and lengths of the individual features are noted. This is done for the entire length of the molecule. This manual approach is tedious and prone to errors. In addition, in order to get statistically valid measurements, 100 to 1000 molecules should be analyzed. The results should then be incorporated in an occupancy histogram to determine the relative frequency and the implicit stability of the features. An example of manually produced secondary structure maps is shown in Fig. 2 and an occupancy histogram of these maps is shown in Fig. 3.

It is intended in this section to demonstrate a methodology by which secondary structure maps can be semi-automatically produced from electron micrographs of nucleic acid molecules. The techniques described will handle molecules that consist of distinguishable single stranded and double stranded regions in which the 5′- and 3′-ends of the molecule are discernable. At the present time only features consisting of 'hairpin' loops are handled.

2.1. *Data acquisition*

The process of data acquisition and subsequent processing consists of several steps directed toward a specific goal. This is to obtain a reasonably noise-free, closed boundary outline of a molecule from a two-dimensional digital image in turn derived from a photographic original. The data acquisition process, as defined here, consists of two parts. The first being video acquisition and enhancement of the initial two-dimensional grey scale digital image. This pre-processing then facilitates the second stage: extraction of a contour representing the molecular outline of the nucleic acids.

2.1.1. *Acquisition of a working two-dimensional image*

Nucleic acid molecules are prepared for visualization by spreading them onto a plastic film substrate which rests on a mesh grid. To improve the contrast and to enlarge their size, the molecules are stained and shadowed. This process makes the features more

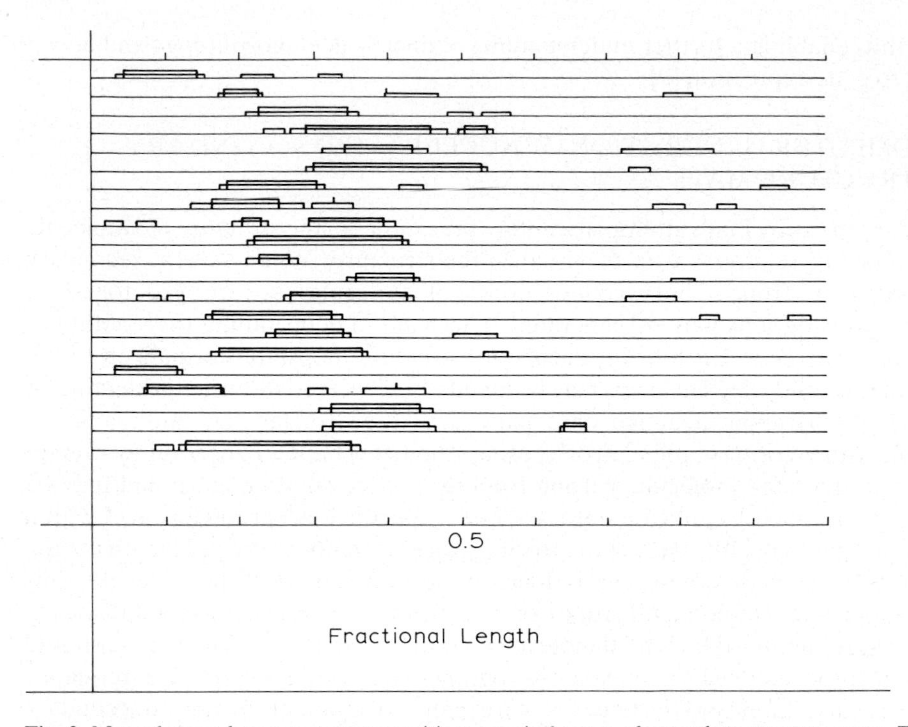

Fig. 2. Manual secondary structure recognition: a typical group of secondary structure maps. Each line represents the mapping of a molecule from a single nucleic acid preparation. The molecular features (secondary structures) are denoted by the raised areas.

distinguishable under 10,000× magnification [11]. A special high resolution film is used to photograph the specimen. This is either in the form of a 4″ by 3¼″ negative or 35 mm film strip which is then viewed through a Quantimet 720 film reader with a Plumbicon TV camera. A 256 × 256 pixel image is then acquired into a digital picture memory in 1/10 of a second using an A–D converter within the Real Time Picture Processor (RTPP) [12–17]. The effective magnification of the image is 53 Angstroms per pixel (with the 4″ by 3¼″ negatives). The picture memory stores up to 8 bits of grey scale per pixel (256 grey levels). A typical photograph of a picture memory image from the Quantimet display is shown in Fig. 4. Upon examination it will be noted that this image has a bad shading problem that must be dealt with before any useful processing can be accomplished. Shading correction is necessary to simplify the extraction of the molecular contours. Because of the shading, the extraction is impossible to do with a single global grey scale threshold. A grey scale threshold of a digital image defines a new image by the following operation

$$g'(x,y) = \begin{cases} g(x,y) & \text{if } g(x,y) \geq t \\ 0 & \text{otherwise,} \end{cases} \tag{1}$$

where t is a specified grey level and $g(x,y)$ is the original digital grey scale image where $g(x,y)$ defines the grey level (0–255) of a pixel at coordinate x,y. Figure 5 shows the effects of applying a global threshold to a poorly shaded image.

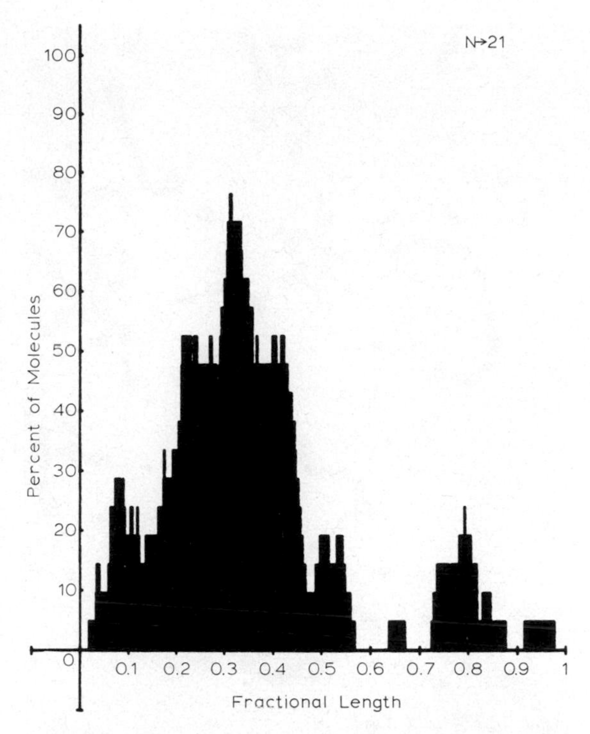

Fig. 3. Secondary structure map histogram corresponding to the set of maps shown in Fig. 2.

Shading may be defined as a position dependent non-uniformity of the density of the acquired image. These non-homogeneities arise from several sources. One major source is the sample itself. When the sample is prepared, it is almost impossible to produce a uniform substrate, thus causing a non-uniform background density. Local and global heterogeneities in the illuminating electron beam (such as off center alignment, etc.) will also result in local variations in brightness. The camera and film processing techniques as well as the illumination system on the film reader further contribute to shading variations. Finally, the electronic image acquisition process itself, which includes the Plumbicon camera will introduce shading effects. The Plumbicon has an intrinsically non-systematic non-uniformity in light sensitivity across its face.

To correct for these shading difficulties, the effect of all the contributory sources are combined and treated as one, rather than treating each one independently. (Although a pointwise shade correction algorithm is under development in this laboratory which when applied to acquired images will correct for shading distortions introduced from the light source and camera). For viewer convenience and photographic quality some minor modifications to the image are first introduced. Before a shading correction technique is applied, the digitized image is complemented so that the strands are darker than the background. Complementing is accomplished by applying the following operation to the digital image

$$g'(x,y) = 255 - g(x,y). \tag{2}$$

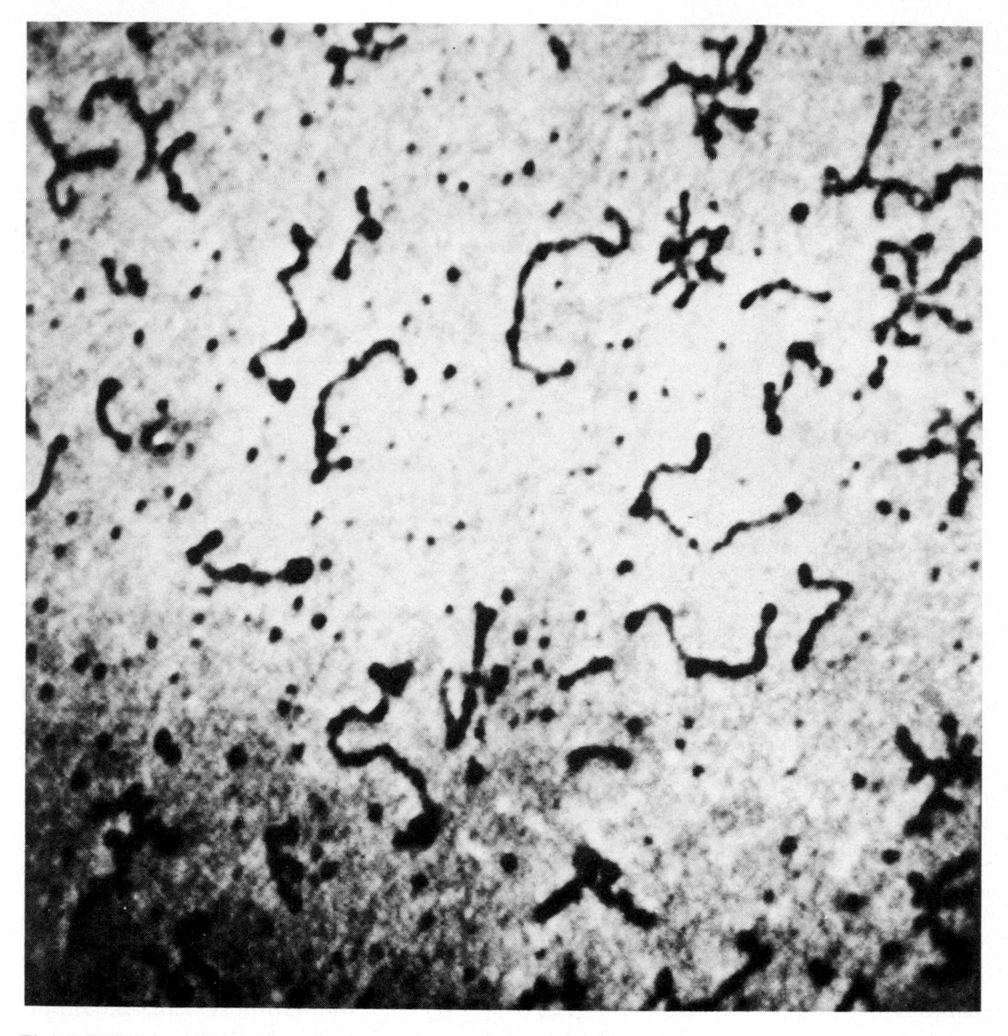

Fig. 4. RTPP image of non-uniformly shaded adenovirus type 2 messenger RNA.

The entire grey level range of the image is then biased toward one end of the dynamic grey scale range of the image processing system since it consists mostly of background. The image is therefore shifted toward the middle of the range by a scalar pointwise subtraction from the picture matrix. This causes the grey level distribution of the complemented image to start at approximately grey level 50 (0 is white and 255 is black).

2.1.2. The DC notch filer

The images consist of a relatively small number of dark strands on a relatively light background, as is evident from Fig. 4. This fact is used in designing a digital filter to shift individual pixels belonging to the background to a uniform level of background. Similarly, pixels belonging to the nucleic acid strands, dark artifacts and molecular

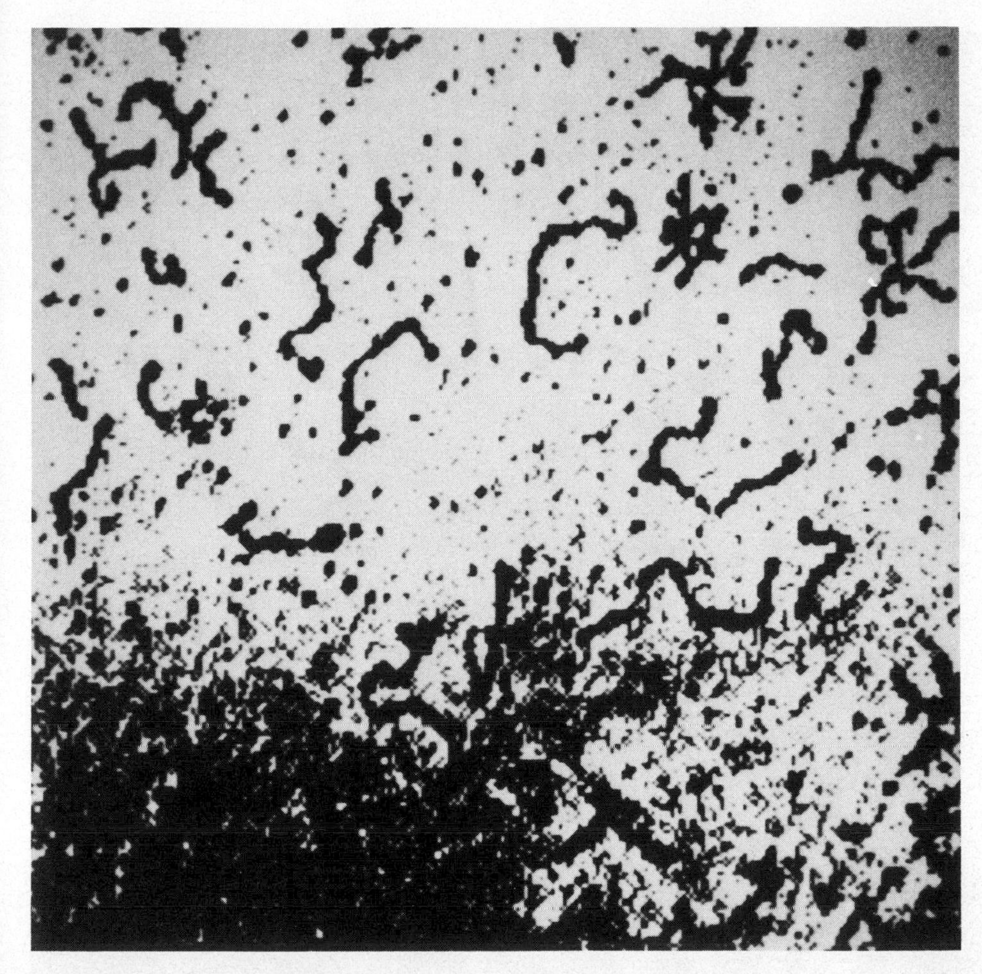

Fig. 5. Global threshold applied to the poorly shaded image of Fig. 4. The threshold process is defined in the text. It can be seen that molecules are washed out due to the non-homogeneity of the background density.

fragments are shifted to a uniform darker level. To accomplish this, the complemented shifted image described in Section 2.1.1. is shade corrected by application of a digital DC notch filter [18–19]. This is a signal analysis transformation designed to supress low spatial frequencies from a signal. An $n \times n$ pixel sampling window is moved through the image and its average is subtracted from the center point of the window for each point in the image. For the electron micrographs used here, strands appear to be 2–3 pixels wide, and a value for n of 32 gives reasonable results. This has the effect of supressing the low spatial frequencies corresponding to the image shading. A DC offset value, b, is added to the value computed by the filter at each point. This requires two passes of the algorithm (the first being used to compute b). The filter is computed in (3).

$$g_j(x,y) = gi(x,y) - \text{AVG}n \times n(x,y) + b. \tag{3}$$

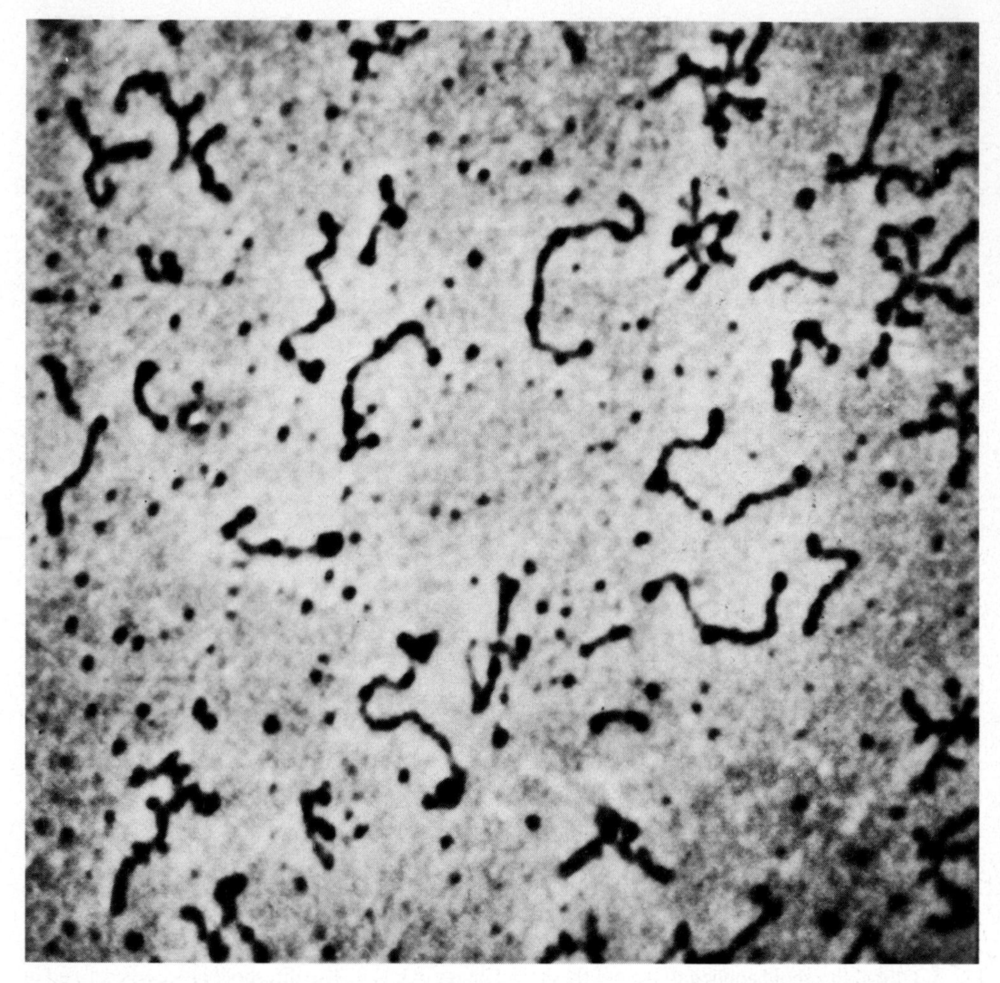

Fig. 6. DC notch filter applied to the image depicted in Fig. 4. An averaging window of 32 × 32 pixels was used.

Let

$$n1 = (n/2) - 1. \tag{4}$$

Then,

$$\text{AVG}n \times n(c,r) = \frac{1}{n*n} \sum_{x=c-n1}^{c+n/2} \sum_{y=r-n1}^{r+n/2} gi'(x,y). \tag{5}$$

To handle the edge conditions, the image is reflected about the boundaries using (6)

$$gi'(x,y) = gi(R(x),R(y)). \tag{6}$$

$R(\cdot)$ is the reflection function and computes the reflected coordinate value on the edges at 0 and 255, under the following conditions in (7).

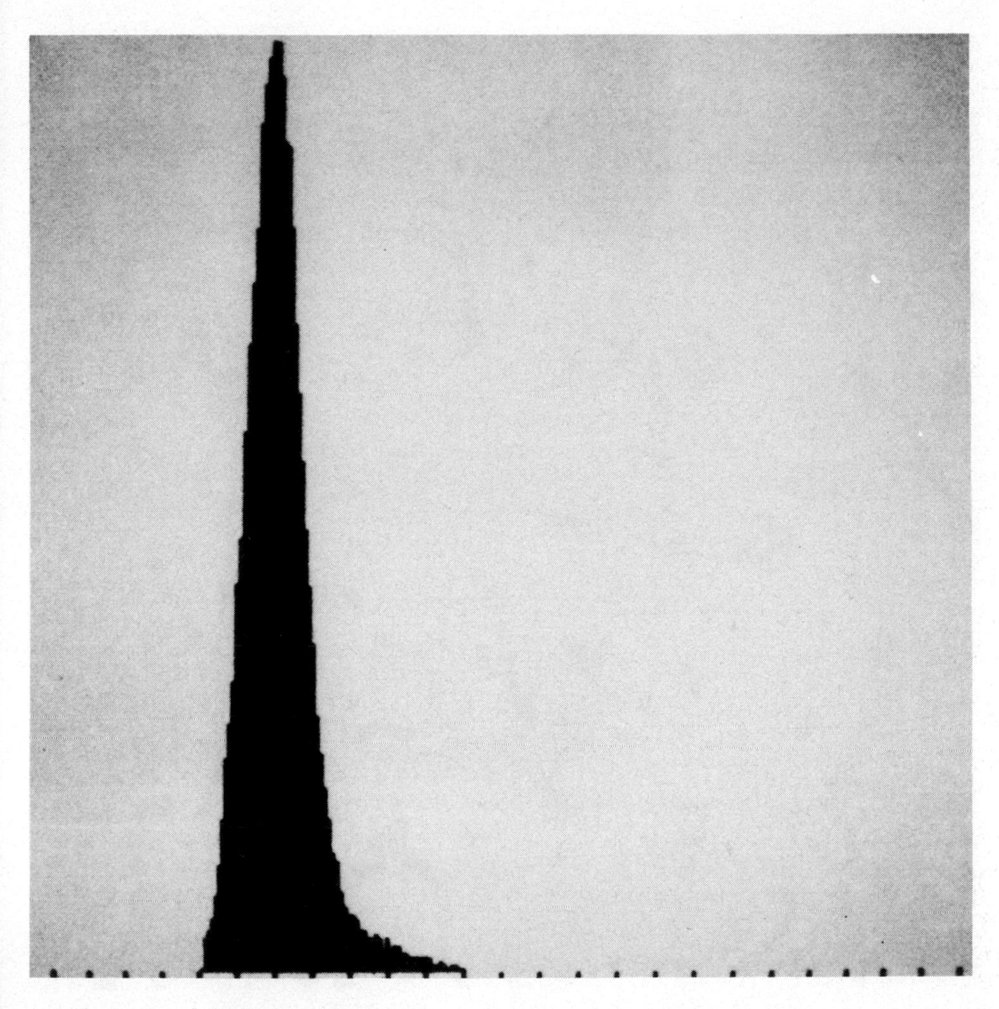

Fig. 7. Histogram of the DC notch filtered image depicted in Fig. 6. The horizontal axis indicates the grey levels (0–255) and the vertical axis indicates the number of pixels in the image.

$$\begin{aligned} R(p) &= -p, & p < 0, \\ &= 255 - (p - 255), & p > 225, \\ &= p, & \text{otherwise}. \end{aligned} \tag{7}$$

During the first pass b is computed in (8). The final output image is computed during the second pass

$$b = -\min_{\text{All } x,y} (gi(x,y) - \text{AVG}n \times n, i(x,y)). \tag{8}$$

To increase the efficiency, AVG$n \times n,i$ is computed iteratively in (9) and (10)

$$\text{AVG}n \times n(0,r) = \frac{1}{n*n} \sum_{y=r-n1}^{r+n/2} \text{COLSUM}(x,r) \tag{9}$$

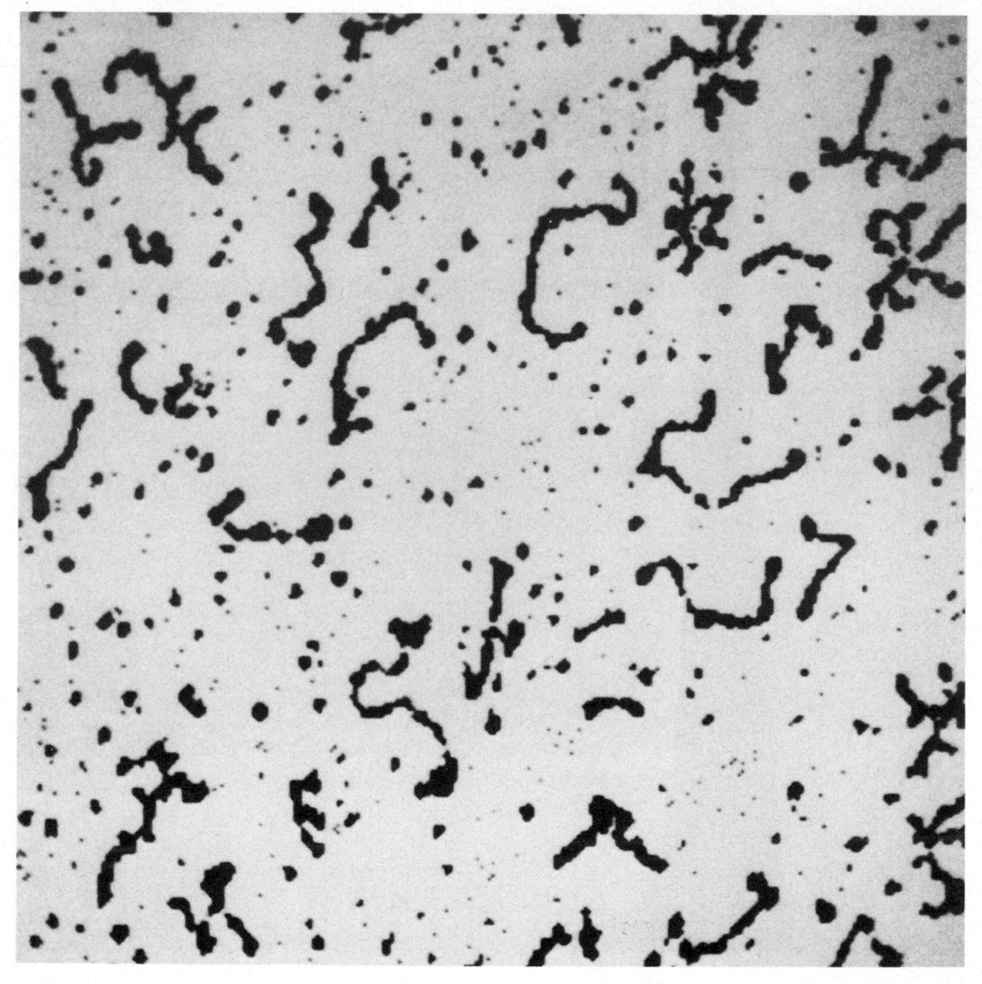

Fig. 8. Threshold of 18 levels above peak histogram value applied to DC notch filtered image depicted in Fig. 6.

and,

$$\text{AVG}n \times n(c,r) = \text{AVG}n \times n(c-1,r) - \text{COLSUM}\,(c-n/2,r) + \text{COLSUM}(c+n/2,r), \tag{10}$$

$$\text{COLSUM}\,(c,r) = \sum_{y=r-n1}^{r+n/2} gi'(c,y). \tag{11}$$

Although this transformation is specified here for even n, it may easily be extended to odd n taking the symmetry of the sampling window into account.

Figure 6 illustrates the results of applying the DC notch filter to a typical electron micrograph. Some slight shading is visible on the outer edges of the figure. This is due to the non-homogeneities in the display as opposed to the actual data. For more detail on the problems of shading and segmentation of electron micrographs of nucleic acids see [20–21].

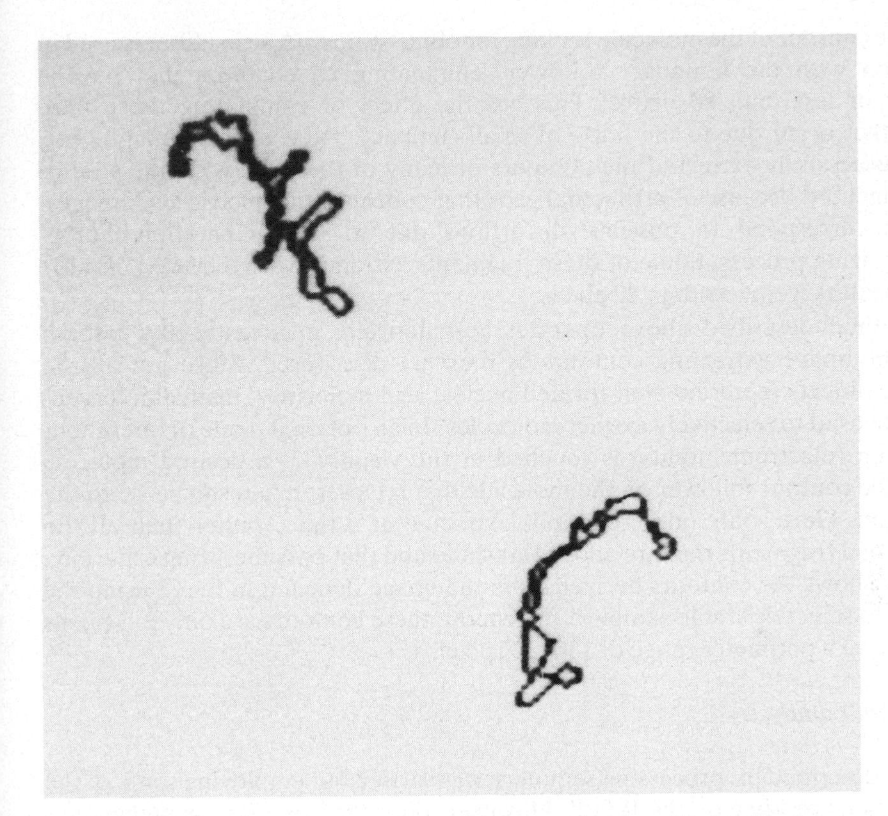

Fig. 9. Molecular contours extracted from just two molecules with the aid of the graph pen using molecules shown in Fig. 8. The use of the stylus allows the interactive choice of the molecules of interest. A cursor just has to be moved to the vicinity of the molecule of interest and a boundary follower will then automatically trace the contour of the molecule. The *x,y* coordinates of these boundaries are then stored for later processing.

2.1.3. *Contour extraction*

After the shade correction process has been applied, the image is now ready for the extraction of the molecular contours. To accomplish this several steps are necessary. A global threshold is determined by computing the grey scale histogram of the complemented, level adjusted, DC notch filtered image. Because of the unimodal nature of the histogram, it is difficult to automatically select an appropriate threshold to separate the object from the background. For this reason, the peak of the histogram is determined and an empirical value of n levels above this peak frequency is used to threshold the image. The value of n is currently chosen between 8 and 18 grey levels depending upon the degree of contrast that exists in the image. The empirical value produces a reasonable separation of the molecule from the background. A typical histogram of a complemented, level adjusted DC notch filtered image is shown in Fig. 7. Figure 8 illustrates the result of applying the determined grey scale threshold to the complemented, level adjusted, notch filtered image.

A boundary follower algorithm similar to the one described in [22] is applied to the thresholded image. The algorithm moves a 3×3 neighborhood around the contour of the molecule looking for object points. A sizing filter is used in conjunction with the boundary follower, eliminating all contours that have a perimeter of less than 80 pixels. This has the effect of eliminating those false contours that occur due to the noise of small compact artifacts in the image. This method successfully extracted the contours of many of the strands. Some strands were fragmented because of artifactual gaps that appear in the nucleic acid images. The gaps correspond to possible distortions due to the preparation/photographic/digitizing process. Some of these fragmented strands were removed because their perimeters were less than 80 pixels.

The method described above operates automatically in a raster-like fashion through the image extracting contours as they are discovered. Since not all the strand-like objects represent well-formed nucleic acid molecules, manual intervention may be used to selectively extract molecules. In an optional mode of operation, a graph pen (electronic stylus) is touched in the vicinity of a desired molecule initiating the contour follower on the molecule that is closest in a raster sense to the pen position. Here, only one contour is extracted at a time, rather than all the molecules and fragments that are above threshold and that pass the sizing criterion.

Figure 9 shows the contours derived from the image depicted in Fig. 8 using the graph pen to select desirable samples. In general, these contours are only 2–3 pixels wide and have a perimeter range of 130–220 pixels.

2.2. *Contour analysis*

The above described preprocessing sequence was carried out employing some of the set of functions residing on the RTPP. However, once the boundary sequences are extracted from the molecules they are processed by programs written in SAIL [23] resident on the DECSYSTEM-10. The shape of the contours are then analyzed to determine the position and size of the 'hairpin' loops; the features that are being explored to generate the secondary structure maps. These contour analysis techniques are now discussed in some detail.

2.2.1. Contour approximation and smoothing

Given a set of extracted contours such as those in Fig. 9, it is desirable to smooth the contours so that the noise introduced by the data acquisition process may be minimized. The more noise is present in the contour, the more noise will be introduced into the secondary structure maps. However, care must be taken to ensure that smoothing does not eliminate significant information.

To accomplish this smoothing, a recursive contour fitting process described in [21] is used with the circle transform [21,24]. Circular arc segments are fit to the original contour. Each segment is subdivided into smaller segments until an error bound is satisfied. The error is defined as the difference between the radius of the circular arc segment and the distance of the point of maximal deviation on the original contour from the center of the circular segment. The region within which this maximum may occur is defined by the endpoints of the circular segment. If the error bound is not satisfied, the point p on the boundary having the maximal deviation is used as a new end point of a new circular segment. A new point is also chosen midway

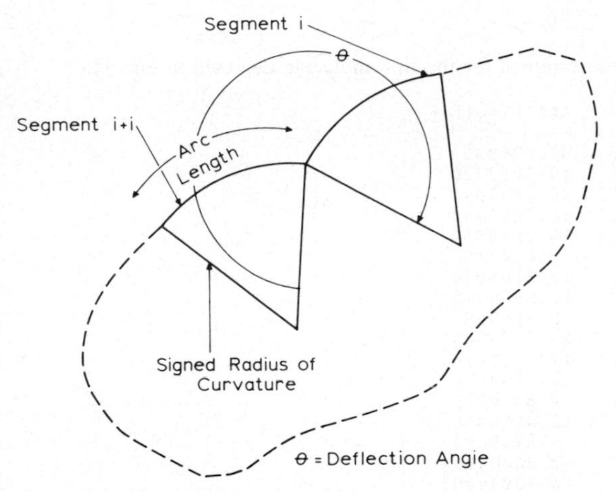

Fig. 10. Components of a circle transform triple.

between p and one of the other end points, thus producing the three points necessary for a new circular segment. The circle transform fitting process is initiated using a sampling distance of one-third the total number of boundary points on the original contour, thus defining the initial three points that determine a circle. The effect of this process is to produce a larger number of short arc segments in regions of high curvature while producing larger arc segments in regions of the contour that are not quite as convoluted. This makes the fitting process information dependent.

As a result of the circle transform, a list of triples is generated, each triple representing a circular arc segment (see Fig. 10). A triple contains four pieces of explicit information, namely: (1) the radius of curvature, indicating the radius of the circular arc that was fit to this region of the original boundary; (2) its sign, a plus sign indicating a concave arc and a minus sign indicating a convex arc; (3) the deflection angle, indicating the degree of cornering that takes place between two adjacent arcs; and (4) the arc length for the given segment.

The list of triples is then scaled up by a factor of 3 to facilitate viewing of the smoothed molecular contours and to place the triples' values in a range suitable for later analysis. Once this is done the triples are used as input to the inverse circle transform (see [21,24]) to synthesize a smoothed, scaled up image of these contours. At this point, however, the synthesized contour and its representation contain corners, i.e., points on the curve where there are discontinuities in the first derivative. These are points where the adjacent segments meet. A corner removing process is applied to the synthesized contour. This accomplishes two things. First, the corners are removed by replacing a corner by a circular arc segment that is tangent to the two arc segments that produce the corner, thus smoothing the contour further. Second, a standard form of the triples is produced that will facilitate further analysis. That is, all segments of the boundary will be composed of circular arcs without any discontinuities in the first derivative. This eliminates the need for considering special cases when analyzing the triples. Thus, the deflection angle becomes superfluous as a parameter.

A point 10% or less of an arc length in from a corner is used as one point of

TABLE 1

List of circle transform triples before corner removal for the first molecule depicted in Fig. 11a.

	Radius	Def Angle	Arc Length
[1]	-29.4204170,	6.2476275,	61.8799950
[2]	-6.5192620,	.2725954,	16.2853720
[3]	13.1573350,	1.5332988,	30.6801670
[4]	17.1038300,	5.2592093,	11.0055440
[5]	-23.7171070,	4.7679102,	15.2619550
[6]	7.6488345,	1.4464781,	12.0141150
[7]	2996.6534000,	.9075592,	13.4164020
[8]	46.5241770,	5.5363420,	73.0800210
[9]	-2997.0015000,	1.2092530,	5.9999995
[10]	16.3469870,	1.2822975,	29.7433420
[11]	-17.1020980,	4.4869536,	17.7575420
[12]	-11.4236370,	1.7094633,	27.1952570
[13]	-10.6068740,	1.1902987,	9.8354075
[14]	17.1005740,	3.5496794,	11.0055940
[15]	-10.6074340,	1.4463500,	9.8355094
[16]	53.4962610,	1.8295307,	53.8600900
[17]	57.1147380,	.6857484,	16.2098440
[18]	-2995.7603000,	3.4059934,	8.4852966
[19]	-5.3032507,	.6449076,	23.4858840
[20]	2995.7581000,	3.7843832,	4.2426563
[21]	17.1026670,	5.7633319,	11.0055620
[22]	11.3843980,	5.0440818,	22.8620680
[23]	59.0785380,	.8853091,	25.8374270
[24]	11.4236720,	6.0183957,	17.9442360
[25]	-6.0467033,	3.4221325,	24.6033750
[26]	9.4868873,	2.3007027,	14.9017510
[27]	28.8530580,	.1651669,	18.5669950
[28]	-27.6586100,	4.0068198,	31.3499060

tangency. The other point of tangency in the adjacent arc is then determined (see [21]). The new arc is spliced into the triple list and the arcs to which it is tangent are shortened accordingly. The result of this operation is barely visible but the uniformity of approach that results from this technique makes it useful.

The corner removal process alters all the triples which represent the contour. These triples are then used to resynthesize the contour using the inverse circle transform. The resynthesized contour and its triple description are now used for the analysis of the shape of the molecule for the 'hairpin' features. Figure 11a shows some typical original molecular contours. Figure 11b shows the synthesized contours without corner smoothing. Figure 11c shows the resynthesized boundaries with corner removal. Table 1 shows a list of circle transform triples before corner smoothing for the first molecule depicted in Fig. 11a. Table 2 shows a list of circle transform triples with the corners smoothed using the first molecule depicted in Fig. 11a.

At this stage a smoothed contour of the molecule has been produced which preserves the essential features of the shape. A description, a list of triples, has also been generated which is now further analyzed to determine the location of specific features.

2.2.2. *Finding the tips of hairpin features*

The features of interest appear as 'hairpin' turns or arms emanating from the molecule. To produce the secondary structure maps, the tips of the arms must be detected. This is accomplished by first merging those adjoining circular segments

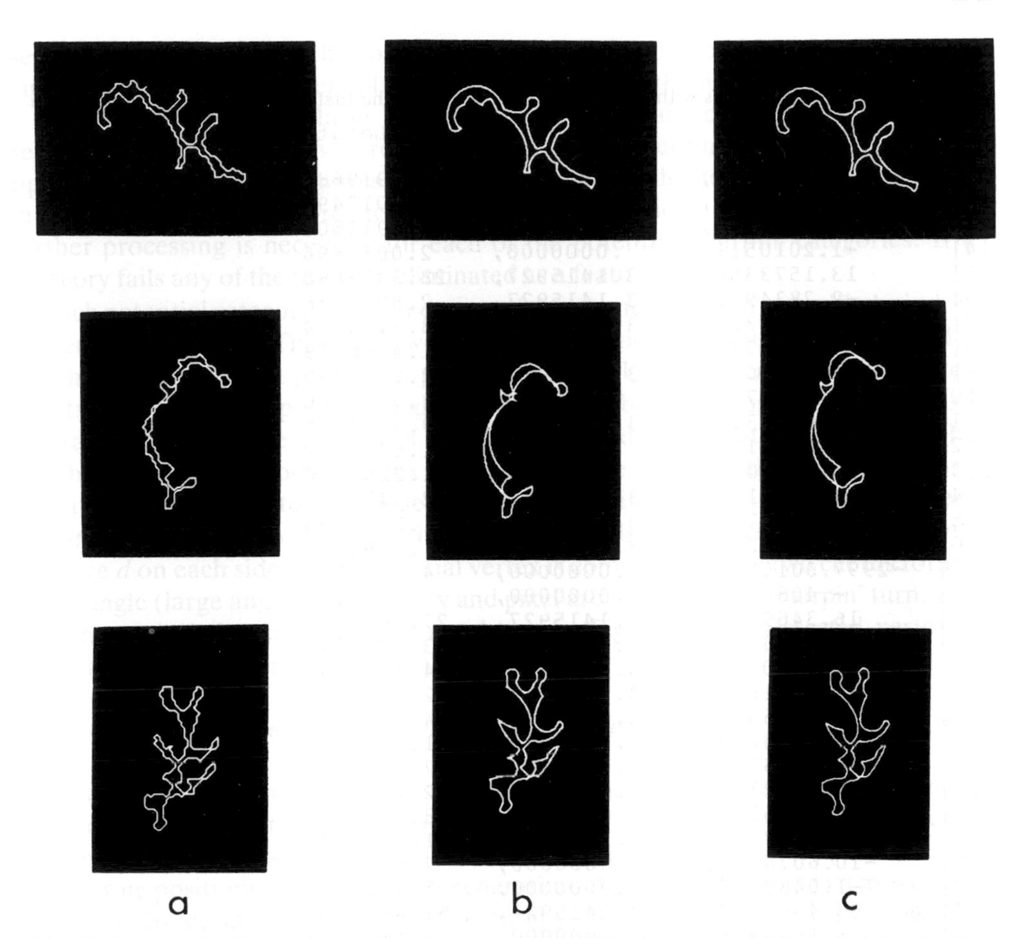

Fig. 11. Extracted and smoothed molecular contours: (a) original contours; (b) inverse circle transform synthesized smoothed contours with corners; and (c) inverse circle transform resynthesized molecular contours with corners removed. All contours used an error value of 1.7 and a one-third contour length initial sampling distance. A corner removal value of 0.1 was used.

with similar characteristics, followed by analysis of those merged segments that appear to be vertices (feature tips). Noise and/or sharp bends in the molecule must not be accepted as legitimate features, thus, the vertex detection must be robust enough to eliminate these regions while accepting actual features. The details of these processes are discussed in [21]. However, a summary of the algorithms used will be presented here.

The merging of adjoining segments consists of testing the adjoining segments for similarity of curvature. Segments are considered similar if they are: (1) all concave or convex; (2) lie within a specified range of curvature; and (3) there is a smooth transition from one to the other as is the case when the corner removal algorithm is applied. If the segments meet the above criteria, the merged segments are placed into the same category. The merging continues until all the triples have been examined and categorized. The category of interest, when looking for the 'hairpin' turn in the molecule is 'convex sharp curve'. This is defined as the set of all convex

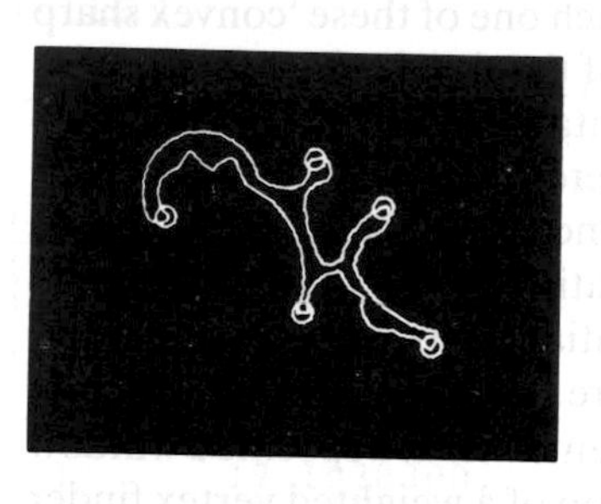

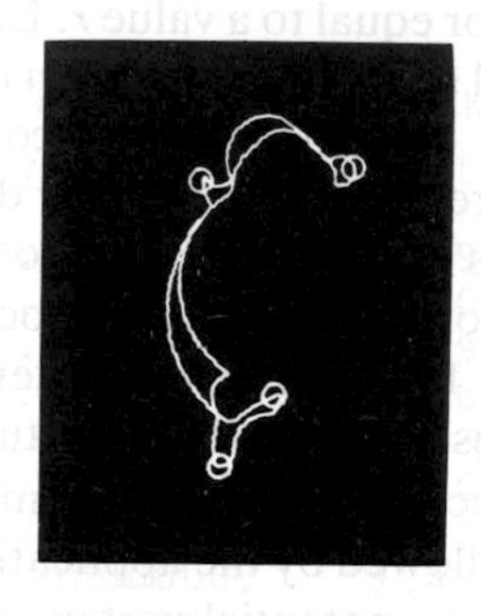

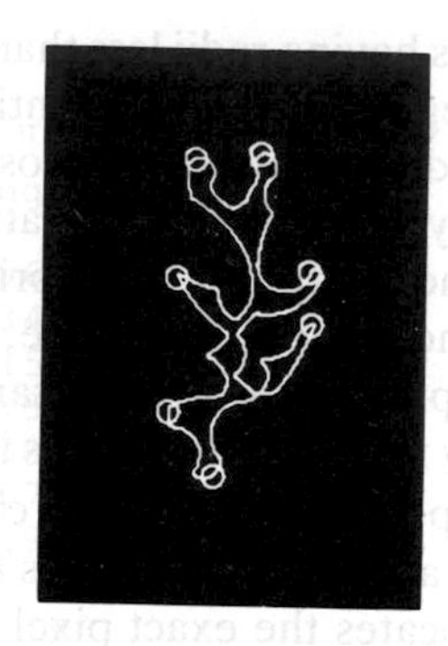

Fig. 12. Illustration of vertex detection. Small circles are centered on the locations of discovered vertices that correspond to the tips of the 'hairpin' turns. It should be noted that those vertices that appear to have been missed, did not satisfy the criteria for feature detection. These features because of their size and shape are considered to be noise by the feature detection process.

connectivity array indicating which pieces of the skeleton are connected to other pieces is also produced. Some typical molecular contours and their corresponding skeletons are shown in Fig. 13.

2.4. *Generating secondary structure maps*

The aim of this section has been to show how to generate secondary structure maps for nucleic acid molecules. A secondary structure map has been defined as a trace along the molecule (preferrably starting from the 5′-end when known) noting the positions and lengths of features as one traverses the entire length.

A secondary structure map is derivable once the skeleton for the molecule has been produced. Another piece of information is required however, namely, those two vertices which represent the 5′-and 3′-ends of the molecule. The determination of these points may at times be difficult. These strands are quite thin to begin with and undergo several error inducing steps before analysis can commence (e.g., those errors imparted due to the data acquisition process), so that the 5′- and 3′-ends may become obscure. Obviously, if some point other than the 5′ and 3′ positions were chosen, the secondary structure maps would not be correct. The determination of the correct points is aided by some simple heuristics. For example, the backbone (single stranded portion) usually constitutes the linear segment of greatest length across the molecule, i.e., the path along the molecule from one vertex to another excluding any intervening features. Another helpful heuristic is that double stranded regions are thicker than single stranded regions. At present the two backbone vertices (5′ and 3′) are defined manually from the raw image and are supplied to the routine that generates the secondary structure maps.

These two end points are used in finding a path connecting them by using the

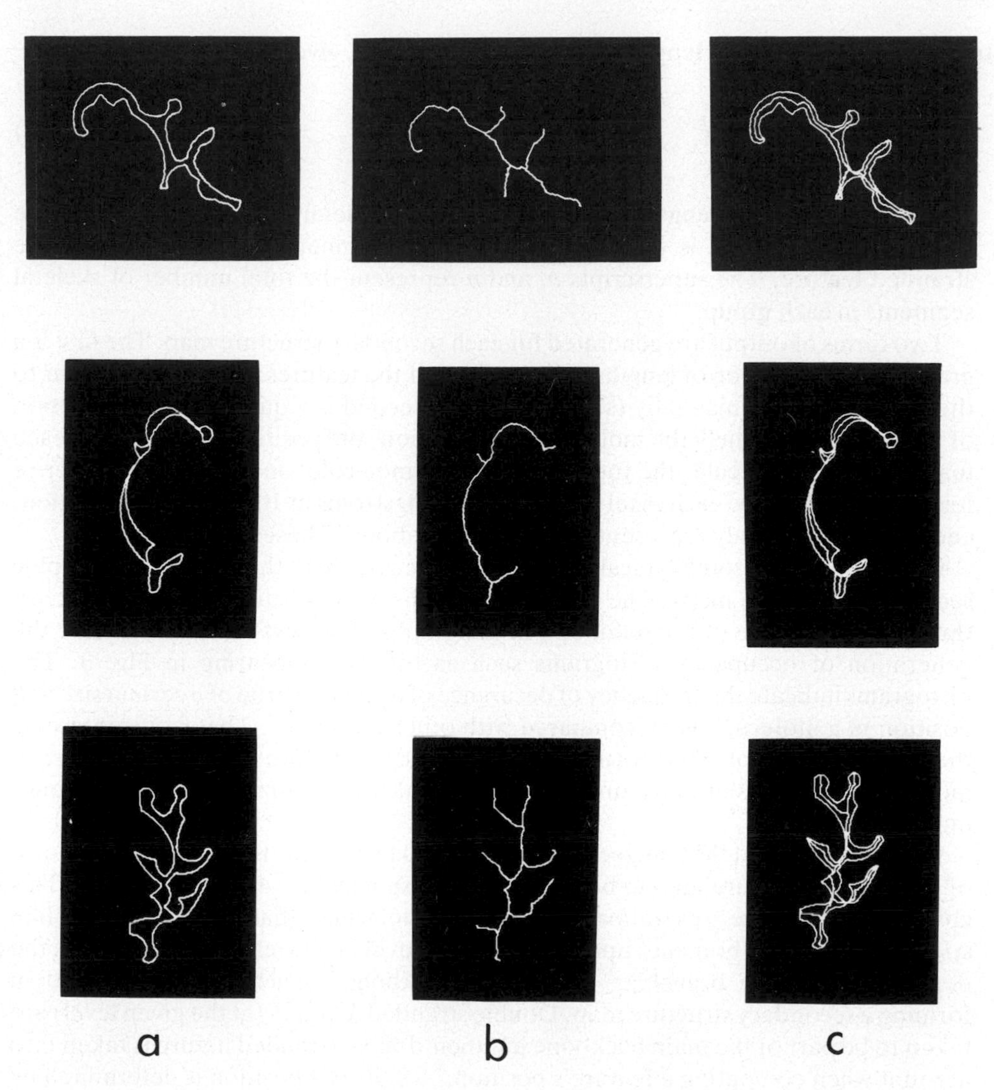

Fig. 13. Some typical molecular contours and their skeletons: (a) resynthesized contours; (b) skeletons; and (c) skeletons with superimposed contours.

connectivity array produced from the skeleton algorithm. This connected path, without any intervening features, constitutes the backbone of the molecule. Its size is just the length of the backbone skeleton. Next, all the non-backbone features are traversed. Since these represent double stranded portions of the molecule, their size is doubled. Their sizes and the positions at which these non-backbone features intersect the backbones are noted, thus, making it possible to construct a secondary structure map. The total length of the molecule is then given by

$$T = \sum_{i=1}^{m} Bi + 2 \sum_{i=1}^{n} Si \tag{12}$$

where each Bi is the length of a single stranded skeletal segment comprising the backbone and each Si is the length of a skeletal segment representing a double stranded feature. The superscripts m and n represent the total number of skeletal segments in each group.

Two forms of output are generated for each secondary structure map. The first is a graphic representation of lengths and positions of the features. These look similar to the maps produced manually (see Fig. 2). The second is a quantitative description of the molecule, namely the molecule identification, the position and length of each feature on the molecule, the total length of the molecule, and the total number of features found. Since each pixel represents 53 Angstroms at 10,000× magnification, each pixel in this study represents a sequence of about 17 bases.

Figure 14 depicts some typical adenovirus molecules with their associated graphic secondary structure maps. The molecular lengths were normalized to themselves, thus the appearance of the number 1 in the graphs. This was done to facilitate the generation of occupancy histograms such as the one appearing in Fig. 3. The histograms indicate the frequency of occurence of a given featrue of a certain size and position in a molecule when compared with other molecules. That is, it represents the combination of the normalized secondary structure maps from several molecules. Table 3 shows the unnormalized quantitative output for the corresponding maps.

It should be noted that the technique described here permits a more general form of secondary structure map to be produced. These may be called 'level' maps. This concept permits the generation of maps for molecules that have very tree-like structures, namely, branches upon branches. A level in a level map is defined as the depth of backbone branching (or pseudo-backbone branching) that is used in forming a secondary structure map. Double stranded features (at the given level) are taken to be part of the main backbone but their double stranded nature is taken into account when computing a feature's position. A feature's position is determined by the distance from the 5′ vertex (or that vertex which is chosen as the 5′ vertex) to the junction point of the feature to the current pseudo-backbone branch. Note that there may be several pseudo-backbone branches existing for a given level if there are several features emanating from the main backbone. A feature's size is determined by all the skeletal segments comprising the feature branch at the given level and above. A feature, by convention is always agreed to emanate from that side of the branch that has the lowest base position relative to the 5′ vertex. Thus, a level one map for molecules of the type discussed in this section would correspond to the standard secondary structure map. Maps with levels greater than one would be empty. With more general molecules however, maps with levels greater than one, say two, would look like a standard secondary structure map except that those features at level two in branching would appear in the level two map, with their

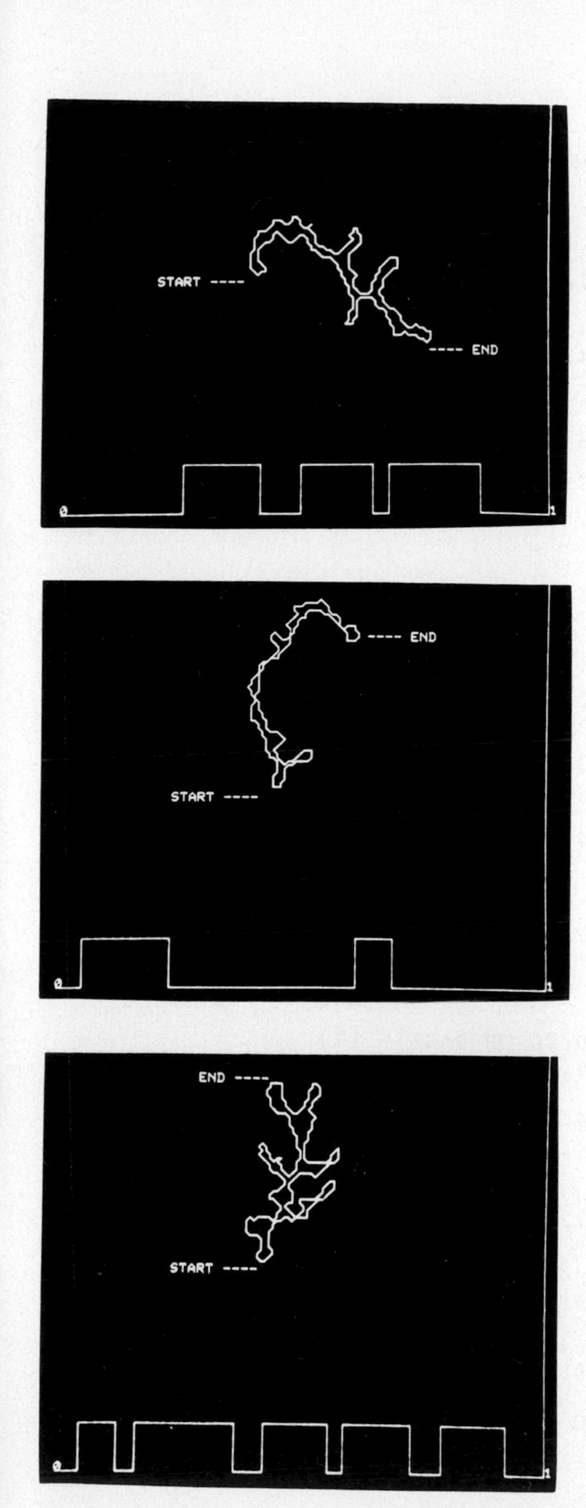

Fig. 14. Molecular contours and their corresponding normalized graphic secondary structure maps. The direction of mapping and the backbone endpoints are indicated by the labels START and END.

TABLE 3

Quantitative secondary structure maps for the molecules depicted in Fig. 14.

```
MOLECULE ID: RNA3.BD

FEATURE      1 LOCATED AT  40.57 OR BASE 716
                 LENGTH IS  24.88 PIXELS  OR  439 BASES

FEATURE      2 LOCATED AT  78.76 OR BASE  1391
                 LENGTH IS  23.38 PIXELS  OR  413 BASES

FEATURE      3 LOCATED AT  107.28 OR BASE  1895
                 LENGTH IS  30.05 PIXELS  OR  530 BASES

TOTAL LENGTH:  159.79 PIXELS  OR  2823 BASES

NUMBER FEATURES:      3

------------------------------------------------------------

MOLECULE ID: RNA2.BD

FEATURE      1 LOCATED AT  6.94 OR BASE  122
                 LENGTH IS  20.32 PIXELS  OR  359 BASES

FEATURE      2 LOCATED AT  70.05 OR BASE  1237
                 LENGTH IS  8.44 PIXELS   OR   149 BASES

TOTAL LENGTH:  114.09 PIXELS  OR  2015 BASES

NUMBER FEATURES:      2

------------------------------------------------------------

MOLECULE ID: RNA4.BD

FEATURE      1 LOCATED AT  10.16 OR BASE    179
                 LENGTH IS  16.00 PIXELS  OR  282 BASES

FEATURE      2 LOCATED AT  34.24 OR BASE  605
                 LENGTH IS  42.42 PIXELS  OR  749 BASES

FEATURE      3 LOCATED AT  88.88 OR BASE  1570
                 LENGTH IS  27.04 PIXELS  OR  477 BASES

FEATURE      4 LOCATED AT  122.69 OR BASE  2167
                 LENGTH IS  28.37 PIXELS  OR  501 BASES

FEATURE      5 LOCATED AT  164.14 OR BASE  2900
                 LENGTH IS  26.99 PIXELS  OR  476 BASES

TOTAL LENGTH:  207.78 PIXELS OR 3671 BASES

NUMBER FEATURES:      5
```

Note: Data are presented in the same order as the molecules appear in Fig. 14.

distance from an end indicated and their size. Thus, a description of the molecule is produced by generating a secondary structure map for several levels. Comparisons amongst several molecules may be accomplished by using the occupancy histograms except that one would be generated for each level.

One problem exists with the level map approach and that is determining which side a given feature may exist on in a branch (this is reflected in determining the correct distance from the 5′-end) since this is not resolvable in an electron micrographic image. That is, a feature may appear to be emanating from one side of a branch due to the random nature of the spreading techniques while in reality it may actually emanate from the other side. The resolution of a problem such as this will now be discussed in the ensuing section.

3. COMPUTER PREDICTION OF SECONDARY STRUCTURE

As is indicated in the previous sections there are molecular configurations that are visibly recurrent amongst a homogeneous group of molecules. A converse problem in nucleic acid morphology is the computation of the molecular structure that is most energetically stable, given sequence data, and which should correspond to those structures that are visible in the electron microscope. To accomplish this requires the solution to a highly combinatoric problem.

3.1. Paired base region generation rules

Given sequence data for a particular nucleic acid, it is possible to construct rules for the different types of base pairings that can arise (see [26]). These are listed below.

(a) G (guanine) may bond to C (cytosine) and vice-versa.
(b) A (adenine) may bond to U (uracil for RNA) and vice-versa.
(c) G may bond to U and vice-versa except when the bond appears as a terminator to a region.

Once these rules have been established a program is used to generate all the 'potential' base-paired regions of a sequence. A region is defined as an uninterrupted run of consecutive complementary pairs of bases. A region is represented by a 4-tuple consisting of: (a) the 5′ starting base position: (b) the 3′ ending base position (this base bonds to the base specified in 'a'; (c) the size of the region; and (d) the stabilizing energy of the region (see Table 4). It should be noted that base positions are ordered by increasing number (starting at 1) from the 5′ side of the molecule. Empirically, the number of possible regions of size two or greater that can be generated from a sequence n nucleotides long is approximately $0.027\ n^2$. Thus, a sequence of around 600 bases in length will produce about 9000 'potential' regions. The problem then becomes one of finding a very small subset of these possible

TABLE 4

Stacking energies of base pairs

5' \ 3'	G	C	A	U	X	Y
G	-4.8	-4.3	-2.1	-2.1	-1.3	-1.3
C	-3.0	-4.8	-2.1	-2.1	-1.3	-1.3
A	-2.1	-2.1	-1.2	-1.8	-0.3	-0.3
U	-2.1	-2.1	-1.8	-1.2	-0.3	-0.3
X	-1.3	-1.3	-0.3	-0.3	-0.3	-0.3
Y	-1.3	-1.3	-0.3	-0.3	-0.3	-0.3

Note: The bases represented on the axes of the table are adjacent to each other, polarized as indicated. They base pair in the standard Watson–Crick way except for X which represents a G that only pairs to a U and Y which represents a U that only pairs to a G. Terminal GU's are not permitted. The values given are in units of kcal (see Salser [26]).

regions which when taken together will produce the most energetically stable configuration. It is now apparent why the combinatorics of this problem are quite large.

3.2. *Energy rules*

Recent experiments attempt to show the free energy contributions of various nucleic acid morphologies ('hairpin' loops, 'bulge' loops, 'internal' loops, see [27–31]). That is, bonds which form regions that follow the rules mentioned in Section 3.1 tend to stabilize a molecular structure, while single stranded loops have a tendency to destabilize the structure. The degree of stabilization or destabilization also appears to be a function of the context in which the individual bases appear within the regions. Those bases that are adjacent to each other, those that are the opening and/or closing bases of loops, as well as the size of loops all contribute significantly to the overall configuration energy [26]. These values are determined by table look-up (see Tables 4 and 5). It is therefore possible to incorporate these rules within a program designed to find that set of regions, from amongst the 'potential' set of regions, that produce the most stable structure.

3.3. *Graphic matrix*

All the generated 'potential' regions may be visualized with what may be called a 'graphic matrix', a two-dimensional plot of all paired bases. Along the horizontal axis (left to right) is the base sequence data ordered from the 5′-end of the molecule to the 3′-end. Along the vertical axis (top to bottom) is the reverse complement of the sequence. The reverse complement orders the molecule from the 3′-end to the 5′-end and all the bases in the sequence are complemented. The origin of the matrix, the (1,1) position, is in the upper left corner. A dot is placed at each position in the

TABLE 5

Loop destabilizing energies

LOOP SIZE	HAIR(G)	HAIR(A)	INT(GG)	INT(AA)	INT(GA)	BULGE
1	–	–	–	–	–	2.80
2	–	–	0.10	1.80	0.95	3.90
3	8.40	8.00	0.90	2.60	1.75	4.45
4	5.90	7.50	1.60	3.30	2.45	5.00
5	4.10	6.90	2.10	3.80	2.95	5.15
6	4.30	6.40	2.50	4.20	3.35	5.30
7	4.50	6.60	2.62	4.32	3.47	5.45
8	4.60	6.80	2.72	4.42	3.57	5.60
9	4.80	6.90	2.82	4.51	3.67	5.69
10	4.89	7.00	2.90	4.60	3.75	5.78
11	4.96	7.07	2.98	4.68	3.83	5.86
12	5.03	7.13	3.05	4.75	3.90	5.93
13	5.10	7.19	3.12	4.82	3.97	5.99
14	5.16	7.25	3.18	4.88	4.03	6.05
15	5.22	7.31	3.24	4.94	4.09	6.11
16	5.27	7.36	3.29	4.99	4.14	6.16
17	5.32	7.41	3.34	5.04	4.19	6.21
18	5.37	7.46	3.38	5.08	4.23	6.26
19	5.42	7.51	3.43	5.13	4.28	6.31
20	5.46	7.55	3.47	5.17	4.32	6.35
21	5.50	7.59	3.51	5.21	4.36	6.39
22	5.54	7.63	3.55	5.25	4.40	6.43
23	5.57	7.66	3.58	5.28	4.43	6.46
24	5.61	7.70	3.62	5.32	4.47	6.50
25	5.65	7.74	3.66	5.36	4.51	6.54
26	5.70	7.77	3.70	5.40	4.55	6.57
27	5.74	7.80	3.74	5.44	4.59	6.60
28	5.79	7.84	3.77	5.47	4.62	6.64
29	5.83	7.87	3.81	5.51	4.66	6.67
30	5.88	7.90	3.85	5.55	4.70	6.70

Note: The letters adjacent to some of the column headings represent one of the complementary bases that close the loop, e.g., HAIR(G) – hairpin loop closed by a GC; INT(GA) – internal loop closed by a GC on one side and an AU on the other. The value given for bulge loops must be refined by taking into account the stabilizing influence of the stacking energy across the bulge (see Table 4). Since no experimental data exists as yet for multibranch loops, the scheme used by Saler [26] is used for computing their believed destabilizing influence. INT(GA) is used if the loop has GC and AU closures, INT(GG) is used if the loop has only GC closures and INT(AA) is used if the loop has only AU closures. Portions of strands that are not in loops or double stranded regions are assumed to have free energy 0. Table entries are in the units of kcal. Some of the entries in the table were linearly interpolated from Salser [26].

matrix where the base represented by the horizontal coordinate is equal to the base represented by the vertical coordinate. (Since we are dealing with a complemented sequence along the vertical axis, equality represents complementarity.) Thus, all the 'potential' regions appear as diagonal lines in the matrix, the length of a diagonal depicts the length of a bonding run or the size of a region, and the position of a diagonal indicates what bases in the sequence are actually partaking in the pairing. An added visualization feature exists with the aid of the RTPP's picture memory, namely, the display of the diagonal region lines in grey scale where the grey scale represents the energetic stability of a given region; the darker the line the more stable the region (see Fig. 15). It should be noted that only the upper half of the matrix is displayed since the upper and lower halves of the matrix are symmetrical.

Once the graphic matrix is displayed it then becomes possible for a user to interactively select or delete regions with a tablet. Each region chosen lights up on

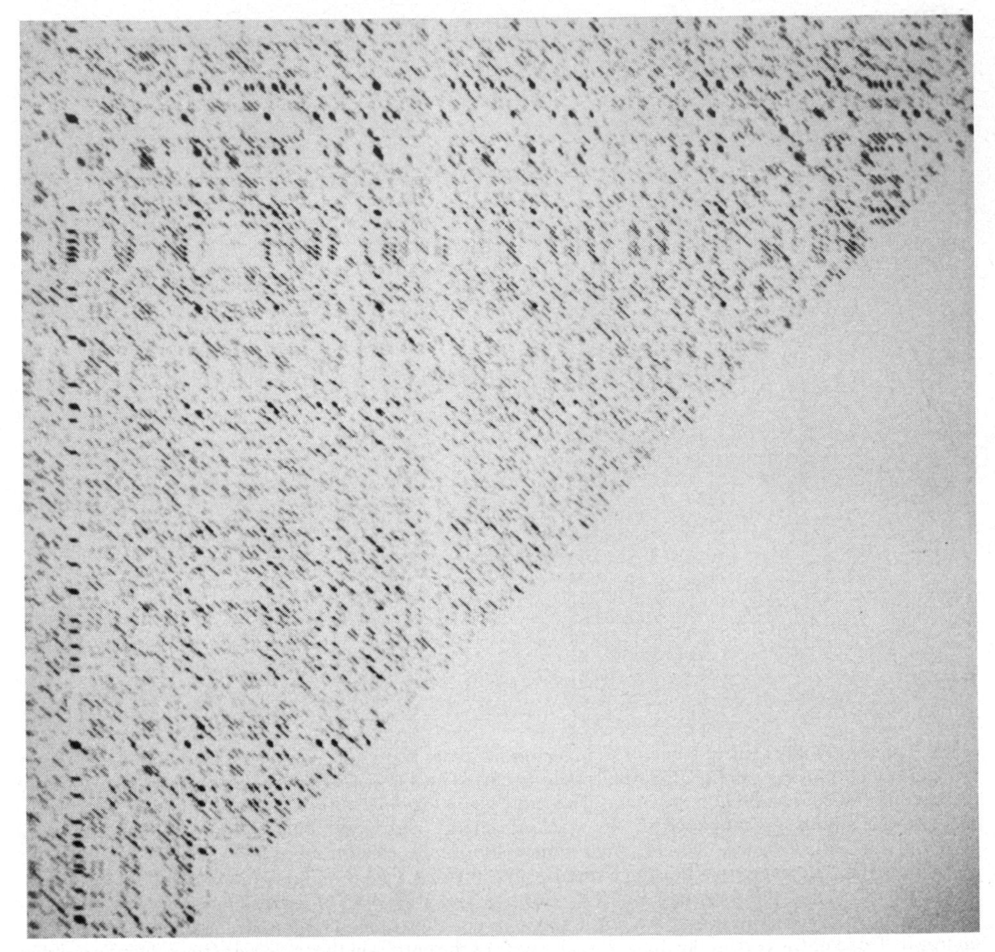

Fig. 15. A 512 × 512 RTPP image of a graphic matrix of rabbit beta-globin mRNA (591 bases). Sequence axis runs left to right across the top (ordered 5′ to 3′) and reverse complement axis runs top to bottom along the left side of the figure. Diagonal runs represent potential double stranded regions and grey scale represents region's energetic stability, the darker the more stable (see Table 4).

the screen and the total current energy of the molecule is computed and is available to the user on a computer terminal along with the current configuration (positions of 'hairpin', 'bulge', 'internal' and 'multibranch' loops, see Fig. 16). In general, attempting to line up selected regions in near diagonals tends to produce more stable configurations than a random selection. This is because diagonal runs tend to minimize the destabilizing influences of loops. Internal loops are indicated in the matrix by gaps between chosen regions. The larger the gap the larger the loop. The more displaced from a diagonal two consecutive regions are, the more unequal are the two sides comprising the internal loop. Hairpin loops are indicated by the gap that exists between the last region diagonal and the matrix diagonal running from the lower left corner to the upper right corner.

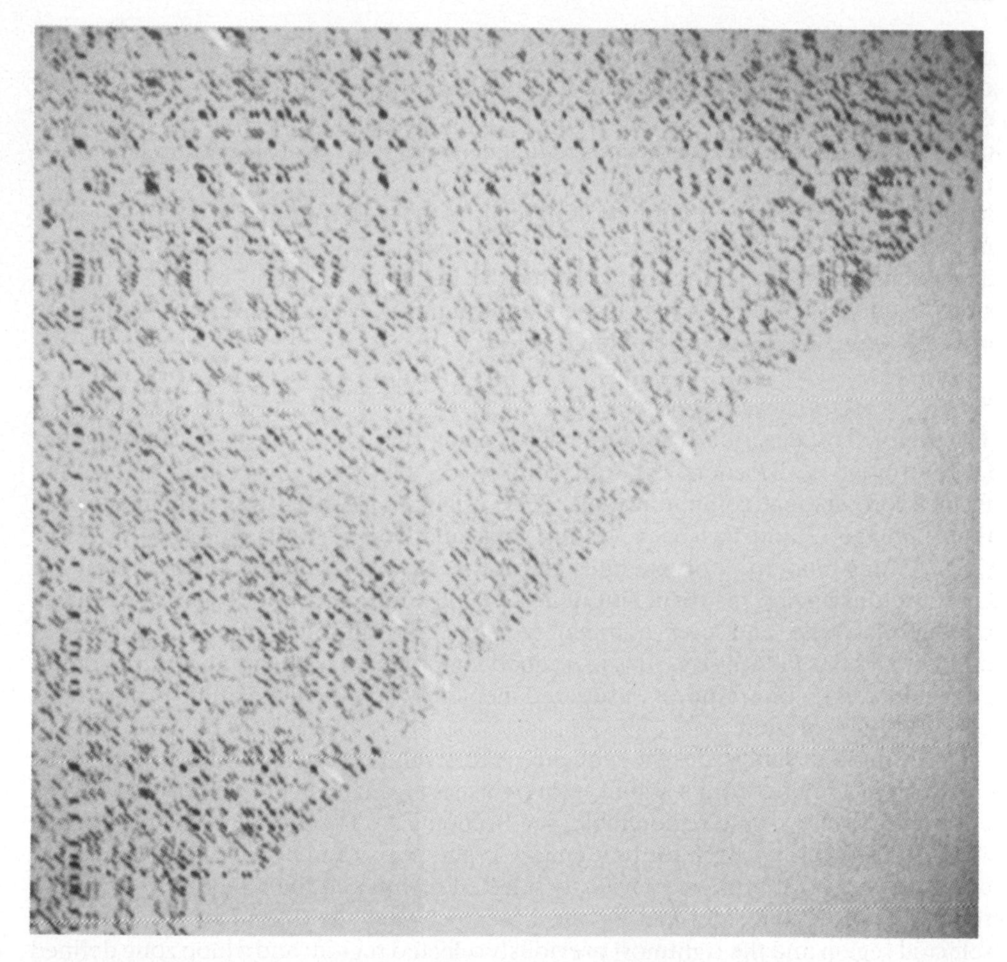

Fig. 16. The graphic matrix shown in Fig. 15 with regions selected (bright diagonals) using the modified Martinez algorithm (see Section 3.4.1.) and user interaction with the tablet (see Section 3.4.2.).

It should be noted that pairs of regions that overlap are not permitted since they would produce structures which allow bases to pair to more than one other base. These overlapping structures are indicated in the matrix by the intersection of the regions' vertical and/or horizontal projections. Also, knot-like structures (free bases in a loop pairing to free bases outside a loop) are not permitted since they further increase the combinatorics of an already very combinatoric problem and in reality it is believed that they contribute minimally to stabilizing the molecule. Knot-like structures are indicated in the matrix by those areas in which the vertical (horizontal) projection of a gap intersects the vertical (horizontal) projection of a region and this region's horizontal (vertical) projection lies outside the gap.

3.4. *Algorithmic approaches to finding the best energy structure*

As was mentioned earlier, the problem of finding the best energy structure for a given sequence is highly combinatoric. One aspect of developing an algorithm is to define a mechanism that generates all possible configurations without duplicating effort or without generating impossible configurations. The problem then becomes one of accomplishing this in the most efficient manner. Several algorithms have been developed [32–36] which generally fall into two categories. One works from the assumption that the best configuration can be derived by using a region table as initial input [32–33] and these regions are explored in various ways. Another approach involves working directly with the sequence data and building up from small fragments of sequence to the entire strand in a systematic way [34–36]. Both of these approaches will be illustrated below.

3.4.1. A region based algorithm

The algorithm described here is based upon that developed by H. Martinez (Department of Biochemistry and Biophysics, University of California at San Francisco, personal communication). The original algorithm was implemented in the language C and has been written and modified by this laboratory in SAIL [23]. At this juncture loop energies are not incorporated in the algorithm but are computed using the resultant solution. The Martinez algorithm appears to run in exponential time and uses minimal memory, the algorithm described here as modified by this laboratory, runs in n cubed time where n is the number of regions in the region table, but requires m squared memory where m is the number of regions used to find a solution.

The algorithm subdivides the sequence recursively. A subdivision is defined as the selection of a region from a region table (which is arranged in increasing order based upon the 5′ entry of the region table, see Section 3.1). Each subdivision produces 3 zones over which a search for best energy is pursued. The left zone, defined by the bases between the leftmost previously selected region and the 5′-end of the current region; a right zone, defined by the bases between the 3′-end of the currently selected region and the rightmost previously selected region; and a loop zone defined by the bases between the 5′- and 3′-ends of the currently chosen region. The general idea of the algorithm is to find the best energy configuration within each of these 3 zones. New subdivisions are determined by taking the first 5′ region in one of these zones and then taking alternate regions that either form knots with the first 5′ prime region or overlap with the first 5′ region. This process is recursively repeated, generating all possible non-knotting and non-overlapping configurations (see Fig. 17).

The algorithm may be made more efficient by recording the ‘best’ energy discovered for each of these zones in a matrix. This eliminates redundant trials of already traversed paths. The algorithm may be further improved by eliminating the attempted traversal of many paths by accepting as an input parameter the largest zone of contiguous free bases likely to be found, and eliminating any paths that may produce configurations with larger zones. Thus, for example, if potential configurations have maximum free base zones of 40, terminate the exploration of other regions at this point of recursion. Because the regions are ordered from the 5′-end any other alternatives at this point would be greater than or equal to 40.

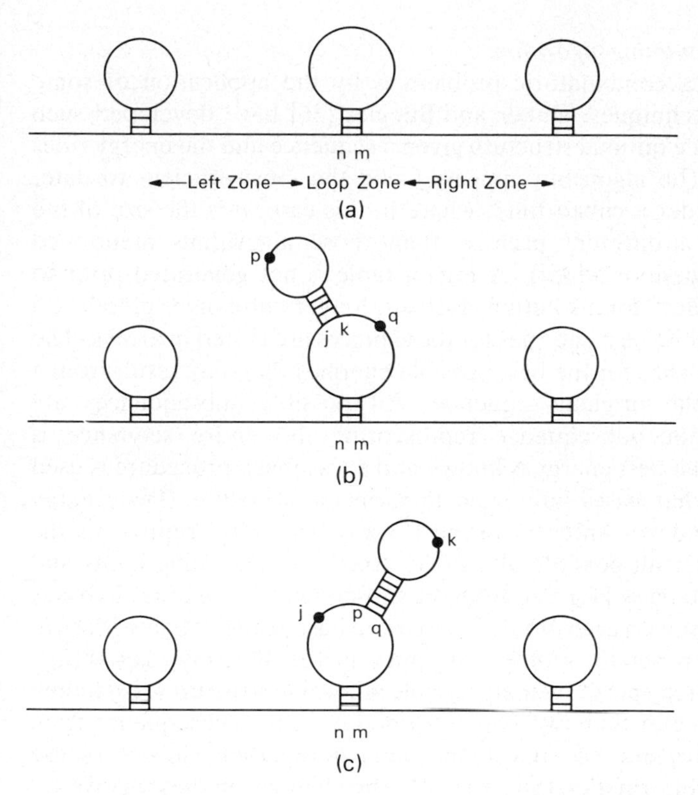

Fig. 17. Example of the Martinez algorithm. (a) A looped region between base positions m and n in which an optimal energy is to be determined. (b) The region between j and k represents the first 5′ region forming a loop that is used to compute an optimal energy for this configuration. (c) The next region used for testing for optimality lies between p and q. This region was chosen next since it knots with the region between j and k. The algorithm cointinues recursively in this fashion, computing optimal energies for new left, right and loop zones as new competing regions are chosen.

3.4.2. Combining the algorithmic and graphic matrix approaches

Because even relatively small sequences generate large numbers of regions, it becomes combinatorically difficult to run the algorithm with all the regions. An alternative approach (that does not guarantee the finding of the 'best' energy) is to threshold the region table accepting only those regions that have energy values better than the threshold. Once the algorithm determines a solution from this limited set, the solution may be displayed in the picture memory. One may then proceed manually to fill in the gaps to attempt to improve upon the answer, interactively acquiring total energy values including the destabilizing effects of loops. A solution for a molecule consisting of about 600 bases (over 9000 regions) may contain less than 100 regions, thus indicating the relatively small number of regions that partake in a solution compared with the total number that one began with. Using this technique a better solution was discovered for a beta globin mRNA (Shapiro, Maizel, Lipkin – this configuration is illustrated in Fig. 19) than that previously published in the literature [26].

3.4.3. A dynamic programming algorithm

Another approach to this combinatoric problem is by the application of some dynamic programming techniques. Zuker and Stiegler [36] have developed such an algorithm. This finds the optimal structure given a sequence and the energy rules defined in Section 3.2. The algorithm appears to be the most efficient to date, finding the solution in order n cubed time, where in this case, n is the size of the sequence. It works on a different premise than those algorithms mentioned previously (except for Nussinov's [35]). A region table is not generated prior to initiation of the actual search for a solution. Instead, longer and longer pieces of a sequence are examined while best energies for these pieces are stored in arrays. The array elements contain entries for the best possible energies that may result from a given subsequence of the original sequence. All possible subsequences are examined until finally the subsequence representing the entire sequence is evaluated. At this point the best energy is known and a traceback procedure is used to determine precisely what bases interacted to form the structure. Two energy arrays are required, V and W. An entry in the V array, say $V(i,j)$ represents the minimum free energy over all possible allowable structures (excluding knots and bondings of more than two bases) formed from the subsequence from base i to base j where base i and base j pair with each other. If base i and base j can not pair with each other then this element is set to infinity. An entry in the W array, say $W(i,j)$, represents the minimum free energy over all possible allowable structures (excluding knots and bondings of greater than two bases) formed from the subsequence from based i to base j where base i and base j may or may not pair together. Thus, in a sense W contains more general information than array V. The elements in these arrays are computed recursively, i.e., the current entries in the arrays may be computed from prior entries.

There are basically three different types of configurations that must be accounted for when attempting to find the minimum energy for a particular $V(i,j)$. A particular $V(i,j)$ entry may be given by,

$$V(i,j) = \min \{E1, E2, E3\}, \tag{13}$$

where $E1$, $E2$, and $E3$ represent these different possible configurations and are described below.

($E1$) Bases i and j pair together, and the interventing bases form a hairpin loop. In this case, $E1 = E(H(i,j))$ where E is the energy function (see Section 3.2) and H is the hairpin loop between i and j.

($E2$) $E2$ may be defined as

$$E2 = \min_{i<i'<j'<j} \{E(L(i,j,i',j')+V(i',j')\}. \tag{14}$$

Here, L represents either a bulge loop, internal loop or stacking region between base pairs i and j and i' and j' and $V(i',j')$ represents the previously computed best energy value for the structure formed when bases i' and j' paired. Note that this value is computed from the minimum over all i' and j' such that $i<i'<j'<j$ (see comment below).

($E3$) $E3$ may be defined as

$$E3 = \min_{i+1<i'<j-2} \{W(i+1,i') + W(i'+1,j-1)\}. \quad (15)$$

The above equation represents the case of a multibranch loop, where i and j represent the currently paired bases, i' represents the $5'$ base position of the next stem in the loop such that $i + 1<i'$. Thus, the portion of the loop after i is broken into two portions. Again, this minimum is taken over all i' such that $i + 1<i'<j - 2$.

The above defines the V array. But, this definition requires the definition of the W array. Here, i and j do not necessarily have to base pair. Again there are basically three possibilities.

(1) The structure has at least one dangling end, i.e., either base i or base j or both do not pair with another base. Therefore,

$$W(i,j) = W(i+1,j) \text{ or } W(i,j) = W(i,j-1). \quad (16)$$

(2) Bases i and j pair with each other. This has already been covered in the definition of $V(i,j)$. Thus,

$$W(i,j) = V(i,j). \quad (17)$$

(3) Both i and j base pair, but not with each other. Let us assume that base i pairs with base i' and base j pairs with base j' where $i<i'<j<j'$, then we have

$$W(i,j) = W(i,i') + W(i'+1,j). \quad (18)$$

Thus, the best energy in this case is

$$E4 = \min_{i<i'<j-1} \{W(i,j') + W(i'+1,j)\}. \quad (19)$$

The best energy for the entry $W(i,j)$ is

$$\min\{W(i+1,j), W(i,j-1), V(i,j), E4\}. \quad (20)$$

The algorithm computes by constantly increasing the size of the subsequence by one and determining what the best energy is at each step, filling in the appropriate positions in the V and W arrays. Eventually when the position $W(1,n)$ is filled in, a solution has been determined. The regions that partake in this solution may be determined from a trace back procedure which starts with the position $W(1,n)$ and works through the W array finding local minima.

The Zuker–Stiegler algorithm as described essentially computes in n to the fourth time. A heuristic has been used in the computation of $E2$ above, which terminates the examination of the i' and j' bases when it is determined that any further examination will not result in an improvement of a $V(i,j)$ value. This allows the algorithm to compute in time n cubed.

The W and V arrays may each be stored in half arrays to save space. However, another array INC is used in the program to aid in the computation of loop energies to keep track of those bases that form the loops. Thus, the total storage of this

another array, PAIRS, is filled in the corresponding base positions with the base number that pairs to it. Now that all double stranded regions emanating from a loop have been accounted for in the array SEQANGLE, the angle representing the regular polygon forming the loop is now computed.

The number of vertices in the polygon forming the loop is equal to the number of free bases in the loop plus two times the number of double stranded regions emanating from the loop. It can therefore be shown that the angle α between two unit vectors in the polygonal loop may be given by,

$$\alpha = (n - 2)\pi/n \tag{21}$$

where n is the total number of bases (vertices) in the loop. The appropriate positions in the SEQANGLE array are then filled in with the angle α for all the unbound bases in the loop.

At this point all that remains to be entered, for this loop structure, in SEQANGLE, are the angles for those base positions that correspond to the transition points from the double stranded regions to the loop. These points correspond to the starting and terminating base pairs of a region. The transition point falls into two categories. Either the transition point is part of a region which consists of a single base pair region or it is part of a region of more than one base pair. For a point satisfying the latter condition the SEQANGLE entry is just $\pi/2 + \alpha$. For single base pair regions the entry is the sum of the prior bend angle (an α from a previously processed loop on one side of the base pair) plus the current angle α. If this point is being processed for the first time (a previously processed loop does not exist at this time) the SEQANGLE entry is just α. By continuing this process for all regions and all loops, the SEQANGLE array is filled in. If it is desired for free bases that do not reside within loops to have a bend associated with them to make them more visible, the SEQANGLE array should be initialized to a value slightly more than π. Since free base positions are not processed in the above description the SEQANGLE array for them will not be altered and a bend will be imparted. At this stage, one may traverse the SEQANGLE array reading the bend angles and generating the x,y coordinates of the base positions by producing unit vectors that are at the bend angles with respect to each other as is indicated.

3.5.2. *Interactive modification of molecular drawing*

As is indicated in Fig. 18, the resultant drawing may contain portions of the molecule overlaying each other making it difficult to discern important features of the structure. To alleviate this problem, an interactive mechanism has been incorporated into the algorithm that enables one to unravel the structure, eliminating overlaps. This is currently accomplished using the crosshair facility on the Tektronix 4012 graphics terminal. All transition points (points which go from a double stranded region to a loop) become potential pivot points. The crosshair may be placed over one of these privot points and by mapping the crosshair position into a base position entry in the SEQANGLE array one can then alter the appropriate entries in this array to pivot a stem relative to its associated loop. Currently, the pivot angle is .5 radians per attempt and is in the direction which decreases the angle

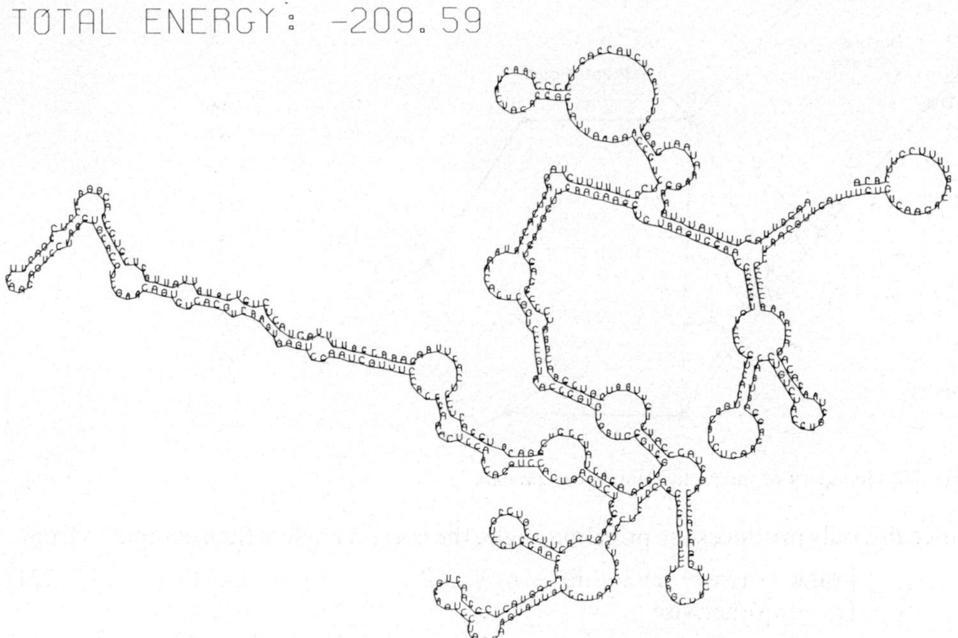

Fig. 19. Interactively untangled molecular drawing of molecule shown in Fig. 18 making molecular configuration more legible.

between the stem and the associated loop. If the other side of the stem (corresponding paired base) is used, the effect is to undo any previous pivots by .5 radians. This interactive process may be done on several pivot points before redisplaying the altered molecule which may again be altered in a similar manner. Eventually, the untangled molecule may be displayed and/or plotted (see Fig. 19).

The pivoting process may be understood in the following way. The main goal is to keep the bases in a double stranded stem opposite each other i.e., no skewing. When a pivot occurs, the angle between the unit vectors emanating from the pivot point decreases. To keep the bases on the stem opposite each other, it is necessary to lengthen the unit vector emanating from the point opposite the pivot point (this point may be easily found from the PAIRS array). Appropriate angles must also be computed to maintain the consistency of the drawing, namely angles GBA, BAC and ACF (see Fig. 20). It can be shown, using some trigonometry and geometry that the size of the lengthened vector can be computed from,

$$\mathrm{d1} = \mathrm{SQRT}(1 + 8\mathrm{SIN}(\alpha/2)\mathrm{SIN}(\Theta/2)\mathrm{COS}((\alpha - \Theta)/2)) \tag{22}$$

where α is the bend angle as defined in equation (21) and Θ is the total angular pivot. Once d1 is computed, some other angles must also be determined. The angle γ may be written as,

$$\gamma = \mathrm{ARCSIN}(2\mathrm{SIN}(\alpha/2)\mathrm{COS}(\Theta - \alpha/2)/\mathrm{d1}) \tag{23}$$

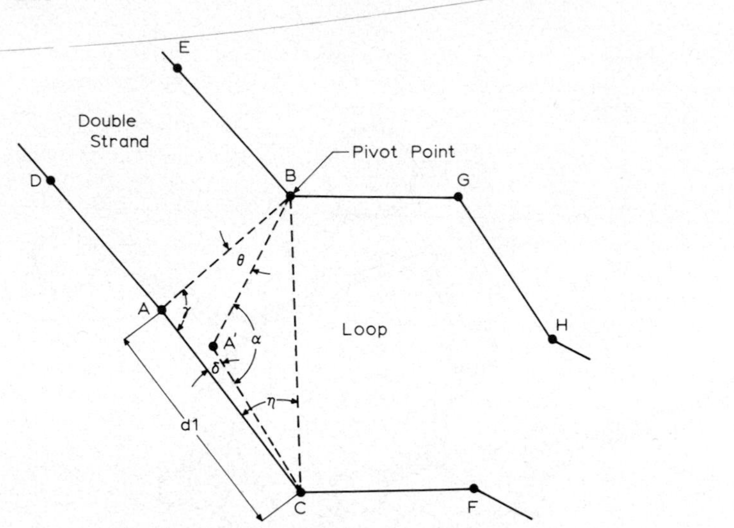

Fig. 20. Geometry of interactive untangling process.

since this only produces the principle angle, the correct angle is then computed from,

$$\gamma = \begin{vmatrix} \gamma \text{ if } \pi - ((\pi - \alpha)/2 + \Theta + \eta) < \pi/2 \\ \pi - \alpha \text{ otherwise} \end{vmatrix} \tag{24}$$

where η can be shown to be

$$\eta = \text{ARCSIN}(\text{COS}(\Theta - \alpha/2)/\text{d1}). \tag{25}$$

The angle at BAC may now be computed to be $\gamma + \pi/2$. The angle at GBA may be computed to the current bend angle plus $d\Theta$($d\Theta$ being the increment of bend, i.e., .5 radians). One more angle must be determined, this is ACF. To do this, the incremental change in the angle A′CF must be determined. The incremental change will be given by δ where,

$$\delta = \text{ARCCOS}(\text{d1}^2 + \text{d2}^2 - 2(1 - \text{COS}(d\Theta)))/2\text{d1 d2})) \tag{26}$$

where d1 is the current length of the lengthened side (this will be 1 if it is the first pivot) and d2 is the new length of the lengthened side as computed from equation (22). Again, only the principle angle δ is computed. To determine the correct angle we have,

$$\delta = \begin{vmatrix} -\delta \text{ if } \eta' > \eta \\ +\delta, \text{otherwise} \end{vmatrix} \tag{27}$$

where η' is the previous angle η which is

$$\eta' = (\text{COS}(\Theta' - \alpha/2)/\text{d1}), \tag{28}$$

where Θ' is the previous angle Θ. Thus, the new angle at ACF is equal to the current angle at A′CF + δ. One more piece of information must be associated with this pivot point, namley the length d1 of the lengthened side. This is done so that it may be used later in equation (26).

Having done the above, the array SEQANGLE has now been updated and may be used to compute the new x,y coordinates of the configuration or may be further updated by more pivoting. It should be noted that if an undoing of a previously pivoted point is desired, all that is necessary, is for the crosshair to be placed on the base opposite the previoulsy pivoted base and the above equations may be used by just swapping the pivot point with its opposite point and replacing $d\Theta$ by $-d\Theta$.

One additional feature should be noted. For convenience, one may rotate the molecule to get a good aspect ratio for publication purposes.

4. DISCUSSION

We have presented some techniques that may be applied to the study of nucleic acid morphology, in particular nucleic acid secondary structure. There are however, some problems remaining to be solved which will provide a more general applicability. The class of images that are being analyzed, while they superficially appear to be quite simple are in reality quite complex. The difference between the molecules and the background is relatively small. In many cases it is difficult to distinguish between the noise generated in the substrate by the preparative technique and the actual molecule. The DC notch filter mentioned in Section 2.1.2 appears to be effective in solving some of these problems, however, it is fairly expensive with respect to timc. It takcs about cight minutes to filter an image on the RTPP. Some other experiments have been performed using localized variable thresholding on an unfiltered image which appear to give positive results. These experiments also indicate that it may be possible to automate the threshold selection process.

The ability to distinguish easily between the single and double stranded portions of the molecule as well as determining the positions of the 5′- and 3′-ends of the molecule would greatly facilitate the automatic mapping process. This problem can be pursued from at least two directions. One way involves an improved preparation technique that enhances the differences between the single and double stranded regions. Some of the methods available to do this have tendency to obscure the smaller secondary structures [37]. The other direction involves improvement of the image acquisition process, thereby reducing some of the inherent losses incurred at this level. Some heuristics (which are not guaranteed to be correct) may also be used, such as choosing the longest direct path between two fcature points and using this as the main backbone of the molecule.

Following the gap that may exist artifactually between molecular fragments is another problem that needs to be resolved. The problem of bridging these gaps may be solvable by using heuristics incorporating continuity or curvature, expand and shrink techniques [38], and/or minimal search distance criteria. Some of these methods are also discussed in [20,39].

Another area of concern involves the analysis of loop-like structures. Some of these structures are not actually loops, but cross-overs which can easily be mistaken for loops. There are no hard fast rules for distinguishing the differences. A heuristic using continuity or curvature representing the cross-over might be used, since loops do not usually have this characteristic while cross-overs do.

Finally, from the analysis standpoint, there is a class of nucleic acid images that are currently not measured manually. These include those molecules that form extremely tangled structures, i.e., spaghetti-like. These molecules occur either because of the preparative techniques, e.g., denaturation did not separate many of the small regions, or may actually be due to the presence of many strongly energetically stable close tight loops.

Some of the problems described above may be able to be solved by the synthetic discussed in the latter half of this chapter. This may be a viable approach because of the predictive capability inherent in the process. This type of situation is one of the prime motivations for the merger of the analytic and synthetic approaches. In this way, one may be able to clarify the ambiguities in the other.

The synthetic algorithms, at present, are still quite combinatoric and require large amounts of computer memory. Currently, relatively small sequences (up to 1000 bases) can be handled by some machines that allow large virtual memory configurations. Those algorithms that require an array of the order of the square of the length of the sequence (such as the dynamic programming algorithm described in this chapter) may be able to generate optimal solutions, but at present must deal with the entire sequence to do so. Those algorithms that deal with region tables directly may at present be more inefficient and may not find the optimal solution but, may more easily reduce the space requirements if one can determine a set of heuristics that permits the usage of only a small portion of the entire region table.

The interactive graphic matrix approach also suggests some interesting possibilities. It permits the user to select regions based upon some external criteria (including visual feedback from the matrix) or heuristics (such as diagonal runs). It also allows the user to specify particular areas of the graphic matrix to search or not to include in the search thereby reducing the combinatorics.

Ultimately, some improved significant biological heuristics are going to be required to shorten both the processing time and size of these programs. Biochemical data relating to secondary structure configurations and/or energy rules should be quite significant in this respect. Merging the analytic and synthetic approaches should also greatly improve both processes one serving to speed up and verify the other. Combining these aspects with human intervention holds the potential for significant advances in the understanding of nucleic acid morphology.

ACKNOWLEDGEMENTS

The authors wish to express their deepest appreciation to their collaborator Jacob Maizel of the Molecular Structure Section, National Institute of Child Health and Human Development for having provided many significant suggestions and ideas. The authors also wish to thank Peter Lemkin, Earl Smith, Morton Schultz and Marta Wade of the Image Processing Section for their invaluable support. Portions of this paper were derived from [8].

REFERENCES

1. Wollenzien, P., Hearst, J.E., Thammana, P. and Cantor, C.R. (1979) J. Mol. Biol. 135, 255–269.
2. Thammana, P., Cantor, C.R., Wollenzien, P.L. and Hearst, J.E. (1979) J. Mol. Biol. 135, 271–283.
3. Rabin, D. and Crothers, D.M. (1979) Nucleic Acids Res. 7, 689–703.
4. Appel, B., Erdmann, V.A., Stulz, J. and Ackerman, Th. (1979) Nucleic Acids Res. 7, 1043–1057.
5. Cantor, C.R., Wollenzien, P.L. and Hearst, J.E. (1980) Nucleic Acids Res. 8, 1855–1872.
6. Boyle, J., Robillard, G.T. and Kim, S. (1980) J. Mol. Biol. 139, 601–625.
7. Noller, H.F. and Woese, C.R. (1981) Science 212, 403–411.
8. Shapiro, B., Lipkin, L. and Maizel, J. (1979) Comp. Biomed. res. 12, 545–568.
9. Shapiro, B., Maizel, J. and Lipkin, L. (1981) Annals of the World Association for Medical Informatics, 4th Meeting, pp. 93–98.
10. Bender, W. and Davidson, N. (1976) Cell 7, 595–607.
11. Eron, L. and Westphal, H. (1974) Proc. Natl. Acad. Sci. USA 71, 3385.
12. Lemkin, P., Carmen, G., Lipkin, L., Shapiro, B., Schultz, M. and Kaiser, P. (1974) J. Histochem. Cytochem. 22, 725–731.
13. Carmen, G., Lemkin, P., Lipkin, L., Shapiro, B., Schultz, M. and Kaiser, P. (1974) J. Histochem. Cytochem. 22, 732–740.
14. Lemkin, P., Shapiro, B., Gordon, R. and Lipkin L. (1976) PROC10-An Image Processing System for the PDP10. NCI/IP Tech. Rep. No. 8, Nat. Tech. Info. Serv. PB261535/AS.
15. Lemkin, P. (1978) Buffer Memory Monitor System for Interactive Image Processing. NCI/IP Tech. Rep. No. 21b, Nat. Tech. Info. Serv. PB278789.
16. Lemkin, P., Carmen, G., Lipkin, L., Shapiro, B. and Schultz, M. (1977) Real Time Picture Processor – Description and Specification. NCI/IP Tech. Rep. No. 7a, Nat. Tech., Info. Serv. PB269600/AS.
17. Lemkin, P. and Lipkin, L. (1980) Comput. Prog. Biomed. 11, 21–42.
18. Schwartz, A.A. and Soha, J.M. (1977) Appl. Opt. 16, 1779.
19. Lipkin, L., Lemkin, P., Shapiro, B. and Sklansky, J. (1979) Comput. Biomed. Res. 12, 279–289.
20. Ito, T. and Sato, K. (1976) In: Digital Processing of Biomedical Images (Preston, K. and Onoe, M., eds.) pp. 89–100, Plenum, New York.
21. Shapiro, B.A. (1978) Shape Description Using Boundary Sequences, Ph.D. dissertation, University of Maryland.
22. Rosenfeld, A. and Kak, A.C. (1976) Digital Picture Processing, pp. 341–347, Academic Press, New York.
23. Reiser, J.F. (1976) SAIL User Manual, Stanford University Artificial Intelligence Laboratory memo AIM-289. Also available from U.S. Dept. Commerce. Nat. Tech. Info. Serv. No. AD-A045-102, Springfield, Va.
24. Shapiro, B. and Lipkin, L. (1977) Comput. Biomed. Res. 10, 511–528.
25. Shapiro, B., Pisa, J. and Sklansky, J. (1981) Comput. Graph. Image Process. 15, 136–153.
26. Salser, W. (1977) Cold Spring Harbor Symp. 42, 985–1002.
27. Tinoco, I., Uhlenbeck, O.C. and Levine, M.D. (1971) Nature 230, 362–367.
28. Uhlenbeck, O.C., Borer, P.N., Dengler, B. and Tinoco, I. (1973) J. Mol. Biol. 73, 483–496.
29. Gralla, J. and Crothers, D.M. (1973) J. Mol. Biol. 73, 497–511.
30. Gralla, J. and Crothers, D.M. (1973) J. Mol. Biol. 78, 301–319.
31. Tinoco, I., Borer, P.N., Dengler, B., Levine, M.D., Uhlenbeck, O.C., Crothers, D.M. and Gralla, J. (1973) Nature New Biol. 246, 40–41.
32. Pipas, J.M. and McMahon, J.E. (1975) Proc. Natl. Acad. Sci. USA 72, 2017–2021.
33. Studnicka, G.M., Rahn, G.M., Cummings, I.W. and Salser, W.A. (1978) Nucleic Acids Res. 5, 3365–3387.
34. Waterman, M.S. and Smith, T.F. (1978) Math. Biosc. 42, 257–266.
35. Nussinov, R. and Jacobson, A.B. (1980) Proc. Natl. Acad. Sci USA 77, 6309–6313.
36. Zuker, M. and Stiegler, P. (1981) Nucleic Acids Res. 9, 133–148.
37. Evenson, D.P. (1977) In: Methods in Virology (Maramorosch, M. and Kaprowski, H. eds.) Vol. 6, pp. 219–263, Academic Press, New York.
38. Lemkin, P., Shapiro, B., Lipkin, L., Maizel, J., Sklansky, J. and Schultz, M. (1979) Comput. Biomed. Res. 12, 615–630.
39. Peacocke, R.D. (1976) A Region-Based Model of Pictoral Data for Pattern Recognition, Ph.D. dissertation, University of Toronto.

Geisow & Barrett (eds.) Computing in biological science

Two-dimensional and three-dimensional reconstruction in electron microscopy

P. J. SHAW

1. INTRODUCTION

The basic concern of this chapter is a description of some of the computing methods which have been used to extract information about structure at the molecular level from electron micrographs of biological specimens. Various analogue techniques such as optical filtering and photographic averaging, have been in use for a number of years. However, these techniques are all somewhat limited, and computer image processing can match or improve upon all the analogue techniques as well as opening up many new possibilities, of which the most notable example is three-dimensional object reconstruction. Klug [1] has recently reviewed many of the developments in biological image processing, and Misell [2] has given a full account of many areas of electron microscopic image analysis. Although the subject of the volume is computing, the theoretical and practical aspects of processing these electron microscopic images are very closely linked with the experimental procedures used in the electron microscope. Some consideration of the electron microscope is therefore essential in order to give a coherent account of this field, although space will allow only a very cursory description.

At first sight the electron microscope does not appear to be one of the most appropriate tools for studying biological structure. Specimens must be placed in the high vacuum of the microscope and then subjected to extremely high levels of radiation in the form of a focussed beam of electrons. This treatment has disastrous effects on biological macromolecules. They are for the most part substantially hydrated and therefore withstand the vacuum conditions very poorly. They are also mainly composed of light atoms. This has two consequences for their interaction with electrons: firstly their scattering of electrons is fairly weak compared to that of elements of high atomic number; secondly the ratio of inelastic to elastic scattering is large compared to that of heavy elements. Since it is only elastically scattered electrons that can be accurately focussed, it is mainly these electrons that provide the high resolution image. On the other hand, it is the inelastically scattered electrons that impart their energy to the specimen, thus damaging it by various processes. The ratio of damage to information is therefore very high for unstained biological material, and until very recently this made direct examination of such specimens in the microscope impossible.

The consideration that outweighs all these problems, of course, is that when an image can be observed it is more or less interpretable as an image of the electron scattering potential of the specimen, which in turn is directly related to its atomic structure. Electron optics have been developed to the stage where the instrumental limit of resolution in modern microscopes is about 0.3 nm, and this resolution is routinely achieved in the examination of inorganic materials. The extent to which biological studies will be able to take advantage of the resolution of the electron optics depends entirely on the extent to which the problems of vacuum and radiation damage and lack of contrast can be overcome. The importance of obtaining structural information at this resolution for biological macromolecules has been amply demonstrated by the results of the x-ray crystallographic analysis of proteins.

The vast majority of biological specimens for electron microscopy have been and continue to be produced by the familiar techniques of thin sectioning. The specimen is cross-linked, dehydrated, embedded in resin and stained with various heavy metal compounds. Although this approach has revolutionised our understanding of sub-cellular morphology, it almost always fails to give any meaningful information below 4–5 nm. The reasons are fairly clear; the fixing and sectioning procedures may be expected to completely disrupt the detailed molecular structure of the specimen. In any case the positively stained image that is observed is the result of a series of chemical reactions of the specimen with the heavy metal compounds, whose products one hopes bear some geometrical relationship to its reactants. (However, a recently reported sectioning technique using tannic acid fixation [3] has been shown to give images comparable to negative staining in resolution.) The more recently introduced techniques of metal shadowing and negative staining routinely provide higher resolution images. These methods produce an electron dense replica of the specimen. With shadowing, a beam of metal particles, generally produced by heating a metal or alloy in a vacuum, is directed onto the specimen. A layer of metal is deposited on the specimen surface whose thickness is a function of the surface relief. Negative staining, on the other hand, produces a volume replica; a solution of a heavy metal salt is added to the specimen, and on drying forms an amorphous electron dense layer. The original specimen is revealed by the regions of low electron scattering from which the stain has been excluded by the substructures in the specimen. The modified specimens, besides being much more easily visible in the microscope, are considerably less susceptible to damage by radiation.

However, there are considerable problems with these replica techniques at resolutions approaching molecular dimensions. It is difficult to know how faithfully the shadow or stain follows the surface of the macromolecules. Stain exclusion by the molecular envelope is liable to be confused with stain exclusion for any other reason, for example, simple inaccessibility of a pocket to stain. Metal shadow layers tend to be grainy, being composed of clusters of metal atoms rather than a smooth layer. Also the conventional metal evaporation techniques produce sizeable globules of hot vapourised metal which can cause considerable local heating and therefore damage at the specimen. The directionally shadowed image is not easily directly interpreted in terms of the specimen surface, since the depth of shadow depends on the inclination of the surface to the beam of incident metal. The commonly used technique of rotation of the specimen during shadowing so as to coat it equally from all directions further degrades the resolution and makes detailed interpretation even more suspect.

In negatively stained specimens the heavy atom salts, which for ideal negative staining should form an amorphous glass, in fact form small crystallites 1–2 nm in diameter. The negatively stained image cannot therefore reflect the molecular envelope to a higher resolution than ~2 nm and the stain distribution also tends to be somewhat uneven. The resulting images at high magnification appear very granular and distinguishing the meaningful from the spurious detail is not easy. Although negatively stained specimens are more resistant to radiation damage than the biological material itself, they are by no means entirely so. The electron doses used in normal elecron microscopy (several tens of thousand electrons per nm^2) cause significant changes in the distribution of the heavy metal atoms of the stain. It appears that oxides of the metals are usually formed, for example, uranium oxide from uranyl acetate and uranyl formate. The oxides have a higher density than the original salts and so a contraction of the stain takes place. This is often accompanied by other rearrangements of the stain. In order to obtain reliable structural information, therefore, it is necessary to reduce the electron dose to a level at which it does not cause significant damage. The critical dose for negative stained specimens seems to be 500–1000 e/nm^2. (The dose sustainable by unstained proteins, at least in crystalline arrays, is about an order of magnitude less than this [8,11].) Unfortunately at such low doses a further factor limits the resolution. This is the so-called quantum noise and arises because the electron image is produced by the statistical distribution of the discrete particles hitting the film. At low electron doses not enough electrons pass through the specimen to the film to define high resolution features with any reasonable certainty.

A typical low dose picture is shown in Fig. 1a. This is a micrograph of the cell wall of the alga *Lobomonas piriformis*, which has a two-dimensionally crystalline structure. Little more than the basic unit cell repeats can be seen in these micrographs, corresponding to about 20 nm. (Fig. 1b shows a normal dose micrograph of the same cell wall.) However, the problem has been reduced by low dose microscopy from that of irreparable specimen damage to the more tractable one of the ratio of signal to noise.

The problem of a small signal obscured by randomly distributed noise is a very

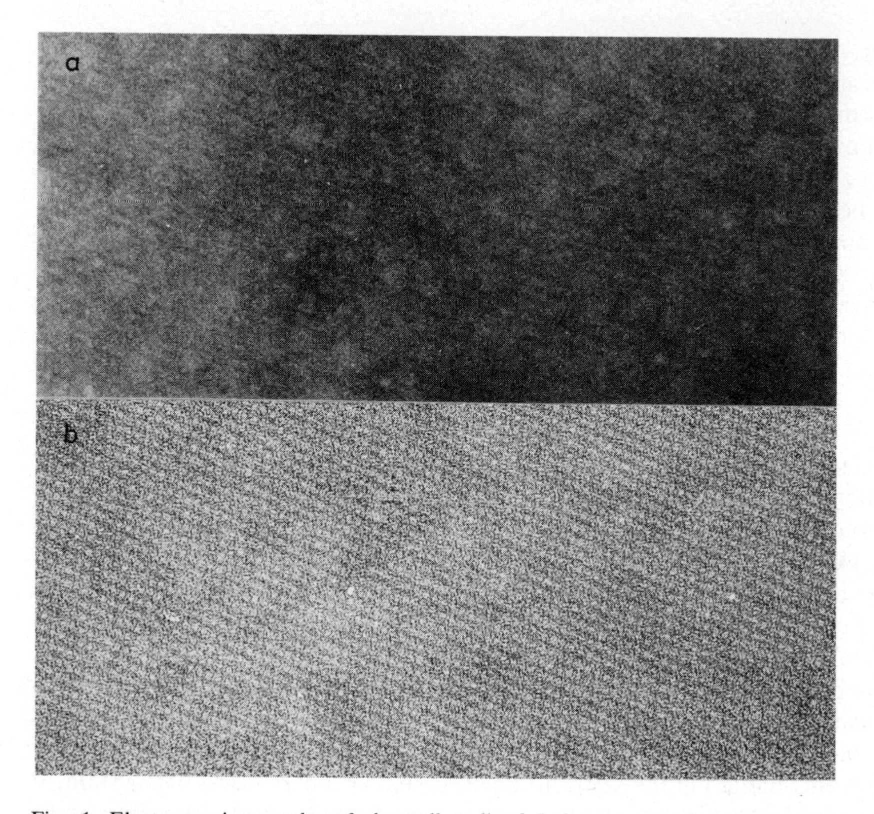

Fig. 1. Electron micrographs of the cell wall of Lobomonas piriformis negatively stained with methylamine tungstate. (a) Low dose micrograph (~50 e/nm^2). (b) Normal dose micrograph (~5000 e/nm^2).

common one in many fields, and the solution to it is well known; many measurements are made and added together. The noise, if truly random, increases as the square root of the number of measurements and the signal directly as the number of measurements. Thus the ratio of signal to noise increases by a factor proportional to the square root of the number of measurements added. In the same way many noisy images of a structural motif may be averaged and the relative contribution of the noise reduced. With electron micrographs this reduces not only the quantum noise, but also the noise from the other sources such as granularity of shadow and crystallites in stain. It is important to remember, however, that it is an average structure that is produced; if there are true differences between the different motifs averaged together this information will be lost.

The main problem to be solved in this approach is the spatial correlation of the different images of the repeated structural motif. In general the relative parameters that have to be determined between different images of an isolated particle are magnification, three-dimensional orientation and translation. This is a formidable problem even in theory with no prior information about the particle, and it is not clear yet whether it is possible to solve it. However, some practical studies have used situations where there is a common axial orientation between the different particles. The correlation problem is considerably simplified for particles with extensive internal symmetry, such as helical and icosahedral particles [4,5]. The symmetry axes define the relative origins of different particles. Correlation becomes trivial, however, if the specimen is crystalline in two dimensions. The repeating motif is simply the crystal unit cell, and the powerful techniques of Fourier analysis may be used to extract the repeating motif from the image. It is also true for a very coherent crystalline image that useful information may be extracted at noise levels which would make correlation of the individual motifs impossible [6,7]. A striking example of this was the work of Unwin and Henderson [8] on the unstained protein of the purple membrane of *Halobacterium halobium*. It seems unlikely with current techniques of transmission electron microscopy that a similar procedure could be carried out on unstained single particles.

Although the only way of obtaining information beyond 2 nm about biological macromolecules in the electron microscope must be by observing them in an unstained state, the experimental procedures for unstained biological electron microscopy are still in their infancy, and only a few successful studies of this kind have yet been reported. However, the image processing techniques which were originally applied to unstained macromolecular crystalline arrays to produce noise-filtered two-dimensional and three-dimensional reconstructions [8,9] are now being widely applied to negatively stained and metal shadowed crystals. This is allowing much more reliable information to be deduced from their electron microscopic analysis than has previously been possible, albeit at the lower resolutions inherent in these specimen preparation techniques. Many important biological systems either can be found naturally in ordered arrays or may be induced to crystallize in vitro. The convenience and power of determining structures by means of the methods that may be applied to crystalline arrays is the justification for devoting most of this chapter to a description of techniques which have been used. The approach is illustrated primarily by considering the structure determination of the regular glycoprotein cell wall shown in Fig. 1.

2. PRACTICAL CONSIDERATIONS

A description of the practical aspects of image analysis in electron microscopy would logically begin with a description of the electron microscopical techniques used. However, only two points will be mentioned here: firstly, images should be recorded with as small an electron dose as practicable, and preferably with a 'minimal dose' technique [8,10], so that the images represent a structure as little damaged by the electron beam as possible. For electron microscopy of unstained specimens to have any chance of success a low electron dose is of critical importance. The second point relates to focussing of the microscope. Defocussing and objective lens aberrations, primarily spherical aberration, produce a non-linear imaging system [12]. The characteristics of the microscope as a frequency filter may be displayed in Fourier space by the contrast transfer function (c.t.f.). This is a real-valued function which oscillates between +1 and −1, and the position of its zeros depend on the objective lens aberrations and the defocus value. Because of the form of this function some frequency components in the object appear in the image with reversed contrast (where the c.t.f. is less than 0). This is the origin of the great differences in a high resolution image that may be produced by altering the focussing. This non-linearity in the spatial frequency response of the electron microscope is also the only reason why many biological specimens produce any contrast at all in the image. They are often weakly scattering objects which merely introduce small phase changes into the incident electron beam, so-called 'weak phase' objects. In a perfect imaging system they would not give rise to any amplitude contast in the image and thus would be invisible in the same way that phase objects are invisible in high quality light microscopes. The defocus and spherical aberration in the electron microscope fortuitously produce amplitude variations in the image from the phase variations in the specimen in an analogous way to that intentionally employed in the phase contrast light microscope. There is still some dispute as to what types of specimen actually are weak phase objects; the heavy metals used in positive and negative staining and in shadowing probably also introduce actual amplitude contrast as well.

Ideally, one should aim to achieve a focus value which transmits all significant frequency components with their correct relative phases. This is perfectly feasible with negative stain images, which contain significant information to only ~1.5 nm, with a moderate degree of defocus-of the order of a 100 nm underfocus. For the higher resolution they obtained from unstained protein specimen Unwin and Henderson [8] used larger degrees of defocus, so as to increase the contrast of what were very weak phase objects. In this case it was necessary to correct the alterations thus introduced into the phases of the spatial frequency components by subsequent processing of the Fourier transforms of the resulting micrographs. Micrographs at different defocus values also had to be combined in order to fill in the regions of the transform where the c.t.f. was close to zero. An optical diffractometer is virtually a necessity for this type of work. It enables a rapid assessment to be made of the degree of focussing and extent of astigmatism. This is because the carbon film usually used as a support on the electron microscope grid is essentially an amorphous specimen, and produces in the Fourier transform of its image a noise spectrum which depends only on the modulus of the two dimensional contrast transfer function. A series of concentric dark rings are observed, the so called 'Thon rings' [13], whose diameters depend on the defocus. Astigmatism is seen as a deviation of the rings from circular to elliptical. Other defects in the image and the performance of the microscope may also be detected (for example specimen drift). A crystalline object in the image, of course, superimposes its own transform on the diffraction pattern, giving a series of spikes or spots lying on a lattice reciprocal to that of the crystal. It is thus very simple to detect the presence of crystalline order in a specimen and to assess its suitability for analysis. The resolution to which spots are present indicates the limit to which reliable information can be extracted, at least by Fourier techniques, and it is simple to ensure that only images are used which have a suitable degree of underfocus, such that all the diffraction spots lie within the first ring of the c.t.f. A simple practical design for an optical diffractometer has been given by Horne and Markham [14], although this can be further simplified if only the diffraction patterns are required, and no optical filtering is envisaged.

2.1. *Image digitisation*

The first requirement for processing images by computer is obviously a method of converting the images into a form accessible to a computer, in other words an array of numbers. Electronic image intensifiers and detectors are now being developed and may be directly interfaced to a computer. It has been shown that electronic image recorders can provide a high detection efficiency for low dose images (see, for example, Hermann et al. [15]). However, there are still severe technical problems with these devices and the number of image elements that can be recorded is a great limitation for recording images of regular arrays. Most computer processing of biological images has used the conventional photographic records of the images. The specular optical density response of photographic emulsions is a linear function of incident electron intensity, up to one quarter to one half of the film saturation density [16] and thus provides a direct record of the electron image. Although photographic emulsions can be highly efficient detectors of electrons, other factors are also important in their suitability for low dose work. Chiu and Glaeser [17] have examined several high sensitivity films for use in low dose microscopy, and have demonstrated marked differences between different emulsions, which they attribute to differing background fog levels as well as to differing electron sensitivities. Photographic film has the advantage of being relatively cheap. Standard practice with low dose microscopy is to take many exposures, more or less 'blind', and select those for analysis after development (discard rates of 99% are not uncommon).

In order to produce a representation of the film, therefore, the logarithm of the attenuation of a parallel beam of light as it passes through a small area of the film is measured at a series of positions generally lying on a square grid. This grid is termed the scanning raster. It is usually best to set the scanning aperture, that is, the aperture defining the area of film whose average optical density is measured at each raster point, to the scanning raster interval, In this way every point in the scanned area is included exactly once. Sampling theory shows that the scanning interval should be at most half the minimum spatial period to be recorded. However, this is a necessary, rather than a sufficient condition. Consider, for example, measuring the sine wave illustrated in Fig. 2 at a raster equal to one half of its period. The average value is recorded at each raster position. If the raster is as shown in Fig. 2a this will yield a (reduced) value for the signal. However, if measured at the positions in Fig. 2b, the signal will not be detected at all. Also if there are significant spatial frequencies at less than the scanning interval, these will be 'aliased' by lower frequencies (see, for example, Bracewell [18]). For these reasons, the safest approach is to scan at an interval between one-third and one-quarter of the minimum significant spatial period present in the image. Thus with a maximum resolution of ~2.0 nm at a magnification of 30,000, the scanning interval should be $\sim 2.0 \times 30{,}000 \times 1/3 = 20\ \mu m$.

be determined by individual circumstances. Some applications, such as interactive analysis and handling of images, necessitate a local, more-or-less dedicated machine. On the other hand, some groups have used large mainframe machines with success. The types of image manipulations described in this chapter are quite well suited to 'batch' processing. However, image processing requires a fairly large allocation of resources, and input and output of large amounts of data. This means that a local machine, whatever size, is a necessity in practice. Also the immediate access to a 'hands-on' minicomputer simplifies much of the work. Some degree of interaction is very helpful at certain stages even with a 'batch' oriented programme system.

2.2.2. Software

The type of computer which is available will determine some features of the software for image analysis, and some of the peripheral devices, such as for grey scale output, will need special programmes or routines. This means that no software system can be entirely portable. Many of the techniques described in this chapter, and many of the original programmes implementing them were developed at the MRC Laboratory of Molecular Biology, Cambridge, U.K. The extensive set of programmes in use there are all independent, 'stand alone' programmes. The only point of contact between different programmes in the suite is the data file, and a common format and accessing protocol is used for the image files. Although this approach gives some limitations to the flexibility of operations, they are not very serious for the crystallographic type of image analysis decribed below, and it may often be the only feasible approach for a mainframe computer.

However, in many image processing applications it is desirable to be able to specify sequences of image manipulations and to include new operations with the minimum of effort. With this in mind, therefore, some attempts have been made to write systems which are as widely useful and portable as possible. One example is SEMPER [24]. This system, written in standard FORTRAN, was developed initially on an extended DEC PDP 8, but has since been implemented on several other, larger minicomputers. The basic idea was to develop an image processing 'super-language', and thus allow new procedures to be put together from a set of basic routines or operations in as flexible a way as possible. The commands are interpreted by the SEMPER monitor routine which calls other routines to carry out the specified operations. The use can be either interactive where the user specifies operations by commands at the terminal, or 'batch', where the monitor reads commands from a prepared file. Provisions are made for input and output of parameter values and for branching and conditional testing and setting. Another specific aim was to make incorporation of new commands as easy as possible. All images are stored by SEMPER in one large FORTRAN direct access file; the SEMPER system maintains its own directory of the images contained in this file. This system was originally developed with the investigation of the physics of high resolution microscopy in mind, but it has recently been applied to various biological problems [25].

A system which is similar in its fundamental design has been developed by Frank and his colleagues [26]. This system, SPIDER, has been developed on a PDP 11/45 computer, and uses some features which are quite specific to the DEC RSX 11 operating system. It is therefore not as easily transportable to other operating environments, although the general system facilities required are not unusual in small computer operating systems. The general strategy, again, is to have a 'friendly' monitor routine with which the user communicates, either interactively or through a batch file. The monitor programme then calls slave programmes to carry out the commands, communicating with them via a message sending and receiving protocol.

A third biological image processing system was written by Smith and his colleagues [27]. This system, written for a large mainframe computer, consists of a single large FORTRAN programme, rather than smaller intercommunicating programmes. The user interface is more rudimentary than with the previous two systems; the user types a two letter code to specify an operation, and all necessary parameters are then requested by the relevant routine. Addition to, and modification of, this system would appear to be considerably more difficult, since any new routine would have to be explicitly included in and compiled with the entire existing system. A fairly detailed knowledge of the internal workings of the programme would therefore be necessary. This system has also recently been implemented on an DEC minicomputer (A. C. Steven, personal communication). A compromise between isolated programmes and an integrated system was adopted in this laboratory which took advantage of two features of the operating system in use (RSX 11M). The first is that programmes can be 'installed', that is, run by referencing the programme name directly as a command. Furthermore, the entire command line can be passed to the programme. Routines were written to parse the command line and extract file name specifications and numerical and character arguments associated with specified keywords. The syntax for the command line was designed to be as close as possible to that used by the system programmes. The general form for an image processing command line in this system is:

PROCEDURE   file/key1:*x*/key2:*y* etc.,
for an operation on a single image or
PROCEDURE   outfile = infile/key1:*x*/key2:*y* etc.,

for an operation involving an output and an input file, where key1, key2 etc are keywords, and *x*,*y*, etc., are parameters to be associated with them.

Most image processing programmes are thus written as independent tasks which take their file specifications and any necessary parameters from the command line. Common routines are used to access image files for input and output. The operator may therefore use what appear to be image processing commands. The second useful feature of the operating system is the facility for using indirect command files. This means that a file of commands to the system monitor can be set up and executed. These commands may of course be a string of image processing commands. There is a provision in the command processor for inputting arguments, and for conditional branching and looping. The main programmes and keyword parameters for them are listed in the appendix, as a guide to what we have found to be useful. Image file allocation and deletion requires some care, since the large image files can soon accumulate and fill all available space on the disc. The following convention has therefore been adopted in the command line interpreting routines: file names which

TABLE 1

Examples of command files to digitise, transform and output an image.

(a) Image File stored as LOB153, transform as LOB153FT:	
SCAN	LOB153/LENGTH:512/WIDTH:512/RASTER:24
SHADE	LOB153
FFT	LOB153FT = LOB153
SHADE	LOB153FT
(b) Image file deleted, transforms file stored as LOB153FT:	
SCAN	/LENGTH:512/WIDTH:512/RASTER:24
SHADE	PYPLYN
FFT	LOB153FT =
SHADE	LOB153FT

are omitted are given a default name (PYPLYN). A defaulted input file is deleted on completion of a programme, whereas a defaulted output file is not. This means that the intermediate files needed in a string of processing commands may be ignored; unwanted files will automatically be deleted.

As a simple example, Table 1a shows a command file to scan an image, calculate its Fourier transform and plot out a grey scale representation of it. In Table 1b the same operations are performed, but only the final transform is stored. Although this implementation is very much tailored to the particular operating system, the general facilities required of an operating system for a similar approach are quite commonly available. Indirect command file processors are a part of many current operating systems, and it is generally possible to arrange for the transmission of a command line to a programme. The advantage of this type of system is that it is very simple to set up and that all the system utility programmes can be used alongside the image programmes, yet some of the power and flexibility of an integrated image processing system is obtained. There are however limitations compared to systems such as SEMPER and SPIDER. Mainly these lie in the ease with which different programmes can communicate with each other. However, for the sorts of applications discussed in this chapter this arrangement has proved very simple and convenient to use.

Most of the basic programmes needed for the manipulations described below are quite straightforward. Handling the large amount of data contained in the images requires some thought so as not to arrive at impossibly slow algorithms; the smaller the computer being used, the more care is required. The calculation of the large two-dimensional Fourier transforms requires particular care. The fast Fourier transform algorithm [28] (FFT) is now very widely known, and indeed without this algorithm much image processing by Fourier methods would not be feasible computationally. What is not so widely known is that the FTT algorithm is not limited to arrays whose dimensions are powers of two. The Cooley–Tukey factorisation can be applied to any integral factor of the array dimension, the greatest savings in computation being for factors of two, three and four. A 'mixed radix' routine determines the factors in an array dimension and uses the relevant routine for each factor. This relaxes almost all the restrictions on the array dimensions, although practical programs generally restrict the highest prime factor to be handled

TABLE 2

Blocking algorithm for two-dimensional Fourier transformation

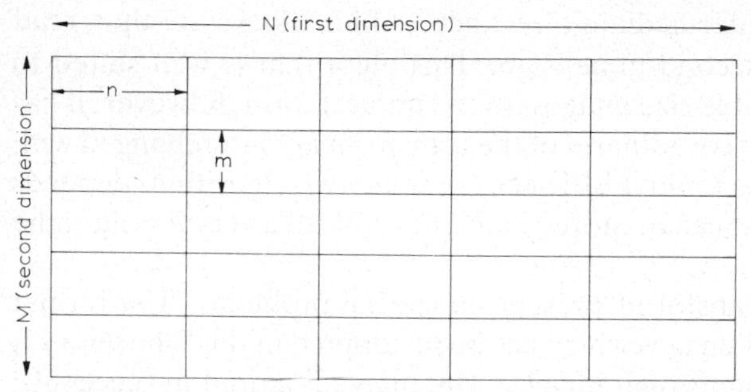

1. Read in a strip of image m rows wide.
2. Transform in first dimension.
3. Write out to intermediate random access file as blocks $m \times n$.
4. Goto 1 until all M blocks are finished.
5. Read in blocks to form a strip of the intermediate image n columns wide.
6. Transform in second dimension.
7. Write out by column.
8. Goto 5 until all N columns are finished.

in the interests of economy. A mixed radix routine is a great help in image processing, since it enables the best choice to be made about the area for analaysis without reference to the number of digitised points contained in it. If a radix 2 routine is used the required area must be 'padded' with zeroes to the next highest power of two in each dimension. It is generally then necessary to 'float' the image (see below) to avoid large origin peaks arising from the edge between the image and the surrounding padding. This procedure also increases the size of subsequent Fourier transformation, which can be a significant consideration on a small computer. The FFT program used in this laboratory is based on subroutines written by Ten Eyck [29]. Since the initial images are real, the transforms have Hermitian symmetry. This means that only half the transform needs to be calculated and stored, and the transform thus occupies the same amount of storage as the original image. The other half of the transform is generated as required, for example for plotted or printed output.

The simplest format for storing images and transforms is as a direct access file where each record contains one image scan line or transform line. The two-dimensional Fourier transform is calculated as a series of one-dimensional transforms, firstly on each line in turn, then on each column. The main problem is that it will almost certainly be impossible to have the whole image in memory at once, and it may only be possible to have a very few images lines in memory. One solution is to calculate the transforms for each row in turn, then to calculate the column transforms in parallel, reading in each row as it is required and carrying out the relevant operations on each element of the rows in turn. Thus the rows in this algorithm take the place of the individual elements in the basic FFT algorithm. The main disadvantage with this algorithm, apart from some additional programming complexity, is that it is not well suited to mixed radix transforms; as many image lines must be resident in memory as the maximum factor to be allowed. This is not always possible for large images. An alternative strategy is illustrated in Table 2. The image is divided into strips containing as many image lines as can be acccommodated in the available memory. After transformation in one dimension the strips are written out in blocks to be an intermediate direct access file. Blocks are then read back to build up strips in the second dimension. This algorithm is well suited to mixed-radix transforms for variable size images. In its simplest form, however, it has the disadvantage that the x and y coordinates of the transform are interchanged with respect to the original image. The relative efficiency of these two algorithms depends on the number of image lines that can be stored in memory. With a very few lines the former is more efficient.

Three-dimensional Fourier transforms present no special problems. The format for storage of three-dimensional images which has been adopted in this laboratory is an extension of that for two-dimensional images. The image is stored in successive planes (by convention perpendicular to z); each plane is equivalent to a two-dimensional image. For the reconstructions of single layer crystals it is advisable considerably to oversample the z^* lattice lines (see below). This means that the thickness of the layer is considerably less than the complete z interval. To take advantage of this therefore, the three-dimensional Fourier program written in this laboratory calculates first the z^*/z transform on each z^* lattice line, and stores only the values for the required z planes. The calculation is completed by a series of two-dimensional transforms on each required z plane in turn.

3. TWO-DIMENSIONAL RECONSTRUCTION

Consider first an image with two-dimensional periodicity. This type of ordering in an image is easily detected and assessed by means of the optical diffraction pattern. A well-ordered crystal shows sharp, unsplit spots, and the resolution to which the spots extend gives an objective assessment of the maximum amount of periodically repeating information present in the image. It is a logical step next to extract this information from the transform discarding the background noise, which is randomly distributed. Although it is possible to do this optically, it is much simpler and more reliable to do it by computer processing. For subsequent processing of the transform, such as symmetry averaging, computer methods are a necessity.

It is worth taking considerable care when selecting areas of micrographs for analysis. One should aim to find an area which is as coherently crystalline and evenly stained as possible. If the digitised area is not entirely filled by a coherent crystal, the analysis of its Fourier transform given below is not valid. A smooth variation in background intensity in the image can often be removed by fitting a plane to the data, so as to give the same average optical density all over the image. In any case such a background variation is not usually very troublesome. However, a variation in image contrast or staining can lead to phase gradients and other deformation in the transform spots. It is often tempting to overestimate the size of image area which is coherent enough to be safely used.

The mathematics of the Fourier transform of a periodic array are very familiar to crystallographers. The crystal may be considered as being formed by the convolution of the unit cell function with a set of regularly spaced delta functions representing the crystal lattice. The Fourier transform of the crystal is then the product of the unit cell transform with the Fourier transform of the delta function set. In the case of an infinite crystal, the transform of the delta function set is another set of delta functions representing a lattice reciprocal to the crystal lattice; i.e., if the crystal unit cell vectors are

$$\mathbf{a}\ ,\ \mathbf{b}$$

the reciprocal lattice vectors are

$$\mathbf{a}^*\ ,\ \mathbf{b}^*$$

where

$$\mathbf{a}^* \cdot \mathbf{a} = \mathbf{b}^* \cdot \mathbf{b} = 1$$
$$\mathbf{a}^* \cdot \mathbf{b} = \mathbf{a} \cdot \mathbf{b}^* = 0$$

if the unit cell vectors have magnitude a and b and are separated by an angle γ, we have:

$$a^* = \frac{1}{a \sin\gamma}$$

$$b^* = \frac{1}{b \sin\gamma}$$

$$\gamma^* = 180 - \gamma$$

for the magnitude a^*, b^* and interaxial angle γ^* of the reciprocal set.

Many of the same considerations apply to interpretation of optical or computed 'diffraction patterns' as to those measured in X-ray crystallography. The images must show plane group symmetry (see International Tables for X-ray Crystallography [30] for enumeration of plane groups) and the diffraction pattern must therefore show two-dimensional point group symmetry. However, in the case of a computed transform the phases are also available, and the phase symmetry, if any, may be determined. Thus the full symmetry of the image plane group is available (see [31]).

In practice, however, the images derive from real two-dimensional crystals composed of three-dimensional objects and thus having a finite thickness, therefore the full three-dimensional symmetry is described by a two-sided plane group [32]. Since biological molecules are enantiomorphous only the plane groups not containing mirror or glide planes or inversion centres are strictly allowed. However, this condition is not as strong with the comparatively low resolution data typical of negatively stained crystals as it is with X-ray data extending to atomic resolution. It is perfectly possible, in principle, to have a symmetry relation obeyed at 2 nm, such as a mirror plane, which breaks down at higher resolution. Nevertheless, it would seem safest to regard such apparent symmetry, if it is ever observed, as non-crystallographic, and to attribute it either to the low resolution, or to statistical disordering. The seventeen allowed two-sided plane groups are listed by Fuller et al. [33]. In some cases the (untilted) projected structure perpendicular to z is sufficient to determine the two sided plane group, in others the three-dimensional transform must be examined.

For reconstruction of the average unit cell by Fourier methods it is necessary to extract the amplitudes and phases of the unit cell transform at each reciprocal lattice point. This requires some consideration of the effect of replacing the infinite, continuous transform with a finite, sampled Fourier series. The effect of sampling at the discrete series of points of the scanning raster has already been considered. To see the effects of finiteness consider, (in one dimension for simplicity) an image function g_i sampled at N points (i.e., $i = 0, N - 1$). We assume in what follows that the image completely fills the sample area (i.e., no zero padding). The Fourier transform calculation, using for example the FFT algorithm, yields frequency components at spatial frequency values J/N, $J = 0,1,\ldots,N - 1$. The crystal reciprocal lattice can only be exactly represented by a subset of these points if the unit call is an integral multiple of the sampling period. This is not possible for a two-dimensional crystal with a non-rectangular unit cell scanned on the normal square sampling lattice, and is very difficult to achieve with sufficient accuracy even for a crystal with square unit cell.

Now consider the contribution of a spatial frequency S with (complex) amplitude $\tilde{F}_S$ to the m'th point of g.

$$g_m = F_S \exp(-2\pi i\ Sm/N)$$

thus in the Fourier transform, at point K:

$$G_K = \sum_{m=0}^{m=N-1} F_S \exp(2\pi i(K - S)m/N)$$

We may sum the R.H.S. as a geometric series to give:

$$G_K = F_S \frac{\exp(2\pi i(K - S)(N - 1)/N) - 1}{\exp(2\pi i(K - S)/N) - 1}$$

$$= F_S P(K - S)$$

i.e., The amplitude is multiplied by a profile function $P(K - S)$.

We have so far defined the origin at zero, which is the definition most practical FFT progammes take. However, transforming the origin to the centre of the sampled interval shows the form of the profile function more clearly:

$$P'(K - S) = P(K - S)\exp(-\pi i(K - S))$$

$$= \frac{\sin(\pi(K - S)(N - 1)/N)}{\sin(\pi(K - S)/N)}$$

$$\simeq N\ \mathrm{sinc}(K - S)\ (\text{if } K - S \ll N \text{ and } N \text{ is large}) \qquad (1)$$

similarly in two dimensions:

$$G_{K,L} = F_{S,T} N^2 \mathrm{sinc}(K - S)\,\mathrm{sinc}(L - T) \qquad (2)$$

for a spot position S, T.

The transform points around each (non-integral) lattice point thus have complex amplitudes which are proportional to the unit cell transform at the point multiplied by a sinc function centred on the lattice point and sampled at the transform points. When an integral number of unit cell repeats is contained in each direction (i.e., S and T are integral) the sinc function is sampled at its maximum and zeros, and the unit cell transform may be directly extracted from the transform values at these points. In general the function is sampled at an arbitary series of points separated by one transform unit, and the spot is spread over several transform points. There is a problem therefore as to how best to extract the information contained in the transform spots. Aebi et al. [34] have approached this problem by resampling the image by means of interpolation such that the new sampling raster is integrally related to the unit cell, and then trimming the image to an integral number of unit cells. The reciprocal lattice points are then present in the calculated transform points, and may be directly read from it. There arc two disadvantages to this procedure. Firstly, it is time consuming; an initial transform must be calculated to derive accurate lattice vectors, then the image must be interpolated onto the new sample raster. This can be an extremely lengthy process, especially on a small computer. Finally, a second large Fourier transform must be calculated. The second disadvantage is that a good deal of the image may have to be discarded in trimming to an integral number of unit cells.

A more efficient procedure is to extract the values from the first transform by using equation (2). The values for the unit cell transform may be derived by direct inspection of the computer printout of the intensities and phases of the Fourier transform [8]. The summed intensity of the points surrounding the true lattice point should be equal to the intensity of the sinc profile maximum, and the phase should be constant across the four points surrounding the spot maximum. An alternative method, discussed by Roberts et al. [35], is to carry out a least-squares fit of the sinc profile around each spot. Some typical printed listings of windows of the data round transform spots are shown in Fig. 3.

Determination of accurate parameters for the reciprocal lattice is very important, whatever method is used. We routinely use an interactive programme to determine approximate lattice vectors, followed by a programme which systematically searches the region of each lattice point within a specified resolution limit on the transform for the peak centres using a search procedure which incorporates the form of the profile function. All significant spots are then used to determine the best lattice vectors.

Once the amplitudes and phases have been extracted from the transfrom, they may be handled as index lists in much the same way as X-ray reflection data, except that the phases are immediately available. Simply applying the inverse Fourier transform to the data generates the averaged, noise-free unit cell [34], which may then be displayed in exactly the same way as the original image. Various image manipulations are more easily carried out on the Fourier transform, or rather the list of indices, amplitudes and phases, than on images. For example, if there is apparent symmetry in the reconstructed image, this may be imposed exactly by carrying out averaging over the symmetry elements. In order to do this, if the symmetry defines an unique origin for the image, as in the case of a rotation axis perpendicular to the plane of the crystal, the origin of the observed data should be set to this position so as to show the phase symmetry between related spots clearly. The simplest way to do this is by use of the Fourier phase shift theorem. One way of stating this theorem is

SPOT -7 4
REFINED SPOT CENTRE ... 78.327 27.814 INTENSITY AND PHASE FROM PROFILE ... 0.18944E+11 -29.59

INTENSITY SCALE FACTOR 0.100E+09 PHASES IN TEN DEGREE INTERVALS

INTENSITY

	75	76	77	78	79	80	81	82	83
23	1.	1.	3.	1.	4.	0.	1.	2.	2.
24	0.	0.	7.	0.	0.	1.	0.	4.	1.
25	1.	0.	2.	0.	6.	2.	0.	2.	0.
26	2.	1.	0.	1.	1.	3.	1.	4.	2.
27	1.	2.	2.	8.	2.	2.	4.	4.	1.
28	1.	10.	12.	12[illegible]	34.	5.	1.	1.	1.
29	2.	3.	0.	14.	6.	7.	6.	9.	2.
30	0.	2.	0.	2.	9.	5.	1.	2.	2.
31	5.	0.	0.	0.	2.	2.	8.	5.	0.

PHASE

	75	76	77	78	79	80	81	82	83
23	20.	32.	26.	3.	2.	7.	29.	16.	3.
24	8.	9.	8.	14.	15.	4.	34.	33.	31.
25	21.	36.	29.	31.	25.	20.	28.	26.	30.
26	18.	16.	7.	6.	36.	20.	12.	36.	24.
27	15.	5.	17.	31.	32.	26.	15.	9.	22.
28	24.	1.	18.	3[illegible]	31.	21.	1.	19.	8.
29	29.	18.	25.	20.	28.	8.	18.	10.	30.
30	6.	16.	28.	26.	5.	30.	25.	33.	20.
31	28.	20.	36.	3.	18.	21.	11.	30.	18.

SPOT -4 4
REFINED SPOT CENTRE ... 49.474 44.191 INTENSITY AND PHASE FROM PROFILE ... 0.22896E+11 -75.30

INTENSITY SCALE FACTOR 0.100E+09 PHASES IN TEN DEGREE INTERVALS

INTENSITY

	45	46	47	48	49	50	51	52	53
40	1.	2.	2.	2.	10.	3.	2.	0.	0.
41	0.	4.	1.	4.	0.	9.	1.	0.	2.
42	2.	0.	9.	5.	2.	3.	2.	3.	2.
43	14.	3.	6.	0.	0.	12.	5.	3.	2.
44	2.	5.	14.	2.	117.	[illegible]6.	11.	6.	6.
45	1.	1.	1.	1.	9.	9.	0.	1.	2.
46	1.	2.	1.	2.	2.	0.	5.	1.	1.
47	4.	3.	0.	8.	1.	8.	0.	7.	4.
48	3.	3.	2.	1.	7.	0.	4.	3.	1.

PHASE

	45	46	47	48	49	50	51	52	53
40	15.	24.	24.	20.	20.	15.	18.	3.	13.
41	5.	15.	6.	19.	4.	29.	23.	3.	29.
42	30.	1.	24.	7.	29.	15.	20.	25.	7.
43	29.	31.	21.	30.	21.	4.	24.	35.	9.
44	28.	10.	29.	13.	29.	[illegible]9.	16.	27.	16.
45	34.	34.	28.	9.	26.	26.	35.	20.	4.
46	10.	27.	33.	24.	31.	11.	12.	31.	32.
47	26.	7.	9.	28.	15.	13.	7.	33.	23.
48	10.	26.	36.	18.	21.	2.	17.	26.	36.

SPOT -5 1
REFINED SPOT CENTRE ... 50.908 -10.553 INTENSITY AND PHASE FROM PROFILE ... 0.13323E+12 94.95

INTENSITY SCALE FACTOR 0.100E+09 PHASES IN TEN DEGREE INTERVALS

INTENSITY

	48	49	50	51	52	53	54	55	56
-16	0.	0.	0.	1.	1.	2.	1.	0.	0.
-15	1.	2.	5.	17.	2.	4.	2.	1.	0.
-14	2.	2.	2.	14.	3.	7.	3.	15.	3.
-13	1.	1.	1.	31.	0.	1.	3.	2.	2.
-12	0.	0.	2.	65.	1.	1.	3.	5.	0.
-11	5.	11.	3.	700.	3.	5.	6.	1.	1.
-10	9.	8.	1.	348.	6.	6.	3.	1.	2.
-9	6.	0.	0.	79.	2.	0.	3.	15.	0.
-8	4.	1.	0.	26.	0.	2.	6.	4.	1.

PHASE

	48	49	50	51	52	53	54	55	56
-16	7.	29.	20.	34.	28.	5.	1.	6.	5.
-15	14.	32.	9.	12.	24.	11.	22.	32.	34.
-14	24.	10.	32.	31.	15.	22.	34.	15.	11.
-13	4.	35.	2.	11.	27.	5.	29.	15.	23.
-12	35.	26.	27.	30.	32.	22.	25.	11.	7.
-11	8.	25.	5.	[illegible]	1.	7.	25.	15.	20.
-10	12.	30.	4.	[illegible]	33.	5.	22.	5.	22.
-9	6.	21.	12.	29.	4.	31.	34.	29.	11.
-8	5.	30.	9.	13.	36.	35.	3.	21.	21.

Fig. 3. Listings of a Fourier transform in windows around some spot positions. The refined spot centre has been indicated by a cross.

that shift in the image of $(\Delta x, \Delta y)$ is equivalent to a phase-shift in the (h,k) spot equal to $2\pi(h\Delta x + k\Delta y)$. In the case of a two-fold axis, for example, we would carry out a phase search for the best two-fold axis by shifting to successive positions over the unit cell by applying the phase shifts $2\pi(h\Delta x + k\Delta y)$ to all the spot phases. At each point a measure is calculated of the discrepancy between the shifted phases and their ideal values, in the case of a two-fold axis 0 or π. One possible measure is a simple unweighted R factor.

$$R = \frac{1}{N} \sum_{\text{all } h,k} |\phi^0_{hk} - \phi'_{hk}|$$

where ϕ'_{hk} is the shifted phase: i.e., $\phi'_{hk} = \phi_{hk} + 2\pi(h\Delta x + k\Delta y)$ and ϕ^0_{hk} is the ideal phase nearest the shifted value. This search should show a minimum for the best phase origin. Another possible measure is an R factor weighted by the amplitude F_{hk}.

$$R_{\phi} = \frac{\sum\limits_{\text{all}\,|h,k} F_{hk}|\phi^{\circ}_{hk} - \phi'_{hk}|}{\sum\limits_{\text{all}\,h,k} F_{hk}}$$

The same general procedure may be used for other types of symmetry. When the best phase origin has been determined, symmetry averaging is equivalent merely to setting each phase to its ideal value. Spot data from several films may also be averaged. First of all a common phase origin for all the films is determined, and then the phases from the different films may be averaged. Amplitude scaling is also necessary, after which the amplitudes may be averaged.

The values of the various statistics resulting from the study of Roberts et al. [35] are shown in Table 3, and are similar to those obtained with other negatively stained specimens by other workers. In this case, the specimen was the Lobomonas cell wall. Images similar to the low dose image shown in Fig. 1a were analysed. A typical computed Fourier transform is shown in Fig. 4 and the image reconstructed from the data averaged from three films in Fig. 5. Comparison of the phase discrepancy between symmetry related reflections in one film, or in this case the discrepancy from 0° or 180°, with the discrepancies between different films, gives an indication of the extent to which the supposed symmetry is obeyed. If the symmetry discrepancy is significantly greater than the interfilm discrepancy the symmetry is obviously only approximate. Compared to typical X-ray data, the amplitude agreements in Table 3 are rather poor, whereas the phase agreements are extremely good. This is a fairly general observation, at least for negatively stained specimens. The reason for this is broadly that amplitudes derive from the contrast in the image whereas phases derive from shapes in the image. The degree of staining of different images can vary, although the overall shapes delineated are quite reproducible. This type of analysis gives an interesting reversal of the normal situation crystallographers find themselves in, namely that of having well determined amplitudes but poorly determined or undetermined phases. There is little doubt that good phases are preferable to good amplitudes.

Analysing the degree of fit of the actual spot profiles provides an assessment of the coherence of the crystalline area. If the residual in the least squares fit is large compared to the background levels of the transform this is an indication of disordering of some sort in the crystal lattice. In this case the procedures for extracting the unit cell transform are not valid and will give spurious results. It is necessary to examine the transform very critically before employing Fourier processing.

The question then arises of what may be done with an image which contains some disordering. Limited Fourier processing may still be usefully employed. An example is shown in Fig. 6. This specimen is a large tube of the coat protein subunits from Turnip Yellow Mosaic Virus [36]. The tube has flattened down on drying, and a pattern produced by the superposition of the two sides is seen. This pattern is very difficult to interpret. The two sides of the tube are clearly distinguishable in the

TABLE 3

Statistics on the phase and amplitude agreement between the images averaged of negatively stained cell walls of *Lobomanas piriformis* cell wall

Three images averaged	Weighted	Unweighted
R_{Φ_o}* before averaging	9.9°	20.1°
R_F**	0.106	0.148
R_Φ***	6.4°	10.1°
R_{Φ_o} after averaging	7.2°	11.1°

(130 independent spots were included in the final reconstruction)

$$^*R_{\Phi_o}\text{ (weighted)} = \frac{\Sigma F_{hk}|\phi_{ohk} - \phi_{hk}|}{\Sigma F_{hk}}$$

$$R_{\Phi_o}\text{ (unweighted)} = 1/N\ \Sigma|\phi_{ohk} - \phi_{hk}|$$

where ϕ_{hk} is the measured phase of spot (h,k) shifted to the best phase origin, and ϕ_{ohk} is the two-fold symmetric phase (0° or 180°) closest to this value.

$$^{**}R_F\text{ (weighted)} = \frac{\Sigma \bar{F}_{hk}|\bar{F}_{hk} - F_{ihk}|}{\Sigma \bar{F}_{hk}{}^2}$$

$$R_F\text{ (unweighted)} = \frac{\Sigma|\bar{F}_{hk} - F_{ihk}|}{\Sigma \bar{F}_{hk}{}^2}$$

where $\bar{F}_{hk}$ is the average amplitude for spot (h,k) taken over all the images, and F_{ihk} is the amplitude for spot (h,k) in image i.

$$^{***}R_\Phi\text{ (weighted)} = \frac{\Sigma \bar{F}_{hk}|\bar{\phi}_{hk} - \phi_{ihk}|}{\Sigma \bar{F}_{hk}}$$

$$R_\Phi\text{ (unweighted)} = 1/N\ \Sigma|\bar{\phi}_{hk} - \phi_{ihk}|$$

where $\bar{\phi}_{hk}$ is the average phase for spot (h,k) taken over all the images, and ϕ_{ihk} is the phase for spot (h,k) in image i.

Fourier transform (Fig. 7) where there are two reciprocal lattices — one from each side of the tube. The methods described for extracting amplitudes and phases fail with this image, however, because there is still a residual curvature in the lattice. The best that can be done is to place a 'mask' on the transform by setting the transform to zero at all positions except for small windows around the points of one or the other reciprocal lattice. This has the effect of averaging over smaller areas of the image rather than over the whole of the crystalline array, and thus allowing for some distortion. Figures 8 and 9 show the two sides reconstructed. However, the reconstructed unit cells show some variation with this type of filtering, and this can lead to ambiguity in interpretation.

Severe distortions or disordering can make Fourier filtering, even with a large window, meaningless. An approach to this problem which is potentially very useful, is to abandon Fourier methods, and to use instead real-space correlation techniques.

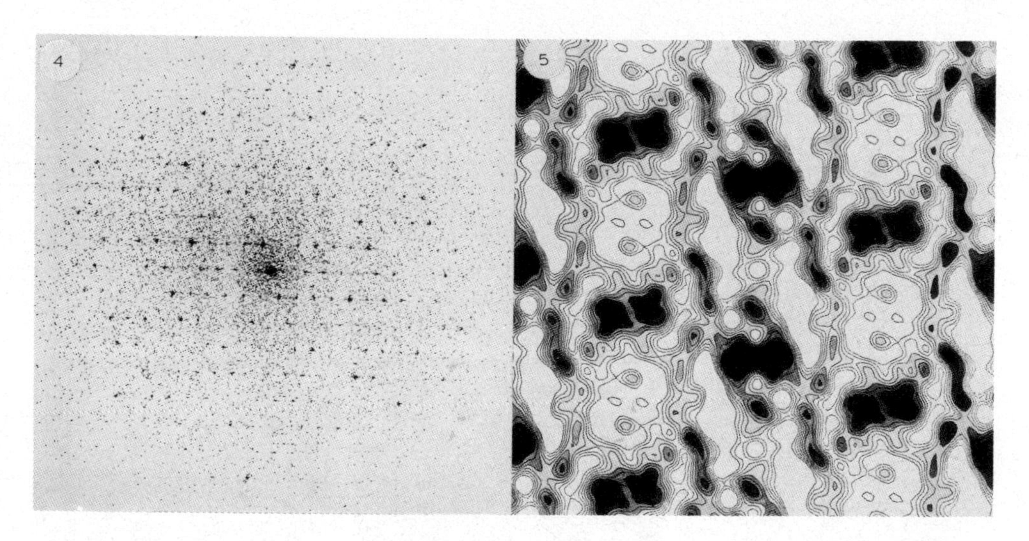

Fig. 4. Computed Fourier transform from a typical minimal dose micrograph of the Lobomonas cell wall. An example of SHADE output. Each transform point is represented by a dot whose size is proportional to the intensity.

Fig. 5. Two-dimensional reconstruction of the Lobomonas cell wall. An example of CONTOUR output. The most negative contour levels have been omitted and the picture has been subsequently shaded by hand to improve clarity. Dark shading indicates stain exclusion.

This method was originally suggested by Frank [6] with completely aperiodic images in mind. It is only really feasible for images of particles which are aligned with at least a common axis, as was the case for the membrane-bound acetyl choline receptors [37]. However, a disordered crystal provides many slightly disoriented particles and thus an ideal object for correlation analysis. A recent example has been published of this type of analysis [38]. In this paper it is shown that for negatively stained tubulin sheets, the correlation averaging is at least equal to Fourier processing for the most perfect crystals, and considerably superior for slightly deformed crystals.

4. THREE-DIMENSIONAL RECONSTRUCTION

Provided the specimen thickness is small compared to the depth of focus of the microscope and provided multiple scattering can be neglected, the image produced by the electron microscope may be regarded as a projection of the scattering electron potential in the plane normal to the direction of the electron beam. Once all the necessary corrections and averaging have been made to this image, the problem is how to relate this two-dimensional projection to the original three-dimensional specimen. If there is enough prior information about the object, or if it is a special structure such as a helix, the three-dimensional structure can sometimes be uniquely deduced from its two-dimensional projection. In general, however, there is no

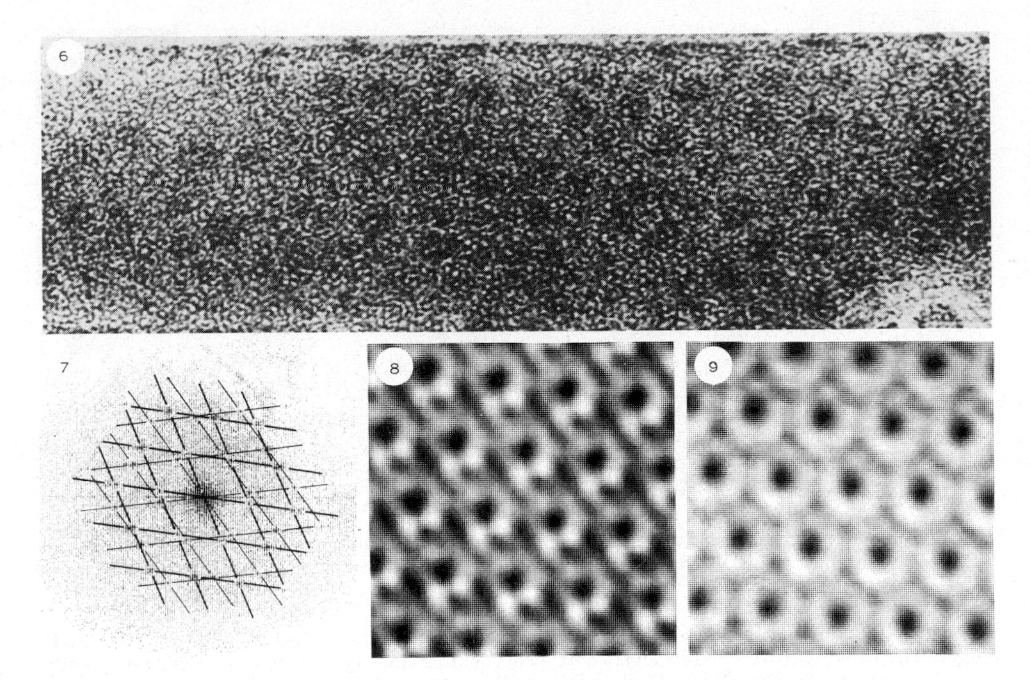

Fig. 6. Micrograph of a collapsed tube of Turnip Yellow Mosaic Virus subunits. A superposition pattern of the two sides is produced, which is not easily interpreted.

Fig. 7. Computed Fourier transform of a portion of the image shown in Fig. 6. The two reciprocal lattices have been drawn on.

Figs. 8 and 9. Reconstructed images from each of the two sides, obtained by use of a mask passing onc or other of the lattices in Fig. 7. An example of SHADE output of an image.

unique solution. The initital attempts at the solution of protein structures by X-ray crystallography were carried out using two-dimensional projections. These proved almost completely uninterpretable, and it has been a fairly general observation since that high resolution maps of these complex macromolecules are not interpretable in two-dimensional projection. As far as electron microscopy is concerned, the only real answer is to obtain several projections in different directions and to attempt to fit the different projections together to form the three-dimensional structure. Electron microscopists have for many years produced galleries of different views of structures, and, by inspection, proposed three-dimensional models. With computer techniques, it is possible to attempt to combine the data from different projections in a more objective, and mathematically precise manner.

The general problem of reconstructing a function from a finite number of its projections is a common one in several areas of science, and many different solutions have been used. A review of the different methods has been given by Gordon [39]. The most successful techniques so far used in electron microscopy have been Fourier space reconstruction methods, which have evolved as extensions of the two-dimensional Fourier reconstruction techniques. The present discussion has therefore been limited to this method, and as an example of three-dimensional analysis, we shall use the algal cell wall described in the previous section. The simplifications in analysis with a coherent crystalline specimen are just as great in three-dimensional analysis as in two-dimensions.

A crystal consisting of copies of a finite three-dimensional object arranged in a two-dimensional lattice may, as before, be considered as the convolution of the object with the lattice. The Fourier transform of the crystal is then the object transform multiplied by the lattice transform. However, the three-dimensional transform of a two-dimensional lattice consists of a set of lines perpendicular to the lattice. The three-dimensional transform of the crystal is thus the three-dimensional object transform sampled along the lattice lines. The two-dimensional transform of each projection gives a central section through the three-dimensional transform:

$$F(x,y,z) = \iiint_{-\infty}^{+\infty} f(x,y,z) \exp(2\pi i(Xx + Yy + Zz))\, dxdydz$$

for the Fourier transform F of a function f.

The central section (perpendicular to z) of the transform is given by:

$$F(X,Y,0) = \iint_{-\infty}^{+\infty} \{\int_{-\infty}^{+\infty} f(x,y,z)\, \mathrm{d}z\} \exp(2\pi i(Xx + Yy))\, \mathrm{d}x\mathrm{d}y$$

$$= \iint_{-\infty}^{+\infty} P_z(x,y) \exp(2\pi i(Xx + Yy))\, \mathrm{d}x\mathrm{d}y$$

where $P_z(x,y) = \int_{-\infty}^{+\infty} f(x,y,z)\, dz$

i.e., P_z is the projection into the x,y plane of $f(x,y,z)$, and its transform is the central section perpendicular to z of $F(X,Y,Z)$.

The various projected views give different central sections through the three-dimensional transform. In order to invert the transform computationally to give a reconstructed image, we need to have estimates of the transform over the whole of a regular three-dimensional grid. It is therefore necessary to interpolate from the points for which measurements are available onto the required grid. Crowther et al [40] have discussed the best way of achieving this interpolation, incorporating the constraint of finiteness of the object.

Achieving the best three-dimensional interpolation from an arbitrary series of central sections results is quite a formidable computing problem. For the least-squares analysis given by Crowther et al. [40] one must invert a matrix whose order corresponds to the number of points in the three-dimensional grid. The problem is considerably simplified for a crystal, however, since the lattice lines are independent of one another and the interpolation needs to be only one-dimensional. Smith et al. [41] have shown furthermore that if the data are oversampled the method of interpolation is in practice unimportant. It becomes very important if a limited number of views or a restricted angular range of views is used. However, it is better to measure more than enough data points than to rely on mathematical procedures to make up for missing data. With a crystalline specimen this is not difficult to achieve.

With a suitable specimen stage it is possible to record tilted views up to an angle of 60°, or more. When many views tilted about various axes are combined the whole transform except for a conical region around the z^* axis may be sampled. Omitting this core of data will decrease the resolution in the z direction of the reconstruction. This means that the contrast variation of the reconstructed image in this direction may be inaccurate, but the overall effect may be expected to be fairly small; the volume of the transform left unsampled is a small fraction of the total volume. Some studies [42,43] have included data about the contrast in the z direction in the form of values for the $(00z^*)$ line derived from sectioned specimens or from micrographs where the layers may be seen edge-on at a bend. The effect of including these data seems to be small. Sampling theory shows that each lattice-line is uniquely determined, in the ideal, noise-free case, by a series of values along it spaced by $1/T$ where T is the thickness of the crystalline specimen in the z direction. This distance on the highest resolution lines gives an expression for the spacing in tilt angle between successive views (Δ) needed to achieve this sampling

i.e., $\tan \Delta = D/T$

where D is the resolution limit (in the same units as T). This figure should, however, be regarded only as a rough guide. It is best, as already mentioned, considerably to oversample the lattice lines. Tilted pictures are most easily recorded using a eucentric goniometer specimen stage. These devices enable the specimen to be tilted, generally through angles up to ±60°, about a point which is always the centre of the illuminated area. With a non-eucentric goniometer the field of view changes as the specimen is tilted, and the required area can easily be lost. For obtaining different views of a single particle therefore an eucentric goniometer is in practice a necessity. However, with a crystalline specimen it is possible to use a fixed angle tilt-holder. The holder is set to a given angle of tilt and several micrographs of different areas are recorded. Since the crystals will be oriented in arbitrary directions on the grid, this will result in projections about a number of different tilt axes. If minimal dose pictures are required this is really the only feasible approach. A tilting holder for a Siemens electron microscope has been constructed in this laboratory [44]. It is also possible to extend the range of tilt angles obtainable simply by bending the specimen grid [43].

The advantage of this approach with crystalline specimens is that it is possible to calculate the tilt axes directly from the micrographs. The tilted and untilted unit cell dimensions are related geometrically (see Fig. 10). We have given the details of the calculation elsewhere [44]. Alternatively, the tilt parameters can be determined by means of the contrast transfer function. The focus changes across the image of a tilted specimen, since each part of the specimen is a different distance from the focal plane. Thus as different parts of the micrograph are moved through the light beam in the optical diffractometer, the diameters of the Thon rings alter. When the micrograph is moved along an axis parallel to the tilt axis, however, there is no change in the ring diameter, since by definition all points along the tilt axis are at the same height. The rate of change in ring diameter in the direction perpendicular to this axis can be related to the tilt angle, if the magnification and objective lens parameters which determine the c.t.f. are known.

In the study of the cell wall of Lobomonas pirifomis, we collected approximately

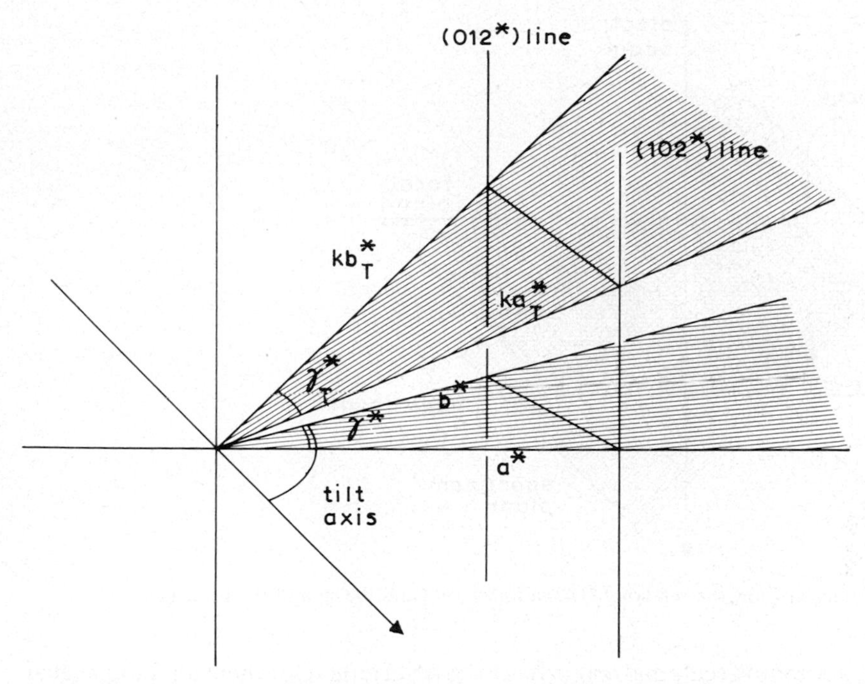

Fig. 10. Geometrical construction relating to the tilted reciprocal unit cell parameters (a^*_T, b^*_T and γ^*_T) to the untilted ones (a^*, b^* and γ^*). k is a constant to account for any difference in magnification between the tilted and untilted micrographs.

forty micrographs at each of several tilt angles ranging from 0° to about 70° in steps of approximately 10°. The four or five best images from each angle were selected by observing their optical diffraction patterns. They were digitised, their Fourier transforms calculated and accurate reciprocal lattice parameters were calculated. The direction of the tilt axis was then also determined for each micrograph using the optical diffractometer. The angle was then related to the crystal lattice vectors by construction on enlarged prints. (It should be remembered that the angle thus determined is the angle from the tilt axis to the tilted lattice vectors, not to the untilted vectors.) It was also possible to determine the absolute sign of the tilt angle by observing how the focus changed across the micrograph (Fig. 11). An example of this is shown in Fig. 12.

Indexing the transform spots was quite straight forward for the lower angle tilted pictures. A few strong lines were followed through the tilted pictures; it is a reasonable assumption that the spot intensities will not change very rapidly for the very strong spots. This process did not work for the very high angle tilted pictures, however. The ambiguities were resolved by taking all possible indexing systems and calculating the tilt angle and axis based on each indexing. In all cases the correct indexing was the only one that gave good agreement with the tilt axis measured by the c.t.f. method. In this way data sets were accumulated, each consisting of a set of indices, amplitudes and phases, at a known tilt angle and axis.

Before data sets like this can be combined to give the three-dimensional

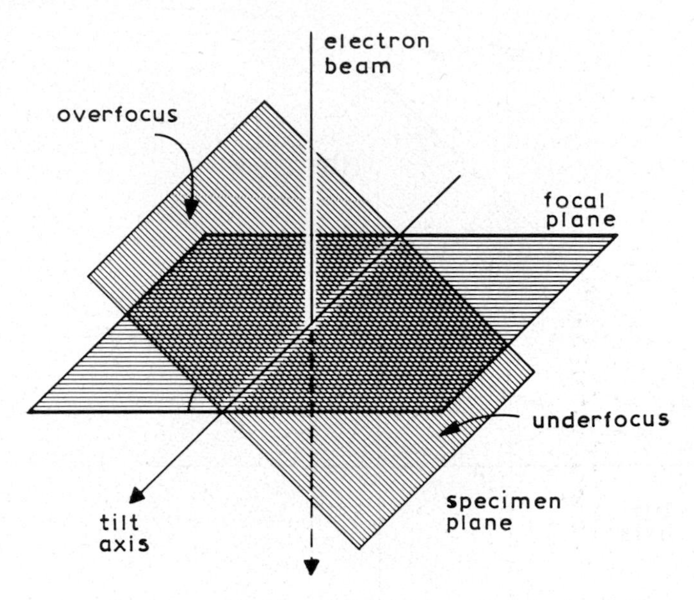

Fig. 11. Diagram to show the relation between focus level and the sign of the tilt angle.

transform, a common scale and origin has to be determined for them all. Hoppe et al. [45] have presented an argument to show that the cross correlation function between two projected views of an object should have a peak corresponding to their relative origin shift provided the angle between the projections is small. One might therefore envisage starting with a zero tilt picture and correlating the lowest tilt angle projections with it. The next highest tilt pictures would then be correlated with the lower angles, and so on. The problem with this procedure is that each image is compared separately as a whole with one or more images. It seems all too likely that errors in the origin, especially in z, would build up between the zero tilt pictures and the highest angle tilt pictures. Henderson and Unwin [9] used a slightly different procedure, and their example has been followed in several subsequent studies, including the one described here. This relies on determining origins in the Fourier transforms by phase shifts, as for two-dimensional origin refinement. In this case the residual whose minimum is searched for is a 'cross-residual'.

$$R_{\phi_x} = \sum_{\text{all refs}} |\phi'_{ihk} - \phi_{jhk}|$$

where ϕ'_{ihk} is the shifted phase for the (h,k) spot on the film whose origin is being refined, and ϕ_{jhk} is the phase for the (h,k) spot on a film whose origin has already been determined. Reverting briefly to one dimension for simplicity, the cross correlation between $f_1(x)$ and $f_2(x)$ is equivalent to the Fourier transform of the product of the transform of $f_1(x)$ (i.e., $F_1(X)$), with the conjugate of the transform of $f_2(x)$ (i.e., $F_2{}^*(X)$).

$$\text{i.e.,} \quad \int_{-\infty}^{+\infty} F_1(X)\, F_2{}^*(X) \exp(2\pi iXx)\, dx$$

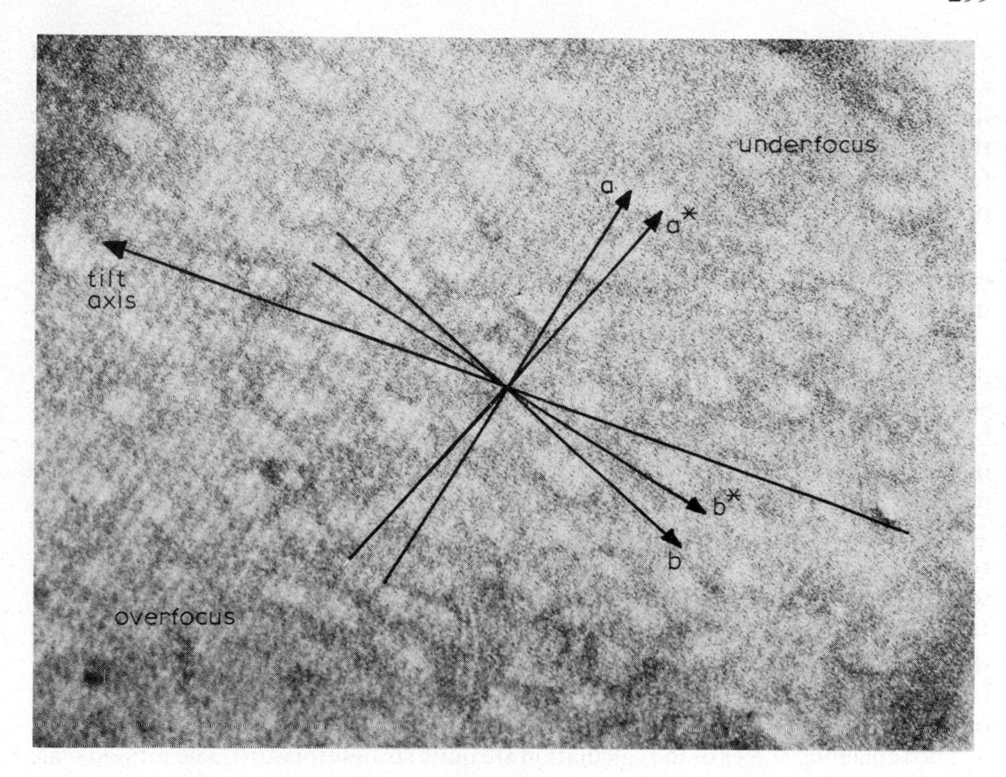

Fig. 12. An example of the determination of the sign of the tilt angle. The direction and sign of the tilt axis is calculated from the refined reciprocal lattice parameters. Determining which side of the axis is overfocus and which underfocus determines the sign of the tilt angle.

or putting

$$F_1(X) = |F_1(X)| \exp(i\phi_1(X))$$
$$F_2^*(X) = |F_2(X)| \exp(-i\phi_2(X))$$

this becomes

$$\int_{-\infty}^{+\infty} |F_1(X)|\,|F_2(X)| \exp(i(2\pi Xx + \phi_1(X) - \phi_2(X))\,\mathrm{d}X$$

Searching for the phase shift ($2\pi x$) which minimises the summed phase differences would be in principle similar to finding the maximum of the cross correlation function if the amplitudes $|F_1|$ and $|F_2|$ were all equal and constant. In practice amplitude weighting of the phase residual seems to have no effect on the peak position [4].

However, the phase search method has the advantage that it allows a correlation of each image with all the preceding images together. Contributions to the residual are added for each line from all points from the other films with similar values of z^*. As noted before the continuous z^* lattice line is determined by its values at points separated by $1/T$. It might therefore be reasonable to include in the phase comparison only points to $\Delta z^* = 1/T$. In practice it is often necessary to experiment with different values for Δz^*, especially since the thickness, T, is often not known before three-dimensional analysis. An alternative possibility might be to include comparison with all other points on the lattice line, but weighted inversely with their distance from the z^* value for the image being refined. The amplitude scale factors may be calculated simply by comparison with the closest amplitudes to the points on the lattice line.

The chief disadvantage to this procedure is that the reference data is initially limited and is constantly changing as new images are added. With care, starting with a zero-tilt image and gradually adding lower angle, then higher angle tilt images works satisfactorily. However, errors in the first few images can completely upset the process; it is very much a "bootstraps" method and errors are not necessarily self-correcting. It is also necessary to iterate the process. After a few of the images have been added, it is advisable to re-refine the first images against the later images. Finally, in the Lobomonas work each image was refined in turn against all the other images until only very small changes occurred. There is in fact no guarantee that this process will converge. In this case there were no significant changes in origin after two cycles of refinement, but there was a slight decrease in amplitude scale factors. The amplitude scaling problem is similar to the layer scaling problem in X-ray crystallography [46]. In this case, also, it is possible to get inconsistencies by scaling each film individually to the rest. It is, however, difficult to see how to apply the approach suggested by Hamilton et al. [46] and, later Fox and Holmes [47], to the present scaling problem.

The remaining stages of the calculation are quite straightforward. The refined scale factors and phase shifts are applied to all the images, which are then merged and sorted into a list of h, k, z^*, amplitude and phase. The measurements of amplitude and phase for each lattice line may then be plotted out as a function of z^* and a smooth curve interpolated by hand [9]. This can become very tedious for a large number of lattice lines. In the Lobomonas work values were interpolated by computation. The following algorithm was found to work quite well. The experimental points were joined together with a linear interpolation between each of the points. The resulting jagged curve was then smoothed by low pass Fourier filtering. The computed curves and the data points were plotted out for each lattice line (see Fig. 13). In nearly all cases the curves agreed well with those that would have been drawn by hand through the points. The computation was regarded as merely a convenience, and visual assessment was relied on to justify the algorithm. Regularly spaced values (spaced at about $1/3T$ to $1/4T$) along each lattice line were written out for input to a three-dimensional Fourier programme. The output of this programme, a three-dimensional image in sections of constant z, was contoured at equal, arbitrary levels onto a stack of transparent sheets (Fig. 14).

Such a stack of contoured sections is probably the simplest and best way of examining and interpreting this type of three-dimensional reconstruction. Displaying the results in a form more suitable for publication can be a problem. In some cases, simply photographing the stack can give an idea of the map. Stereoscopic pairs of photographs can also help. Another possibility is to construct a wooden model of the envelope of the density. This has the advantage that the contours can be smoothed so

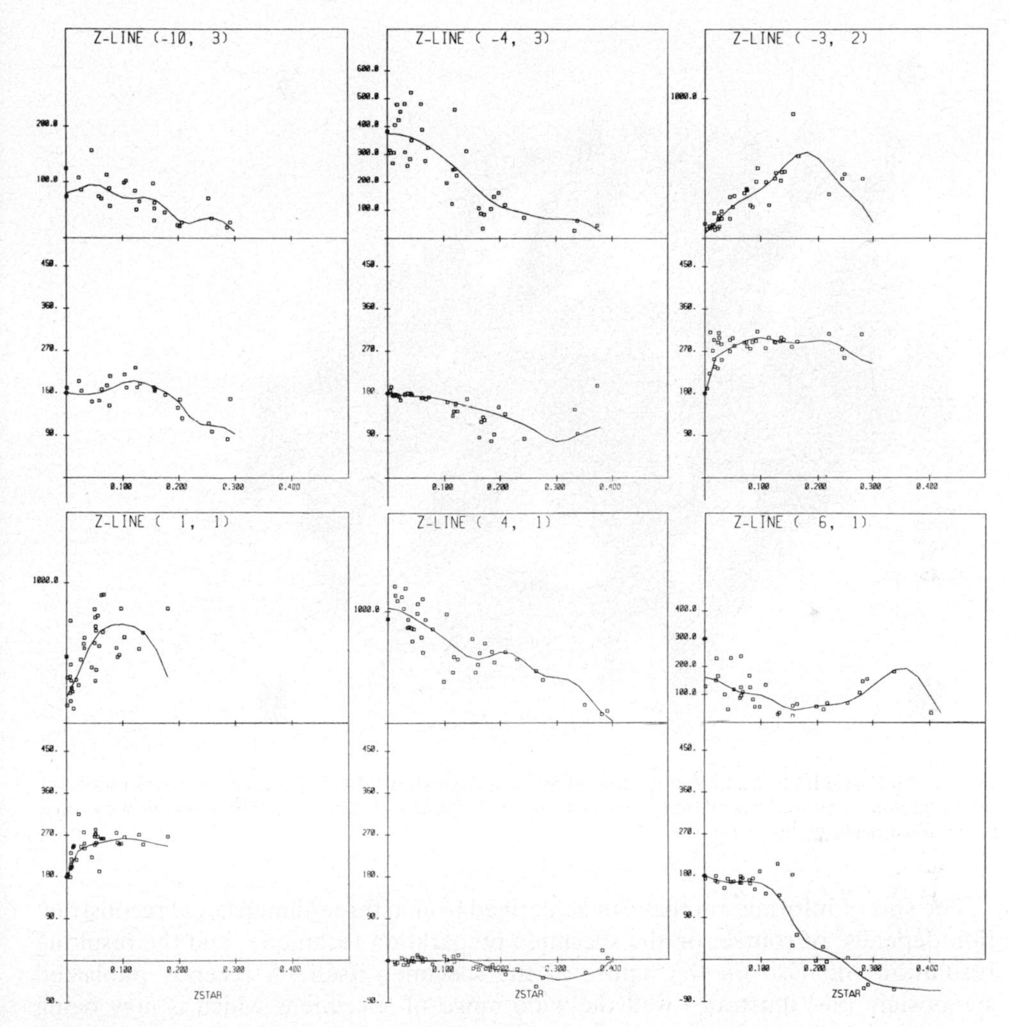

Fig. 13. Plots of the measured values and smoothed lines along a selection of z^*-lines, plotted as amplitudes (arbitrary units) and phases (in degrees).

as to avoid the misleading impressions that sectioning can give (Fig. 15). Also, if an interpretation can be made about subunit structure, the model can be taken apart to demonstrate this. The disadvantage inherent in this type of model is that all the information about the different densities within the envelope is lost and this can often be very important information in interpreting molecular and subunit boundaries.

Although computer graphic facilities are rapidly becoming more sophisticated and more widely available, they are probably not a great deal of help in displaying and interpreting low resolution maps such as these. Unless an accurate molecular model can be displayed and manipulated, as with fitting a polypeptide chain to high resolution X-ray protein maps, there is no compensating advantage for the loss of a three dimensional representation such as a stack of contoured sections provides.

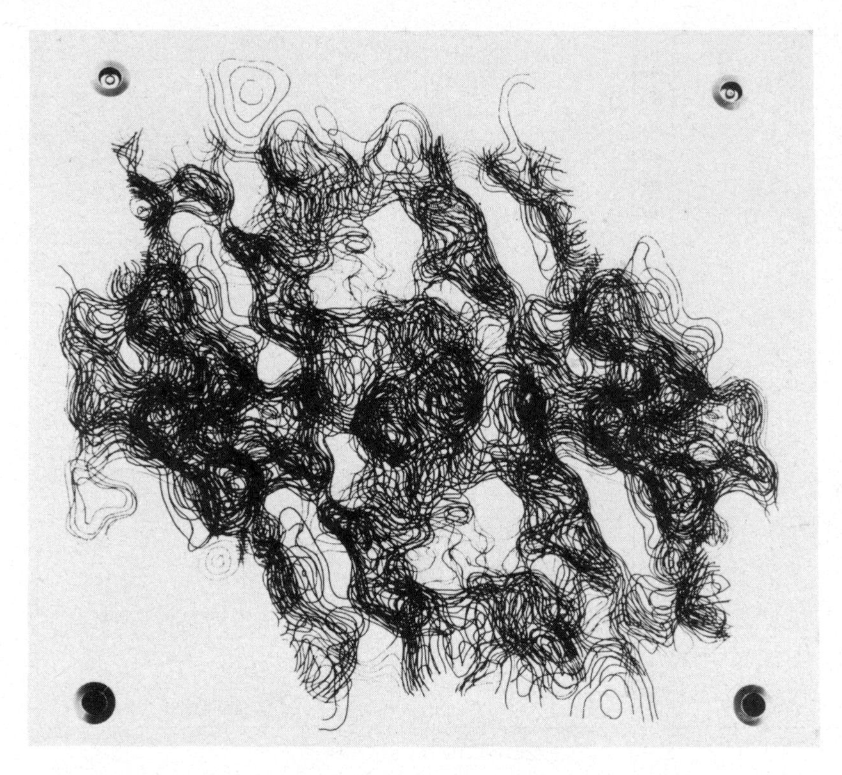

Fig. 14. Photograph of the stack of contoured sections through the three-dimensional reconstruction of the Lobomonas cell wall traced on to 'perspex' sheets. Contours at equal, arbitrary intervals with some of the most negative levels omitted.

The sort of information that can be derived from a three-dimensional reconstruction depends, of course, on the specimen preparation technique, and the resulting resolution and also on the nature of the specimen itself. A recently published symposium [48] illustrates well the wide range of specimens which is now being studied, and the sort of results which are emerging. In the study of unstained bacteriorhodopsin it was possible [9] to interpret a map at 0.8 nm resolution in terms of the protein secondary structure. Structural determinations of proteins by X-ray methods have shown, however, that secondary structure cannot generally be unambiguously delineated until the resolution approaches 0.3 to 0.4 nm. With negatively stained specimens no information about the interior of the macromolecules is present, and the reconstruction gives a three-dimensional envelope of the molecule at about 2.0 nm resolution. Even at this resolution, however, two-dimensional projections can be very misleading and a three-dimensional reconstruction gives much more reliable information about the molecular shapes. For example, in the study of Lobomonas, we interpreted the structure from the two-dimensional work as containing two dimeric structural units with fibrillar interconnections (see Fig. 5). The three-dimensional picture gives quite a different interpretation. We still find two major structural units, but they are much more intricate than they appear in

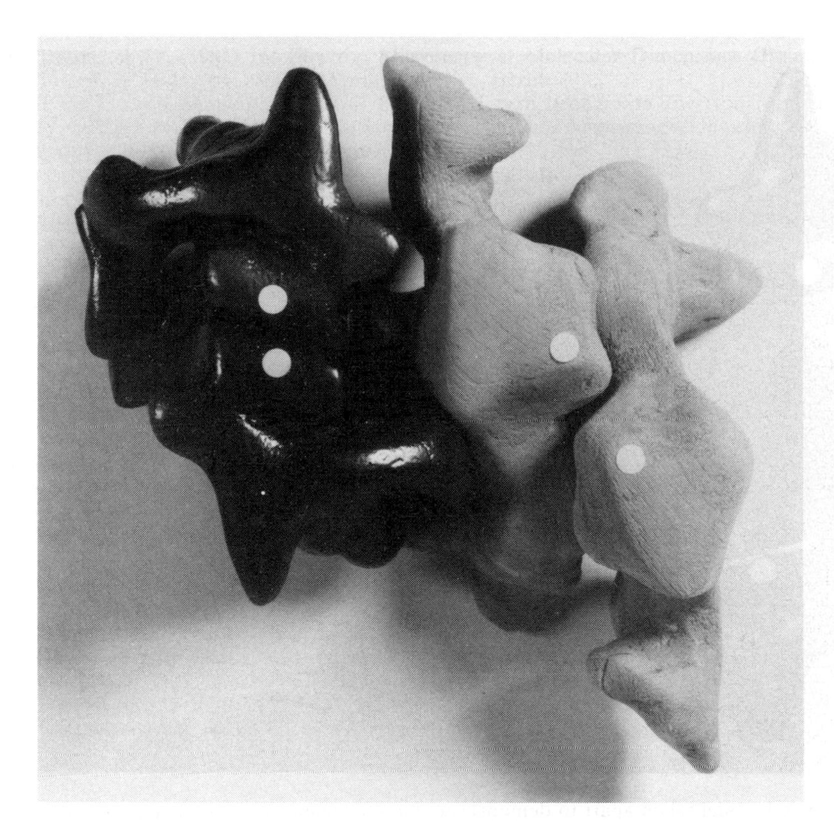

Fig. 15. Balsa wood model of the density envelope of the three-dimensional reconstruction taken at the lowest contour level shown in Fig. 14. The model has been smoothed to remove the misleading impression contour levels can give.

projection and the delineation of their boundaries is not as it appears in two-dimensions. The 'fibrillar' regions of the projections are not, in fact, real features of the three-dimensional structure, but are formed by various pieces of the somewhat elongated structural units overlapping in projection. Thus for any serious study of molecular structure, even by negative staining, three-dimensional analysis seems essential. The higher the resolution, the more, proportionately, is information lost by projection. If the current hopes for extending the scope of unstained specimen preparation techniques are realised we may hope to see many more structure determinations at a resolution of less than 1.0 nm. Three-dimensional structure determination would seem to be essential at these resolutions.

REFERENCES

1. Klug, A. (1978) Chemica Scripta 14, 245–256.
2. Misell, D.L. (1978) Image Analysis, Enhancement and Interpretation in Practical Methods in Electron Microscopy (Glauert, A.M., Ed.) Vol. 7, Elsevier, Amsterdam.
3. Akey, C.W., Moffat, K., Wharton, D.C. and Edelstein, S.J. (1989) J. Mol. Biol. 136, 19–43.

x,y denote floating point parameters.
c denotes a character string parameter.
image, imagein and imageout denote image format files.
Imageft denotes an image format file containing a Fourier transform
spotlist, splotlist1 etc. denote formatted ASCII files containing index list data.
listfile is a formatted ASCII file containing output data.

SCAN image
Operates microdensitometer to produce digitised image, file name *image*.
Options:
/LENGTH:n number of samples in scan line.
/WIDTH:n number of scan lines in image
/RASTER:n raster size in μ (multiples of 12 μm).
TVSCAN image identical to *SCAN*, but image is displayed on Tektronix 4014 as it is being scanned

INTERPOLATE imageout = imagein
Interpolates image on to new sample raster.
Options:
/XANGLE:x angle of new *X* axis with respect to old *X* axis in degrees – default 90°,
/YANGLE:x angle of new *Y* axis with respect to old *X* axis – default 0°.
/SAMPLES:n number of pixels along new *X* axis.
/YSAMPLES:n number of pixels along new *Y* axis.
/TRIM trim to maximum parallelogram contained in old image – otherwise new image is padded out with zeros.
/EQUALISE equalise *X* and *Y* sampling.
/MATRIX:c take interpolation matrix from ASCII file named *c*.

TRIM imageout = imagein
Extracts a rectangular area from *imagein*.
Options:
/XLOW:n
/XHIGH:n
/YLOW:n
/YHIGH:n
Limits of required rectangle in *X* and *Y*.

FLOAT imageout = imagein
Fits a plane to image by least squares to eliminate contrast gradients.
Options:
/CUT:x replace any optical density values greater than *X* by the average OD (to eliminate dirt, etc., from image).

FFT imageft = image
Calculates 2D Fast Fourier Transform in either forward or reverse direction according to whether the contents of the file are an image or a transform.

Options:
NONE

SHADE image or imageft
Produces grey scale picture of image or Fourier transform on Versatec 1200A.
Options:

/LEVELS:n	number of grey levels – up to 20.
/Bits:n	number of bits/pixel (between 2 and 8) e.g.,
/BITS:5	would represent each pixel as a 5 × 5 box with the number of bits set as black in proportion to the grey level.
/NOEXP	No expansion of Fourier transform. Only half the transform is stored, since it has Hermitian symmetry. It is normally expanded on output to give the full transform.
/REF	reflect the image
/NEG	contrast reverse the image
/FACTOR:x	multiply grey level thresholds by *x*. Normally grey levels are determined by dividing the interval between the maximum and minimum value in the image into the required number of intervals.

CONTOUR image
Produces contour plot of image (or a series of plots for each section of a three-dimensional image).

/LEVELS:n	number of contour levels – up to 32.
/ZERO:n	take zero to be level *n* – default is number of levels/2.
/NONEG	no negative levels to be plotted, otherwise negative levels are plotted as dotted lines.
/SCALE:x	Geometrical scale factor for size of plot.
/GRID:n	*n* grid lines across plot in each direction – default none.
/REF	reflect image
/NEG	reverse contrast

SKETCH image or imageft
Produces grey scale picture of image of Fourier transform on Tektronix 4014 graphics terminal.
Options:

/REF	
/NEG	as for shade
/FACTOR:z	

REFINE imageft
Refines reciprocal lattice vectors by interactively centreing spots in the transform.
Options:
Program requests approximate starting values for vectors and size of search window.

SPTLST imageft

N COLUMNS WIDE.

produces printed listing of windows of the transform around each reciprocal lattice point.
Options:
Program requests parameters such as reciprocal lattice vectors, window size, etc.

SPTRPR listfile = imageft
Refines reciprocal lattice vectors by profile fitting method, searching the whole transform to a specified resolution cut-off for significant spots.
Options:
Program requests starting values for lattice vectors, resolution cut-off, etc.

PROFILE spotlist = imageft
Extracts list of indices, amplitudes and phases for spots in transform. Either using spot profile analysis or by taking a weighted average phase, and integrated intensity, as requested.
Options:
Program requests values for reciprocal lattice vectors, etc.

BLDFT imageft = spotlist
Builds 2D list of indices amplitudes and phases into image format file in preparation for *FFT*.
Options:

/NACELL:n	number of unit cells in a direction in reconstructed image.
/NBCELL:n	number of unit cells in b direction in reconstructed image
/NASAMPLE:n	number of pixels/unit cell in *a* direction.
/NBSAMPLE:n	number of pixels/unit cell in *b* direction.

ORIGIN listfile = spotlist
Finds best two-fold origin for image and lists two-fold symmetrised phases along with the indices, amplitudes and measured phases.
Options:

/STEP:n	step size for phase origin search in degrees shifted for an index of 1 – default 3°.
/NSTEP:n	number of steps searched either side of starting origin – default 30.
/ORX:x	starting origin in *x* in fractional unit cells – default 0.0
/ORY:x	starting origin in *y* in fractional unit cells – default 0.0.

SCALEN
Correlates phase origins of several *spotlist* files, and produces list of average phases and average amplitudes. Program requests file names.

ORD3DIM @filename
Carries out 3D phase origin correlation of a new *spotlist* file with previously refined *spotlist* files. *Filename* is the name of a file containing the parameters required by *OR3DIM* in the following form:

listfile/key1:x/key2:y.........
spotlist1/keya:x/keyb:y.........
spotlist2/keya:x/keyb:y.........
.
.
.
spotlistn/keya:x/keyb:y.........

where *listfile* is the listing file for the output.

key1, key2, etc., are as follows:

/ASTAR:x	a^*value
/BSTAR:x	b^* value
/GAMMASTAR:x	γ^*value
/BIGSTEP:x	starting step size in degrees for origin search
/LITTLESTEP:x	final step size in degrees for origin search
/ZRANGE:x	range of z^* (in the same units as a^* and b^*) for phase and amplitude comparison.

spotlist1, spotlist2, etc., are files of spot data. The first file is refined against all the subsequent files.
keya, keyb, etc., are as follows, for each file.

/AZIMUTH:x	tilt axis for film.
/TILT:x	tilt angles for film
/SCALE:x	amplitude scale factor for film
/ORX:z	origin in x for film.
/ORY:x	origin in y for film.

AZIMUTH and *TILT* must be present for each data set. *SCALE, ORX* and *ORY* must be present for a data set used to refine against, but are ignored and may be omitted for the refined data set (i.e., the first).

ME3DIM @filename
Applies amplitude scale factors and phase origins to all data sets, and produces a merged list of h,k,z^*, amplitude and phase, sorted on h,k and z^* with z^* least significant (i.e., sorted into z^* lines).
Filename is a parameter file in the same format as that for *OR3DIM*.

ZLINE
Carries out smoothing of z^* lines and produces interpolated z^* lines. Plots each line together with its data points. Program requests filenames and parameters.

F3D imageout = spotlist
Calculates 3D Fourier transform from 3D list of indices, amplitudes and phases.
Options:
Additional parameters – such as which z sections required – are requested by the program at the operator's terminal.

Numerical analysis

Geisow & Barrett (eds.) Computing in biological science

9

Computer analysis of transient kinetic data

G. MARIUS CLORE

1. INTRODUCTION

Chemical kinetics deal with the study of the mechanism whereby a system in one state is converted into another state, and the rate at which the processes involved take place. Thus kinetics provides a dynamic picture of a system in contrast to the static one of thermodynamics which concerns itself purely with initial and final states.

The mechanism of an enzyme reaction may be defined as the description of the reaction pathway in terms of the number of intermediates, their sequence of formation and degradation, and the values of the rate constants for their interconversion. To achieve a complete understanding of an enzyme system in terms of an explanation for the catalytic process and of the profile of forces in it, the dynamic properties of the system must be related to all the static structural information that can be obtained. However, the biological significance of any system is its dynamic behaviour so that in the description of biological systems, kinetics can often be considered as an end in itself. Nevertheless, the use of rapid spectroscopic techniques coupled to rapid reaction techniques enables one to obtain structural as well as purely kinetic information.

Abbreviations used: ODE, ordinary differential equation; PDE, partial differential equation; SD, standard deviation; SD_{ln}, standard deviation of the natural logarithm of an optimized parameter; EPR, electron paramagnetic resonance.

Essentially two approaches have been used in the study of enzyme kinetics: the steady state and transient state approaches. Although there has been and continues to be a tremendous amount of work in the literature using the steady state approach (see [1,2] for recent reviews), it must be emphasised that steady state kinetics produce only extremely limited information about enzyme mechanisms. The reason for this is three-fold:

(1) The only parameter of the reaction that can be monitored is the overall rate of substrate utilisation and/or product formation because the steady state is defined by the constancy of the concentrations of the intermediates.

(2) Steady state kinetics only give information about the intermediate which precedes the rate limiting step. Often one step is clearly rate limiting and all the enzyme is in essentially one form in the steady state. Because very small concentrations of enzyme (10^{-7} to 10^{-10} M) must be used in order to maintain the steady state for a prolonged period and to reduce the rate of substrate utilisation and product formation to a measurable rate (owing to the relationship $V_{max} = k_{cat}E_0$ where V_{max} is the maximum velocity of the steady state for a concentration of E_0 of enzyme with a turnover number of k_{cat}), even the predominant steady state intermediate is not susceptible to investigation by spectroscopic techniques (because its concentration is below the limit of detectability).

(3) Although exact steady state rate laws are easy to derive, the kinetic constants extracted from the steady state data are complicated functions of individual rate constants (see [3] for a striking example of this), so that information about the rates of individual steps cannot be obtained for any realistic mechanism.

The principal disadvantage of the steady state approach, (namely the absence of any direct information about the states occurring between the binding of substrate(s) and the release of products), can be overcome by employing higher concentrations of enzyme (10^{-6} to 10^{-3} M), thus enabling one to detect individual intermediates and measure their rates of formation and degradation. Under these conditions, enzyme reactions usually occur very rapidly so that special experimental techniques must be employed: e.g., stopped flow, temperature jump, cryoenzymology, rapid spectroscopic techniques (see [4–10] for reviews). Further, under most circumstances the solution of the kinetic equations can no longer be obtained analytically due to their non-linearity, so that special numerical techniques must be employed for data analysis.

The general approach used in the analysis of transient kinetic data to elucidate reaction pathways involves comparing experimental kinetic data with hypothetical kinetic mechanisms. If the experimental data are sufficiently extensive and accurate, it is generally the case that most of the hypothetical mechanisms can be rejected, and the number of possibilities can be narrowed down to one under most circumstances. This model, however, can only be considered as the minimum mechanism. This is because, unlike in X-ray crystallography, for example, there is no unique solution in kinetics, such that further data using other techniques may lead to the discovery of further intermediates, branching pathways, and so on.

In order to compare experimental progress curves with the behaviour of kinetic models, one must integrate the coupled simultaneous ordinary differential equations (ODE's) describing a particular model. Unless every elementary process in the mechanism is first order in a single reactant species, the ODE's cannot be solved analytically. Further, for any mechanism involving more than two steps, analytical solutions (where they can be obtained) are difficult and tedious to derive. One must therefore resort to numerical techniques to solve the coupled ODE's. Because the numerical solution of the ODE's encountered in enzyme kinetics is technically difficult, owing to a problem known as stiffness (see Section 2.1.4), and involves the use of large scale digital computers, it is only in recent years that extensive use of this approach has been made [11–19].

This chapter deals with the principles and methodology involved in the complete analysis of transient kinetic data which essentially encompasses three areas:

(1) Numerical integration of large sets of coupled ODE's.
(2) Non-linear optimization of unknown parameters in the kinetic model (e.g., rate constants, extinction coefficients, activation energies).
(3) Error analysis evaluation which involves the process of decision making in respect to acceptability or non-acceptability of a given kinetic model to account for the experimental data.

The importance of incorporating all three areas into the analysis of transient kinetic data cannot be over-emphasized as it is only by using such an approach that the mechanisms of complex reaction systems can be unravelled clearly and unambiguously. Given the computational complexity of such an approach it is essential to have a software package incorporating both sophisticated numerical techniques and a problem oriented high level language making user programming of complex problems both easy and highly efficient. In this respect particular emphasis will be placed on the FACSIMILE package [20] which incorporates all these facilities.

Section 2 deals with the problems and methodology of numerical integration; Section 3 with non-linear optimization; Section 4 with numerical software (placing particular emphasis on the FACSIMILE package); and Section 5 with the criteria for adequacy of fit of a model to a set of experimental data. In Section 6 the general approach and philosophy towards the analysis of transient kinetic data is discussed and illustrated by examples of the computer analysis of transient kinetic data.

2. NUMERICAL INTEGRATION OF ODE's IN CHEMICAL KINETICS

The ODE's describing the kinetic behaviour of a chemical reaction system can be written in the general form (using vector notation)

$$d\mathbf{y}/dt = \mathbf{f}(\mathbf{y}), \quad \mathbf{y}(0) = \mathbf{y}_0 \tag{2.1}$$

where $\mathbf{y}$ and $\mathbf{f}$ are in dimensional vectors. y_i is the concentration of the ith chemical species; f_i is a function containing terms representing the fluxes of the various reactions which affect the concentration of y_i. The expression for a reaction flux, relation to the time scale T of evolution of the solution. In other words, the non-dimensional product $T\,Re(-\lambda)$ is large for this eigenvalue so that perturbations of the solution of Eq. (2.1) in the direction of the corresponding eigenvector would be extremely rapidly damped in relation to T.

In terms of a chemical reaction system, stiffness may arise in one of three ways:

(1) One or more of the rate constants may have very much larger values than those of the others.

(2) The interacting chemical species may be present in concentrations differing by orders of magnitude.

(3) Certain of the species may be subject to high flux concentrations approximate to an equilibrium during the time course of the reaction. For example, consider the equation

$$\mathrm{d}y/\mathrm{d}t = p - ry \tag{2.7}$$

where $\mathrm{d}y/\mathrm{d}t$ is the difference between two nearly equal terms, a concentration independent production term p and a removal term ry which is nearly proportional to the concentration of y. Under these conditions $\mathbf{J}$ will have an eigenvalue close to $-r$ while T may be of the order $|y/(\mathrm{d}y/\mathrm{d}t)|$, and we have

$$-\lambda T = |ry/(p - ry)| \gg 1 \tag{2.8}$$

Classical explicit forward difference methods (e.g., Runge–Kutta) for the numerical solution of Eq. (2.1) either fail dismally due to their intrinsic instability, or at best are extremely inefficient in the stiff case, because the step size h is limited by $|\lambda h| \leqslant 1$ in order to maintain stability. In many systems of interest at least one time constant is of the order of 10 μs. To maintain stability, h must therefore be equal to or less than 10 μs so that the right hand side of Eq. (2.1) would then have to be evaluated 10^5 times for a value of T of 10 s.

Prior to the pioneering work of Gear [22] in 1967, stiffness presented a serious problem to the numerical solution of the initial value problem defined by Eq. (2.1). However, today, largely because of the introduction of implicit backward difference methods, chiefly developed by Gear [22–25], even extremely stiff problems, both large (> 1000 ODE's) and small are routinely solved. In these implicit methods the solution vector $\mathbf{y}_r$ at the end of the integration step r is obtained by solving one or more set of equations, whose Jacobian matrix is related to $\mathbf{J}$. Newton iteration is necessary in solving these generally non-linear equations in order to avoid re-introducing the severe step size limitation of methods applicable only to non-stiff problems. In general, by using an implicit method especially developed for stiff problems the number of times the right hand side of Eq. (2.1) has to be evaluated is reduced by a factor as large as 10^6 or even 10^8 relative to methods for non-stiff problems. A large study [26] carried out using the FACSIMILE program [20] indicates that an enormous reduction in computing time of the order of 10^{15} may be possible in certain cases using special methods for stiff problems while still retaining control of the error in each variable. It should also be noted that the methods developed for stiff problems are also applicable to non-stiff problems.

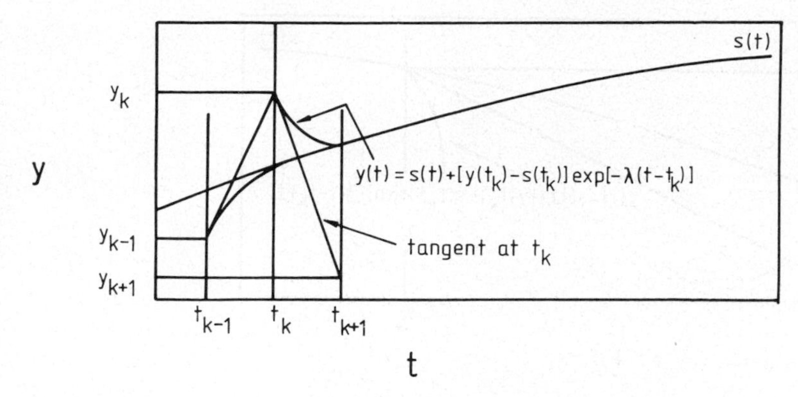

Fig. 1. Behaviour of the explicit first order forward Euler point slope.

2.2. *Stable methods for stiff ODE's*

To illustrate the difficulties encountered in the numerical solution of stiff ODE's and the way these may be overcome, we consider the simple linear ODE

$$\frac{\mathrm{d}y}{\mathrm{d}t} + \lambda y = \frac{\mathrm{d}s}{\mathrm{d}t} + \lambda s(t) \tag{2.9}$$

where λ corresponds to an eigenvalue of $\mathbf{J}$, and the required solution is the known function $s(t)$. If $s(t)$ changes slowly in a time δt of order $1/\lambda$, Eq. (2.9) is stiff. The most general solution of Eq. (2.9) is:

$$y(t) = s(t) + c\,\exp(-\lambda t) \tag{2.10}$$

For the initial value problem $y(0) = s(0)$, we have $c = 0$; round-off errors in the numerical integration, however, make c non-zero.

2.2.1. *Behaviour of classical methods*

All classical higher order methods (e.g., Runge–Kutta, Burlisch–Stoer, Adams methods) behave analogously to the *explicit* first order *forward* Euler point slope method, in which the value of y at time $t_{k+1} = t_k + h$ is defined by

$$y(t_{k+1}) = y(t_k) + h\,\frac{\mathrm{d}y(t_k)}{\mathrm{d}t} \tag{2.11}$$

whose behaviour is illustrated in Fig. 1. The value of the slope supplied to the forward integration equation (2.11) is that of the rapidly decaying perturbation; this means that the error overshoots. The error increases in magnitude and oscillates in sign from one step to the next unless $\lambda h \leqslant 1$. If the latter condition is satisfied, h will be so small that many steps will be needed to cover a range over which the required solution $s(t)$ varies considerably. Thus these methods are only applicable to non-stiff problems where λ has only a small negative real part.

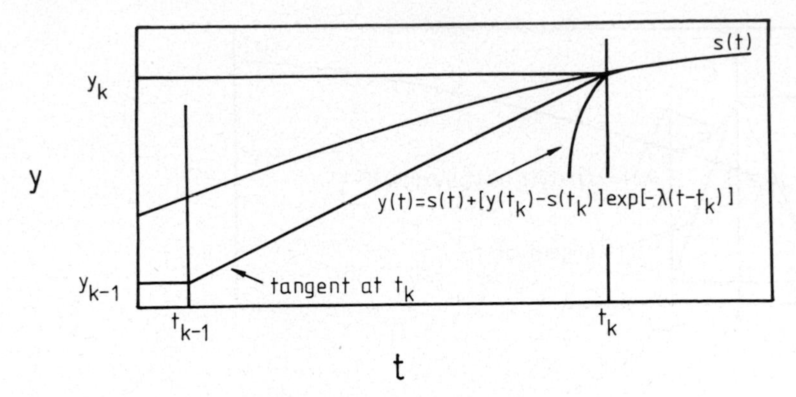

Fig. 2. Behaviour of the implicit first order backward Euler method.

2.2.2. *Behaviour of methods developed for stiff problems*

Implicit higher order *backward* difference methods [22–25,27,28] employed for stiff problems behave analogously to the *implicit* first order backward Euler method defined by

$$y(t_{k+1}) = y(t_k) + h\,\frac{\mathrm{d}y(t_{k+1})}{\mathrm{d}t} \tag{2.12}$$

Eq. (2.12) is implicit because the unknown value of y at t_{k+1} occurs implicitly on the right hand side through its derivative $\mathrm{d}y/\mathrm{d}t$. The behaviour of Eq. (2.12) is illustrated in Fig. 2 and is stable for any value of $\lambda h > 0$.

2.2.3. *The numerical method for stiff ODE's*

In applying Eq. (2.12) to solve Eq. (2.9) we may write

$$(1 + \lambda h)y(t_{k+1}) = y(t_k) + h\,[\frac{\mathrm{d}s(t_{k+1})}{\mathrm{d}t} + \lambda s(t_{k+1})] \tag{2.13}$$

Eq. (2.13) may clearly be solved immediately. In real simulation problems, in place of the simple linear ODE (2.9) in one variable, we usually have to solve a system with two additional complications:

(1) the equations are non-linear, that is to say λ depends on y; and
(2) more than one dependent variable is involved.

The first factor imposes on Eq. (2.12) a method of solution appropriate to a non-linear system; in general, an iterative method. The second factor means we must regard y as a vector of variables, and λ as a matrix.

It turns out that in linearising the non-linear vector equation

$$\mathbf{y}(t_{k+1}) - h\mathbf{f}(\mathbf{y}(t_{k+1})) = \mathbf{y}(t_k) \tag{2.14}$$

which results from applying Eq. (2.12) to (2.1), a central part is played by the matrix $(\mathbf{I} - h\mathbf{J})$ where $\mathbf{I}$ is the identity matrix, h the step length and $\mathbf{J}$ the Jacobian matrix of the system. As stated in Section 2.1.2, the Jacobian matrix changes relatively slowly in practical situations. This is fortunate, since one could not afford to recalculate the matrix $(\mathbf{I} - h\mathbf{J})$ for every iteration needed to solve Eq. (2.14). In practice one recalculates $\mathbf{J}$ automatically, when the iterations of the solution procedure for Eq. (2.14) fail to converge.

A further useful property of the Jacobian matrix is its sparseness. A not uncommon experience is that on average only about five elements per row are non-zero, and since the number of variables may easily be of the order of 100 in practical problems (see Section 6), it is of great importance to be able to exploit its sparseness. By doing so, one can economise not only on the amount of computer storage needed, but also on the computer time taken in operations with the matrix.

2.2.4. *Higher order methods*

So far we have been talking about the use of the backward Euler method defined by Eq. (2.12). This is only a first order method, that is to say the error committed on a step of length h is of the order h^2. Fortunately Gear [22–25] was able to find higher order methods which have the same general properties. A typical Gear method of order $q \geqslant 1$ is defined by

$$\mathbf{y}(t_{k+1}) = \sum_{j=1}^{q} a_{qj}\mathbf{y}(t_{k+1-j}) + G_q h \frac{\mathrm{d}\mathbf{y}(t_{k+1})}{\mathrm{d}t} \tag{2.15}$$

where the numerical coefficients a_{qj} and G_q are chosen to make the method of order q, that is the error on a step length h is of the order of h^{q+1}. For $q = 1$ this is identical with Eq. (2.12) since then $a_{11} = 1 = G_1$. Gear proposed methods of order $q = 1$ to 6, but the stability properties for $q = 6$ leave something to be desired, and experience with biochemical systems has shown that it is adequate to restrict the methods to $q \leqslant 4$. By analogy with Eq. (2.14), to solve Eq. (2.15) requires an iterative approach using the matrix $(\mathbf{I} - G_q h\mathbf{J})$, where previously we had $G_q = 1$. A number of refinements over Gear's original method have been developed by Curtis [28]. First, the matrix $\mathbf{J}$ is stored as well as the decomposed form of the matrix $(I - G_q h\mathbf{J})$; this has the advantage that when h or q is changed it is not necessary to recompute $\mathbf{J}$. Second, the rate of convergence of the iterations for $\mathbf{y}(t_{k+1})$ is monitored so that it is possible to predict when the next correction will fall below the preset limit, and thus not waste time in calculating it; this saves nearly one whole iteration per step on average. Third, in order to obtain values of the vector $\mathbf{y}$ of variables at a specified time, which may be required for printed or graphical output, it is not necessary in Curtis' method [28] to take a time step terminating at that time; instead an interpolation process within a step is used. Integration is performed until the output point is passed, and then interpolation is carried out within the previous step to get the values at the required time. Finally, the procedure for starting a range of integration, or for restarting after some discontinuous change has been made in the system, has been greatly improved. By differentiating Eq. (2.1) with respect to time, we obtain equations which can be written in vector notation

$$\frac{d^2\mathbf{y}}{dt^2} = \mathbf{J}\,\frac{d\mathbf{y}}{dt} \tag{2.16}$$

and since in any case one needs to compute the matrix **J**, Eq. (2.16) is used to estimate the truncation error in starting with $q = 1$. By this means, one is able to choose an acceptable step length h, and experience shows that this avoids wasting much computer time in finding the best value for h for starting purposes.

2.2.5. *Behaviour of Curtis' method* [28]

Generally speaking, biochemical systems exhibit a small extremely rapid initial transient of no particular interest; this is due to slightly inaccurate choice of initial values. This transient is followed faithfully by the integration method. During this process, the order increases rapidly from $q = 1$ to about $q = 4$, under control of an order choosing strategy built into Curtis' method [28]. At the same time, h increases even more rapidly, often by several orders of magnitude as the transient dies away. Later h continues to increase while q may fall off again as the system approaches a steady state, when a lower order integration method is adequate. In our experience h can easily reach values of the order of 10^6 times the shortest time constant in the system, so that quite a moderate number of steps are able to advance the integration a long way. In general, therefore, much of the work is done following the initial transient, and for this reason we have found that the computer time taken is not critically dependent on the total time for which the system has to be simulated; it depends much more strongly on how many initial transients are caused by discontinuous changes imposed upon it.

3. NON-LINEAR OPTIMIZATION*

The analysis of a complex kinetic problem requires one not only to be able to integrate the ODE's describing the system but also to optimize the parameters in the ODE's in order to fit a particular kinetic model to a set of experimental data. The parameters to be optimized in kinetic problems are generally rate constants, activation energies and extinction coefficients. Because, in general, the parameters occur non-linearly in the ODE's describing a chemical system, we require non-linear methods of optimization. It should also be noted that non-linear optimization is required not only to determine parameters in kinetic systems but also in many other systems of biological and chemical interest (e.g., to determine binding constants in protein–ligand equilibria; parameters of the Gaussian bands in complex circular dichroism spectra; spin-lattice relaxation rates, rate constants and equilibrium constants by lineshape analysis of dynamic nuclear magnetic resonance spectra).

---

* The words optimize and minimize are regarded as synonymous. However, we generally employ the term minimize to the function for which a minimum is sought, and the term optimize to the parameters in the function (i.e., the parameters are optimized in order to minimize the function).

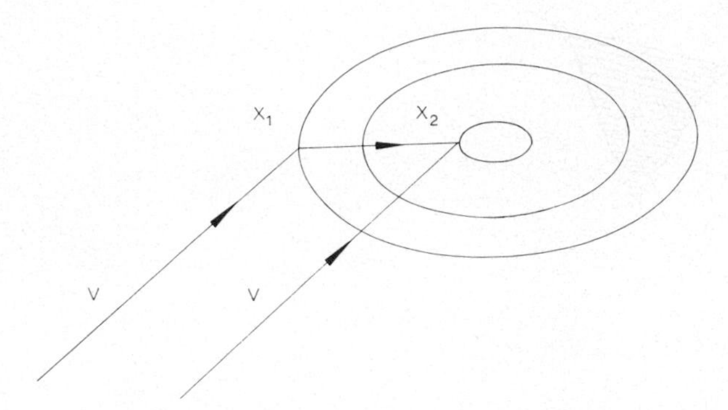

Fig. 3. The principle of Powell's method of optimization which doesn't employ partial derivatives. The method is based on the observation that if the minimum of a positive definite quadratic form is sought in the direction $\mathbf{v}$ from each of two distinct points, then the vector joining the resulting minima $\mathbf{x}_1$ and $\mathbf{x}_2$ is conjugate to $\mathbf{v}$.

3.1. *Methods of unconstrained non-linear optimization*

The problem of unconstrained minimization can be stated as follows:

$$\text{Minimize } F(\mathbf{x}) \quad \text{for} \quad \mathbf{x} \in R^n \tag{3.1}$$

where $F(\mathbf{x})$ is a non-linear function in n parameters, $\mathbf{x}$ is an n-dimensional vector, and R^n is the n-dimensional Elucidean space. A point $\mathbf{x}^*$ is said to be a solution of the unconstrained minimization problem if

$$F(\mathbf{x}^*) < F(\mathbf{x}) \quad \text{for} \quad \mathbf{x} \in R^n \tag{3.2}$$

and $\mathbf{x}^*$ does not have to satisfy any constraints.

Given a function $F(\mathbf{x})$ the task of finding a global minimum is in general a non-trivial problem. The computational difficulties may be classified into three broad categories:

(1) Convergence properties of the method.
(2) Sensitivity of the method to initial guesses in the parameters to be optimized (the less sensitive the better).
(3) Computational cost.

There are essentially two categories of non-linear optimization methods: (1) Direct Search methods, and (2) Descent techniques. The theory and methods behind these techniques will only be discussed briefly here as they are extensively described in a number of books solely devoted to non-linear optimization (the reader is referred to [29] and references therein for further details). These two methods and their relationship to the computational aspects are discussed below.

3.1.1. *Direct search methods*

Direct Search methods are based on the sequential evaluation of the function $F(\mathbf{x})$ using trial values of the parameters to be optimized, which by successive comparisons give an indication for a further searching procedure towards a minimum. These methods do *not* give a rapid rate of convergence and are *inefficient* for finding a minimum with high precision [29].

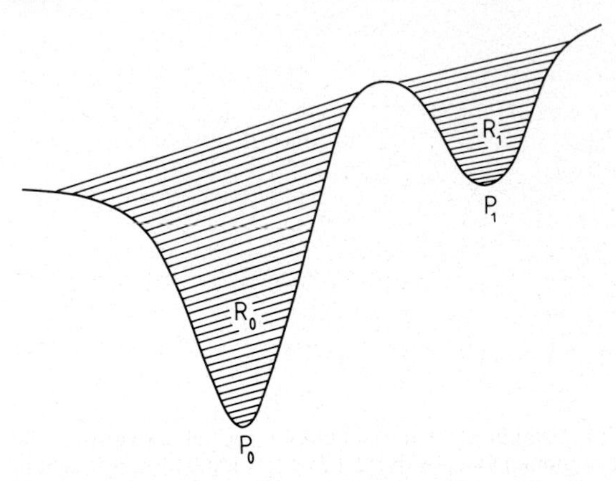

Fig. 4. Global versus local minima. P_0 is the global minimum and P_1 a local minimum. The region R_0 and R_1 are associated with P_0 and P_1 respectively, and do not intersect with each other. Further optimization starting in R_0 and R_1 will end up at P_0 and P_1 respectively.

3.1.2. *Descent techniques*

By using additional information about the function to be minimized, efficiency of optimization is increased. In the Descent techniques, the gradient of $F(\mathbf{x})$, $\nabla F(\mathbf{x},)$ which is the vector normal to the local contour surface is used. The components of $\nabla F(\mathbf{x})$ are the first order partial derivatives of the differentiable function $F(\mathbf{x})$. The most important Descent methods from a computational point of view are the Conjugate Direction methods which assume that in a neighbourhood of the minimum the function $F(\mathbf{x})$ can be closely approximated by a positive definite quadratic form. Examples of the Descent technique include the Newton–Raphson method, the method of steepest descent, the Gauss–Newton procedure and Marquardt's method. These methods all require the explicit calculation of the partial derivatives of $F(\mathbf{x})$.

3.1.3. *Descent methods not requiring the computation of partial derivatives*

As in general, the expression for $\nabla F(\mathbf{x})$ is at least as complicated as that for $F(\mathbf{x})$, this means that effectively $(n + 1)$ function values must be evaluated at each step. Therefore, a method which avoids the calculation of partial derivatives has the possibility of not only being more efficient but also the considerable advantage of being much more convenient to use. Probably the best Descent method which does not use partial derivatives is that of Powell [30–31] which is based on the observation that if the minimum of a positive definite quadratic form is sought in the direction $\mathbf{v}$ from each of two distinct points, then the vector joining the resulting minima is conjugate to $\mathbf{v}$. This is illustrated in Fig. 3. It should also be noted that Powell's method [30–31] appears to be the least sensitive to initial guesses in the parameters to be optimized [32].

3.2. *Choice of function to be minimized*

The function to be minimized, $F(\mathbf{x})$, may be either a general function of the parameters or a sum of squares. In general, it is best to minimize the sum of squares and optimize the residuals at the same time as a number of definite advantages follow. Efficient methods for functions which depend smoothly on the parameters which are being varied, use the function values supplied to them to estimate, either explicitly or implicitly, derivatives with respect to the parameters. The accuracy with which this can be done depends on how nearly linear these functions are. However, the sum of squares is highly non-linear in the region of the optimum as it has a minimum there. The residuals, on the other hand, are more likely to be approximately linear functions of the parameters in the region of the minimum so that numerical differentiation of them will be more accurate. A number of optimization methods need to be provided with both the residuals and the sum of squares, but only Powell's method [30–31] does not require the evaluation of partial derivatives.

3.3. *Global versus local minima*

In certain cases, not only is there a global minimum but also a number of subsidiary local minima such that there is a risk of finding sets of parameter values which, while locally optimum, do not give the best global fit. In terms of the optimization process, the problem of local minima is illustrated in Fig. 4 and may be represented as follows. We consider an N-dimensional space in which the N parameters being optimized are the N coordinates. There will exist a point P_0 which is the global minimum, and a set of points $P_1, P_2 \ldots, P_j$ which are local minima. Associated with each point P_i is a region R_i which has the properties that (1) it does not intersect with any other region R_j, and (2) optimization starting at any region R_i will end up at P_i. Further, there may be another region Z_i with the property that the optimization procedure will diverge when starting from any point within it.

Therefore, if one finds that a particular model which one would expect to fit the data from qualitative considerations yet fails to do so, it is wise to try a series of initial estimates for the parameters being varied, in order to exclude the possibility of local minima.

4. NUMERICAL SOFTWARE

The user of software for solving stiff ODE and non-linear optimization problems may not be a numerical analyst or a mathematician, but rather a chemist, biochemist or physicist interested in solving a particular problem. It therefore follows that any software package of this type must satisfy four requirements:

(1) Reliability.

(2) Efficiency of the numerical methods employed.

In this respect it follows that there are a number of matters which need consideration in software design, but in which the user is not interested. Ideally, the software should be so designed that the user is unaware of these matters; if this is impracticable within the state of the art, care should be taken such as to minimize the demands on the user and to guide him in making choices which the software is unable to make for itself.

(3) Efficiency of user programming.

In this respect the use of a high level user oriented language is vital to the success of any software package.

(4) Broad applicability.

The software package must be sufficiently flexible to enable relatively simple programming (from the user point of view) of a very large number of problems, not necessarily involving the solution of ODE's as in transient kinetics, but also of partial differential equations as found in diffusion problems, and even of complex arithmetic problems (which don't require the use of numerical integration).

At the present time there is only one package, in the author's experience, which adequately fulfils the above four criteria. This is the FACSIMILE program developed by A.R. Curtis and colleagues at Harwell [20]. FACSIMILE is a large computer program for the automatic solution of initial value problems (especially stiff ones) arising as systems of differential equations. It also incorporates an optimization package for minimization problems. In addition, it offers the user a powerful, high level program description language which has been designed for maximum ease of use and programming efficiency.

4.1. The FACSIMILE package

4.1.1. Numerical algorithms

Numerical integration is carried out using Curtis' modified version [28] of Gear's backward difference method [22–25], together with spare matrix handling subroutines, automatic sparsity pattern determination, automatic initial step size selection and automatic error control [20,33,34,35]. (The latter two can be over-written by the user who has the freedom to make his own choice of initial step size and of the tolerance value on the local truncation error.)

FACSIMILE can carry out a single integration run on a set of ODE's with a single set of initial values, a succession of runs with different initial values, or a number of runs with different initial values simultaneously by using its array facilities. As an important extension of this flexibility it can also execute parameter fitting runs by carrying out successive simulation runs under control of a powerful non-linear least squares optimization routine, varying the values of specified parameters to obtain the best fit between calculated and observed values. Optimization is carried out using Powell's method [30–31] which does not require the computation of partial derivatives. Further, by use of array facilities, a large number of curves of experimental data may be fitted simultaneously. Both numerical and graphical output can be obtained on a line printer or terminal; and numerical values of optimized parameters may be stored on disc files to be written out for input to a later run.

4.1.2. The FACSIMILE language

The FACSIMILE language, in which problems presented to the program for solution are described, consists of a sequence of statements of which there are five types:

(i) end-of-list statement consisting of **; which is used to terminate a list of statements forming part of a more complex structure;

TABLE 1

The main six FACSIMILE code blocks.

Code block name	Execution storage
GENERAL	Whenever values of variables change (to compute parameters)
EQUATIONS	Whenever time derivatives of variables are needed.
INITIAL	On a BEGIN command, to set initial values.
RESID	Before computing parameter fitting residuals.
INSTANT	Immediately after compilation (this block is then deleted).
TESTOUT	During WHEN testing, at each trial value.

(ii) comment statements which must begin with * (equivalent to C in FORTRAN);

(iii) declarations which define the characteristics of the variables and parameters to be used in the problem (e.g., dimensions, initial values, etc. ...);

(iv) commands which specify an action to be carried out immediately (some of these introduce a structure of statements terminated by a **;);

(v) instructions, which results in the storing of code for execution at various points during a simulation run; these occur in code blocks introduced by a COMPILE command.

There are fifteen standard FACSIMILE code blocks which are executed at set stages during the running of the program (the main ones, of which there are six, are summarized in Table 1). In addition, the user may name his own code blocks for execution at any particular point(s) he desires.

A generalized flow chart of the FACSIMILE program is given in Fig. 5. Two examples of FACSIMILE programming are given in Figs. 6 and 7 which serve to illustrate the ease with which problems can be coded in the FACSIMILE language.

4.1.3. Status and implementation of FACSIMILE

FACSIMILE is coded in Fortran for IBM 360 and 370 computers and can be compiled by FORTRAN G1, H or H extended (X) compilers. At present, some features are special to IBM FORTRAN so that conversion for a different make of computer is not entirely trivial, but in most cases practicable. In addition there is a version already available for use on the ICL 2900 series computers.

The present version of FACSIMILE contains about 150 subroutines and is about 14,000 card images in length. The program is divided into five files (see Table 2).

There are eight array size parameters whose sizes can be chosen to suit one's requirements. These are given in Table 3 together with the number of extra bytes of memory required if the size of each array is increased by unity. The total memory requirement in kilobytes can be calculated approximately from:

$$\text{Total memory requirement} = \{[122\text{NVARS} + 8(\text{NPARS} + \text{NVARY}) + 10\text{NJAC} + 12\text{NLU} + 32\text{NCV} + 4(\text{NOBS} + \text{NWSLIM})]/1024\} + \text{NPROG}$$

where NPROG has the following values

150 (overlaid, H or X with OPT = 1 or 2)
275 (not overlaid, H or X with OPT = 1 or 2)

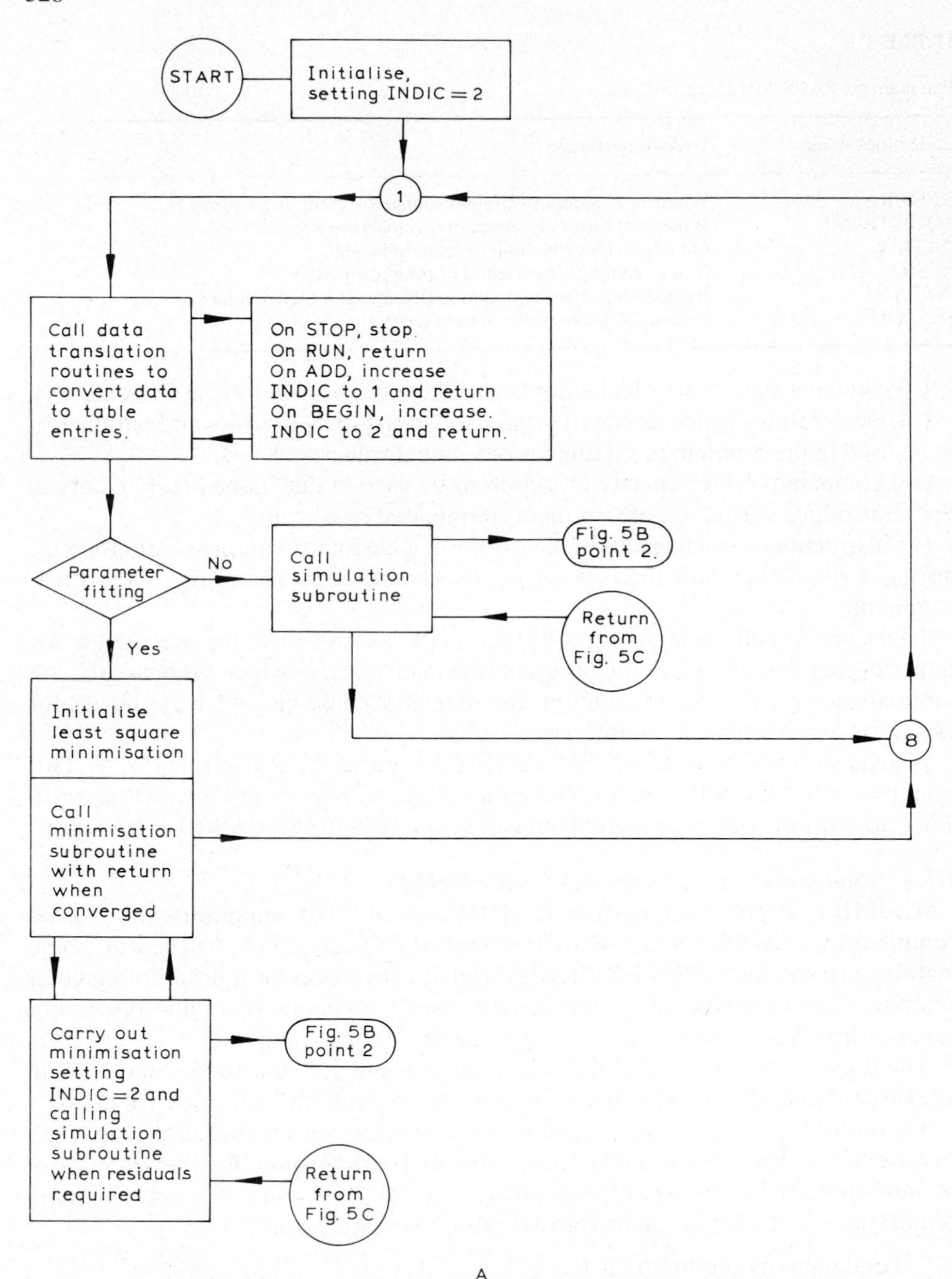

A

Simplified control flowchart for FACSIMILE

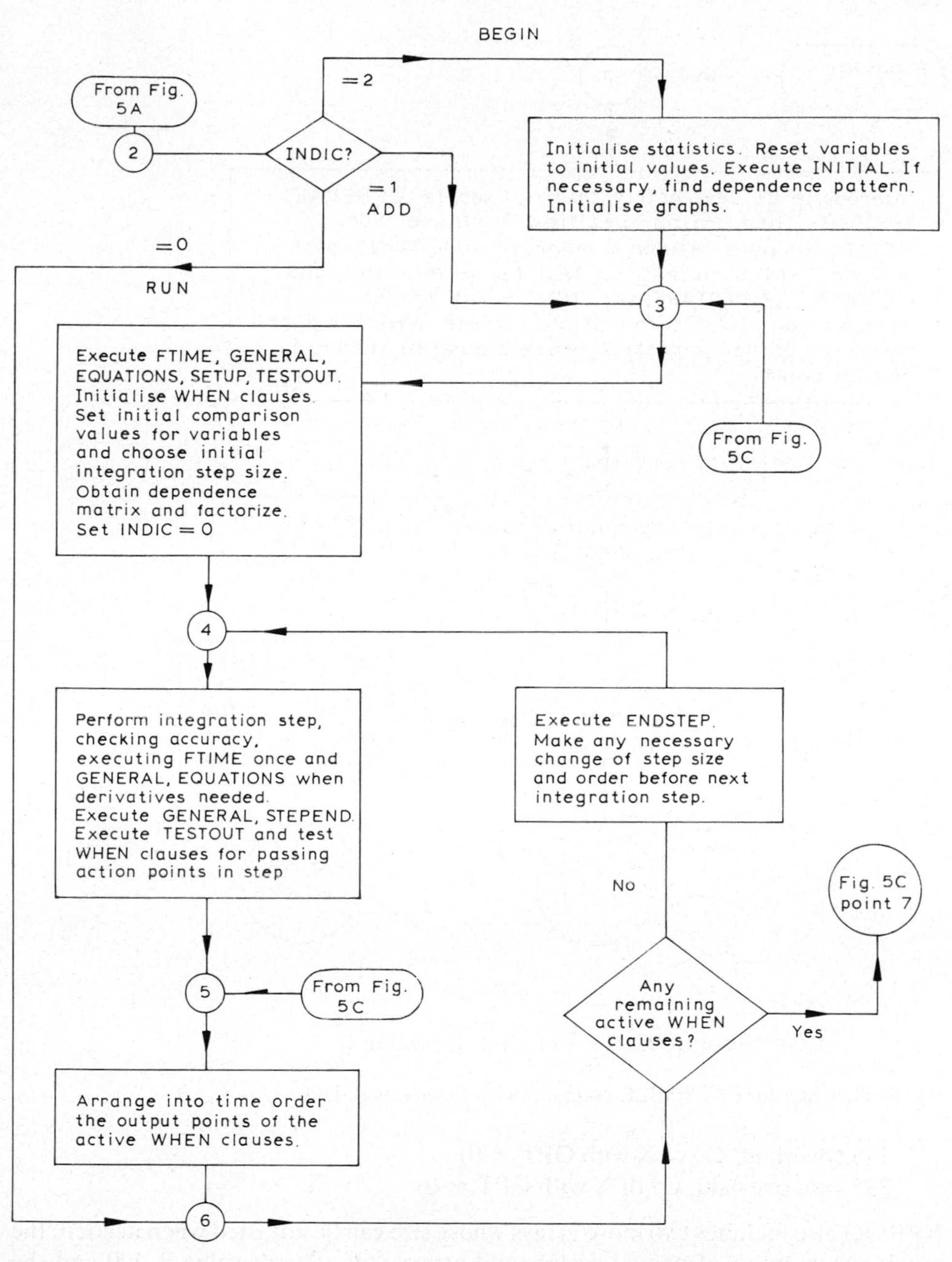
BEGIN
=2
From Fig. 5A
2
INDIC?
=1
ADD
=0
RUN
Initialise statistics. Reset variables to initial values. Execute INITIAL. If necessary, find dependence pattern. Initialise graphs.
3
From Fig. 5C
Execute FTIME, GENERAL, EQUATIONS, SETUP, TESTOUT. Initialise WHEN clauses. Set initial comparison values for variables and choose initial integration step size. Obtain dependence matrix and factorize. Set INDIC = 0
4
Perform integration step, checking accuracy, executing FTIME once and GENERAL, EQUATIONS when derivatives needed. Execute GENERAL, STEPEND. Execute TESTOUT and test WHEN clauses for passing action points in step
Execute ENDSTEP. Make any necessary change of step size and order before next integration step.
No
Fig. 5C point 7
5
From Fig. 5C
Any remaining active WHEN clauses?
Yes
Arrange into time order the output points of the active WHEN clauses.
6

B
Simulation code simplified flowchart

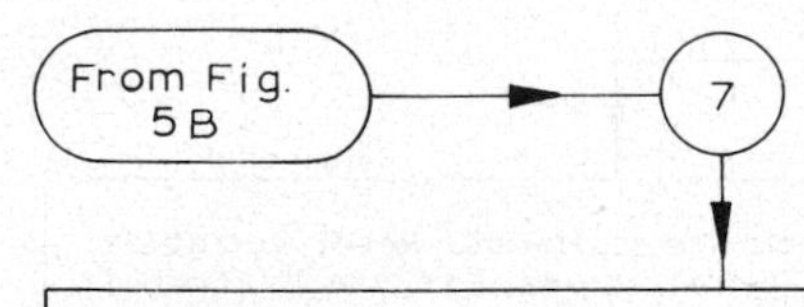

Interpolate at next action point. Execute GENERAL, TESTOUT. Take action specified in clause. For RESID, compute residuals, execute TIMEPOINT, plot and/or print if necessary. Test for termination and RESTART: if RESTART, set INDIC = −1 for no termination, 1 for termination. Delete WHEN if right-hand end of range passed. Move clause to its next action point.

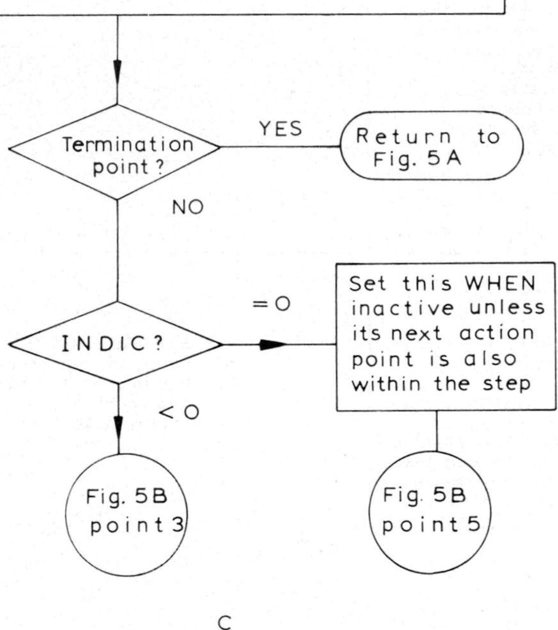

C

WHEN/WHENEVER simplified flowchart

Fig. 5. Flowchart for FACSIMILE program (from Chance et al. [20]).

185 (overlaid, G1 or X with OPT = 0)
335 (not overlaid, G1 or X with OPT = 0)

NPROG also includes two more arrays whose size can be adjusted when needed: the maximum number of named scalars and arrays (the default value is 400 and the increment is 16 bytes per unity); and the maximum number of locations for the pseudo-code into which the user's code blocks are translated (the default value is 6000 and the increment is 3 bytes per unity). Table 4 summarises the memory requirements for a number of combinations of array sizes.

For details of implementation and conversion for other computers the reader is referred to [20,36].

```
FACSIMILE CODE                                  COMMENTS

PERMIT -;

VARIA E 5E-6 S 2ØE-6;                           Variables and their initial
VARIA ES EP P;                                  values defined

PARAM K1 1E3 K-1 1 K2 5;                        Parameters and their values
PARAM K-2 3 K3 1Ø K-3 1 E2;                     defined

COMPILE EQUATIONS;

|K1| K-1 : E + S = ES;                          Chemical equations defined

|K2| K-2 : ES = EP;

|K3| K-3 : EP = E + P;

**;

GSTREAM 1  1  5Ø;                               Defines graphical output
E 5E-6 1;
ES 5E-6 2;
EP 5E-6 3;
S 2ØE-6 S;
P 2ØE-6 P;

**;

PSTREAM 1 2;                                    Defines printed output
TIME;
E ES EP S P;

**;

COMPILE GRAPH;                                  User defined code block
                                                to be called in WHEN Command
GSTREAM 1;
PSTREAM 1;

**;

WHEN TIME = Ø + Ø.Ø1*1ØØ | CALL GRAPH;          Time points at which the block
                                                GRAPH is to be executed

BEGIN;                                          Starts integration

STOP;                                           End of program
```

Fig. 6. FACSIMILE code for the simulation of the simple kinetic scheme

$$E + S \underset{k_{-1}}{\overset{k_1}{\rightleftharpoons}} ES \underset{k_{-2}}{\overset{k_2}{\rightleftharpoons}} EP \underset{k_{-2}}{\overset{k_3}{\rightleftharpoons}} E + P$$

at a single concentration of E and S. Most of the comments are self-explanatory. However, the following require further comment. PERMIT -: allows one to use - symbolically as in K-1 without carrying out a subtraction providing there are no spaces between K,- and 1; 1E3: this is equivalent to 10^3 (similarly 1E-3 would be equivalent to 10^{-3}); GSREAM i j k: i is the identifier, j the FORTRAN output stream and k the width of the graph (full scale on a line printer is 100); scalar name i j in GSTREAM command: i is the range of the dependent variable and j the symbol used in the graphical output; PSTREAM i j; i and j have the same meaning as i and j in GSTREAM command. in WHEN command indicates time at which integration is stopped (in this example l s).

```
FACSIMILE CODE                               COMMENTS

PERMIT -;
EXEC #1 = 2;                                 Set value of #1 index to 2
EXEC #2 = 5*#1;                              Set value of #2 index
PARAM #2 WS;                                 Workspace parameter for array code
VARIA <#1> E 5E-6 5E-6;                      Variables and their initial
VARIA <#1> O2 1ØØE-6 5ØØ-6;                  values defined
VARIA <#1> I  IIA  IIB  III;
PARAM K1 8Ø.9 K-1 4E-3 K2 5.61E-3;           Parameters and their initial
PARAM K3 8.Ø5E-3 K4 9.88E-4;                 values defined
PARAM <#1> A59Ø G2;
PARAM X1 1.71  X2 2.59  X3 1.25;
PARAM EI 5E-6;
EQUIV A590: A59ØA  A59Ø;                     Equivalence array elements
EQUIV  G2:  G2A  G2B;                        to scalar names for plotting

COMPILE EQUATIONS;
ARRAY <#1> WS;
|K1 | K-1: E + O2 = I;
|K2:  I = IIA;                               Chemical equations defined
|K3:  I =  IIB;
|K4: IIA = III;
ARRAY;
**;
COMPILE RESID;
ARRAY <#1> WS;                               Defines A59Ø and G2 which
A59Ø = (X1*I + X2*IIA + X3*III)/EI;          will be compared to observed
G2 = (IIB + III)/EI;                         data
ARRAY;
**;

GSTREAM 1 1 5Ø;                              For graph of observed versus
**;                                          calculated data
GSTREAM 2 2 5Ø;                              For graph of residuals versus
**;                                          time
DATA .Ø2;                                    Data
TIME  A59Ø  A59ØB  G2A  G2B ;
RANGE  1      1      1    1 ;
  Data follows
**;
VARY K1 K-1 K2 K3 K4 X1 X2 X3;               Parameters to be optimized
BEGIN;                                       Starts optimization and
                                             integration
GSTREAM  3  3  5Ø;                           Defines graphical output
A590A    1     A ;                           for simulation
A590B    1     B ;
G2A      1     1 ;
G2B      1     2 ;
**;

COMPILE GRAPH;                               User defined code block to be
CALL RESID;                                  called in WHEN command.  Will
GSTREAM 3;                                   compute values of A59Ø and G2
** ;                                         using optimized values of
                                             parameters.
WHEN TIME = Ø + 5*1ØØ |CALL GRAPH;           Time points at which the block
** ;                                         graph is to be executed

BEGIN;                                       Starts integration using the
                                             optimized values of the parameters
END;                                         End of program
```

5. CRITERIA FOR ADEQUACY OF FIT OF A MODEL TO A SET OF EXPERIMENTAL DATA

Unlike the refinement of structural models using X-ray crystallographic data, where it is usually the case that a unique solution can be obtained, in kinetics no unique solution ever exists. This is because it is always possible to increase the complexity of a given model and still produce an adequate fit of the data on the basis of the SD of the fit. It is therefore essential to have a set of rigorous quantitative criteria on which to base one's choice of model. These have been developed by Clore and Chance [11] and consist of the following triple requirement.

(1) a SD of the fit within the standard error of the data;
(2) good determination of the optimised parameters;
(3) a random distribution of residuals.

This triple requirement greatly decreases the number of models available. Indeed, in initial value problems involving stiff non-linear ODE's it is usually the case that only a single model will satisfy this triple requirement. Thus, for a given set of data, although there may be many models with a SD within the standard error of the data, models with too many degrees of freedom will fail such an analysis because of under-determination, whereas models with too few degrees of freedom will fail such an analysis as a result of the introduction of systematic errors in the distribution of residuals.

5.1. *Estimation of the standard error of the data*

As a first step to data analysis, the standard error of the data should be estimated. There are a number of ways of doing this. One fairly simple method developed by Clore and Chance [11] makes use of orthogonal polynomials and is suitable for implementation on a minicomputer or microprocessor during data acquisition. The method proceeds as follows. For any curve, we observe that over a short length H the curvature is small, and can be considered to approximate to a straight line. So

Fig. 7. FACSIMILE code for fitting the scheme C

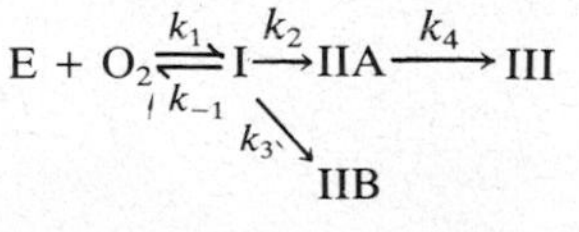

$$E + O_2 \underset{k_{-1}}{\overset{k_1}{\rightleftharpoons}} I \overset{k_2}{\longrightarrow} IIA \overset{k_4}{\longrightarrow} III$$
$$I \overset{k_3}{\searrow} IIB$$

simultaneously to data at two O_2 concentrations. Most of the comments are self-explanatory. However, the following require further comment. The ARRAY instruction switches the mode of interpretation of assignment and reaction instructions in its range so that they act on corresponding elements of arrays, all of the same dimensions specified in this case by # 1 (which has been set to 2); the ARRAY instruction has the same function as the DO loop in FORTRAN but unlike the latter does not require the indexing of each element within its range, thus making the code simpler. (It should also be noted that DO loop facilities exist in FACSIMILE but their syntax is slightly different from that in FORTRAN.) The data: the data is started by a statement DATA i where i specifies the standard error of the data, a statement listing the independent and dependent variables, and a statement giving the range of the dependent variables; the data is then typed in columns i,j,k,l,m, where i is the independent variable and j,k,l,m, the dependent variables.

TABLE 2

Composition of the FACSIMILE package.

File No.	File name	Content
1.	TAILOR	FORTRAN source for the pre-processor TAILOR (approx. 800 card images in length).
2.	FACSE	TAILOR control cards to set up the size of arrays, followed by a generalized COMMON in which arrays are dimensioned symbolically and finishing with the generalized FACSIMILE source containing reference to COMMON which are to be inserted by TAILOR (approx. 12000 card images).
3.	MODFILE	TAILOR data cards supplying editing information, constituting the second part of the data check.
4.	OVERLAY	IBM linkage editor control cards used to produce overlaid modules.
5.	TEST1+	FACSIMILE test problems.

TABLE 3

Array size parameters.

Name of array size parameter	Memory increment per unity (bytes)	Description of effect
NVARS	122	Limit on the total number of variable scalars and array elements (including time).
NPARS	8	Limit on the total number of parameter scalars and elements.
NJAC	10	Limit on the total number of non-zeros in the Jacobian (dependence) matrix **J**.
NLU	12	Limit on the total number of non-zeros in the Gaussian decomposition of the convergence matrix $(\mathbf{I} - G_q h\mathbf{J})$.
NVARY	8	Limit on the number of parameters which can be varied during an optimization run (n_v).
NCV	32	Limit on the number of data curves which can be fitted simultaneously in an optimization run.
NOBS	4	Limit on the number of observations which can be used in an optimization run (n_c).
NWSLIM	4	Limit on workspace requirement. $(r_w = n_v(2n_v + 5) + 2n_0(n_v + 1)$ optimization runs).

consider a regression line fitted to the length H

$$y = ax + b \tag{5.1}$$

where the true regression line would be

$$y = a_0x + b_0 \tag{5.2}$$

The residuals in the length H are

$$r_j = y_j - (ax_j + b) \tag{5.3}$$

for Eq. (5.1), instead of

$$r_{0j} = y_j - (a_0x_j + b_0) \tag{5.4}$$

for Eq. (5.2) as they should be. Assuming that r_{0j} is Gaussian with mean zero and

TABLE 4

Memory requirement of FACSIMILE for various combinations of array size parameter.

NVARS	NPARS	NJAC	NLU	NVARY	NCV	NOBS	NWSLIM	Memory requirement (Kilobytes) Overlaid	Not overlaid
101	2,000	1,000	2,000	30	50	300	2,500	220	345
401	5,000	2,000	3,000	30	400	400	4,500	320	445
801	8,000	7,500	20,000	30	50	300	2,500	630	750
901	10,000	9,000	35,000	1	1	1	1	835	960
1,101	12,000	9,000	20,000	1	1	1	1	700	825

variance σ^2, we construct the polynomials (P_i) orthogonal on the points x_j

$$P_0(x) = 1 \tag{5.5}$$

$$P_1(x) = x - \frac{1}{J}\sum_j x_j \tag{5.6}$$

where J is the number of points in the interval H.

$$\begin{aligned} P_2(x) &= (x - \alpha_2)P_1(x) - \beta_2 P_0(x) \\ &= \sum_j P_1(x_j)P_2(x_j) \\ &= \sum_j x_j P_1^2(x_j) - \alpha_2 \sum_j P_1^2(x_j) \end{aligned} \tag{5.7}$$

$$\therefore \alpha_2 = \sum_j x_j P_1^2(x_j) / \sum_j P_1^2(x_j) \tag{5.8}$$

Now

$$\begin{aligned} 0 &= \sum_j P_0(x_j)P_2(x_j) \\ &= \sum_j P(x_j) + \frac{1}{J}\sum_j (x_j - \alpha_2)P_1(x_j) - \beta_2 \sum_j P_0^2(x_j) \\ &= \sum_j P_1^2(x_j) - J\beta_2 \end{aligned} \tag{5.9}$$

$$\therefore \beta_2 = \frac{1}{J}\sum_j P_1^2(x_j) \tag{5.10}$$

We define

$$\begin{aligned} R &= \sum_j r_j P_2(x_j) \\ &= \sum_j r_{0j} P_2(x_j) \end{aligned} \tag{5.11}$$

because P_2 is orthogonal to the straight-line difference. Then

$$E(R^2) = \sum_j P_2^2(x_j)\sigma^2 \left(\frac{J-2}{J}\right) \tag{5.12}$$

because two degrees of freedom have to be taken from J in fitting the true regression:

$$\therefore \sigma^2 = \frac{J}{J-2} R^2 / \sum_j P_2^2(x_j) \tag{5.13}$$

The choice of the points x_j must satisfy the following three conditions:

(1) there must be enough of them (say $J \geqslant 10$);
(2) they must be uniformly distributed in the interval H;
(3) they must be chosen without reference to the peaks and troughs of the noise.

To obtain the true regression line ($y = a_0x + b_0$) from the fitted one we proceed as follows.
Let

$$x_j = jh \tag{5.14}$$

where h is the interval between x_{j+1} and x_j, and

$$r_{0j} = y_j - \alpha_0 P_1(x_j) - P_0(x_j) \tag{5.15}$$

We define

$$Y_0 = \sum_j y_j P_0(x_j)$$
$$= \sum_j y_j \tag{5.16}$$
$$Y_1 = \sum_j y_j P_1(x_j) \tag{5.17}$$

Then since for the true regression:

$$0 = \sum_j r_{0j} P_0(x_j)$$
$$= \sum_j r_0 P_1(x_j) \tag{5.18}$$

we have

$$Y_0 = \beta_0 \sum_j P_0^2(x_j)$$
$$= J\beta_0 \tag{5.19}$$
$$Y_1 = \alpha_0 \sum_j P_1^2(x_j)$$
$$= \frac{J(J+1)(J-1)h^2\alpha_0}{n} \tag{5.20}$$

$$\therefore \quad \alpha_0 = nY_1/[J(J^2-1)h^2] \tag{5.21}$$
$$\beta_0 = Y_0/J \tag{5.22}$$

The variance V of the fitted value at x is then

$$V = E[\alpha_0 P_1(x) + \beta_0 P_0(x) - y(x))]^2 \tag{5.23}$$

where $y(x)$ is the 'true unknown value'. By standard statistical theory

$$V = P_1^2(x)\mathrm{Var}(\alpha_0^2) + 2P_1(x)P_0(x)\mathrm{Covar}(\alpha_0\beta_0) + P_0^2(x)\mathrm{Var}(\beta_0^2)$$
$$= \left(\frac{P_0^2(x)}{J} + \frac{nP_1(x)}{J(J^2-1)h^2}\right)\sum_j \frac{{r_{oj}}^2}{J-2} \tag{5.24}$$

The mean variance $\bar{V}$ for $x = 0$ to Jh is

$$\bar{V} = \frac{1}{(J-1)h}\int_0^{(J-1)h} V\,dx$$

$$= \frac{\sum_j r_{0j}^2}{J(J-2)}\left(1 + \frac{(J-1)}{(J+1)}\right)$$

$$= \frac{2\sum_j r_{0j}^2}{(J+1)(J-2)} \tag{5.25}$$

The square root of $\bar{V}$ is the standard error of the curve. The fractional error in the estimation of $\bar{V}$, assuming a Gaussian distribution, is given by $1/(2n-2)^{1/2}$ where n is the number of points x_j used to estimate $\bar{V}$.

5.2. *Error analysis evaluation*

Clearly there are a number of ways of going about estimating the parameters involved in the triple requirement of Clore and Chance [11], namely the SD of the fit, the determination of the optimised parameters and the distribution of residuals. The method used in the FACSIMILE program employed by the author is described below.

We define the residuals R_{ij}:

$$R_{ij} = (v_{ij} - u_{ij})/\sigma_i \tag{5.26}$$

where j identifies the time point and i the data curve, v_{ij} the observed values, u_{ij} the corresponding calculated values, and σ_i the standard error of curve i.

5.2.1. *Computation of the overall SD of the fit*

We minimise the residual sum squares RSQ

$$\mathrm{RSQ} = \sum_{i=1}^{n}\sum_{j=1}^{m} R_{ij}^2 \tag{5.27}$$

At the minimum RSQ is equal to the normalised χ^2. From the RSQ we calculate the overall SD of the fit, expressed as a percentage

$$\mathrm{SD}(\%) = \Phi[\mathrm{RSQ}/(d-p)]^{1/2} \tag{5.28}$$

where d is the total number of experimental points, p is the number of optimised parameters and Φ the overall standard error of the data, expressed as a percentage

$$\Phi(\%) = 100 \sum_{i=1} \sigma_i r_i / \sum_{j=1} r_i \tag{5.29}$$

where r_i is the range of curve i.

5.2.2. *Estimation of the determination of the optimized parameters*

A quantitative measure of how accurately an unknown paramater P_ℓ is determined by optimisation is given by the standard deviation of the natural logarithm (SD_{ln}) of P_ℓ. Because rate constants and other parameters need to be varied over a very large range of values, we vary $\ln(P_\ell)$ rather than P_ℓ as this is computationally both more efficient and more reliable. When the minimum residual sum of squares is reached, a sensitivity matrix, expressing the dependence of each residual R_{ij} on each parameter $\ln(P_\ell)$ (at its optimum value), is determined by adding 0.2 to each $\ln(P_\ell)$ in turn and examining the effect on each R_{ij}. The normal matrix is then calculated from the sensitivity matrix and inverted to obtain the variance–covariance matrix which refers to $\ln(P_\ell)$. The variance $(SD_{ln})^2$ of each $\ln(P_\ell)$ is given by the diagonal elements of this matrix and used to convert it into a correlation matrix, whose components $r_{k\ell}$ is the correlation coefficient between $\ln(P_k)$ and $\ln(P_\ell)$. Because of the linearity of logarithms less than 0.2, the fractional error $\Delta x/x$ of P is approximately equal $\pm$ SD_{ln} and is considered to have a well determined minimum in multidimensional parameter space. For larger values of SD_{ln}, up to 1 in magnitude, the parameter is determined to within a factor of $e \sim 2.72$, and so its order of magnitude is known. Values of SD_{ln} significantly greater than 1 show that the observations are inadequate to determine the parameter for the particular model being fitted. The 5 and 95% confidence limits, $P_{5,\ell}$ and P_{95}, are calculated from:

$$\begin{aligned} P_{5,\ell} &= P_\ell \exp(-\ 1.645\ SD_{1n}) \\ P_{95,\ell} &= P_\ell \exp(1.645\ SD_{1n}) \end{aligned} \tag{5.30}$$

and represent values for which the 'true' value of P_ℓ would lie between 5 and 95% probability respectively, on the assumption that $\ln(P_\ell)$ has a Gaussian distribution.

5.2.3. Estimation of the nature of the distribution of residuals

We use one of two statistics to examine the nature of the distribution of residuals. These are, the correlation index and the normalised auto-correlation index.

The correlation index, C_i, for curve i is given by

$$C_i = \sum_{j=1}^{m} R_{ij}/\surd \mathrm{RSQ}_i \tag{5.31}$$

(note $\mathrm{RSQ}_i = \sum_{j-1}^{m} R_{ij}^2$), and the mean absolute correlation index, $\bar{C}$, is given by:

$$\bar{C} = \frac{1}{n} \sum_{i=1}^{n} |C_i| \tag{5.32}$$

For a value of $|C_i| < 1$ or $\bar{C} < 1$ (the expected root mean square value of $|C_i|$ and $\bar{C}$ if the residuals for each curve were all independent random variables of zero mean and the same variance), the residuals are as likely to be positive as negative, and the distribution of residuals can be considered random; a value of $|C_i| \gg 1$ or $\bar{C} \gg 1$ indicates that the residuals are one-sided, and that the departures between calculated and observed values are systematic [11].

The normalised auto-correlation index, A_i, for curve i is given by:

$$A_i = \sum_{j=1}^{m} R_{ij}R_{i,j-1}/\mathrm{RSQ}_i \tag{5.33}$$

and the expected value of the normalised auto-correlation index $E(A_i)$, if the residuals for curve i were all independent random variables of zero mean and the same variance, is given by:

$$E(A_i) = (m_i - 1)^{1/2}/m_i \tag{5.34}$$

where m_i is the number of observed data points for curve i.

If $|A_i| \lesssim E(A_i)$, the distribution of residuals is random; if $|A_i| \gg E(A_i)$ the residuals show some systematic behaviour.

6. GENERAL APPROACH AND PHILOSOPHY TOWARDS THE ANALYSIS OF TRANSIENT KINETIC DATA

In order to extract the most information and establish with as much certainty as possible the mechanism of a particular reaction, it is essential to follow a number of guidelines which are summarised below:

(1) The data should cover as wide a time range as possible. It is particularly important that neither the start (where possible within the time resolution of one's monitoring technique) nor the end of the reaction are truncated. In this respect it should be noted that information resides not only in the number and rate of the kinetic phases but also in their amplitudes.

(2) The experiments should be carried out under as wide a set of conditions as is feasible. This may include, for example, variations in the concentrations of the reactants and in the temperature at which the reaction is allowed to take place.

(3) The reaction should be monitored by several techniques. This includes looking at different features within a single spectroscopic technique (e.g., different wavelengths in optical absorption spectroscopy), using different spectroscopic techniques (e.g., optical absorption spectroscopy, EPR, etc. ...), or using different kinetic techniques (e.g., stopped flow, temperature jump, etc. ...).

(4) All the data acquired on a particular reaction should be fitted simultaneously so as to impose as high a level of constraint as possible on the models being fitted to the data.

(5) Any proposed model for a reaction must satisfy the triple requirement of a SD within the standard error of the data, good determination of the optimised parameters and a random distribution of residuals [11] (see Section 5).

In the following sub-sections, four examples will be given which illustrate the above guidelines. In all four cases, the FACSIMILE package [20] was used to analyse the data.

6.1. Examples of analysis of transient kinetic data

6.1.1. The reaction of fully reduced and mixed valence state cytochrome oxidase with CO at low temperatures

This example illustrates the importance of analysing all the available data simultaneously in order not to arrive at incorrect conclusions.

Clore and Chance [13,15] examined the reactions of fully reduced and mixed valence state cytochrome oxidase with CO in the frozen state at six temperatures

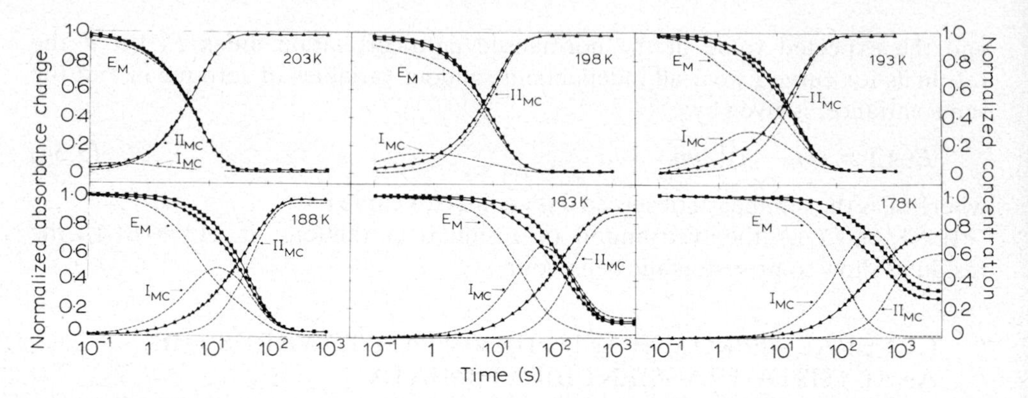

Fig. 8. Comparison of the observed kinetics of the reaction of mixed valence state cytochrome oxidase with CO in the 178–203 K range with the best fit computed kinetics for the scheme given by Eq. (6.1). The experimental points are shown as: ■, 444–463 nm; ▲, 590–630 nm; ●, 608–630 nm. The theoretical normalized absorbance changes are shown as continuous lines (———), and the computed time courses of species E, I and II as interrupted lines (----). The overall SD of the fit is 1.95% compared to the overall standard error of the data of 2 ± 0.36%. The distribution of residuals is random and the optimized parameters are well determined with a relative error $|\Delta x/x| < 0.15$. Initial conditions (i.e., immediately after flash photolysis at $t = 0$) are: μM unliganded mixed valence state cytochrome oxidase (E_M) in the presence of 1.2 mM CO. Species II is the final CO complex. Experimental conditions: 21 mg/ml beef heart mitochondria containing 7 μM cytochrome oxidase, 30% (v/v) ethylene glycol, 0.1 M mannitol, 50 mM sodium phosphate buffer pH 7.2, 5 mM succinate, 1 mM potassium ferricyanide and 1.2 mM CO. (From Clore and Chance [15].)

between 178 and 203 K by dual wavelength multichannel spectroscopy and flash photolysis. They found that the normalised absorbance changes at three wavelength pairs (444–463, 590–630 and 608–630 nm) were different for both reactions (see Fig. 8), indicating, on the basis of Beer's law, the presence of a minimum of three optical species. Three schemes were examined:

$$\mathrm{E + CO} \underset{k_{-1}}{\overset{k_1}{\rightleftharpoons}} \mathrm{I} \underset{k_{-2}}{\overset{k_2}{\rightleftharpoons}} \mathrm{II} \quad (\mathrm{II} \xrightarrow{h\nu} \mathrm{E}) \qquad \textit{Scheme 1} \qquad (6.1)$$

$$\mathrm{E + CO} \underset{k_{-1}}{\overset{k_1}{\rightleftharpoons}} \mathrm{I} \underset{k_{-2}}{\overset{k_2}{\rightleftharpoons}} \mathrm{II} \quad (\mathrm{II} \xrightarrow{h\nu} \mathrm{I}) \qquad \textit{Scheme 2} \qquad (6.2)$$

$$\mathrm{E + CO} \underset{k_{-1}}{\overset{k_1}{\rightleftharpoons}} \mathrm{I} \quad (\mathrm{I} \xrightarrow{h\nu} \mathrm{E}) \qquad \textit{Scheme 3} \qquad (6.3)$$

The data at the three wavelength pairs and six temperatures were analysed *simultaneously* by optimising the rate constants at a reference temperature (chosen as 188 K), the corresponding activation energies, and in the case of Schemes 1 and 2, the relative extinction coefficients of species I and E respectively. Only Scheme 1 was found to satisfy the triple requirement of Clore and Chance [11]. The fits to the data and the time courses of the intermediates for the mixed valence state cytochrome oxidase–CO reaction are shown in Fig. 8.

In a later paper De Fonseka and Chance [37] analysed data at two temperatures *separately* and found that some of the parameters for Scheme 1 were poorly determined. At the two temperatures examined the differences in the normalized

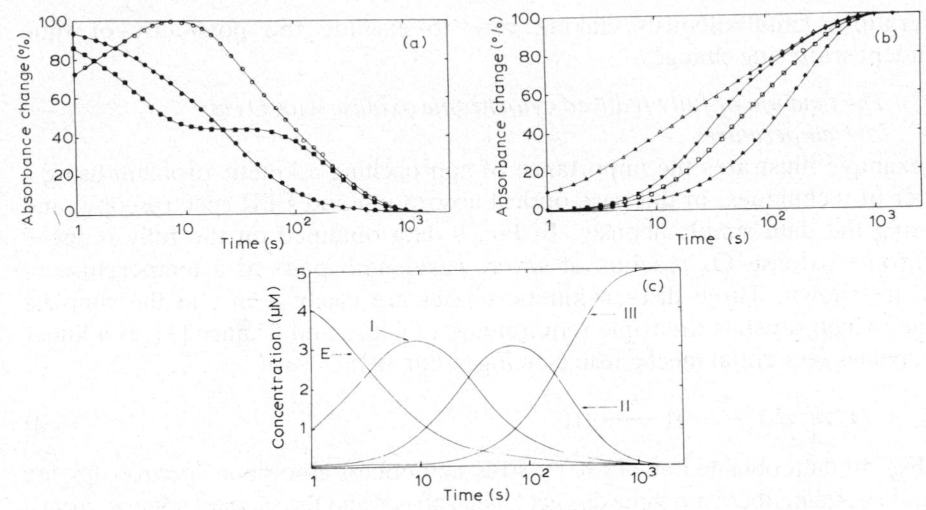

Fig. 9. Comparison of the observed kinetics of the reaction of fully reduced cytochrome oxidase with O_2 at 176 K as measured at seven wavelength pairs simultaneously with the best fit computed kinetics for the linear three species sequential mechanism given by Eq. (6.4). Symbols: △, 430–463 nm; ●, 444–463 nm; ○, 590–630 nm; ■, 608–630 nm; ▲, 740–940 nm; □, 790–940 nm; ▼, 830–940 nm. The normalized computed absorbance changes are shown as continuous lines. The computed kinetics of the individual intermediate are shown in (C). The overall SD of the fit is 1.95% compared to the overall standard error of the data of 2.0 ± 0.2%. The distribution of residuals is random and the optimized parameters are well determined with a relative error $|\Delta x/x| \lesssim 0.15$. Initial conditions: 5 μM fully reduced cytochrome oxidase (E) in the presence of 750 μM O_2. Experimental conditions: beef heart mitochondria containing 5 μM cytochrome oxidase, 30% (v/v) ethylene glycol, 0.1 M mannitol, 50 mM sodium phosphate buffer pH 7.2, 5 mM succinate, 0.6 mM CO and 750 μM O_2. The reaction is initiated by photolysis of the fully reduced cytochrome oxidase–CO complex, and CO does not recombine to a detectable extent in the presence of the relatively high O_2 concentration used, as shown by control experiments where repeated flashes over the course of the experiment only produces approximately 1% further photolysis of the CO complex, the O_2 intermediates not being susceptible to photolysis at the flash intensity used. (From Clore and Chance [14].)

absorbance changes at the three wavelength pairs were small enough to enable them to obtain a fit for Scheme 3 with a SD within the standard error of data and good determination of the optimized parameters. However, there were systematic errors in the distribution of residuals at both temperatures. In the belief that poorly determined parameters are a more important cause for rejection of a kinetic model than systematic errors in the distribution of residuals, De Fonseka and Chance [37] concluded that Scheme 3 was the best model representing the reaction of both fully reduced and mixed valence state cytochrome oxidase with CO, and they attributed the small differences in the normalized absorbance changes at the three wavelength pairs to time and wavelength dependent scattering changes [38]. The failure of De Fonseka and Chance [37] to determine all the parameters for Scheme 1, however, is not surprising, and is simply due to the use of a small data-information base (three wavelength pairs and two temperatures analysed individually) compared to the much larger data-information base (three wavelength pairs and six temperatures analysed simultaneously) used by Clore and Chance [13,15]. In a further study, Clore [39] confirmed the findings of Clore and Chance [13,15] by analysing data at a single wavelength pair (444–463 nm) and six temperatures simultaneously, taking care to exclude the possibility of time dependent scattering changes.

6.1.2. *The reaction of fully reduced cytochrome oxidase with O_2 at low temperatures*

This example illustrates the importance of approaching a kinetic problem using a number of techniques, in this case optical absorption and EPR spectroscopy, and analysing the data simultaneously. In Fig. 9 data obtained on the fully reduced cytochrome oxidase–O_2 reaction at seven wavelength pairs at a temperature of 176 K are shown. Three distinct kinetic phases are easily seen and the simplest scheme, which satisfies the triple requirement of Clore and Chance [11] is a linear three species sequential mechanism (see Fig. 9 for fit):

$$E + O_2 \underset{k_{-1}}{\overset{k_1}{\rightleftharpoons}} I \xrightarrow{k_2} II \xrightarrow{k_3} III \tag{6.4}$$

In Fig. 10 data obtained at 173 K by EPR and optical absorption spectroscopy are shown [17]. Again, there are three distinct kinetic phases and the simplest scheme, from a conceptual point of view, required to fit the data, is the linear three species sequential mechanism (6.4). However, it can be seen that at the end of the reaction, whereas 100% of the EPR detectable copper, Cu_A, (g 2 EPR signal), is oxidised, only 40% of the low-spin cytochrome *a* (g 3 EPR signal) is oxidised. As a result, in order to fit Scheme 6.4 to the data requires one to assume that the contribution of the EPR detectable copper (g 2) and low-spin ferric cytochrome *a* (g 3) EPR signals to species II and III are non-integer. Such an assumption is completely unjustified because each signal arises from only a single metal centre [17]. A number of schemes were tested, and the only scheme which satisfied the triple requirement of Clore and Chance [11] and accounted for the data without any assumptions, was a branching mechanism (see Fig. 10 for fit):

$$\begin{array}{ccl} & & IIA \xrightarrow{k_4} III \\ & \overset{k_2}{\nearrow} & \\ E + O_2 \underset{k_{-1}}{\overset{k_1}{\rightleftharpoons}} I & & \\ & \underset{k_3}{\searrow} & \\ & & IIB \end{array} \tag{6.5}$$

Species E and I contain no EPR detectable components. The EPR detectable components in species IIA, IIB and III are as follows: for species IIA, cytochrome a^{3+}; for species IIB, Cu_A^{2+}; and for species III, cytochrome a^{3+} and Cu_A^{2+}. Thus, using a combination of EPR and optical absorption spectroscopy, and analysing the complete data set simultaneously, it was possible to demonstrate the presence of a branching pathway and assign unambiguously the valence states of cytochrome *a* and Cu_A in each of the intermediates. It should be noted, however, that had the optical absorption and EPR data been analysed separately, this would not have been possible.

6.1.3. *The reaction of aquo Fe(III) myoglobin with H_2O_2 at pH 8.0*

This example illustrates the importance of analysing kinetic data under conditions where the largest amount of information is available.

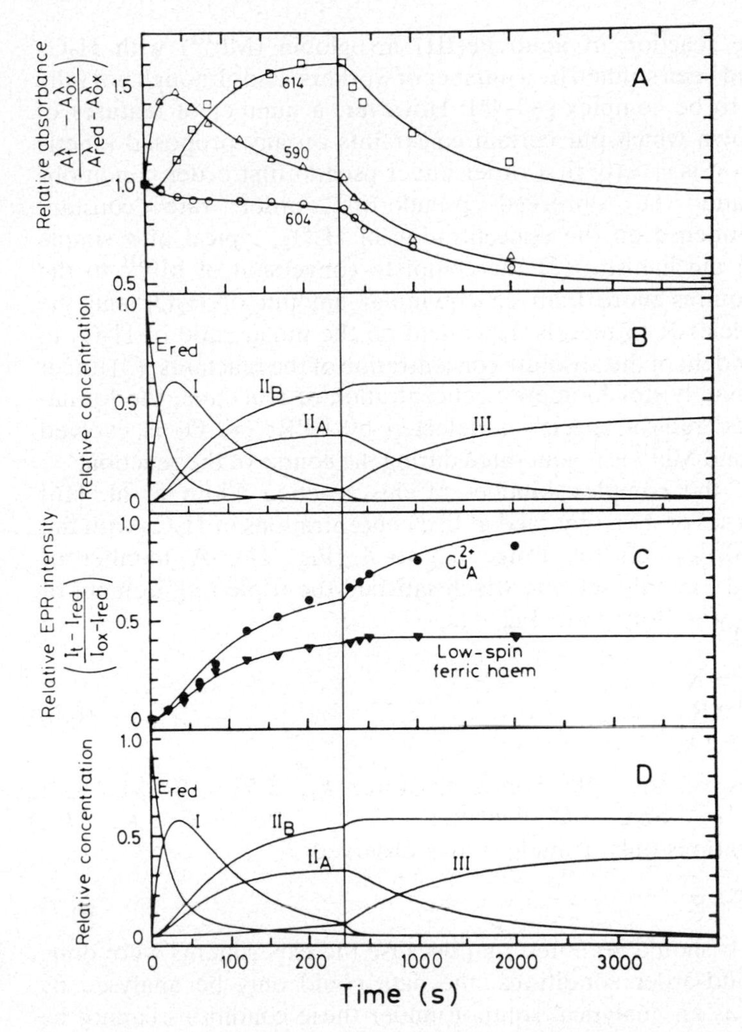

Fig. 10. Comparison of the observed kinetics of the reaction of fully reduced cytochrome oxidase with 0_2 at 173 K as measured by optical absorption and EPR spectroscopy with the best fit computed kinetics for the branching scheme given by Eq. (6.5). Symbols: △, 590–630 nm; ○, 604–630 nm; □, 614–640 nm; ●, Cu_A^{2+} EPR signal; ▼, low-spin cytochrome a^{3+} EPR signal. The computed curves are shown as continuous lines. The initial conditions are: 10 μM fully reduced cytochrome oxidase (E) and 0.67 mM O_2 for the optical data (a and b); 106 μM fully reduced cytochrome oxidase and 0.67 mM O_2 for the EPR data (c and d). The overall SD of the fit is 2.3% compared to the overall standard error of the data of 2.5 ± 0.2% The distribution of residuals is random and the optimized parameters are well determined with a relative error $|\Delta x/x| \lesssim 0.15$. The absorbance changes are digitized relative to the difference in absorbance between fully reduced minus fully oxidized cytochrome oxidase normalized to 1.0; the EPR intensity changes are digitized relative to the difference in intensity between fully oxidized minus fully reduced cytochrome oxidase normalized to 1.0. Experimental conditions: (a) 10 μM soluble cytochrome oxidase, 1.6 μM phenazine methosulphate. 0.3 mM NADH, 50 mM sodium phosphate buffer pH 7.4, 30% (v/v) ethylene glycol, 0.8 mM CO and 0.67 mM O_2; (b) as in (a) except for 107 μM cytochrome oxidase, 2.7 μM phenazine methosulphate and 0.6 mM NADH. (From Clore et al. [17].)

The kinetics of the reaction of aquo Fe(III) myoglobin (Mb^{III}) with H_2O_2 between pH 8 and 9 had been studied by a number of workers, and although not fully characterized, shown to be complex [40–44]. However, a number of features of the reaction were known which put certain constraints on any proposed kinetic scheme: (1) the reaction is strictly first order under pseudo-first order conditions ($[H_2O_2] \gg [Mb^{III}]$) and the observed pseudo-first order rate constant exhibits a linear dependence on the concentration of H_2O_2, typical of a simple second order reaction mechanism; (2) the complete conversion of Mb^{III} to the product species R requires more than an equimolar amount of H_2O_2 and the relative amount of species R formed is dependent on the molar ratio of H_2O_2 to myoglobin and independent of the absolute concentration of the reactants; (3) under conditions of approximately stoichiometric concentration of reactants, the formation of a transient free radical species is detected by EPR; (4) O_2 is evolved following a lag phase, and Mb^{III} is regenerated during the course of the reaction.

In order to resolve the complex kinetics of this reaction Clore et al. [16] proceeded to analyse a set of data obtained at five concentrations of H_2O_2 with the molar ratio $[H_2O_2]/[Mb]_{total}$ in the range 0.3 to 4 (Fig. 11). A number of schemes were tried and the only scheme which satisfied the triple requirement of Clore and Chance [11] was (for fit see Fig. 11):

$$\begin{aligned} Mb^{III} + H_2O_2 &\xrightarrow{k_1} X \\ X + H_2O_2 &\xrightarrow{k_2} R \\ R &\xrightarrow{k_3} Mb^{III} + O_2 \end{aligned} \tag{6.6}$$

The optimized values of the rate constants were: k_1, 3.53×10^2 M^{-1} s^{-1}; k_2, 3.40×10^3 M^{-1} s^{-1}; and k_3, 9.60×10^{-5} s^{-1}. Because $k_2 \gg k_1$, at high H_2O_2 concentrations only a single step is observed

$$Mb^{III} + H_2O_2 \xrightarrow{k_{app}} R \tag{6.7}$$

with k_{app} equal to k_1. It should be noted that because the experiments were done under non-linear second-order conditions, the data could only be analysed by numerical integration as an analytical solution under these conditions cannot be obtained.

6.1.4. *The reaction of ethanolamine ammonia-lyase with L-2-aminopropanol: determination of the number of functional active sites per enzyme molecule*

Ethanolamine ammonia-lyase is an adenosylcobalamine (AdoCbl) dependent enzyme which catalyzes the formation of propionaldehyde and ammonia from L-2-aminopropanol in a reaction involving the migration of an amino-group between adjacent substrate carbon atoms [45]. Kinetic studies [46] using rapid scanning stopped flow spectrophotometry have demonstrated that the reaction proceeds by a three-step mechanism

$$EC + S \underset{k_{-1}}{\overset{k_1}{\rightleftharpoons}} ECS \xrightarrow{k_2} ERP \xrightarrow{k_3} EC + P \tag{6.8}$$

where EC is the enzyme–AdoCbl complex, ECS the enzyme–substrate complex, ERP the Cob(II) alamin intermediate, and P the primary product 1-aminopropanol, which dissociates rapidly to propionaldehyde and ammonia. In species EC and ECS

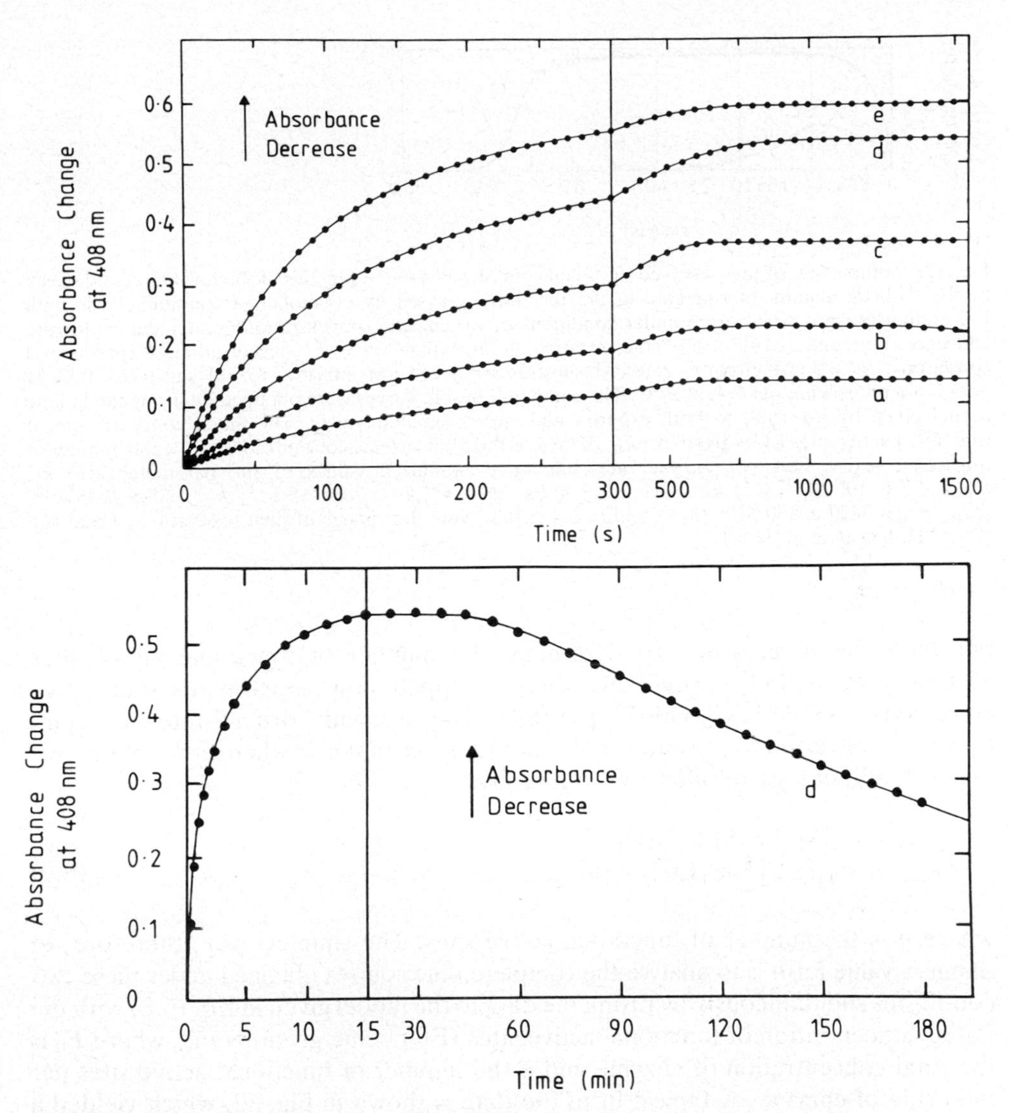

Fig. 11. Comparison of the observed (●) kinetics at 408 nm for the reaction of Mb^{III} with approximately stoichiometric concentrations of H_2O_2 with the best fit computed time courses (———) for the scheme given by Eq. (6.6). Experimental conditions: 8.49 μM (final concentration) Mb^{III} was dissolved in 0.05 M tris-(hydroxymethyl)amine methane buffer, pH 8.0, at 20°C and reacted with (a) 2.78 μM, (b) 5.55 μM, (c) 11.1 μM, (d) 20.0 μM, and (e) 35.1 μM (final concentrations) H_2O_2. The overall S.D. of the fit is 1.2% compared to the overall standard error of the data of 1.5 ± 0.2%. The distribution of residuals is random and the optimized parameters are well determined with a relative error $|\Delta x/x| \lesssim 0.05$ (from Clore et al. [16]).

the cobalt atom of the coenzyme is in the oxidised Co(III) state, and in the ERP intermediate in the reduced Co(II) state. The molecular weight of the enzyme is 520,000 [47] and is composed of two types of subunits (I and II) of molecular weight 51,000 and 36,000 indicating a subunit structure I_6II_6 for the native enzyme [48]. Given this subunit structure it is likely that there is more than one active site

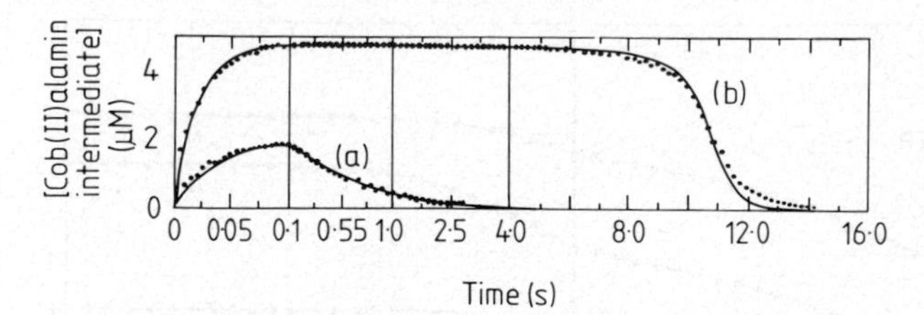

Fig. 12. Comparison of the observed (○) and computed (———) kinetics of the formation and decay of the Cob(II) alamin intermediate in the reaction catalysed by ethanolamine ammonia-lyase with L-2-aminopropanol as substrate under conditions of (a) enzyme concentration greater than substrate concentration, and (b) substrate concentration greater than enzyme concentration. Experimental conditions: (a) 3.5 μM enzyme, 2.18 μM substrate; (b) 0.88 μM enzyme, 87 μM substrate; 0.01 M potassium phosphate, pH 7.4, at 27°C. The computed best fit curves were obtained by fitting the kinetic model given by Eq. (6.8) to both experimental curves simultaneously. The initial concentration of functional active sites EC is given by nE_0 where e_0 is the total enzyme concentration and n the number of functional active sites per enzyme molecule. The optimized values of the parameter are: k_1, $1.6 \pm 0.7 \times 10^8$ M^{-1} s^{-1}; k_{-1}, $8.7 \pm 3.7 \times 10^3$ M^{-1} s^{-1};$k_{\ 2}$, 78 ± 15 s^{-1}; k_3, 1.55 ± 0.35 s^{-1}; (ε_{ERP}–ε_{EC}), 3820 ± 180 M^{+1} cm^{-1}; and n, 5.4 ± 0.3. Note the change of time scale at 0.1, 1 and 4 s. (From Hollaway et al. [49].)

per molecule of enzyme. To determine the number of functional active sites Hollaway et al. [49] carried out kinetic stopped flow experiments under two conditions: $[S] \gg [EC]$ and $[EC] \gg [S]$. The second order rate constant k_1 can be converted to a pseudo-first order rate constant k_{obs} when either of the two above conditions are fulfilled, and will be given by

$$\begin{aligned} k_{obs} &= k_1[S] \text{ for } [S] \gg [EC] \\ k_{obs} &= nk_1[EC] \text{ for } [EC] \gg [S] \end{aligned} \tag{6.9}$$

where n is the number of functional active sites. The simplest way, therefore, to obtain a value for n is to analyse the complete time courses obtained under these two conditions simultaneously by fitting the data to the model given in Eq. (6.8) with the initial concentration of functional active sites (EC) being given by nE_0 where E_0 is the total concentration of enzyme and n the number of functional active sites per molecule of enzyme. A typical fit of the data is shown in Fig. 12, which yielded a value of $n = 5.4 \pm 0.3$ from which it was deduced that there were six functional active sites per enzyme molecule.

ACKNOWLEDGEMENTS

I am indebted to Drs. E.M. Chance and M.R. Hollaway for introducing me to and alighting my interest in the analysis of transient kinetic data.

REFERENCES

1. Cornish-Bowden, A. (1976) Principles of Enzyme Kinetics, Butterworths, London.
2. Ainsworth, S. (1977) Steady State Enzyme Kinetics, Macmillan, London.
3. Bardsely, W.G., Leff, P., Kavanach, J. and Waight, R.D. (1980) Biochem. J. 187, 739–765.
4. Chance, B. (1973) In: Techniques of Chemistry (G.C. Hammes, Ed.), Vol. 6, part 2, pp. 5–62; Wiley, New York.
5. Bernasconi, C.F. (1976) Relaxation Kinetics. Academic Press, New York.
6. Douzou, P. (1977) Cryobiochemistry: an introduction. Academic Press, New York.
7. Hammes, G.G. (1973) Techniques of Chemistry: Investigation of Elementary Reaction Steps in Solution and Very Fast Reactions, Wiley, New York.
8. Hollaway, M.R. and White, H.A. (1975) Biochem. J. 149, 221–231.
9. Porter, G. and West, M.A. (1973) In: Techniques of Chemistry (G.C. Hammes, Ed.) Vol. 6, Part 2, pp. 367–462, Wiley, New York.
10. Gibson, Q.H. (1966) Ann. Rev. Biochem. 35, 435–456.
11. Clore, G.M. and Chance, E.M. (1978) Biochem. J. 173, 799–810.
12. Clore, G.M. and Chance, E.M. (1978) Biochem. J. 173, 811–820.
13. Clore, G.M. and Chance, E.M. (1978) Biochem. J. 175, 709–725.
14. Clore, G.M. and Chance, E.M. (1979) Biochem. J. 177, 613–621.
15. Clore, G.M. and Chance, E.M. (1980) Biochim. Biophys. Acta 590, 34–49.
16. Clore, G.M., Lane, A.N. and Hollaway, M.R. (1980) Inorg. Chim. Acta. 46, 139–
17. Clore, G.M., Andreasson, L.E., Karrlson, B., Aasa, R. and Malström, B.G. (1980) Biochem. J. 185, 139–154.
18. Clore, G.M., Andreasson, L.E., Karrlson, B., Aasa, R. and Malström, B.G. (1980) Biochem. J. 185, 155–167.
19. Hollaway, M.R., Ticho, T. and Clarke, P.H. (1980) Biochem. J. 191, 811–826.
20. Chance, E.M., Curtis, A.R., Jones, I.P. and Kirby, C.R. (1979) FACSIMILE: a computer program for flow and chemistry simulation, and general initial value problems. A.E.R.E. Report No. R-8775, Harwell.
21. Rodiguin, N.M. and Rodiguina, E.N. (1964) Consecutive Chemical Reactions: mathematical analysis and development. D. Van Nostrand Company, Princeton, New Jersey.
22. Gear, C.W. (1967) Math. Comp. 21, 146–156.
23. Gear, C.W. (1967) Numerical Integration of Stiff Ordinary Differential Equations. Department of Computer Science, University of Illinois, Report No. 221.
24. Gear, C.W. (1969) In: Information Processing 68 (Morell, A.H.J., Ed.), pp. 197–19, Elsevier, Amsterdam.
25. Gear, C.W. (1971) Comm. A.C.M. 14, 176–179.
26. Derwent, R.G., Eggleton, A.E.J. and Curtis, A.R. (1976) J. Inst. Math. Its Appl. 8, 344–353.
27. Chance, E.M. and Curtis, A.R. (1970) FEBS Lett. 7. 47–50.
28. Curtis, A.R. (1979) The FACSIMILE Numerical Integrator for Stiff Initial Value problems. A.E.R.E. Report No. R.9352. Harwell.
29. Kowalik, J. and Obsorne, M.R. (1968) Methods for Unconstrained Optimization Problems. Elsevier, New York, and references therein.
30. Powell, M.J.D. (1965) Comp. J. 7, 303–307.
31. Powell, M.J.D. (1972) On Search Directions for Minimization Algoriths., A.E.R.E. Report No. TP.492, Harwell.
32. Curtis, A.R. and Edsberg, L. (1973) Some Investigations into Data Requirements for Rate Constant Estimation. A.E.R.E. Report No. TP.554, Harwell.
33. Curtis, A.R. and Reid, J.K. (1971) J. Inst. Math. Its Appl. 8. 344–353.
34. Curtis, A.R., Powell, M.J.D. and Reid, J.K. (1974) J. Inst. Math. Its Appl. 13, 117–119.
35. Duff, I.S. and Reid, J.K. (1977) The Estimation of Sparse Jacobian Matrices. A.E.R.E. Report No. CSS 48, Harwell.
36. Kirby, C.R. (1980) FACSIMILE Implementation Manual, A.E.R.E. Report No. R 9557, Harwell.
37. De Fonseka, K. and Chance, B. (1978) Biochem. J. 175. 1137–1138.
38. De Fonseka, K. and Chance, B. (1979) Biochem. J. 183. 375–379.
39. Clore, G.M. (1981) Biochim. Biophys. Acta 634, 129–139.
40. George, P. and Irvine, D.H. (1952) Biochem. J. 52, 511–517.
41. Fox, J.B., Nicholas, R.A., Acherman, S.A. and Swift, C.E. (1974) Biochemistry 13, 5178–5186.
42. Yonetani, T. and Schleyer, H. (1967) J. Biol. Chem. 242, 1974–1979.

43. King, N.K., Looney, F.D. and Windfield, M.E. (1967) Biochim. Biophys. Acta 133, 65–82.
44. King, N.K. and Windfield, M.E. (1963) J. Biol. Chem. 238, 1520–1528.
45. Babior, B.M. (1975) In: Cobalamin: Biochemistry and Pathology (Babier, B.M., Ed.), pp. 141–212, John Wiley, New York.
46. Hollaway, M.R., White, H.A., Joblin, K.N., Johnson, A.W., Lappert, M.F. and Wallis, O.C. (1978) Eur. J. Biochem. 82, 143–154.
47. Kaplan, B.H. and Stadtman, E.R. (1968) J. Biol. Chem. 243, 1794–1803.
48. Wallis, O.C., Johnson, A.W. and Lappert, M.F. (1979) FEBS Lett. 97, 196–199.
49. Hollaway, M.R., Johnson, A.W., Lappert, M.F. and Wallis, O.C. (1980) Eur. J. Biochem. 111, 177–178.

Geisow & Barrett (eds.) Computing in biological science
© Elsevier Biomedical Press, 1983

The numerical geometry of biological structures

A.L. MACKAY

Nothing exists except atoms and empty space: all else is opinion.
Democritos of Adera (ca. 400BC)

There is nothing in the world except empty curved space. Matter, charge, electromagnetism and other fields are only manifestations of the curvature of space.
John Archibald Wheeler (1957)

This chapter deals with a number of areas of spatial structure where the ready availability of computer power, especially using the programming language BASIC (developed by J. Kemeny and T. Kurtz of Dartmouth College, New Hampshire, USA) and implemented on small computers, has greatly facilitated the transference of geometrical problems into numerical form.

0.1. INTRODUCTION

The basis of function is structure and the basis of structure is geometry, primarily the geometry of the three-dimensional world which is the theatre in which the drama of life unfolds. Other worlds with different geometries may be used to represent various aspects of certain problems. Indeed, in the very small and the very large, things may not be the same as in our middling, everyday world. More than 100 years ago, W.K. Clifford (1845–1879) suspected this, but we will be concerned here only with the classical geometry systematised by Euclid nearly two millenia ago. However, a great change has recently been brought about by the advent of the computer and of computer graphics which have revolutionised geometry by making it possible to calculate rapidly quantities which would be prohibitively difficult or tedious to find algebraically. In computer graphics we have general physical models of indefinitely high accuracy.

On our natural scale the human brain, eyes and hands are marvellously organised and structured by evolution for existence and functioning in the three dimensions of space and the one dimension of time. When we apprehend a problem, we get hold of it in spatial form with our prehensile intellect. The memory, of which we do not yet understand the operation, is not a miniature theatre of the imagination, but it is nevertheless coded to operate as a theatre in which we can 'rehearse' what we are going to do, trying out various scenarios for the future. In this theatre we can perform 'Gedankenexperimente' – thought experiments, and we can emerge with solid conclusions as to how to act in real life. Perhaps those of our ancestors whose spatial and kinaesthetic senses were not so good, just missed their holds when swinging from tree to tree and failed to reproduce themselves. There has been comparatively little biological evolution in the 5000 years of human civilisation and the senses which were evolved for hunting are now used for thinking about the behaviour of the micro-world. It has been found that the facility for visualising space is a primary one on which other faculties are based. Recently, the computer has become an aid to our imagination and the aim of this section is to show how numerical values may be attached to the physical, representational and mental models of spatial structures.

We include a number of computer programs to show how far the subject has come since Euclid. These programs are in BASIC and we wish to show that giant machines are not necessary in order to take a very significant step.

We should emphasise that geometry in three dimensions is not just like that in two but with extra terms in the expressions. New properties and complications emerge. Similarly, four dimensions is different again. We suspect that the ancients were hampered, as we ourselves still are, by the difficulty of making good models for solid geometry, but now computer graphics, with facilities for stereo perception, offer us the prospect of substantial advance.

The main purpose of this chapter is to emphasise that with the advent of computers capable of handling arrays of numbers as easily as the first pocket calculators handle single numbers, solid geometry has been transformed. The algebraic framework, available in textbooks such as D.M.Y. Somerville [1], remains fundamental but operations involving, for example, the repeated calculation of the inverses of large matrices, which were earlier prohibitive, have become practicable. Calculation can often be used as a substitute for algebra. We include some references which have proved generally useful in the field of biological geometry [2].

1. THE NUMERICAL SOLUTION OF EQUATIONS

1.1. An array of N *linear equations for* N *unknowns*

$$a_{11}x_1 + a_{12}x_2 + a_{13}x_3 = h_1$$
$$a_{21}x_1 + a_{22}x_2 + a_{23}x_3 = h_2$$
$$a_{31}x_1 + a_{32}x_2 + a_{33}x_3 = h_3$$

may be written in matrix notation as $[\mathbf{A}][\mathbf{X}] = [\mathbf{H}]$. The solution is $[\mathbf{X}] = [\mathbf{A}]^{-1}[\mathbf{H}]$ where $[\mathbf{A}]^{-1}$ is the matrix inverse to $[\mathbf{A}]$. It is a central feature of BASIC that inversion of a matric is possible with one statement, such as MAT[**B**] = INV[**A**], although implementations on micro-computers do not have this facility and a specially-written segment of program must be inserted.

Thus a set of equations can be solved directly:

```
DIM[A(N,N)], [B(N,N)], [X(N,1)], [H(N,1)]
MAT READ [A]
MAT READ [H]
MAT [B] = INV [(A)]
MAT [X] = [B]*[H]
MAT PRINT [X],
```

If the determinant of [**A**] is zero, the matrix is singular and the equations have no unique solution.

In the first century AD the Chinese book *Nine Chapters of Arithmetic Technique* gave the following problem:

5 sheep + 4 dogs + 3 hens + 2 hares cost 1496 coins
4 sheep + 2 dogs + 6 hens + 3 hares cost 1175 coins
3 sheep + 1 dog + 7 hens + 5 hares cost 958 coins
2 sheep + 3 dogs + 5 hens + 1 hare cost 861 coins.

How much does each kind of animal cost? (The answer is: 177 coins for a sheep, 121 for a dog, 23 for a hen and 29 for a hare.) Linear equations have been known for a long time but their solution is now extremely fast. A program listing is given in Section 7.5.

1.2. The least squares solution of a block of linear equations

[**A**][**X**] = [**H**], a block of N observational equations for M unknowns [**X**], ($N \geq M$), can be solved by the method of least squares. Here we attach dimensions to each array in brackets to help in keeping track of the calculation. $[\mathbf{A}]^T$ is the transpose of the matrix [**A**].

The set of equations is $[\mathbf{A}(N,M)]*[\mathbf{X}(M,1)] = [\mathbf{H}(N,1)]$ but they may not be consistent with each other and it is required to find the best answer. The solution

$$[\mathbf{X}(M,1)] = (\mathbf{A}^T(M,N)]*[\mathbf{A}(N,M)])^{-1}*[\mathbf{A}^T(M,N)]*[\mathbf{H}(N,1)]$$

is the best solution in the sense that the sum of the squares of the residuals $([\mathbf{A}(N,M)]*[\mathbf{X}(M,1)] - [\mathbf{H}(N,1)])^2$, the squares of the 'misfits' of the equations, is a minimum.

A skeleton program in BASIC would go:

```
DIM[X(M,1)], [A(N,M)], [B(M,N)], [H(N,1)], [C(M,M)], [D(M,1)], [E(M,M)]
MAT READ [A]
MAT READ [H]
MAT[B] = TRN[A]
MAT[C] = [B]*[A]
MAT[D] = [B]*[H]
MAT[E] = INV[C]
```

```
MAT[X] = [E]*[D]
MAT PRINT [X],
```

If the N equations are not all of equal weight, then they can be multiplied through by quantities proportional to their desired weights before beginning the calculation.

1.2.1. The best plane through N *points*

The equation of a plane is $lx + my + nz = p$ and, if $l^2 + m^2 + n^2 = 1$, p is the perpendicular distance of the origin from the plane and l,m,n are the direction cosines of the normal. Thus the distance of a point P at x_1,y_1,z_1 from this plane is $lx_1 + my_1 + nz_1 - p$. For the best plane the sum of the squares of these distances is to be minimised. Thus $S = \Sigma_{i=1}^{N} (lx_i + my_i + nz_i - p)^2$ is to be minimised.

Calculating $\partial S/\partial l$, $\partial S/\partial m$, $\partial R/\partial n$ and equating them to zero we have

$$l\Sigma x^2 + m\Sigma xy + n\Sigma xz - p\Sigma x = 0$$
$$l\Sigma xy + m\Sigma y^2 + n\Sigma yz - p\Sigma y = 0$$
$$l\Sigma xz + m\Sigma xy + n\Sigma z^2 - p\Sigma z = 0$$

and the condition $\partial S/\partial p = 0$ gives $l\Sigma x + m\Sigma y + n\Sigma z = Np$ which expresses the condition that the centre of gravity should lie on the best plane.

The matrix equation:

$$\begin{bmatrix} \Sigma x^2 & \Sigma xy & \Sigma xz \\ \Sigma xy & \Sigma y^2 & \Sigma yz \\ \Sigma xz & \Sigma yz & \Sigma z^2 \end{bmatrix} \begin{bmatrix} l/p \\ m/p \\ n/p \end{bmatrix} = \begin{bmatrix} \Sigma x \\ \Sigma y \\ \Sigma z \end{bmatrix}$$

is then solved for l/p, m/p and n/p and using $l^2 + m^2 + n^2 = 1$ this gives l,m,n,p. The extension to weighted points is clear.

1.3. The solution of non-linear equations

Calculation can be used as a substitute for algebra. A triangle requires three quantities for its specification. The most natural, perhaps, are a, b and c, the lengths of the sides, but others could be chosen. As an example of a computer program for solving non-linear equations, let us suppose that we are given the three quantities S, the area of the triangle, R the radius of the circumcircle and $2s$ the perimeter of the triangle. These quantities are homogeneous in a, b and c which have to be determined from $S^2 = s(s - a)(s - b)(s - c)$

$$s = ½(a + b + c) \qquad \text{and} \qquad 4\,RS = abc.$$

The algebra is tedious and would lead to a cubic, but, if we assume tentative values a_1,b_1,c_1 for a, b and c respectively we will have the equations:

$$S^2 - s(s - a_1)(s - b_1)(s - c_1) = E_1$$
$$s - ½(a_1 + b_1 + c_1) = E_2$$
$$4\,RS - a_1b_1c_1 = E_3$$

where E_1, E_2 and E_3 are the discrepancies which must be minimised by the correct choice of a, b and c. We calculate how the discrepancies change with a, b and c. The derivatives are:

$$\partial E_1/\partial a = s(s-b)(s-c) \quad \partial E_1/\partial b = s(s-a)(s-c) \quad \partial E_1/\partial c = s(s-a)(s-b)$$
$$\partial E_2/\partial a = -½ \quad \partial E_2/\partial b = -½ \quad \partial E_2/\partial c = -½$$
$$\partial E_3/\partial a = -bc \quad \partial E_3/\partial b = -ac \quad \partial E_3/\partial c = -ab$$

In this example, analytic expressions for the partial derivatives are readily obtained, but in less tractable cases they could be calculated numerically by changing a by a finite step and noting the corresponding change in E_1, etc., thus if Δa, Δb and Δc are the deviations required to account for the errors E_1, E_2, E_3, then:

$$E_1 = \frac{\partial E_1}{\partial a} \,.\, \Delta a + \frac{\partial E_1}{\partial b} \,.\, \Delta b + \frac{\partial E_1}{\partial c} \,.\, \Delta c$$

$$E_2 = \frac{\partial E_2}{\partial a} \,.\, \Delta a + \frac{\partial E_2}{\partial b} \,.\, \Delta b + \frac{\partial E_2}{\partial c} \,.\, \Delta c$$

$$E_3 = \frac{\partial E_3}{\partial a} \,.\, \Delta a + \frac{\partial E_3}{\partial b} \,.\, \Delta b + \frac{\partial E_3}{\partial c} \,.\, \Delta c$$

Thus, if initial values of a, b and c are assumed, all the coefficients necessary for calculating the corrections $-\Delta a$, $-\Delta b$, $-\Delta c$ can be found and the corrections got by solving the block of linear simultaneous equations. The substantive equations are non-linear and therefore successive approximations are necessary. A listing of a program SOLVE is given in Section 7.1.

1.4. The generalised inverse

An important function, which is not available in BASIC (although it is provided in APL), is that of the generalised inverse of a matrix. The inversion of a matrix is a most valuable feature of BASIC but if the matrix is singular, that is, if its determinant is zero, the operation will fail with the message 'MATRIX NEARLY SINGULAR'.

The generalised inverse $[\mathbf{A}^+]$ of the matrix $[\mathbf{A}]$ provides the best answer $[\mathbf{X}] = [\mathbf{A}^+][\mathbf{H}]$ to the set of equations $[\mathbf{A}][\mathbf{X}] = [\mathbf{H}]$ under most circumstances. **A** may be rectangular or square, and singular or not. For clarity we insert the dimensions of the arrays as $[\mathbf{A}(N,M)] * [\mathbf{X}(M,1)] = [\mathbf{H}(N,1)]$ and have the following cases: (a) If $N = M$ and $[\mathbf{A}]$ is square and non-singular, there are N equations for N unknowns and $[\mathbf{X}(N,1)] = [\mathbf{A}^+(N,N)] * [\mathbf{H}(N,1)]$. Here $[\mathbf{A}^+]$ is the same as the usual inverse and gives the normal correct answer for $[\mathbf{X}]$.
(b) If N is greater than M and there are redundant equations consistent with each other, $[\mathbf{X}(M,1)] = [\mathbf{A}^+(M,\mathrm{N})] * [\mathbf{H}(N,1)]$ gives the correct answer.
(c) If N is greater than M and there are thus more equations than unknowns, the equations representing perhaps experimental observations, $[\mathbf{X}(M,1)] = [\mathbf{A}^+(M,N)] * [\mathbf{H}(N,1)]$ gives the least squares solution (the sum of the squares of the residues of the equations being minimised).

(d) If there are more unknowns than observations, or if the matrix [**A**] is singular and thus it is impossible to solve for [**X**], the expression $[\mathbf{X}(M,1)] = [\mathbf{A}^+(M,N)] * [\mathbf{H}(N,1)]$ gives unique answers for [**X**] with the condition that the residues of the equations are minimised, the actual values of the components of [**X**] being minimised in so far as they are not determined by the equations. Thus, for example, three equations can be solved for five unknowns which may be the corrections to be applied to the movements of a robot. The calculation does not fail and a definite action can be taken in the face of the uncertainty. Since there is not a unique answer to the equations part of the solution remains arbitrary and the full answer is

$$[\mathbf{X}(M,1)] = [\mathbf{A}^+(M,N)] * [\mathbf{H}(N,1)] + ([\mathbf{I}(M,M)] - [\mathbf{A}^+(M,N)] * [\mathbf{A}(N,M)]) * [\mathbf{Z}(M,1)]$$

where [**I**] is the unit matrix (diagonal elements unity) and [**Z**] is an arbitrary vector.

The question then is how to calculate the generalised inverse $[\mathbf{A}^+(M,N)]$ of a matrix $[\mathbf{A}(N,M)]$. This is best done by iteration. If $[\mathbf{B}_k]$ is the kth approximation to $[\mathbf{A}^+]$, then a better approximation is $[\mathbf{B}_{k+1}] = [\mathbf{B}_k(2\mathbf{I} - \mathbf{A} * \mathbf{B}_k)]$. Iteration is stopped when the trace (the sum of the diagonal terms of the matrix $[(\mathbf{A} * \mathbf{B}_k)]$ is close to an integer, the closeness depending on the accuracy of the computer. The method converges if the terms in the starting matrix $[\mathbf{B}_1]$ are taken to be sufficiently small. This method also gives the ordinary matrix if this exists. A specimen program is given in Section 7.5 written in Microsoft Basic, which is a version commonly available for microcomputers. The same program segment will perform the function of the matrix inversion statement found in full versions of BASIC.

There are many text books on the generalised inverse [3].

2. COORDINATE SYSTEMS

There are two basic ways of describing structures in space:

(a) looking from the outside, seeing the structure in an external framework of coordinate axes, and assigning coordinates to each point. This is the method of the observer. Various systems of coordinates are possible, Cartesian, cylindrical and spherical being the commonest. Algebraic relationships between the coordinates can be used to describe planes, lines, spheres, and other geometrical forms.

(b) Looking from the inside, seeing the immediate local surroundings of each point (or atom or molecule or whatever is taken as the unit) and describing the local surroundings. Since the forces which hold biological (and almost all other structures together are the forces between atoms, this is the natural way of describing structures, in that domains of short-range order build up to give long-range order.

When we deal with complex biological structures mathematically they have to be simplified to the points, lines, surfaces and volumes of geometry, in ways familiar to

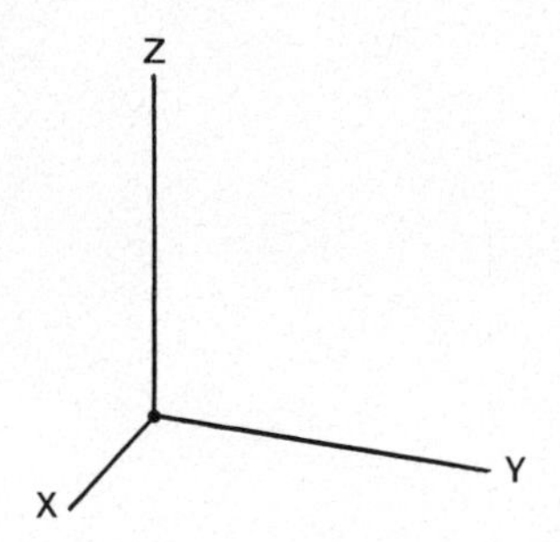

Fig. 1. Right-handed Cartesian axes.

the engineer. Indeed, much of engineering computing is applicable also to biology. It is desirable to be able to convert easily between these two systems of internal and external coordinates and the computer can now do this for us with considerable speed.

In the 19th century there was a movement to express the theorems of geometry in terms which were free of arbitrary coordinates that is, in terms of quantities which were invariant with respect to the choice of coordinate system. The predominant invariants were the distances between pairs of points and the branch was called 'distance geometry'. It is less familiar than Cartesian coordinate geometry. Earlier it was called 'géométrie de position' [4].

2.1. *Coordinate systems*

Usually it is convenient to describe the geometry of a structure, such as a molecule, with respect to a coordinate frame fixed in space. This means, if we are interested only in the internal geometry of the molecule, that the coordinates of each point have no individual physical significance. Moreover, the relationship of the molecule to the frame has to be specified, whether it is significant or not.

Three coordinates must be specified for each point, making $3N$ for a molecule of N atoms, but only $(3N - 6)$ are needed to describe the molecular configuration in isolation. The extra six parameters give the position and orientation of the molecule as a whole with respect to the arbitrary axial system.

2.2. *Cartesian coordinates*

Ordinary Cartesian coordinates x,y,z with respect to a set of three mutually orthogonal axes require no further description. They are conventionally taken to be right-handed the convention being as shown in Fig. 1 with the positive directions of x out of the paper towards us, y in the paper to the right and z in the paper upwards.

Pythagoras' theorem supplies us with the distance d_{ij} between two points i and j as $d_{ij}^2 = (x_i - x_j)^2 + (y_i - y_j)^2 + (z_i - z_j)^2$ and from this everything else flows. Cartesian coordinates are the simplest because there are no cross terms (with products of x and y).

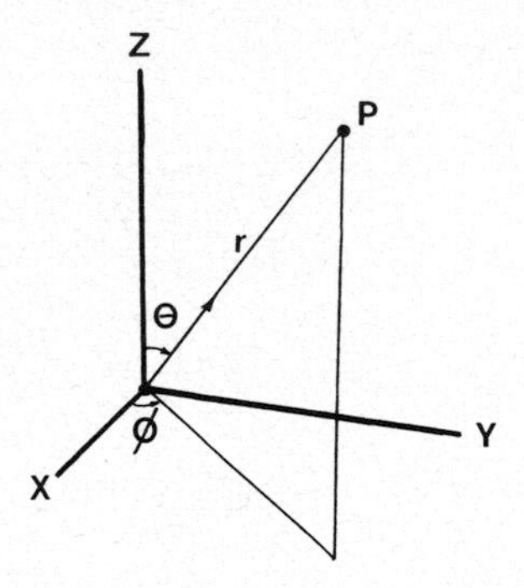

Fig. 2. Cylindrical and spherical axes.

Diagrams of the distances between points often provide in themselves valuable features of structure, as, for example, in the plots of the distances between all pairs of alpha-carbon atoms in a protein molecule used by Go [5] to illustrate domain structure.

2.3. *Cylindrical coordinates and spherical coordinates*

Cylindrical coordinates r,ϕ,z are useful when the system studied (such as a helix of TMV particles) has an axis of rotational symmetry. All other coordinate systems are referred to Cartesian coordinates for definition. The transformations are (Fig. 2):

$$x = r\cos\phi \qquad r = (x^2 + y^2)^{1/2} \qquad (r \leqslant 0)$$
$$y = r\sin\phi \qquad \phi = \arctan\,(y/x) \qquad (0 \leqslant \phi < 360°)$$
$$z = Z$$

The element of volume equivalent to the Cartesian element $dx\ dy\ dz$ is $r\ d\phi\ dr\ dz$.

Spherical coordinates are useful when there is a unique centre such as in an icosahedral virus particle. Here:

$$x = r\sin\theta\cos\phi \qquad r = (x^2 + y^2 + z^2)^{1/2} \qquad (r \leqslant 0)$$
$$y = r\sin\theta\sin\phi \qquad \theta = \arctan((x^2 + y^2)^{1/2}/z) \qquad (0 \leqslant \theta \leqslant 180°)$$
$$z = r\cos\theta \qquad \phi = \arctan(y/x) \qquad (0 \leqslant \phi < 360°)$$

The element of volume here is $r^2 \sin\theta\ dr\ d\theta\ d\phi$.

There is a complete algebraic formalism for each system of coordinates and, for example, density distributions can be expressed as series of orthogonal functions. Thus, a roughly spherical mass can be expressed as a series of spherical harmonics with respect to an origin at its centre.

The choice of coordinates is determined by considerations of what kind of calculations are to be carried out, Cartesian, (orthonormal) coordinates being chosen unless the symmetry of the system dictates otherwise.

2.4. *Area and volume in Cartesian coordinates*

The volume V of a tetrahedron with vertices 1,2,3,4 at x_1,y_1,z_1, etc., is given by the

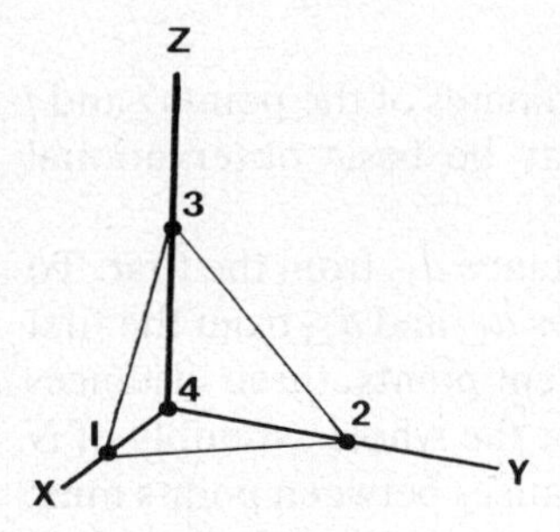

Fig. 3. The sign of the volume of a tetrahedron (in this case positive with respect to right-handed axes.

determinant:

$$V = 1/6 \begin{vmatrix} x_1 & y_1 & z_1 & 1 \\ x_2 & y_2 & z_2 & 1 \\ x_3 & y_3 & z_3 & 1 \\ x_4 & y_4 & z_4 & 1 \end{vmatrix}$$

If the vertices are in the sense indicated in Fig. 3, in relation to the right-handed Cartesian axes, then the volume has a positive sign. If the designations of any two vertices are exchanged then the sign of the volume is reversed. This expression is thus important for calculating on which side of the plane (123) the point (4) lies. Since the volume of a tetrahedron is 1/3 × (area of base) × height, the distance h_4 of the point (4) from the plane (123) is given by $\mathbf{V}_{1234} = (1/3)\mathbf{A}_{123}h_4$ where $\mathbf{A}_{123}$ is the (positive) area of the triangle (123).

The area $\mathbf{A}_{123}$ of the triangle is best calculated from the distances d_{12}, d_{23}, d_{31} using Heron's formula

$$\mathbf{A}_{123}^2 = s(s - d_{12})(s - d_{23})(s - d_{31}) \text{ where } 2s = d_{12} + d_{23} + d_{31}.$$

This can also be expressed as a determinant:

$$-16\mathbf{A}_{123}^2 = \begin{vmatrix} 0 & d_{12}^2 & d_{13}^2 & 1 \\ d_{12}^2 & 0 & d_{23}^2 & 1 \\ d_{13}^2 & d_{23}^2 & 0 & 1 \\ 1 & 1 & 1 & 0 \end{vmatrix}.$$

2.5. *Internal coordinates independent of any coordinate system*

2.5.1. *Distances, bond angles and torsion angles*

To build a structure, such as a molecule, of points joined by lines, three kinds of quantities are needed for describing the configuration in terms of the mutual positions of the points, that is, using internal coordinates. The usual measures are: d_{ij}, θ_{ijk} and ϕ_{ijkl} (distances, bond angles and torsion (or dihedral) angles). Of these only the first, distance, has the dimensions of length, the others being dimensionless ratios, although they may be related to distances, areas and volumes respectively.

2.5.2. *Inter-point distances* d_{ij}

Distances are readily calculated from the Cartesian coordinates of the points i and j as $d_{ij}^2 = (x_i - x_j)^2 + (y_i - y_j)^2 + (z_i - z_j)^2$, or they may be basic observational data.

Building up a framework, the second point lies at a distance d_{12} from the first. To add the third point two measures are needed, the distances d_{13} and d_{23} from the first and second points and for adding the fourth and subsequent points, three distances (or other parameters) must be given making $(3N - 6)$ for the whole assembly of N points. Distances are necessarily positive. Any set of distances between points must obey the triangle inequality — the sum of two sides of a triangle cannot be less than the third side.

2.5.3. *Bond angles* θ_{ijk}

A bond angle is given in terms of distances d_{ij}, d_{jk}, d_{kl} as $d_{ik}^2 = d_{ij}^2 + d_{jk}^2 -$ - $2d_{ij}d_{jk}\cos\theta_{ijk}$ (the cosine rule) so that $\cos\theta_{ijk} = (d_{ij}^2 + d_{jk}^2 - d_{ik}^2)/(2d_{ij}d_{jk})$. The angle must be between 0° and 180° so that it is given uniquely by the cosine.

In terms of vector distances $\mathbf{d}_{ij}$ and $\mathbf{d}_{jk}$, directed from i to j and from j to k respectively, the scalar product $\mathbf{d}_{ij}.\mathbf{d}_{jk}$ equals $|\mathbf{d}_{ij}|\,|\mathbf{d}_{jk}|\,(-\cos\theta_{ijk})$. So that, using Cartesian coordinates

$$-\cos\theta_{ijk} = \frac{(x_j - x_i)(x_k - x_j) + (y_j - y_i)(y_k - y_j) + (z_j - z_i)(z_k - z_j)}{|d_{ij}|\,|d_{jk}|}$$

where

$$|d_{ij}| = ((x_j - x_i)^2 + (y_j - y_i)^2 + (z_j - z_i)^2)^{1/2} \text{ etc.}$$

Similarly, the vector product $\mathbf{d}_{ij} \times \mathbf{d}_{jk}$ is a vector perpendicular to the plane of the two vectors and of magnitude proportional to the area of the parallelogram formed by the two vectors which is $|\mathbf{d}_{ij}|\,|\mathbf{d}_{jk}|\sin\theta_{ijk}$.

Thus,

$$\sin\theta_{ijk} = \frac{|\mathbf{d}_{ij} \times \mathbf{d}_{jk}|}{|\mathbf{d}_{ij}|\,|\mathbf{d}_{jk}|}$$

Since θ is defined as being between 0 and 180° the sign of its sine is not given by the above expression but must be allocated by assessing the handedness of the three vectors $\mathbf{d}_{ij}$, $\mathbf{d}_{jk}$ and their vector product. The hand is shown by Fig. 4. For calculating θ_{ijk} the cosine expression is clearly preferable.

2.5.4. *Tetrahedral bonds*

When four straight lines meet at a point, six bond angles are formed between them. However, only five quantities are required to specify the configuration and thus there is a relationship between the six angles which enables any one to be found in terms of the other five as the solution of a quadratic equation.

The relation is

$$\begin{bmatrix} 1 & \cos\theta_{12} & \cos\theta_{13} & \cos\theta_{14} \\ \cos\theta_{12} & 1 & \cos\theta_{23} & \cos\theta_{24} \\ \cos\theta_{13} & \cos\theta_{23} & 1 & \cos\theta_{34} \\ \cos\theta_{14} & \cos\theta_{24} & \cos\theta_{34} & 1 \end{bmatrix} = 0$$

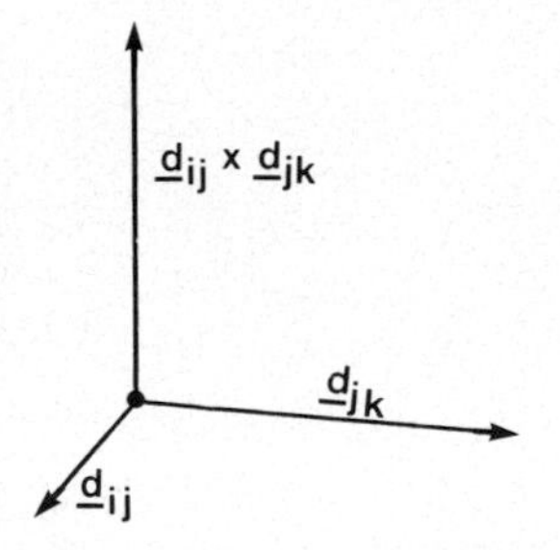

Fig. 4. The sign of the vector product.

It derives from the consideration that the four axes define a four-dimensional unit cell and that, since we are working in three-dimensional space, this four-dimensional content should be zero. This determinant is the square of that content and thus zero.

If more than four lines, for example N lines, meet at a point then the matrix:

$$\begin{bmatrix} 1 & \cos\theta_{12}\ldots & \cos\theta_{1N} \\ \cos\theta_{12} & & \ldots \\ \ldots & & \\ \cos\theta_{1n}\ldots & & 1 \end{bmatrix}$$

should be of rank three and all determinants of order four made from it should be zero. In other words the four-line relationship applies to all combinations of four lines among the N.

2.5.5. Torsion angles

Three points 1,2,3 in a triangle specify a bond angle θ_{123}. When a fourth point (4) is to be added, three further parameters must be specified. These may be distances from the points 1,2,3 or one distance, such as d_{34}, one bond angle θ_{234} and one torsion angle ϕ_{1234}. In all cases it is necessary to know on which side of the plane (123) the point (4) lies. If the volume of the tetrahedron (1234) is calculated from the x,y,z coordinates of the four points, then its sign will be positive if (4) lies on one side of (123) and negative if it lies on the other. The sign requires that the hand of the axial frame should be fixed, so that right-handed axes are always chosen.

The torsion angle ϕ_{1234} (which equals ϕ_{4321}) is defined as the dihedral angle between the planes (123) and (234) and is positive in the case illustrated in Fig. 5. The ϕ value for the cis-configuration is 0° and for the trans-configuration is 180°. ϕ may run from 0° to 360° but it is usually better to make it lie between $-180°$ and $+180°$ since mirror symmetry is then more recognisable. Looking at the tetrahedron of figure the angle ϕ_{1234} is seen to be negative while the volume of the tetrahedron (as given by the formula of Section 2.4) is positive so that V_{1234} and ϕ_{1234} have opposite signs.

ϕ_{1234} is the angle between the normals to the planes (123) and (234) which are parallel to the vector products ($\mathbf{d}_{12} \times \mathbf{d}_{23}$) and ($\mathbf{d}_{23} \times \mathbf{d}_{34}$) respectively. For the coordinate axes $\mathbf{X} \times \mathbf{Y} = \mathbf{Z}$. Thus,

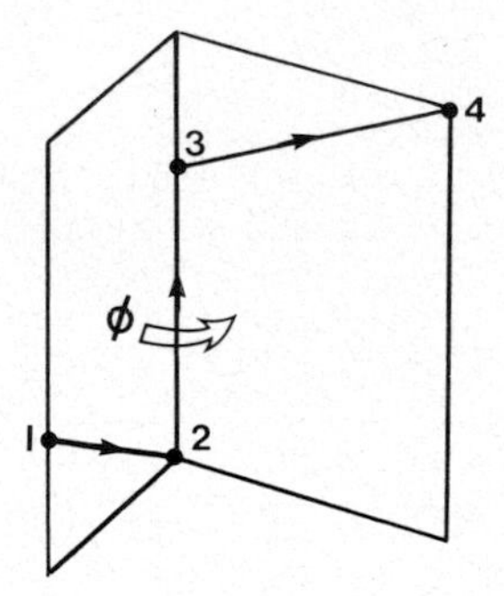

Fig. 5. The sign of the torsion angle ϕ_{1234} is positive.

$$\cos \phi_{1234} = \frac{(\mathbf{d}_{12} \times \mathbf{d}_{23}) \cdot (\mathbf{d}_{23} \times \mathbf{d}_{34})}{|\mathbf{d}_{12} \times \mathbf{d}_{23}|\ |\mathbf{d}_{23} \times \mathbf{d}_{34}|}$$

$$= \frac{\begin{vmatrix} \mathbf{d}_{12}\cdot\mathbf{d}_{23} & \mathbf{d}_{23}\cdot\mathbf{d}_{23} \\ \mathbf{d}_{12}\cdot\mathbf{d}_{34} & \mathbf{d}_{23}\cdot\mathbf{d}_{34} \end{vmatrix}^{1/2}}{\begin{vmatrix} \mathbf{d}_{12}^2 & \mathbf{d}_{12}\cdot\mathbf{d}_{23} \\ \mathbf{d}_{12}\cdot\mathbf{d}_{23} & \mathbf{d}_{23}^2 \end{vmatrix}^{1/2} \begin{vmatrix} \mathbf{d}_{23}^2 & \mathbf{d}_{23}\cdot\mathbf{d}_{34} \\ \mathbf{d}_{23}\cdot\mathbf{d}_{34} & \mathbf{d}_{34}^2 \end{vmatrix}^{1/2}}$$

$$= \frac{\cos \theta_{123}\cdot\cos \theta_{234} - \cos \theta_{(12,34)}}{\sin \theta_{123}\cdot\sin \theta_{234}}$$

(By cancelling $d_{12}d_{34}d_{23}^2$ and using $\mathbf{d}_{12}.\mathbf{d}_{23} = -d_{12}d_{23} \cos \theta_{123}$

$$\cos \theta_{(12,34)} = \left|\frac{\mathbf{d}_{12}\cdot\mathbf{d}_{34}}{|\mathbf{d}_{12}|\ |\mathbf{d}_{34}|}\right|)$$

These scalar and vector products are readily calculated from the orthonormal Cartesian coordinates. For example,

$$\mathbf{d}_{12}.\mathbf{d}_{34} = (x_2 - x_1)(x_4 - x_3) + (y_2 - y_1)(y_4 - y_3) + (z_2 - z_1)(z_4 - z_3)$$
$$|\mathbf{d}_{12}| = [(x_2 - x_1)^2 + (y_2 - y_1)^2 + (z_2 - z_1)^2]^{1/2}$$

but $\cos \phi_{1234}$ can also be expressed in terms of internal parameters as:

$$\cos \phi_{1234} = \frac{d_{13}^2 - d_{14}^2 + d_{24}^2 - d_{23}^2 + 2d_{12}d_{34}\cos \theta_{123}\cos \theta_{234}}{2d_{12}d_{34}\sin \theta_{123}\sin_{234}}$$

When the inverse cosine is taken the resulting angle is between 0° and 180°. The correct sign must then be attached to accord with the hand of the configuration (1234). The hand of the torsion angle (shown in Fig. 5) is positive, the opposite to the sign of the volume as determined from the formula above. The sign of the volume of a tetrahedron can be found by inspection on comparing the configuration with that of the tetrahedron in Fig. 3, which is positive.

A convenient program in BASIC has been given by J.D. Dunitz [6]. This gives distances, bond angle and torsion angles starting from fractional coordinates and unit cell dimensions but is generally applicable.

2.6. *Distance coordinates*

We have seen that bond angles and torsion angles are derived from the mutual distance between sets of three and of four points respectively. It is possible to describe a configuration of points completely, *except* for the hand of the arrangement, by specifying the distances d_{ij} between all pairs.

In fact only $3N - 6$ properly chosen distances need be specified to define an arrangement of N points in space. The remaining distances (bringing the number up to $\frac{1}{2}N(N-1)$) can be found by a generalised form of triangulation for which the program outlined in Section 2.7 can be used. More than the minimum number of distances are, however, useful in deciding the handedness of each tetrahedron.

The basic question is: given the 3 distances of point 4 from points 1,2,3, the distances of point 5 from 1,2,3 and the three distances between 1,2 and 3, find the distance 4–5. There are, in fact two solutions, with points 4 and 5 on the same side or on opposite sides of the plane 1,2,3.

The fundamental relationship is that the five points define a simplex in four dimensions and this should have a zero four-dimensional content. Thus:

$$\begin{vmatrix} 0 & d_{12}^2 & d_{13}^2 & d_{14}^2 & d_{15}^5 & 1 \\ d_{12}^2 & 0 & d_{23}^2 & d_{24}^2 & d_{25}^2 & 1 \\ d_{13}^2 & d_{23}^2 & 0 & d_{34}^2 & d_{35}^2 & 1 \\ d_{14}^2 & d_{24}^2 & d_{34}^2 & 0 & d_{45}^2 & 1 \\ d_{15}^2 & d_{25}^2 & d_{35}^2 & d_{45}^2 & 0 & 1 \\ 1 & 1 & 1 & 1 & 1 & 0 \end{vmatrix} = 0$$

For five points, $3N - 6 = 9$, while there are $\frac{1}{2}N(N-1) = 10$ distances between them, so that among the 10 distances d_{ij} there is one relationship (that given above) which can be used to give any one distance in terms of the others as the solution of a quadratic.

The closeness of this determinant to zero can be used as a check on the consistency of a set of ten distances between five points. In surveying, the accuracy of measures of length (using a tellurometer) has greatly exceeded that of measures of angle, so that distances are now primary data.

If the distance matrix is constructed for a large array of points than five, then all smaller determinants of the type given, connecting five points, will be zero.

2.7. *Conversion of internal coordinates (of a molecule or other structure) to Cartesian coordinates*

L.N.M. Carnot (one of Napoleon's generals, whose epithet was 'the organiser of victory' and who studied geometry when he was out of favour) wrote [4] in 1803: 'In any system whatever of straight lines, in the same plane or not, being given certain of their lengths, or of the angles between them, or of the angles between planes containing them, in number sufficient to determine the figure, find the remainder of these parameters'.

That is to say, given internal coordinates d_{ij}, θ_{ijk}, ϕ_{ijkl} sufficient to determine a structure, find x_i, y_i, z_i, Cartesian coordinates of each of the atoms or points, and thence calculate any other internal coordinates which may be desired.

The problem is solved by the following algorithm. A structure of N points requires $3N - 6$ properly chosen parameters for its unique definition. To specify positions of N atoms requires $3N$ Cartesian coordinates.

(1) Allocate rough coordinates x_i, y_i, z_i to the N points (these are the quantities which are to be refined, so that the closer to the correct values the better).

(2) Calculate values for the M parameters d_{ij}, θ_{ijk}, ϕ_{ijkl} specified and thus find the discrepancy for each parameter.

(3) Calculate the variation of each of the M parameters with respect to each of the $3N$ Cartesian coordinates (by calculating each parameter, adding an increment to the coordinate and re-calculating the parameter).

(4) Thus, set up a block of linear equations to find the corrections to each of the $3N$ coordinates required to reduce the discrepancy between observed and calculated parameters to zero.

(5) Solve this block of equations using the generalised inverse method and repeat steps 2–5 until the coordinates x_i, y_i, z_1 change only inappreciably and yield the observed internal parameters.

(6) From this final configuration calculate whatever other internal coordinates may be required.

We may distinguish three cases:

(a) There are only $3N - 6$ internal coordinates and they determine the configuration exactly. Here a set of $3N$ Cartesian coordinates is obtained but the orientation and position of the axes relative to the structure is indeterminate. J.D. Dunitz [7] gives a program in BASIC which solves a restricted version of this case.

(b) If these are more than $3N - 6$ internal coordinates and they are inconsistent with each other, being 'experimental observations' then the 'least squares solution' minimising the sum of the squares of the discrepancies is obtained. Equations for d, θ and ϕ must be appropriately weighted according to the nature of the experimental observations.

(c) If there are insufficient internal coordinates to determine the structure, then a configuration which fits the data will be obtained, but the solution will not be unique. When calculating the generalised inverse the rank of the matrix is obtained and this can be noted to see whether it is equal to $3N - 6$ or is less.

It has been found a useful device to limit the corrections applied in a refinement by using the inverse tangent function, since $\arctan(x)$ is proportional to x for small values of x but does not exceed $\pi/2$ however large x may be. This prevents the refinement from 'running away' by overcorrection at an early stage. If the indicated correction is X and the maximum which can be sensibly applied is M we may write: $X = (2/\pi)\ M \arctan(X)$.

This program was developed to deal with a problem which looked simple but proved to be very difficult by orthodox methods [8], namely: Given the 12 edge lengths of an irregular octahedron, calculate the lengths of the three remaining diagonals (there are 15 distances between 6 points, given 12 find the other 3). This problem is the simplest case of the geometry of the geodesic domes developed by Buckminster Fuller which becomes extremely difficult if there is no symmetry. The method described enables the stresses in such a structure to be calculated using, for example, the Reciprocal Theorem of Maxwell, Betti and Rayleigh [9].

2.8. *Cartesian coordinates from the distance matrix*

2.8.1. *A procedure for conversion*

Given a table of all the distances d_{ij} between N points, a set of Cartesian coordinates can be obtained in the following way.

(1) Form the Matrix $\begin{bmatrix} 0 & d_{12}^2 & d_{13}^2 \ldots & d_{1N}^2 \\ d_{12}^2 & 0 & & \\ \cdot & & & \\ \cdot & & & \\ \cdot & & & \end{bmatrix}$

(2) From it form the metric matrix for an $N-1$ dimensional simplex with terms $g_{ij} = \mathbf{a}_i.\mathbf{a}_j = \frac{1}{2}(d_{il}^2 + d_{jl}^2 - \mathrm{d}_{ij}^2)$ by subtracting the first row and column from every other row and column. The first point then become the origin of the Cartesian coordinates (0,0,0), the second point is along the x-axis $(x_1,0,0)$, the third point lies in the x–y plane $(x_2,y_2,0)$ and the other points have general values of x_i, y_i, z_i.

(3) This matrix can then be factorised by Choleski's method into a lower triangular matrix and its transpose, but only if the matrix is positive definite. That is the determinants of all minors must be positive. If there are violations of the triangle inequality (i.e., if d_{23} is greater than $d_{12} + d_{23}$) then the method will fail.

If the distances are somewhat inconsistent with each other then the fourth and subsequent columns of the lower triangle matrix obtained will be non-zero and the first three columns can be taken as a set of Cartesian coordinates which can be refined further according to appropriate criteria.

2.8.2. *Lagrange's theorem and the radius of gyration*

It is slightly surprising that the radius of gyration R (the r.m.s. distance of a set of points from their common centre of gravity) can be calculated from a table of their mutual distances. This was proved by Lagrange and follows from the Parallel Axis Theorem for moments of intertia. m_i is the mass of the ith point.

$$R^2 = \frac{\sum_{i=1}^{N} m_i \mathrm{D}_{i0}^2}{\sum_{i=1}^{N} m_i} = \frac{\sum_{i=1}^{N}\sum_{j=1}^{N} m_i m_j d_{ij}^2}{2\left(\sum_{i=1}^{N} m_i\right)^2} \text{ or } \frac{\sum_{i=2}^{N}\sum_{j=1}^{i-1} m_i m_j d_{ij}^2}{\left(\sum_{i=1}^{N} m_i\right)^2}$$

further, the distance D_{i0} of each point from the centre of gravity can be calculated from:

$$D_{io}^2 = \frac{\sum_{j=1}^{N} d_{ij}^2 m_j}{\sum_{j=1}^{N} m_j} - \frac{\sum_{j=2}^{N}\sum_{k=1}^{j-1} d_{jk}^2 m_j m_k}{\left(\sum_{j=1}^{N} m_j\right)^2}$$

In principle it is possible to calculate also the principal moments of inertia of the assembly from the matrix d_{ij}^2 but it is very much simpler to calculate these, and any other such measures, by way of assigning Cartesian coordinates to each of the points first.

2.8.3. *Crippen and Havel's method*

A much improved method of calculating a set of Cartesian coordinates from a table of all the distances d_{ij} between N points has been given by G.M. Crippen and T.F. Havel [10]. The distances may not even satisfy the triangle inequality and may be quite approximate, the idea being that $\frac{1}{2}N(N-1)$ rough measures may allow $3N-6$ parameters (the Cartesian coordinates) to be determined the more accurately the larger N.

Their treatment uses Lagrange's theorem. The steps are as follows:

(1) Calculate the distance of each point from the centre of mass by Lagrange's theorem

$$d_{i0}^2 = \frac{1}{N}\sum_{j=1}^{N} d_{ij}^2 - \frac{1}{N^2}\sum_{j=2}^{N}\sum_{k=1}^{j-1} d_{jk}^2$$

(2) Form the metric matrix $[\mathbf{g}_{ij}]$ where $g_{ij} = \frac{1}{2}(d_{i0}^2 + d_{j0}^2 - d_{ij}^2)$ (The elements of the matrix are all the scalar products

$$\mathbf{d}_{i0}.\mathbf{d}_{j0} = |d_{i0}|\,|d_{j0}| \cos\theta_{i0j} \quad \text{and}$$
$$d_{ij}^2 = d_{i0}^2 + d_{j0}^2 - 2\,|d_{i0}|\,|d_{j0}| \cos\theta_{i0j} \quad \text{so that}$$
$$|d_{i0}|\,|d_{j0}| \cos\theta_{i0j} = d_{i0}^2 + d_{j0}^2 - d_{ij}^2)$$

(3) Find the eigenvalues of $[\mathbf{g}_{ij}]$ (which is symmetric) using, for example, Jacobi's method (see Section 7.4). Take the three largest eigenvalues $\lambda_1,\lambda_2,\lambda_3$ and find the three corresponding eigenvectors given as the three columns of the $N \times 3$ matrix $[\mathbf{W}_{ij}]$.

(4) Calculate coordinates x_{ij} (points $i+1$ to N, axes $j = 1$ to 3) from $x_{ij} = \lambda_j^{1/2}\mathbf{W}_{ij}$. If λ_j is less than 0 due to errors, then use $|\lambda_j|$ in place of λ_j.

The problem is greatly overdetermined if N is large enough and can sustain a heavy load of inaccuracy in the values d_{ij} if we have the a priori knowledge that the points in question really lie in two- or three- (or more) dimensional space. For example a table of the fares in dollars between all pairs of a dozen European cities (66 data) can be used to find approximate relative positions of these cities (21 coordinates) with surprising accuracy. A scaling function may be needed if the measure is not directly proportional to the distance. The method has been used to reconstruct a map of town in ancient Mesopotamia from the frequencies of association of their names on clay tablets [11]. Applications in ecology suggest themselves. The essential pre-requisite is that the vague measures of distance should be labelled as being between two particular places. If the distances are unlabelled the problem is almost impossible to solve.

3. LINES IN SPACE

The next simplest geometrical structures, after arrays of points, are linear continua of points, namely lines. We take first lines in two dimensions.

3.1. *Curved lines in the plane*

At a point, not being a singular point, a plane curve has a tangent, which can be

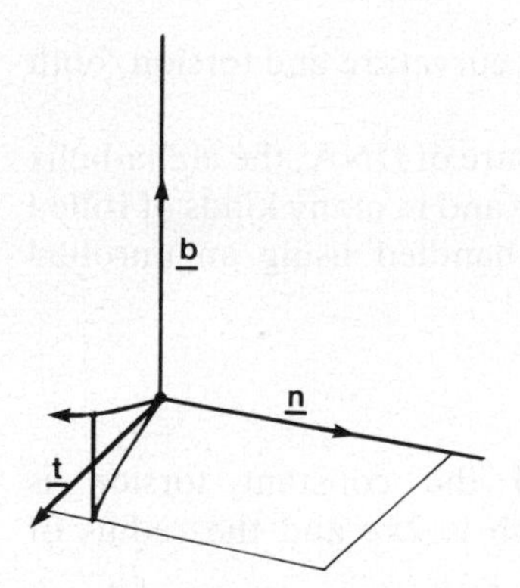

Fig. 6. A curved line in space. t,n is the osculating plane containing the circle of curvature; n,b is the normal plane.

regarded as the limit of the chord joining two points as the two points approach each other. The tangent represents the instantaneous direction of the curve. Similarly, the limit of a circle through three points, as the outer two points approach the third, gives the radius of curvature at the point.

In terms of Cartesian coordinates the length of an element ds of a curve is $\mathrm{d}s = (1 + (\mathrm{d}y/\mathrm{d}x)^2)^{1/2}\mathrm{d}x$ or, if the coordinates x and y are themselves expressed in terms of a parameter t $((\mathrm{d}x/\mathrm{d}t)^2 + (\mathrm{d}y/\mathrm{d}t)^2)^{1/2}\mathrm{d}t$, or, if in polar coordinates, $(r^2 + (\mathrm{d}r/\mathrm{d}\theta)^2)^{1/2}\mathrm{d}\theta$. If the equation of the curve is known ds can be integrated to give the length.

The radius of curvature R (the inverse of the curvature K) is given by $R = (1 + (\mathrm{d}y/\mathrm{d}x)^2)^{3/2}/(\mathrm{d}^2y/\mathrm{d}x^2)$, or in polar coordinates,

$$R = \frac{(r^2 + (\mathrm{d}r/\mathrm{d}\theta)^2)^{3/2}}{r^2 + 2(\mathrm{d}r/\mathrm{d}\theta)^2 - r(\mathrm{d}r/\mathrm{d}\theta)}$$

or, if the curve is given as $F(x,y) = 0$

$$R = \frac{(F_x'^2 + F_y'^2)^{3/2}}{\begin{vmatrix} F''_{xx} & F''_{xy} & F'_x \\ F''_{yx} & F''_{yy} & F'_y \\ F'_x & F'_y & 0 \end{vmatrix}}$$

where F'_x denotes the derivative of the function F with respect to x.

3.2. *Curved lines in space*

At each (non-singular) point, a curve in space, a tortuous curve, has both curvature and tortuosity or torsion. At a point the curve can be considered to lie in a plane (containing its circle of curvature). The curvature is the rate of change of direction in this plane and the torsion is the rate at which the curve departs from the plane (Fig. 6).

This is best seen for a helix composed of a thin tape wound round a cylinder. If there were only curvature and no torsion, the tape would simply form a coil or roll. If there were only torsion and no curvature, the tape would be a straight length of axially twisted ribbon. For a helix, which is a curve of constant curvature and torsion, both are necessary.

The circular helix is, of course, important in the structure of DNA, the alpha-helix of proteins, the rods of TMV, the tails of bacteriophage and in many kinds of rolled sheets of protein particles. Their geometry is best handled using an unrolled cylindrical projection and the parametric equations

$$x = a \cos \theta, \quad y = a \sin \theta, \quad z = c\theta$$

Here the (constant) curvature is $a/(a^2 + c^2)$ and the constant torsion is $c/(a^2 + c^2)$. The pitch angle $\alpha = \arctan(c/a)$, the pitch is $2\pi c$ and the radius of the cylinder is a.

In Fig. 6 t, n is the osculating plane containing the circle of curvature; n, b is the normal plane.

The curvature is

$$\frac{d^2r}{ds^2} = (x''^2 + y''^2 + z''^2)^{1/2} = r'' = K$$

(if the curve is specified by giving x, y and z as functions of a parameter s. Primes denote differentiation by this parameter.) The radius of torsion, which is the reciprocal of the torsion, is

$$\frac{(x''^2 + y''^2 + z''^2)}{\begin{vmatrix} x' & y' & z' \\ x'' & y'' & z'' \\ x''' & y''' & z''' \end{vmatrix}} = \frac{1}{\tau}$$

so that, as the curvature K is $|\mathbf{r}''|$, the triple vector product $[\mathbf{r}'\mathbf{r}''\mathbf{r}'''] = K^2\tau$.

3.2.1. *The regular helix*

As an example, these formulae may be applied to the regular helix where $\mathbf{r}$, the position vector $= (a \cos \theta, a \sin \theta, c\theta)$. The tangent $\mathbf{r}'$, the first derivative with respect to the distance along the curve $= \mathrm{d}\mathbf{r}/\mathrm{d}s = \mathrm{d}\mathbf{r}/\mathrm{d}\theta \,.\, \mathrm{d}\theta/\mathrm{d}s$

$$\mathrm{d}s = (\mathrm{d}x^2 + \mathrm{d}y^2 + \mathrm{d}z^2)^{1/2} = (a^2 \sin^2 \theta + a^2 \cos^2 \theta + c^2)^{1/2}.\mathrm{d}\theta$$

and $\mathrm{d}s/\mathrm{d}\theta = (a^2 + c^2)^{1/2}$.
thus

$$\mathbf{r}' = (-a \sin \theta, a \cos \theta, c)/(a^2 + c^2)^{1/2}$$
$$\mathbf{r}'' = \mathrm{d}\mathbf{r}'/\mathrm{d}s = \mathrm{d}\mathbf{r}'/\mathrm{d}\theta \,.\, \mathrm{d}\theta/\mathrm{d}s = (-a \cos \theta, -a \sin \theta, 0)/(a^2 + c^2)$$

so that the curvature $K = |\mathbf{r}''| = a/(a^2 + c^2)$.
While,

$$\mathbf{r}''' = \mathrm{d}\mathbf{r}''/\mathrm{d}s = \mathrm{d}\mathbf{r}''/\mathrm{d}\theta \,.\, \mathrm{d}\theta/\mathrm{d}s = (a \sin \theta, -a \cos \theta, 0)/(a^2 + c^2)^{3/2}$$

From $[\mathbf{r}\mathbf{r}''\mathbf{r}'''] = K^2\tau$ we have

$$\frac{\begin{vmatrix} -a \sin \theta & a \cos \theta & c \\ -a \cos \theta & -a \sin \theta & 0 \\ +a \sin \theta & -a \cos \theta & 0 \end{vmatrix}}{(a^2 + c^2)^3} = \frac{ca^2}{(a^2 + c^2)^3} = k^2\tau = \frac{a^2}{(a^2 + c^2)^2}\,\tau$$

$$\therefore \tau = \frac{c}{(a^2 + c^2)}$$

3.3. *The cylindrical packing of equal spheres*

The close-packing of spheres, whose centres all lie in the surface of a circular cylinder (and which are all tangent internally and externally to other circular cylinders) has been described by R. Erickson [12]. The actual points of contact will be inside the surface through the centres of the spheres. These helices, on which the spheres appear to be arranged, are called *parastichies*. Spheres on the surface of the cylinder appear to be hexagonally close packed (with six neighbours) but they may in fact be in contact along only two parastichies (rhombic packing) although contact along three (hexagonal packing) is sometimes possible. When the cylinder containing the centres of the spheres is unrolled, a two-dimensional lattice is obtained (periodic in two dircctions).

Fig. 7 (reproduced by kind permission of Science) shows 'representative tubular arrangements of spheres, drawn in parallel projection on to the plane, in side view and as viewed from above. All are triple contact patterns except (1,2) and (3,5)'. The spheres are given numbers in order of increasing z (distance along the cylinder axis) and the symbol for each packing signifies the numbers of those spheres which are in contact with the one labelled 0.

3.4. *Differential geometry and the supercoiling of DNA*

An important relationship applies to the twisting and coiling of a ribbon-like thread (which can store energy on bending and on torsion). The phenomenon will be familiar to anyone who has wound a rubber-powered model aeroplane. If a rubber tube is held at the ends, extended, twisted to store torsional energy, and then the ends are allowed to approach each other, a spur of super-coiled helix appears, a doubled helix standing out like the stalk of the letter Y from the main axis.

The formal relationship was given by J.H. White as $L = T + W$, where L is the linkage number (an integer), T the twist or torsion and W, a new quantity called the writhing number. W is known mathematically as the Gaussian integral of the axis curve [13].

The linkage number L is a topological characteristic describing the linkage of two closed curves which, in a special case, may be the edges of a ribbon with its ends joined (the edge of a Möbus strip would be only one curve). For the two curves to be separable, the linkage number must be zero (but this may not be a sufficient condition). Roughly speaking, L is the number of breaks necessary for the curves to

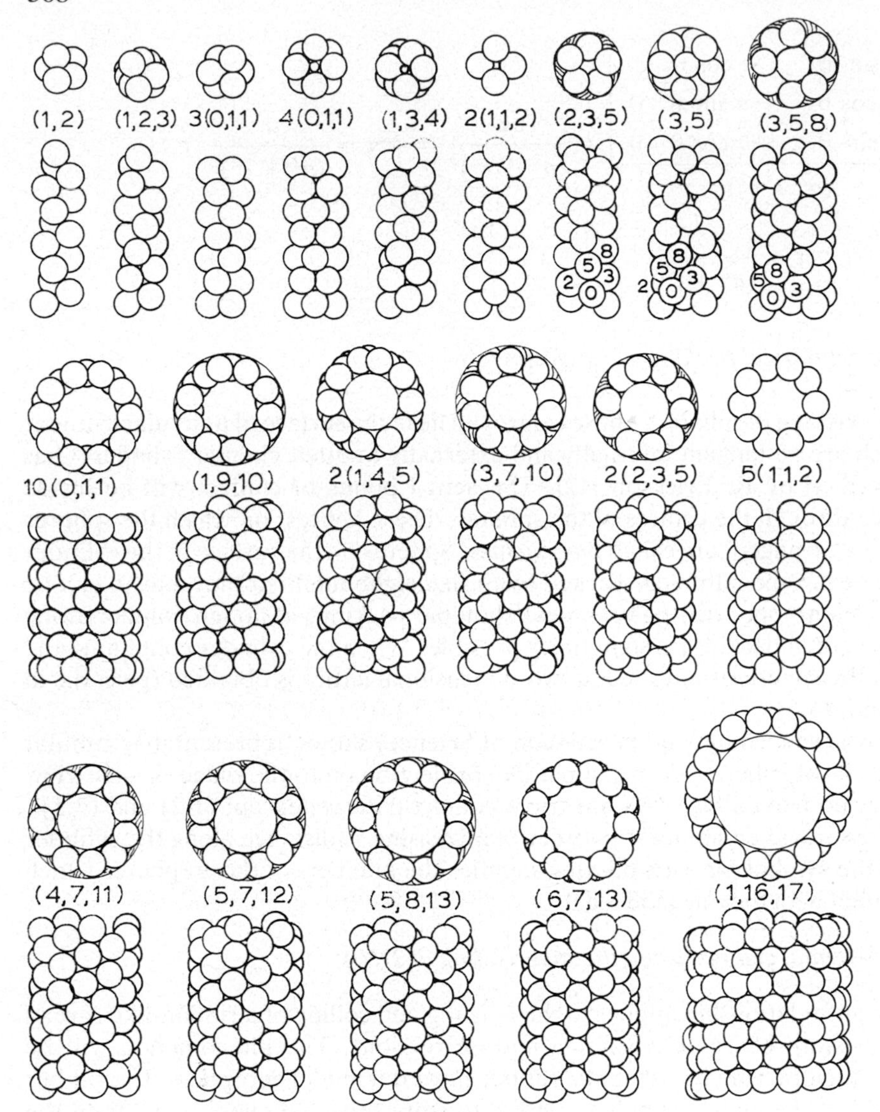

Fig. 7. The possible cylindrical packings of equal spheres. All spheres are at the same distance from the axis and each touches 4 or 6 neighbours (R. Erickson).

be separable by pulling apart without unthreading. L carries a sign if the curves also have identifiable directions (as does a single strand of DNA) and this sign changes on mirror reflection.

The twist T has the conventional meaning and for a flat ribbon is the integral of the rotation of the plane of the ribbon about its axis, summed along the loop (or along a segment) since the twist can be defined for a segment considered in isolation. Twisting into a right-handed helix is measured as positive, so that T also changes sign on reflection. T is measured in turns.

The writhing number W can be computed as the difference $W = L - T$. As a special case, if the axis of the ribbon lies entirely in a plane or on the surface of a sphere, W is zero. W depends entirely on the curve of the axis of the ribbon. For DNA it can be measured directly since W affects the compactness of the loop and thus its rate of sedimentation.

The two papers mentioned [13] give fuller explanation of a concept which it is rather difficult to visualise and still harder to describe without models.

4. SURFACES

A surface in space may be regarded as a continuum in which two parameters are necessary for the specification of a point. It may be described algebraically as a function $F(x,y,z) = 0$ (such as the sphere where $x^2 + y^2 + z^2 - a^2 = 0$ or the plane where $lx + my + nz - p = 0$ (l, m and n being the direction cosines)) or parametrically as $x = X(u,v)$, $y = Y(u,v)$, $z = Z(u,v)$ as a function of two parameters u and v.

In some cases the function can be separated to give z as a function of x and y, so that x and y are themselves these parameters and z is the height of the point above the x–y plane.

A surface may also be broken up into finite elements and specified digitally or numerically by various methods. For example: (a) as an array of points (using Cartesian, spherical or other coordinates); (b) as contour lines in two directions – as links between points; (c) by triangulation, as triangles between triplets of points; (d) as quadrilateral elements where the vectors defining the edges lie in the principle planes X–Y or Y–Z or Z–X of the Cartesian coordinates; (e) as serial sections – outlines parallel to a principal plane.

If the equation can be separated to give $z = F(x,y)$ then the area of such a curved surface, provided it does not become parallel to z, that is the gradients must remain finite, can be calculated as follows: The element of surface area δA for an element $\delta x \times \delta y$ on the ground plane is $\delta x\delta y/\cos\theta$, where θ is the angle between the normal of the element and the vertical 0–Z

$$A = (1 + (\partial F/\partial x)^2 + (\partial F/\partial y)^2)^{1/2}\delta x\delta y.$$

The gradients $\partial F/\partial x$ and $\partial F/\partial y$ should be calculated for the middle of the element.

4.1. The super-ellipsoid

Spheres and ellipsoids are very frequently used in modelling (since many problems in, for example, electrostatics can be solved exactly) but it is often convenient to describe a brick-shaped cell as $(x/a)^n + (y/b)^n + (z/c)^n = 1$, where n is a high power such as 15. (For an ellipsoid $n = 2$.) This automatically handles corners by rounding them off and gives a box approximately $2a \times 2b \times 2c$. If $n = 15$ and $a = b = c = 1$ then the surface goes through the point 0.93, 0.93, 0.93 so that the vertex is within a few percent of reaching to 1,1,1.

The area of the surface of an ellipsoid is not calculable in analytic functions but it can readily be calculated by a computer program given in Section 7.2. The equation $((x^2 + y^2)/a^2)^n + (z/y)^{2n} = 1$ gives a cylindrical 'cannister' shape, similar to that observed in the division of bacteria [14].

4.2. *The random arrangement of points on a sphere*

In simulations it is often necessary to put N points uniformly at random on to the surface of a sphere. It is a theorem (known since Archimedes) that if a sphere is sliced into layers of equal thickness, each slice has the same area of curved surface. The obliquity of the surface compensates for the reduction in radius. Accordingly the following program segment will (in spherical coordinates θ,ϕ) allocate random coordinates to N points.

```
Rem θ and φ in radians
DIM T(100, F(100)
N = 50
P1 = 3.1415926535
REM define inverse cosine (fails at x = −1)
DEF FNC(X) = 2 * ATN(SQR((1 − X)/(1 + X)))
FOR I = 1 TO N
F(I) = 2 * P1 *RND(0)
T(I) = FNC(2 * RND(0) − 1)
NEXT I
```

4.3. *The regular arrangement of points on a sphere*

If it is required to set up a lattice of points to cover a sphere most effectively then one of the icosahedral tessellations might be chosen. In such an arrangement 12 points will have five neighbours and the remainder six. The packing is like a plane hexagonal sheet wrapped round a sphere with 12 dislocations. The number of points is $2 + 10T$ where T is the tessellation number 1,3,4,7,9,12,13, ... ($h^2 + hk + k^2$ with $h,k = 1,2,3,4, \ldots$).

The steps in setting up this packing are as follows:

(1) Taking the centre of the sphere as origin find the coordinates of the 12 vertices of an icosahedron. These are $(1,0,t)$, $(1,0,-t)$, $(-1,0,t)$, $(-1,0,-t)$, $(t,1,0)$, $(-t,1,0)$, $(t,-1,0)$, $(-t,-1,0)$, $(0,1,t)$, $(0,-1,t)$, $(0,1,-t)$, $(0,-1,-t)$ where $t = (1 + 5^{1/2})/2$, the golden number.

(2) The principle is that two adjacent faces of the icosahedron when flattened out make a rhombus which will fit on to a plane hexagonal net with its vertices coinciding with 4 points of the net (Fig. 8). The coordinates of the finer net points can be found as weighted averages of the coordinates of the vertices of the rhombus. When these have been allocated, then all points are normalised to lie at the same distance from the origin. Fig. 9 shows the $T = 7$ tessellation. If A, B and C are the coordinates of the icosahedral vertices, then the coordinates of the three extra points

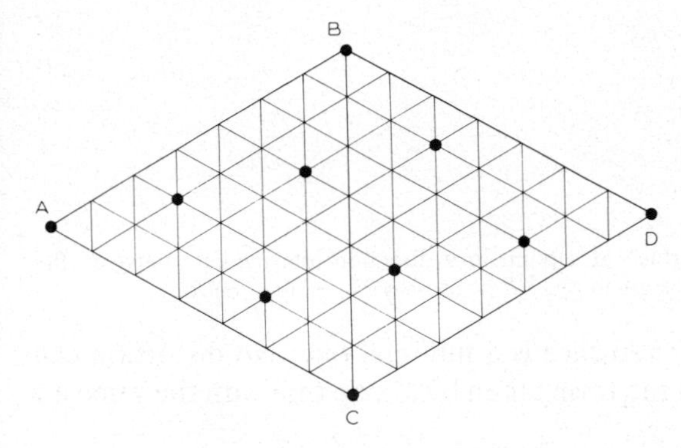

Fig. 8. The $T = 7$ tessellation of a hexagonal lattice. There are 7 points per rhombic cell ABCD and, if the figure is wrapped round an icosahedron, there will be 3 extra points on each triangular face disposed as shown on nodes of a lattice which is seven times finer.

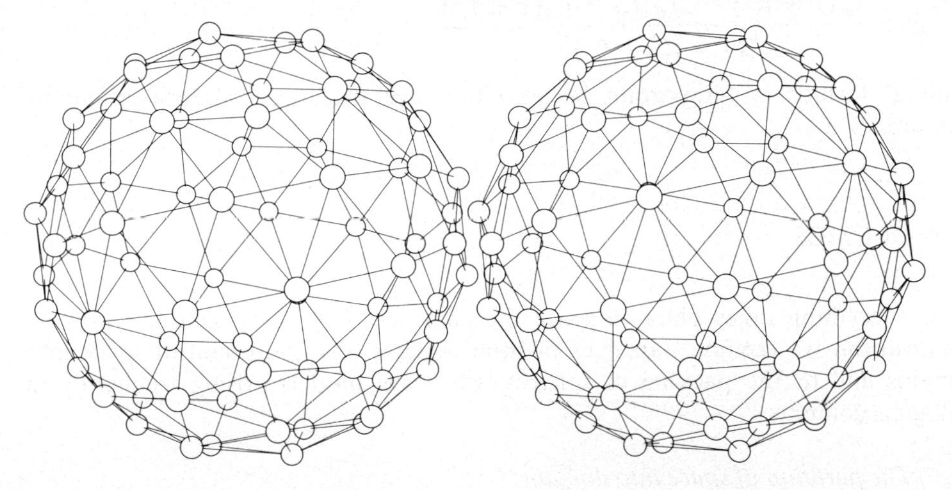

Fig. 9. The tessellation of Fig. 8 applied to an icosahedron and adjusted to make all points the same distance from the centre. (Stereopair arranged for 'cross-eyes viewing'.)

lying on that face have the weights $(4A + 2B + C)/7$; $(A + 4B + 2C)/7$; $(2A + B + 4C)/7$.

4.4. *Minimal surfaces*

At any non-singular point on a surface the vertex of an ellipsoid or hyperboloid can be fitted so that there are two principal radii of curvature R_1 and R_2 in perpendicular directions on the surface. The radius of curvature R in an intermediate direction is given by $1/R = (\cos^2\alpha)/R_1 + (\sin^2\alpha)/R_2$ (Fig. 10).

If the two principal curvatures are of opposite sign then the surface is saddle-shaped. The sum of the principal curvatures at a point gives the mean curvature H as $H = ½(1/R_1 + 1/R_2)$.

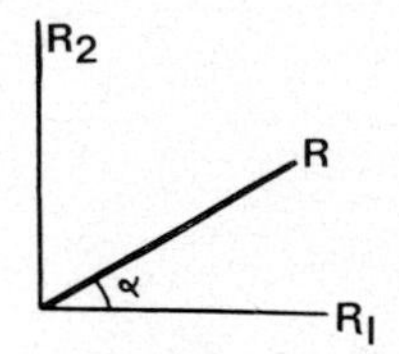

Fig. 10. The radius of curvature R of a surface at a point in a direction intermediate between the directions of principal curvatures R_1 and R_2 is given by: $1/R = (\cos^2 \alpha)/R_1 + (\sin^2 \alpha)/R_2$.

If this is everywhere zero, then the surface is a minimal one. No distortion can make the total surface less and this is the form taken by a soap film with the same air pressure on each side [15].

If a surface can be described in the form $z = f(x,y)$ then the Laplace–Young equation gives the mean curvature H in terms of the derivatives with respect to x and y as:

$$H = \frac{f''_{xx}(1 + f'^2_y) - 2f'_x f'_y f''_{xy} + f''_{yy}(1 + f'^2_x)}{2(1 + f'^2_x + f'^2_y)^{3/2}}$$

Minimal surface are important because they are the forms also taken up by membranes.

5. VOLUMES

Since everything takes place in space the considerations of volume are literally manifold but we confine ourselves to some aspects of the division of space into domains and to the packing of spheres which are models arising frequently in biological contexts.

5.1. *The partition of space into domains*

For ecological, geographical, administrative and many other purposes, it may be necessary to divide space (of 2,3 or indeed of N dimensions), containing a number of points, such as the centres of atoms, the nests of birds or market towns, so that the whole space is allocated to the domain of one or other point, no space being unallocated.

5.1.1. *Voronoi polygons*

The simplest example is the division of a plane containing centres into Voronoi polygons, one surrounding each point, so that every point in a particular (necessarily convex) polygon is nearer to its 'centre' than to any other centre.

The method of achieving this is to draw the perpendicular bisector between every pair of centres and to take as the Voronoi polygon of a particular centre, the inmost segments of these bisecting lines. In this way the whole plane can be divided into convex polygons each containing one centre. Figure 11 gives an example [16].

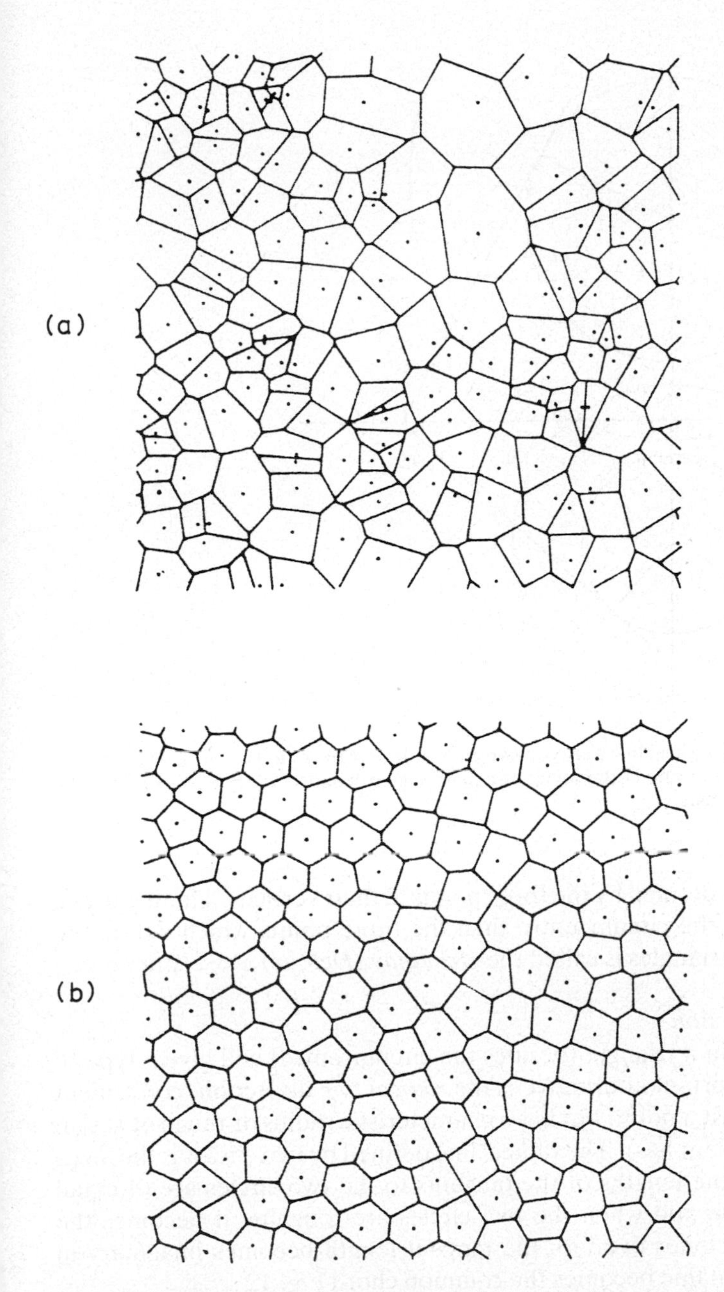

Fig. 11. (a) The Voronoi division of a random distribution of points in two dimensions. (Due to M. Tanemura and M. Hasegawa.) (b) The same pattern after the points have been adjusted to be nearly as possible equidistant from their immediate neighbours.

There are a variety of algorithms for carrying through this process automatically (and a list of relevant papers is given in Section 8) [16].

The basis consists of identifying possible vertices for the polygon which are the

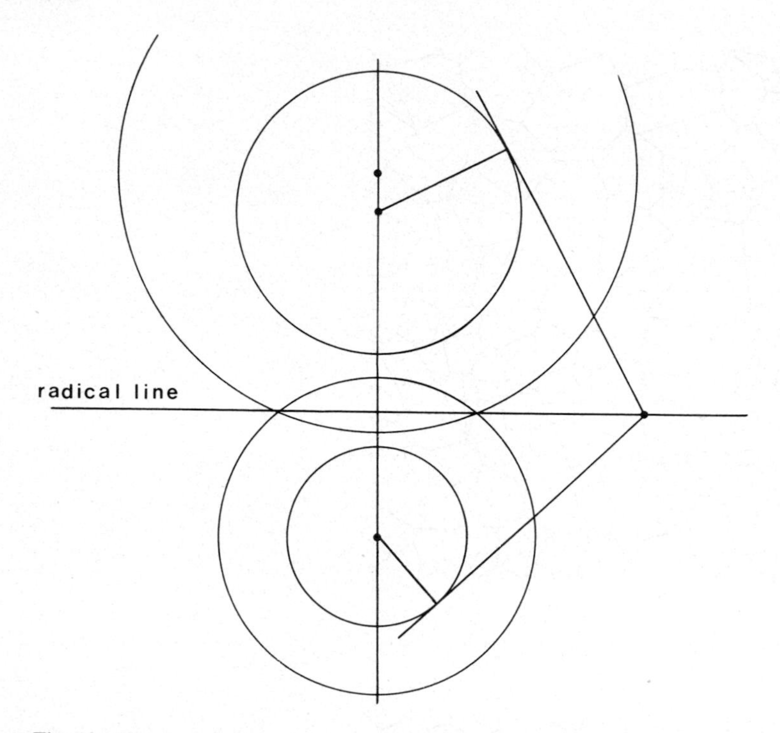

Fig. 12. The construction of the radical line between two circles. If they overlap, the line is the common chord. If they do not overlap, the line is the locus of equal tangent lengths. In three dimensions the radical plane is defined analogously.

circumcentres of triangles defined by the three points at their vertices. There must be no other point nearer to the circumcentre than the three points which define the triangle. The network of triangles is called the *Delaunay (Delone) Tessellation*.

5.1.2. Radical line dissection

In fact any unique point in a triangle (besides the circumcentre) will give a type of dissection. The most important alternative is the *radical line* dissection, convenient when each centre is not just a point, but has a characteristic radius or range of action (as an atom or a city or a plant has). The radical line defined by two circles is the locus of all points from which the lengths of the tangents to the two circles are of equal length. It is a straight line and when the two circles are touching it becomes the common tangent. If the circles overlap, the tangent length becomes imaginary in some places but the radical line becomes the common chord Fig. 12.

5.1.3. The division of space into domains

Given an array, of what we will call atoms, in three dimensions, the space can be divided into Voronoi domains (polyhedra in three dimensions) such that all points in the domain of a given atom are closer to that atom than to any other. At the same time the volume occupied by the array can be divided into simplices, tetrahedra in three dimensions, which have an atom at each vertex, giving the Delaunay tessellation which is the dual of the Voronoi tessellation.

The Voronoi tessellation of a space is unique and, if the arrangement is not degenerate (in having more than four atoms lying on the same sphere) so also is the Delaunay tessellation. A variety of computing algorithms have been developed, see Section 8.

5.1.4. *Properties of the Voronoi tessellation*

(a) All space is allocated so that the total volume is the sum of the volumes of the individual domains.
(b) Every point in a domain is nearer to the centre atom defining the domain than to any other atom.
(c) The partitions between the domains are planar and are the planes perpendicularly bisecting the lines joining pairs of atoms.
(d) The domains are convex.
(e) The partitioning is unique.
(f) The number of odd faces of each domain is even.
(g) Atoms at the surface of a finite cluster may have domains which are unbounded or abnormally large. That is, the procedure for dealing with the surface of a cluster is undefined.
(h) The dissection can be extended for use as a diagram of forces (Maxwell, 1869) [17].

5.1.5. *Properties of the associated Delaunay tessellation*

(a) Fills all space inside the cluster.
(b) Unique, except in cases of high symmetry.
(c) Each region is a tedrahedron with an atom at each vertex.

5.1.6. *Outline of the construction of Voronoi and Delaunay tessellations*

The algorithm described can readily be modified to give the radical plane dissection. It is only one method and, for example, Brostow et al. [16] in 1978 give a very different procedure.

The following is the outline of one algorithm for the Voronoi (or radical plane) dissection:

(a) Each atom is taken in turn, its neighbours within a certain distance (about two atom diameters) – those likely to contribute to defining a vertex – are listed.
(b) All triplets of these neighbours with the centre atom are examined and the position of the circumcentre of each such tetrahedron is calculated (a slightly variant formula deals with the case of the radical plane dissection). Each centre is tested to see whether any other atom is nearer than the four defining atoms.
(c) If more than four atoms lie at the same distance from the centre, then a special procedure must be invoked to deal with the degeneracy. The simplest way is to give each atom a slight random displacement to remove the symmetry.
(d) If the centre passes the test (b) then it is a vertex of the Voronoi polyhedron

about each of the four vertices, and is added to a list of such for the whole system. The tetrahedron of the four vertices is an element of the Delaunay tessellation. The volume of this tetrahedron is calculated and its sign (giving its cheirality) is noted also.
(e) The Voronoi dissection gives an unambiguous definition of coordination, and hence of coordination number. Two points are coordinated if they share in defining a Voronoi vertex. The two Voronoi polyhedra share a common face and its vertices are those which have the corresponding two indices (out of the four) in common.
(f) The geometry of the Voronoi polyhedra can be developed from the coordinates and from the quadruplet numbers of its vertices, together with the signs of the Delaunay tetrahedra which determine their cheirality.

5.1.7. Diagrams of forces

It was first pointed out by Rankin and amplified by J.C. Maxwell, that forces in a spatial framework can be represented by a reciprocal tessellation. 'The simplest case is that of five points in space with their ten connecting lines, forming ten triangular faces enclosing five tetrahedrons. By joining the five points which are the centres of the spheres circumscribing these five tetrahedrons, we have a reciprocal figure of the kind described by Professor Rankin in the Phil. Mag. of Feb. 1864; and forces proportional to the areas of the triangles of one figure, if applied along the corresponding lines of connexion of the other figure, will keep its points in equilibrium' [17].

This is the generalisation of the plane diagrams of forces also developed by Maxwell.

As is evident in a foam of soap bubbles, or in a pile of smooth ball-bearings, the geometry of the structure is, in itself, a diagram of forces. In D'Arcy Thompson's aphorism: 'Form is a diagram of forces' [2].

The geometry of the Voronoi/Delaunay dissections of space is the first step in finding a network to which Kirchoff's or Tellegen's theorems can be applied. The latter connect 'across' and 'through' variables – generalised voltages and currents operating in conformity with the general principles of minimisation of the rate of doing work. Maxwell's work is essentially the foundation of 'finite element analysis' which is the major way in which computers can be applied to structural mechanics. Biological structures are also engineering structures and the various computer packages of programs can be applied to their mechanics.

5.2. The packing of spheres

A standard model for various purposes is the stacking or arrangement of hard spheres in contact with each other. Their geometry can be dealt with by triangulation or distance geometry, but a number of direct relationships between their radii are available.

5.2.1. To find the radius of a sphere inscribed in the cavity between four spheres in mutual contact

If four spheres of radii R_1,R_2,R_3,R_4 are in contact with each other making a tetrahedron, then the radius of the inscribed sphere (just filling the cavity between them) and of the circumscribed sphere are given as the roots of the quadratic equation:

$$(1/R_1 + 1/R_2 + 1/R_3 + 1/R_4 + 1/R_5)^2 = 3(1/R_1^2 + 1/R_2^2 + 1/R_3^2 + 1/R_4^2 + 1/R_5^2).$$

Similarly, for mutually tangent circles in a plane

$$(1/R_1 + 1/R_2 + 1/R_3 + 1/R_4)^2 = 2(1/R_1^2 + 1/R_2^2 + 1/R_3^2 + 1/R_4^2)$$

a formula due to F. Soddy [18], so that

$$1/R_4 = (1/R_1 + 1/R_2 + 1/R_3) \pm 2(1/(R_1R_2) + 1/(R_2R_3) + 1/(R_3R_1))^{1/2}.$$

Since the circles have points of contact on the sides of the triangle made by their centres, the inscribed circle is totally within the triangle. This is not always the case of a sphere inscribed between four other mutually tangent spheres and the inscribed sphere may 'protrude' through the faces of the tetrahedron. Its centre may even not be within the tetrahedron.

5.2.2. *The sphere inscribed between four other spheres*

If the four spheres (discussed in the previous section) do not all touch each other the equations are more complex and, if the radii are R_1, R_2, R_3 and R_4 and the distances between the centres are d_{12}, d_{13}, etc., then R, the radius of the inscribed or circumscribed sphere can be found from the symmetrical determinant

$$\begin{vmatrix} 0 & d_{12}^2 & d_{13}^2 & d_{14}^2 & (R_1+R)^2 & 1 \\ d_{12}^2 & 0 & d_{23}^2 & d_{24}^2 & (R_2+R)^2 & 1 \\ d_{13}^2 & d_{23}^2 & 0 & d_{34}^2 & (R_3+R)^2 & 1 \\ d_{14}^2 & d_{24}^2 & d_{34}^2 & 0 & (R_4+R)^2 & 1 \\ (R_1+R)^2 & (R_2+R)^2 & (R_3+R)^2 & (R_4+R)^2 & 0 & 1 \\ 1 & 1 & 1 & 1 & 1 & 0 \end{vmatrix} = 0$$

This is solved by iteration, beginning with $R = 0$, using a routine for the evaluation of the determinant by pivotal condensation (Section 7.3). The determinant gives the four-dimensional content of the simplex of five points defined by the centres of the spheres and this must be zero [19].

5.2.3. *The sphere inscribed between four other spheres found directly*

If we are working with the coordinates of the centres of the spheres then a much improved algorithm due to Hocquemiller can be used. Here the radius and coordinates of the centre of the inscribed sphere can be found directly without iteration as the solution of a quadratic equation. Langlet [20] has shown how concisely the program can be written in APL.

There are four equations for the four unknowns R,X,Y,Z. These are

$(R + R_i)^2 = (X - X_i)^2 + (Y - Y_i)^2 + (Z - Z_i)^2$ where the radii of the known spheres are R_i and their centres are at X_i, Y_i, Z_i. Subtracting, we obtain three linear equations and one quadratic:

$$2R(R_1 - R_2) + R_1^2 - R_2^2 = -2X(X_1 - X_2) + X_1^2 - X_2^2 \ldots +$$

terms in Y and Z

$R^2 + 2RR_i + R_i^2 = X^2 - 2XX_i + X_i^2$ + terms in Y and Z this is quadratic in R.

Rewrite these three linear equations as a matrix equation:

$$R\begin{bmatrix} R_1 - R_2 \\ R_2 - R_3 \\ R_3 - R_4 \end{bmatrix} = \begin{bmatrix} X_1 - X_2 & Y_1 - Y_2 & Z_1 - Z_2 \\ X_2 - X_3 & Y_2 - Y_3 & Z_2 - Z_3 \\ X_3 - X_4 & Y_3 - Y_4 & Z_3 - Z_4 \end{bmatrix}\begin{bmatrix} X \\ Y \\ Z \end{bmatrix} + \tfrac{1}{2}\begin{bmatrix} (x_1^2 - x_2^2) + () + () - (R_1^2 - R_2^2) \\ (X_2^2 - X_3^2) + () + () - (R_2^2 - R_3^2) \\ (X_3^2 - X_4^2) + () + () - (R_3^2 - R_4^2) \end{bmatrix}$$

Rename these terms as $R[\mathbf{P}] = -[\mathbf{M}][\mathbf{X}] + [\mathbf{S}]$ note $[\mathbf{S}] = \tfrac{1}{2}[\ldots]$.
Therefore, $[\mathbf{X}] = [\mathbf{M}]^{-1}[\mathbf{S}] - R[\mathbf{M}]^{-1}[\mathbf{P}]$ where

$$[\mathbf{X}] = \begin{bmatrix} X \\ Y \\ Z \end{bmatrix}, \quad [X_1] = \begin{bmatrix} X_1 \\ Y_1 \\ Z_1 \end{bmatrix}$$

write $[\mathbf{S}'] = [\mathbf{M}]^{-1}[\mathbf{S}]$ and $[\mathbf{P}'] = [\mathbf{M}]^{-1}[\mathbf{P}]$, $[\mathbf{X}] = [\mathbf{S}'] - R[\mathbf{P}']$ so that the first of the original quadratic equations becomes:

$$[[\mathbf{X}] - [\mathbf{X}_1]]^2 = (R + R_1)^2$$

that is $[\mathbf{X}]^2 - 2[\mathbf{X}]^{\mathrm{T}}[\mathbf{X}_1] + [\mathbf{X}_1]^2 = (R + R_1)^2$
since $[\mathbf{X}]^2 = [\mathbf{X}]^{\mathrm{T}}[\mathbf{X}] = X^2 + Y^2 + Z^2$.
Substitute the value of $[\mathbf{X}]$ derived above and obtain a quadratic for R:

$$[\mathbf{S}']^2 - 2R[\mathbf{S}']^{\mathrm{T}}[\mathbf{P}'] + R^2[\mathbf{P}']^2 - 2[\mathbf{S}']^{\mathrm{T}}[\mathbf{X}_1] + 2R[\mathbf{P}']^{\mathrm{T}}[\mathbf{X}_1]^2 = R^2 + 2RR_1 + R_1^2$$

If the quadratic for R is $aR^2 + bR + c = 0$ then the coefficients collected on the left are:

$$a = [\mathbf{P}']^{\mathrm{T}}[\mathbf{P}'] - 1$$
$$b = -2[\mathbf{S}']^{\mathrm{T}}[\mathbf{P}'] + 2[\mathbf{P}']^{\mathrm{T}}[\mathbf{X}_1] - 2R_1$$
$$c = [\mathbf{S}']^{\mathrm{T}}[\mathbf{S}'] - R_1^2$$

The solutions then are $R = (-b \pm (b^2 - 4ac)^{1/2})/(2a)$ and R can then be substituted into

$$[\mathbf{X}] = [\mathbf{M}]^{-1}[\mathbf{S}] - R[\mathbf{M}]^{-1}[\mathbf{P}]$$

to obtain [**X**] the coordinates of the centre of the sphere.

We thus have an interesting example of a problem which can be solved very simply (previous Section 5.2.2) by an iterative program which can be written quickly if we are using the full version of BASIC which gives the determinant of a matrix in one statement but which can be solved directly without iteration at the cost of much more care and complexity in programming. The latter method (Section 5.2.3) is probably preferable if the process of fitting a sphere between four others has to be done a very large number of times.

6. AFFINE TRANSFORMATIONS

When two structures or distributions are related so that every point x_1,x_2,x_3 in one is transformed into a corresponding point x_1', x_2', x_3' in the other by the same transformation

$$x_1' = a_{11}x_1 + a_{12}x_2 + a_{13}x_3$$
$$x_2' = a_{21}x_1 + a_{22}x_2 + a_{23}x_3$$
$$x_3' = a_{31}x_1 + a_{32}x_2 + a_{33}x_3$$

then this transformation is called an affine transformation.

6.1. *To find the best affine transformation*

By least squares, or by the equivalent generalised inverse procedure we can find the best affine transformation [A] connecting two sets of N points, such as the atoms of two molecules. The two sets of coordinates are first referred each to its centre of mass as origin. One definition of centre of mass is that it is that point for which the weighted sum of the squares of the distances of the mass points is a minimum. This sum is the (radius of gyration)^2 × total mass.

Suppose that $[\mathbf{X}'(N,3)] = [\mathbf{X}(N,3)]\,[\mathbf{A}(3,3)]$, inserting the dimensions of the arrays, then for the best fit the sum of the squares of the discrepancies in this equation is minimised, by applying the least squares procedure (of Section 1.2). $[\mathbf{X}^{\mathrm{T}}]$ is the transpose of the matrix [X].

$$[\mathbf{X}^{\mathrm{T}}(3,N)]\,[\mathbf{X}'(N,3)] = [\mathbf{X}^{\mathrm{T}}(3,n)]\,[\mathbf{X}(N,3)]\,[\mathbf{A}(3,3)]$$

$$\text{so that } [\mathbf{A}(3,3)] = [\mathbf{X}^{\mathrm{T}}(3,N)\;\mathbf{X}(N,3)]^{-1}\,[\mathbf{X}^{\mathrm{T}}(3,N)]\,[\mathbf{X}'(N,3)]$$

The two distributions are then related by a transformation which has nine independent components.

6.2. *To find the best rotation*

There are more complex procedures for finding the best rotation matrix connecting two distributions of points under the restriction that there should be only three adjustable parameters, instead of nine as in the affine method. The three parameters for a rotation correspond to three Euler angles, or to one rotation angle and the direction cosines (2 parameters) of the line about which rotation takes place.

A recent exact treatment is by W. Kabsch [21] 'A solution to the best rotation to relate two sets of vectors'.

6.3. *Decomposition of the affine transformation matrix*

The general affine transformation matrix contains nine independent quantities but it can be expressed as a combination of simpler operations. In each case the determinant of the matrix [**A**] represents the change in volume of an element. A pure rotation thus has a determinant of +1 and a rotation plus an improper operation (a reflection or an inversion in the origin which changes the cheirality of the assembly) has a determinant of −1.

There are a number of special cases

(a) The identity matrix [**I**] $= \begin{bmatrix} 1 & 0 & 0 \\ 0 & 1 & 0 \\ 0 & 0 & 1 \end{bmatrix}$ effects no change.

(b) The matrix $\begin{bmatrix} a & 0 & 0 \\ 0 & b & 0 \\ 0 & 0 & c \end{bmatrix}$ is a dilatation, and a sphere is deformed into an ellipsoid, the principal axes of which coincide with the coordinate axes.

(c) The matrix $\begin{bmatrix} a_{11} & a_{12} & a_{13} \\ a_{12} & a_{22} & a_{23} \\ a_{13} & a_{23} & a_{33} \end{bmatrix}$, which is symmetrical with six independent components, represents a homogeneous dilatation but now along axes which no longer coincide with the coordinate axes. A sphere is transformed into an ellipsoid. The matrix can be diagonalised, turned into the form of (b) above, by a suitable rotation so that [$\mathbf{A}_s$] = [**D**][**R**], where [**D**] is a diagonal matrix and [**R**] is a pure rotation. This can be done most easily by Jacobi's method of iteration (see Section 7.4) although in the case of a 3 × 3 matrix the solution can be found from a cubic equation which can be solved exactly by the standard formulae.

(d) The matrix $\begin{bmatrix} 1 & 0 & 0 \\ 0 & \cos\theta & \sin\theta \\ 0 & -\sin\theta & \cos\theta \end{bmatrix}$ causes a rotation by an angle θ about the *X*-axis.

The reverse rotation, obtained by changing the sign of θ, is the transpose of this matrix and this result is true for the general rotation matrix, for which the transpose is also the inverse. The trace, the sum of the diagonal terms, is equal to

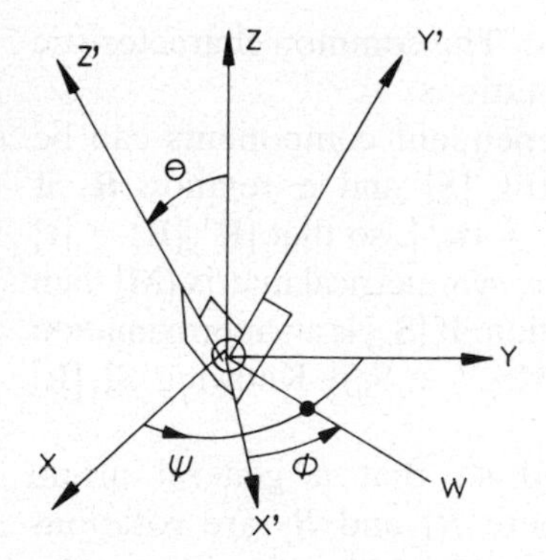

Fig. 13. Euler angles θ, φ, ψ describe the rotation of old axes XYZ to new axes X'Y'Z'. The first rotation is ψ about 0Z'; the second rotation is θ about 0X' and the third rotation is φ about 0Z'.

$1 + 2\cos\theta$ for all rotation matrices.

(e) The general rotation matrix, which represents a rotation through an angle θ about a line with direction cosines l,m,n, is

$$R = \cos\theta \begin{bmatrix} 1 & 0 & 0 \\ 0 & 1 & 0 \\ 0 & 0 & 1 \end{bmatrix} + (1 - \cos\theta) \begin{bmatrix} l^2 & lm & ln \\ lm & m^2 & mn \\ ln & mn & n^2 \end{bmatrix} + \sin\theta \begin{bmatrix} 0 & -n & m \\ n & 0 & -l \\ -m & l & 0 \end{bmatrix}$$

(Reversing this relationship, if we are given a rotation matrix R, then $\cos\theta = \frac{1}{2}(R_{11} + R_{22} + R_{33} - 1)$, $l = (R_{32} - R_{23})/2\sin\theta$, $m = (R_{13} - R_{31})/2\sin\theta$. $n = (R_{21} - R_{12})/2\sin\theta$). This expression enables us to choose a random rotation matrix. The direction of the rotation axis is chosen uniformly over the sphere (see Section 4.2) and the value of θ is taken uniformly in the range $0 \leqslant \theta \searrow 360°$.

(f) *Euler Angles*. The rotation of a frame of coordinates is also expressible by Euler angles. This system is used, for example, in the analysis of the gyroscope. Euler angles are convenient when moving from one orientation to another by successive rotations about the three Cartesian axes. They describe the movement of an asymmetric object in the surface of a sphere. The planes $X0Y$ and $X'0Y'$ intersect in the line $0W$. $Z'0Z = \theta$ = angle of nutation, $0 \leqslant \theta < \pi$ $X\hat{0}A = \psi$ = angle of precession $0 \leqslant \psi < 2\pi$. $X'\hat{0}A = \phi$ = angle of pure rotation $0 \leqslant \phi < 2\pi$.

To move from the old axes XYZ to the new axes $X'Y'Z'$, the first rotation is ψ about $0Z'$; $0X' \rightarrow 0W$ from $0X$, the second rotation is θ about $0X'$; $0Z' \rightarrow 0Z'$ from $0Z$ and the third rotation is φ about $0Z'$; $0X' \rightarrow 0X'$ from $0W$.

The matrices describing the rotations are: $\mathbf{A} = A_3(\psi)A_1(\theta)A_3(\phi)$

$$= \begin{bmatrix} \cos\psi & -\sin\psi & 0 \\ \sin\psi & \cos\psi & 0 \\ 0 & 0 & 1 \end{bmatrix} \begin{bmatrix} 1 & 0 & 0 \\ 0 & \cos\theta & -\sin\theta \\ 0 & \sin\theta & \cos\theta \end{bmatrix} \begin{bmatrix} \cos\phi & -\sin\phi & 0 \\ \sin\phi & \cos\phi & 0 \\ 0 & 0 & 1 \end{bmatrix}$$

Many variants of the Euler system may be encountered. The common characteristic is resolving the transformation into three successive rotations.

(g) A general affine transformation [**A**] with 9 independent components can be decomposed into the product of a symmetrical matrix [**S**] and a rotation **R**. if [**A**] = [**R**] [**S**], then as [**R**] represents a rotation, $[\mathbf{R}^{-1}] = [\mathbf{R}^{T}]$, so that $[\mathbf{R}^{T}][\mathbf{R}] = [\mathbf{I}]$ and $[\mathbf{A}^{T}][\mathbf{A}] = [\mathbf{S}^{T}][\mathbf{R}^{T}][\mathbf{R}][\mathbf{S}] = [\mathbf{S}^{T}][\mathbf{S}]$. If $[\mathbf{A}^{T}][\mathbf{A}]$ is the symmetrical matrix [**M**] then [**S**] is its square root and $[\mathbf{M}] = [\mathbf{S}^{2}]$. [**S**] is found by iteration. If $[\mathbf{S}_n]$ is an approximation to [**S**], then a better approximation $[\mathbf{S}_{n+1}]$ is given by$[\tfrac{1}{2}(\mathbf{MS}_n^{-1} + \mathbf{S}_n)]$. Knowing [**S**], [**R**] is found as $[\mathbf{A}][\mathbf{S}^{-1}]$.

As seen in (c) above, [**S**] can be diagonalised so that a general affine transformation [**A**] can be expressed as $R_1\mathbf{D}R_2$, where R_1 and R_2 are rotations and **D** is a diagonal matrix of dilations.

(h) A general matrix [**A**] can be factorised (by Choleski's method) into the product of upper and lower triangular matrics. For a 3 × 3 matrix this can be written explicitly as:

$$\begin{bmatrix} A_{11} & A_{12} & A_{13} \\ A_{21} & A_{22} & A_{23} \\ A_{31} & A_{32} & A_{33} \end{bmatrix} = \begin{bmatrix} 1 & 1 & 0 \\ A_{21}/B_1 & 1 & 0 \\ A_{31}/B_1 & \mathrm{C} & 1 \end{bmatrix} \begin{bmatrix} B_1 & 0 & 0 \\ 0 & B_2/B_1 & 0 \\ 0 & 0 & B_3/B_2 \end{bmatrix}$$

$$\begin{bmatrix} 1 & A_{12}/B_1 & A_{13}/B_1 \\ 0 & 1 & \mathrm{D} \\ 0 & 0 & 1 \end{bmatrix}$$

$$\mathbf{C} = (A_{11}A_{32} - A_{31}A_{12})/B_2, \quad \mathbf{D} = (A_{11}A_{32} - A_{21}A_{13})/B_2,$$

$$\mathbf{B}_1 = A_{11}, \quad \mathbf{B}_2 = \begin{vmatrix} A_{11} & A_{12} \\ A_{21} & A_{22} \end{vmatrix}, \quad \mathbf{B}_3 = \text{determinant of the matrix } \mathbf{A}.$$

7. COMPUTING IN BASIC

A number of fundamental service programs are given for carrying out operations useful in a variety of contexts. These are:

(1) The solution of a number of non-linear equations by successive approximation. 'SOLVE'.
(2) The calculation of the area of the surface of a 'super-ellipsoid' by integration over two variables. 'ELAREA'.
(3) The evaulation of a determinant by pivotal condensation. 'DETER'.
(4) The diagonalisation of a symmetrical matrix by Jacobi's method. 'JACOBI'.
(5) The calculation of the generalised inverse $[\mathbf{M}^{+}]$ of a matrix [**M**] by iterating $[\mathbf{M}^{+}_{k+1}] = [\mathbf{M}^{+}_{k}][2\mathbf{I} - \mathbf{MM}^{+}_{k}]$. 'GENINV'.

Note on BASIC

BASIC is deficient in a number of respects but some of these deficiencies can be remedied by writing segments into a program which needs them.

(1) It is a useful habit to reserve the variable P1 for π and to write P1 = 3.1415926535.
(2) There are no *arcsine* and *arccosine* functions but these can be written as:

REM DEFINE ARCSINE FUNCTION (−P1/2 < RESULT < P1/2)
DEF FNS(X) = ATN(X/SQR(1 − X * X))

If X = +1 or −1 this will fail by division by zero.

REM DEFINE ARCCOSINE FUNCTION (0 ⩽ RESULT < P1)
DEF FNC(X) = ATN(SQR(1 − X * X)/X) + (L − SGN(X)) * P1/2

again, this will fail when X = 0 when the result should be ½π. An alternative is:

DEF FNC(X) = 2 * ATN(SQR((1 − X)/(1 + X)))

which fails at X = −1 when the result should be π. Similarly, the tangent of half the angle may be used for the arcsin to give a failure only at X = 0.

DEF FNS(X) = 2 * ATN((1 − SQR(1 − X * X))/X)

If it is important to cover the cases which fail, then extra IF statements will have to be included.

7.1. *The solution of non-linear equations*

```
PROGRAM   SOLVE

00100 REM SOLUTION OF EQUATIONS
00110 REM READ IN INITIAL VALUES
00120 REM AREA S, HALF PERIMETER P, RADIUS OF CIRCUMCIRCLE R
00130 READ S,P,R
00140 REM COUNT CYCLES
00150 N=1
00160 REM SET INITIAL VALUES FOR A,B,C
00170 B=2*P/3
00180 A=SQR(4*R*S/B)
00190 C=SQR(A*B)
00200 REM SET UP MATRIX OF COEFFICIENTS
00210 DIM D(3,3), E(3), F(3,3),C(3)
00220 GOSUB 00300
00230 REM SOLVE FOR CORRECTIONS
00240 MAT F=INV(D)
00250 MAT C=F*E
00260 A=A-C(1)
00270 B=B-C(2)
00280 C=C-C(3)
00290 GO TO 00220
00300 REM SUBROUTINE TO CALCULATE ERRORS AND COEFFICIENTS
00310 N=N+1
00320 E(1)=S↑2 -P*(P-A)*(P-B)*(P-C)
00330 E(2)=P -.5*(A+B+C)
00340 E(3)=4*R*S -A*B*C
00350 REM TEST ERRORS
00360 IF ABS(E(1))>1E-9*P THEN 00520
00370 IF ABS(E(2))>1E-9*P THEN 00520
00380 IF ABS(E(3))>1E-9*P THEN 00520
00390 PRINT "RESIDUAL ERRORS";E(1);E(2);E(3)
00400 PRINT "A=";A
```

```
00410 PRINT "B=";B
00420 PRINT "C=";C
00430 REM CHECK
00440 P=.5*(A+B+C)
00450 PRINT "SEMIPERIMETER=";P
00460 S=SQR(P*(P-A)*(P-B)*(P-C))
00470 PRINT "AREA=";S
00480 R=.25*A*B*C/S
00490 PRINT "CIRCUMRADIUS=";R
00500 PRINT "NO. OF CYCLES=";N
00510 STOP
00520 D(1,1)=P*(P-B)*(P-C)
00530 D(1,2)=P*(P-A)*(P-C)
00540 D(1,3)=P*(P-A)*(P-B)
00550 D(2,1)=D(2,2)=D(2,3)=-.5
00560 D(3,1)=-B*C
00570 D(3,2)=-A*C
00580 D(3,3)=-A*B
00590 RETURN
00600 DATA 6,6,2,5

PROGRAM    SOLVE

RESIDUAL ERRORS-9.09495E-13   0   9.09495E-13
A= 3.
B= 4.
C= 5.
SEMIPERIMETER= 6
AREA= 6.
CIRCUMRADIUS= 2.5
NO. OF CYCLES= 11

CP          0.193 SECS.
```

7.2. *To calculate the area of the surface of a 'super-ellipsoid'*

The equation of the figure is $(x/a)^n + (y/b)^n + (z/c)^n = 1$, where, if n is 15 or more, the figure approaches a roughly brick-shaped (or rectangularly parallelepipedal) cell with rounded corners.

In spherical coordinates

$$x = r \cos \phi \sin \theta$$
$$y = r \sin \phi \; 0 \sin \theta$$
$$z = r \cos \theta$$

$r^2 = x^2 + y^2 + z^2$ so that the length of the radius vector to the point x,y,z is given by

$$1/r^n = (\cos \phi \sin \theta/a)^n + (\sin \phi \sin \theta/b)^n + (\cos \theta/c)^n.$$

On the surface of a sphere the element of area is

$$\delta S = r^2 \sin \theta \; \delta\theta\delta\phi \text{ (which integrates up to } S = 4\pi r^2\text{).}$$

On the surface of the figure the corresponding element of area is tilted so that its normal makes an angle ψ with the radius vector from the origin. The area is thus greater by a factor of $1/\cos \psi$.

The direction cosines of the normal to the surface $F(x,y,z) = 0$ at x,y,z are proportional to $\partial F/\partial x : \partial F/\partial y : \partial F/\partial z$ and thus, in this case, to $x^{n-1}/a^n : y^{n-1}/b^n : z^{n-1}/c^n$.

The direction cosines of the radius vector are a $x : y : z$ and thus,

$$1/\cos\psi = (x^2 + y^2 + z^2)^{1/2}(x^{2n-2}/a^{2n} + y^{2n-2}/b^{2n} + z^{2n-2}/c^{2n})^{1/2}$$

therefore,

$$S = 8\int_{\theta=0}^{\pi/2}\int_{\phi=0}^{\pi/2} (1/\cos) r^2 \sin\theta \, d\theta \, d\phi$$

$$\sin\theta = 8\int_{\theta=0}\int_{\phi=0}^{\pi/2} (x^2 + y^2 + z^2)^{1/2}(x^{2n-2}/a^{2n} + y^{2n-2}/ b^{2n}$$

$$+ z^{2n-2}/c^{2n}) r^2 \, d\theta \, d\phi$$

The values for x,y,z and r in terms of θ and ϕ are then substituted and the integral is evaluated numerically. It might be noted that, as here, when evaluating an integral by summing at finite intervals Δx of a variable x, the value of the function at the middle of the interval is taken.

```
LIST


PROGRAM    ELAREA
00100 SET DIGITS(9)
00110 REM PROGRAM NAME ELAREA
00120 REM CALCULATE SURFACE OF SUPER-ELLIPSOID
00130 REM EQUATION (X/A)n +(Y/B)n +(Z/C)n =1
00140 REM FOR ORDINARY ELLIPSOID N=2
00150 REM EXPONENT N
160 N=15
00170 REM M= NUMBER OF DIVISIONS PER RIGHT-ANGLE
00180 P1=3.141592653589793
00190 M=50
00200 D=.5*P1/M
00210 REM SEMI-AXIS
220 A=3
230 R=5
240 C=2
00250 A2=A*A
00260 B2=B*B
00220 C2=C*C
00280 S=0
00290 REM SUM OVER THETA
00300 FOR T=.5*D TO .5*P1 STEP D
00310 S1=SIN(T)
00320 REM SUM OVER PHI
00330 FOR F=.5*D TO .5*P1 STEP D
00340 R1=(COS(F)*SIN(T)/A)**N+(SIN(F)*SIN(T)/B)**N
00350 R1=R1+(COS(T)/C)** N
00360 R1=(1/R1)**(2/N)
00370 X2=R1*(COS(F)*SIN(T))**2
00380 Y2=R1*(SIN(F)*SIN(T))**2
00390 Z2=R1*(COS(T))**2
```

```
00400 R2=SQR((X2/A2)**N/X2+(Y2/B2)**N/Y2+(Z2/C2)**N/Z2)
00410 R3=SQR(X2+Y2+Z2)
00420 S=S+R1*R2*R3*S1
00430 NEXT F
00440 NEXT T
00450 S=S*D*D*8
00460 PRINT "AREA OF SURFACE=".S
00420 PRINT "NUMBER OF INCREMENTS IN RIGHT-ANGLE=";M
00480 PRINT "ESTIMATE FROM 4*P1*(A*B*C) 2/3)"
00490 PRINT "SPHERE"; 4*P1*(A*B*C) 2/3)
00500 PRINT "AREA OF BLOCK=";8*(A*B+B*C+C*A)
00510 PRINT "EXPONENT+";N
READY.
RUN


PROGRAM    ELAREA
vAREA OF SURFACE= 230.74637
NUMBER OF INCREMENTS IN RIGHT-ANGLE= 50
ESTIMATE FROM 4*P1*(A*B*C) 2/3)
SPHERE 121.326974
AREA OF BLOCK= 248
EXPONENT= 15

CP        4.627 SECS

RUN COMPLETE.
```

7.3. *The evaluation of a determinant*

Determinants frequently occur in expressions for areas and volumes and it is necessary to evaluate them expeditiously.

A third-order determinant is most easily evaluated by multiplying it out, to obtain six terms of degree three, but for an Nth order determinant there are $N!$ terms of order N, so that this procedure becomes impracticable if $N > 3$.

For $N = 3$ and $D = \begin{bmatrix} a_{11} & a_{12} & a_{13} \\ a_{21} & a_{22} & a_{23} \\ a_{31} & a_{32} & a_{33} \end{bmatrix}$

$$D = a_{11}a_{22}a_{33} + a_{12}a_{23}a_{31} + a_{13}a_{21}a_{32}$$
$$-a_{11}a_{23}a_{32} - a_{12}a_{21}a_{33} - a_{13}a_{22}a_{31}$$

using the mnemonic shown:

+	+	+	−	−	−
a_{11}	a_{12}	a_{13}	a_{11}	a_{12}	a_{13}
a_{21}	a_{22}	a_{23}	a_{21}	a_{22}	a_{23}
a_{31}	a_{32}	a_{33}	a_{31}	a_{32}	a_{33}

In general $D = \frac{1}{2} \Sigma\, \epsilon_{ijkl} \ldots \epsilon_{pqrs}\, a_{ip}a_{jq}a_{kr}a_{ls} \ldots$ where ϵ_{ijkl} is the permutation tensor which has the value 1 if i,j,k,l ... are an even permutation, -1 if they are an odd permutation and 0 if any two indices are equal. For three suffixes the expression $\epsilon_{ijk} = \frac{1}{2}(j - k)(k - i)(i - j)$ can sometimes be convenient in computing.

Pivotal condensation

The most efficient way to evaluate a determinant of order greater than three is by pivotal condensation. If by subtraction all terms but one in a row (or column) can be reduced to zero, then this remaining term is a factor of D and the order of the determinant can be reduced by one by striking out this row and the corresponding column. This procedure is repeated until the order reaches one. In practice errors are minimised if, each time, the term numerically the greatest is taken as the pivot. An example follows but a program segment is given in an appendix.

$$\begin{bmatrix}1&0&2&3\\8&8&5&1\\1&4&7&6\\3&7&9&8\end{bmatrix} \to \begin{bmatrix}1-\tfrac{2}{3} & 0-\dfrac{2.7}{9} & 2 & 3-\dfrac{2.8}{9}\\ 8-\tfrac{5}{3} & 8-\dfrac{5.7}{9} & 5 & 1-\dfrac{5.8}{9}\\ 1-\tfrac{7}{3} & 4-\dfrac{7.7}{9} & 7 & 6-\dfrac{7.8}{9}\\ 0&0&9&0\end{bmatrix}$$

The first pivot is a_{43}. Multiples of column 3 are subtracted from the other columns to reduce row 4 to zeros.

$$\to 9\begin{bmatrix}1/3 & -14/9 & 11/9\\ 19/3 & 37/9 & 23/9\\ -4/3 & -13/9 & -2/9\end{bmatrix} = 9.1/3.1/9.1/9\begin{bmatrix}1 & -14 & 11\\ 19 & 37 & 32\\ -4 & -13 & -2\end{bmatrix}$$

The next† pivot is a_{22}.

$$\to \tfrac{1}{27}\begin{bmatrix}1+\dfrac{14.19}{37} & -14 & 11+\dfrac{14.23}{37}\\ 0 & 37 & 0\\ -4+\dfrac{13.19}{37} & -13 & -2+\dfrac{13.23}{37}\end{bmatrix} \to \tfrac{37}{27}\begin{bmatrix}1+\dfrac{14.19}{37} & 11+\dfrac{14.23}{37}\\ -4+\dfrac{13.19}{37} & -2+\dfrac{13.23}{37}\end{bmatrix}$$

$$\to -4$$

Subroutine in BASIC for calculating a determinant

```
1000    REM Calculation of determinant by pivotal condensation
1010    REM Det. of order N is in D(N,N). Result in S.
1020    REM Matrix D is destroyed. Uses A(N*N),B,I,J
1030    REM uses A,P,Q,V. N is altered.
1040    S=1
1050    REM begin
1060    B=0
1070    REM find largest element = pivot = D(P,Q)
1080    DIM      A(N*N)
```

```
1090    FOR I=1 TO N
1100    FOR J=1 TO N
1110    A = ABS(D(I,J))
1120    IF A<B  THEN  1160
1130    P=I
1140    Q=J
1150    B=A
1160    NEXT J
1170    NEXT I
1180    REM multiply by largest element
1190    S=S*D(P,Q)*(-1)↑(P+Q)
1200    IF S=0 THEN  1440
1210    V=1
1220    REM reduce other terms by subtr. cols. of pivot to zero
1230    FOR I=1 TO N
1240    IF I=P THEN  1320
1250    FOR J=1 TO N
1260    IF J=Q THEN  1310
1270    D(I,J)=D(I,J)-D(P,J)/D(P,Q) * D(I,Q)
1280    A(V)=D(I,J)
1290    V=V+1
1300    REM run out non-zero elements into det. of lower order
1310    NEXT J
1320    NEXT I
1330    N=N-1
1340    IF N=0 THEN  1440
1350    V=1
1360    FOR I=1 TO N
1370    FOR J=1 TO N
1380    D(I,J)=A(V)
1390    V=V+1
1400    NEXT J
1410    NEXT I
1420    REM repeat until det. has order 1
1430    GO TO 1060
1440    RETURN
```

7.4. *The Jacobi diagonalisation of a symmetrical matrix*

This is an iterative method of diagonalising a symmetrical matrix by successive rotations to reduce the largest off-diagonal terms step by step. It is not the most efficient method but it is not stopped by the presence of repeated or zero eigenvalues. The successive rotations are also applied to a unit matrix which thus become the matrix of unit eigenvectors.

The matrix [**A**] is multiplied by the rotation matrix [**T**] where [**T**] is

$$\begin{bmatrix} 1 & | & | & \\ \text{—} & \cos\theta \text{—} & \sin\theta \text{—} & \\ & | & | & \\ \text{—} & -\sin\theta \text{—} & \cos\theta \text{—} & \\ & | & | & 1 \end{bmatrix}$$

so that its new value becomes $[\mathbf{A}'] = [\mathbf{T}]^{-1}[\mathbf{A}][\mathbf{T}]$ θ is given by $\tan 2\theta = 2\mathbf{A}_{pq}/(\mathbf{A}_{pp} - \mathbf{A}_{qq})$ where $\mathbf{A}_{pq}$ is the largest off-diagonal term. A specimen program is given below.

```
0100  REM JACOBI DIAGONALISATION OF SYMMETRIC MATRIX
00110  REM ORDER=N, MAT IN A(N,N)
00120 DIM A(10,10),T(10,10),R(10,10),W(10,10),U(10,10)
00130  READ N
00140 MAT R=IDN(N,N)
00150 MAT A=ZER(N,N)
00160  MAT  READ A
00170 PRINT 'ORIGINAL MATRIX'
00180 MAT PRINT A;
00190 PRINT
00200 PRINT
00210  REM FIND LARGEST OFF-DIAGONAL TERM
00220  LET A9=0
00230  FOR I=1 TO N
00240  FOR J=1 TO N
00250  IF I=J THEN 00300
00260 IF A9>ABS(A(I,J)) THEN 00300
00270 A9=ABS(A(I,J))
00280  LET I9=I
00290  LET J9=J
00300  NEXT J
00310  NEXT I
00320 IF A9<1.E-10 THEN 00490
00330 MAT T=IDN(N,N)
00340 IF ABS(A(I9,I9)-A(J9,J9))>1.E-12 THEN 00370
00350 T9=3.1415926535*.25
00360 GO TO 00380
00370 T9=.5*ATN(2*A(I9,J9)/(A(I9,I9)-A(J9,J9)))
00380 T(I9,I9)=T(J9,J9)=COS(T9)
00390 T(J9,I9)=SIN(T9)
00400 T(I9,J9)=-SIN(T9)
00410 MAT U=ZER(N,N)
00420 MAT W=ZER(N,N)
00430  MAT U=TRN(T)
00440  MAT W=A*T
00450  MAT A=U*W
00460  MAT W=R*T
00470  MAT R=(1)*W
00480 GO TO 00210
00490  PRINT 'DIAGONALISED MATRIX'
00500 MAT PRINT A,
00510  PRINT
00520  PRINT 'MATRIX TO BE APPLIED FOR DIAGONALISATION'
00530 MAT PRINT R,
00540 DATA 7
```

```
00550 DATA 0,1,3,4,3,1,1
00560 DATA 1,0,1,3,4,3,1
00570 DATA 3,1,0,1,3,4,1
00580 DATA 4,3,1,0,1,3,1,
00590 DATA 3,4,3,1,0,1,1
00600 DATA 1,3,4,3,1,0,1
00610 DATA 1,1,1,1,1,1,0
00620  END
```

7.5. *The generalised inverse of a matrix*

In the example given the program data is from the Chinese problem mentioned in our introduction (1.1) where there are four simultaneous linear equations for four unknowns. The first four data statements (1440, 1450, 1460, 1470) give the relevant figures. However, it is instructive to try the effect of changing the dimensions of the array (using statement 1430). 4,4 gives the normal answer; 6,4 solves six equations by least squares; 7,4 also; 3,4 asks for the values of four unknowns from 3 equations. It gives a consistent answer where the values of the unknowns are also minimised; 5,4 solves five equations where the last is a linear function of the others. It gives the correct answer although the matrix is singular.

```
100 REM GENERALISED INVERSE OF X(N,M)
110 READ N,M
120 DIM H(N),P(M),Q(N)
130 DIM X(N,M), Y(M,N), W(M,N), Z(N,N), A(M,M)
140 REM MATRIX ENTERED IN X AND RETURNED IN Y
150 REM READ IN MATRIX
160 FOR I=1 TO N
170 FOR J=1 TO M
180 READ X(I,J)
190 W(J,I)=X(I,J)
200 REM W IS TRANSPOSE OF X
210 NEXT J
220 REM R.H.S. OF EQUATION
230 READ H(I)
240 NEXT I
250 K=0
260 FOR I=1 TO N
270 FOR J=1 TO N
280 Z(I,J)=0
290 FOR L=1 TO M
300 Z(I,J)=Z(I,J)+X(I,L)*W(L,J)
310 NEXT L
320 K=K+ABS(Z(I,J))
330 NEXT J
340 NEXT I
350 K=1/K
360 PRINT "K=";K
370 REM SMALL CONSTANT
380 D=1E-5
```

```
390 PRINT "CONSTANT FOR INTEGRAL TRACE=";D
400 PRINT "TRACE+2*N"
410 FOR I=1 TO M
420 FOR J=1 TO N
430 REM FIRST APPROXIMATION TO INVERSE
440 Y(I,J)=K*W(I,J)
450 NEXT J
460 NEXT I
470 FOR I=1 TO N
480 FOR J=1 TO N
490 Z(I,J)=0
500 FOR L=1 TO M
510 Z(I,J)=Z(I,J)+X(I,L)*Y(L,J)
520 NEXT L
530 NEXT J
540 NEXT I
550 REM TRACE=T
560 T=0
570 FOR I=1 TO N
580 Z(I,I)=Z(I,I)-2
590 T=T+Z(I,I)
600 NEXT I
610 PRINT 2*N+T
620 FOR I=1 TO M
630 FOR J=1 TO N
640 W(I,J)=0
650 FOR L=1 TO N
660 W(I,J)=W(I,J)+Y(I,L)*Z(L,J)
670 NEXT L
680 NEXT J
690 NEXT I
700 FOR I=1 TO M
710 FOR J=1 TO N
720 Y(I,J)=-W(I,J)
730 NEXT J
740 NEXT I
750 IF ABS(T-INT(T)-1) < D THEN 780
760 IF ABS(T-INT(T)) < D THEN 780
770 GO TO 470
780 REM REPEAT UNTIL T IS AN INTEGER
790 FOR I=1 TO M
800 FOR J=1 TO M
810 A(I,J)=0
820 FOR L=1 TO N
830 A(I,J)=A(I,J)+Y(I,L)*X(L,J)
840 NEXT L
850 NEXT J
860 NEXT I
870 PRINT "RANK OF MATRIX="; 2*N+T
880 REM REMOVE NEXT STATEMENT FOR FULL PRINTOUT
890 GO TO 1240
900 PRINT "GENERALISED INVERSE"
910 PRINT
920 FOR I=1 TO M
930 FOR J=1 TO N
940 PRINT Y(I,J),
950 NEXT J
960 PRINT
970 PRINT
980 NEXT I
990 REM CHECKING PROCEDURE
1000 PRINT
1010 PRINT "ORIGINAL MATRIX"
1020 FOR I=1 TO N
1030 FOR J=1 TO M
1040 PRINT X(I,J),
1050 NEXT J
1060 PRINT
1070 PRINT
1080 NEXT I
1090 PRINT
1100 PRINT "PRODUCTS"
1110 FOR I=1 TO M
1120 FOR J=1 TO M
1130 PRINT A(I,J),
1140 NEXT J
1150 PRINT
1160 NEXT I
1170 PRINT
1180 FOR I=1 TO N
1190 FOR J=1 TO N
1200 PRINT Z(I,J),
1210 NEXT J
1220 PRINT
1230 NEXT I
1240 PRINT "SOLUTIONS TO EQUATIONS"
1250 FOR I=1 TO M
1260 P(I)=0
1270 FOR J=1 TO N
1280 P(I)=P(I)+Y(I,J)*H(J)
1290 NEXT J
1300 PRINT I, P(I)
1310 NEXT I
1320 PRINT "NUMBER OF EQUATIONS=";N
1330 PRINT "NUMBER OF UNKNOWNS=";M
1340 PRINT "CALCULATED AND OBSERVED R.H.S."
1350 FOR I=1 TO N
1360 Q(I)=0
1370 FOR J=1 TO M
1380 Q(I)=Q(I)+X(I,J)*P(J)
1390 NEXT J
1400 PRINT Q(I),H(I),H(I)-Q(I)
1410 NEXT I
1420 REM TEST DATA
1430 DATA 4,4
1440 DATA 5,4,3,2,1496
1450 DATA 4,2,6,3,1175
1460 DATA 3,1,7,5,958
1470 DATA 2,3,5,1,861
1480 DATA 2,3,5,1,861
1490 DATA 2,3,5,1,862
1500 DATA 1,2,3,4,500
```

REFERENCES

1. Somerville, D.M.Y. (1959) Analytic Geometry of Three Dimensions. Cambridge University Press.
2. Williams, R. (1979) The Geometrical Foundation of Natural Structure. Dover, New York.
   Coxeter, H.S.M. (1961) An Introduction to Geometry. Wiley, New York.
   Thompson, D'Arcy W. (1917) On Growth and Form. Cambridge University Press.
   Korn. G.A. and Korn, T.M. (1968) Mathematical Handbook for Scientists and Engineers. McGraw-Hill, New York.
   Klein, F. (1939) Geometry, Dover, New York.
3. Ben Israel, A. and Greville, T. (1974) Generalised Inverses: Theory and Applications. Wiley, New York.
   Rao, C.R. and Mitra, S.K. (1971) Generalised Inverse of Matrices and its Application. Wiley, New York.
   Gupta, N.N. (1971) IEEE Trans. SMC, Jan. 1971, pp. 89–90.
   Mackay, A.L. (1977) Acta Cryst. *A33*, 212–215.
4. Carnot, L.F.M. (1803) Geometrie de Position. Paris.
5. Go, M. (1981) Nature 291, 90.
6. Dunitz, J.D. (1979) X-Ray Analysis and the Structure of Organic Molecules. Cornell University Press, pp. 495–497.
7. Dunitz, J.D. (1979) X-Ray Analysis and the Structure of Organic Molecules. Cornell University Press, pp. 498–501.
8. Mackay, A.L. (1974) Acta Cryst. A30, 440–447.
9. Southwell, R.V. (1936) An Introduction to the Theory of Elasticity. Oxford University Press.
10. Crippen, G.M. and Havel, T.F. (1979) Acta Cryst. A34, 282–284.
11. Tobler, W. and Wineburg, S. (1971) Nature 231, 39–41.
12. Erickson, R. (1973) Science 181, 705–716.
13. Baur, W.R. Crick, F.H.C. and White, J.H. (1980) Sci. Am. 243, 118–133.
    Pohl, W.F. (1980) Math. Intelligencer 3, 20–27.
14. Barrett, A.N. and Burdett, I.D.J. (1981) J. Theor. Biol. 92, 127–139.
15. Isenberg, C. (1978) The Science of Soap Films and Soap Bubbles. Tieto, Clevedon.
16. Bowyer, A. (1981) Comp. J. 24, 162–166.
    Watson, D.F. (1981) Comp. J. 24, 167–172.
    Fischer, W. and Koch, E. (1973) N.Jb.Miner.Mh. 252–273 and 361–380.
    Fischer, W. and Koch, E. (1979) Zeit. f. Krist. 245–260.
    Brostow, W., Dussault, J.-P. and Fox, B.L. (1978) J. Comp. Phys. 29, 81–92.
    Frank, F.C. and Kaspar, J.S. (1958) Acta Cryst. 11, 184–190.
    Finney, J.L. (1879) J. Comp. Phys. 32, 137–143.
    Mackay, A.L. (1972) J. Microsc. (Oxford) 95, 217–227.
    Hasegawa, M. and Tanemura, M. (1980) J. Theor. Biol. 82, 477–496.
    Maxwell, J.C. (1869) Scientific Papers of J.C. Maxwell. pp. 514–525.
    Green, P.J. and Sibson, R. (1978) Comput. J. 21, 168–173.
17. Maxwell, J.C. (1869)
18. Soddy, F. (1936) Nature 137, 1021.
19. Mackay, A.L. (1973) Acta Cryst. A29, 308.
20. Langlet, G. (1979) Acta Cryst. A35, 836–837.
21. Kabsch, W. (1976) Acta Cryst. A32, 922.
22. Kemeny, J.G. and Kurtz, T.E. (1971) Basic Programming. Wiley, New York.

Microprocessing

Geisow & Barrett (ed.) Computing in biological science

The potential of the microcomputer in biological research

ROBERT J. BEYNON

1. OVERVIEW

Despite the content of volumes such as this, the use of computing methods in the life sciences remains an area of specialist application in which relatively few scientists take advantage of the extensive data analysis that is possible. The qualitative or semi-quantitative nature of a large proportion of biological data means that computational analysis is perhaps seen as providing a gloss of sophistication on unworthy experimental data. Rigorous mathematical models of biological processes are often lacking, unsound or require a mathematical knowledge attained by few life scientists. Within this framework it is easier to appreciate that computers have not enjoyed wide acceptance, especially when compounded with the need to acquire a knowledge of the operating procedures of the central minicomputer or mainframe computer. To understand the operation of a computer an effort must be expended that may not, in the eyes of many scientists, justify the benefits that may accrue.

Within the last few years a new development, consequential upon the advances in the manufacture of large scale integrated circuits, has provided an opportunity to alter the attitude of the life scientist to computers in general. This development, the evolution and appearance of the microcomputer in a 'ready to go' marketable form has been hailed as being responsible for a major change in the decentralisation of computing power. It is now possible for individual users to have access to one of a large number of types of small computer that have been the subject of a vast advertising campaign providing descriptions such as 'personal', 'affordable', 'user friendly', 'easy to use' and so on. These descriptions are extremely subjective and do not take into account the abilities, time and experience of the user, the local availability of expert knowledge and the intended application. The low cost and personal nature of the microcomputer have resulted in their acquisition by a large proportion of departments in life sciences where previously the extent of computer usage was minimal. Indeed we may already have reached the point where microcomputer sales have exceeded the total sales of minicomputer and mainframe computers. It may be timely therefore to look at the application of these machines in the life sciences and ask whether they can provide acceptable computational facilities at low cost. The effective use of the microcomputer requires a detailed understanding of some aspects of its electronics and of the way in which information is stored as data and processed. The level of competence needed is not attained without effort, often in the face of limited manufacturer support. The bold claims made for these machines will often not be realised without a considerable investment of time and effort and the naive user may not be in a position to assess what is required. There is a genuine need for introductory, and application-oriented texts, that could reduce the diffuse and extensive learning process consequential to the acquisition of a microcomputer.

This chapter has two primary objectives. Firstly, a brief description of the features that are offered by the current generation of microcomputers will provide a background and glossary for further investigations into this area. Secondly, a summary of some application areas will attempt to emphasise the facilities and limitations that a new user might reasonably expect. The literature pertaining to microcomputer applications in the biological sciences is not extensive but where possible references that are of relevance to the life sciences will be cited.

Finally, it is important to consider the role of the microcomputer in the light of current trends in mainframe computing. The tendency to distributed processing means that the old concept of a computer centre is being superseded by the more nebulous computer network connecting department terminals to a central mainframe. This would seem to dispel any geographical advantage that was offered by the microcomputer. It is therefore essential to begin an evaluation of the advantages that are offered by the microcomputer and to identify those areas in which the microcomputer is most effective.

2. THE MICROCOMPUTER

2.1. Introduction

Perhaps irrelevantly, it is difficult to provide an exact definition of a microcomputer. The simplest definition might be that of Ogdin [1] 'computers based on microprocessors' whilst a more expansive description might be 'a set of system devices, including the microprocessor, memory and input/output elements interconnected for the purpose of performing some well-defined function' [2]. The number of functions that may be accommodated by the microcomputer can vary from a single purpose dedicated machine, often designed to fulfil that one function, to a general machine with a broad application area. The first type of system has often

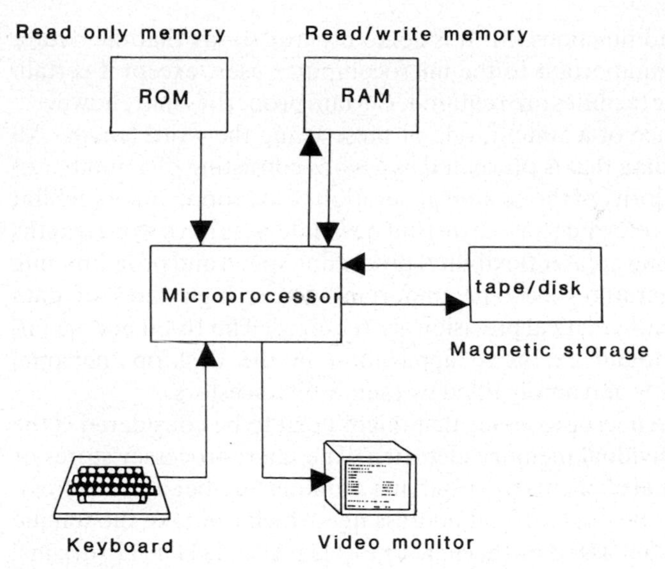

Fig. 1. A block diagram of the components of a typical microcomputer systems. The various connecting lines indicate data movement in the direction shown.

been designed and built at a single or few sites and has limited applicability. The second microcomputer type is represented by the large number (over 250) of user-programmable commercially available microcomputers. The general utility of this type of machine has been a major force in their widespread acceptance and this chapter will concentrate on the applicability of these type of machines.

Certain features will be common to both types of microcomputer (Fig. 1). By definition, they will contain a microprocessor which will be linked to memory devices and facilities for communication, either with the user or with a piece of associated equipment. The tangible electronic components of the microcomputer (in jargon, the 'hardware') can, within the domain of these common features, demonstrate a huge range of computing power. Additional to the hardware is the less tangible series of instructions, or 'software' that dictates the function of the computer. Software may be changed at will to meet new demands and in this respect the microcomputer is no different to a mainframe machine. The differences are primarily in size and scope, such that the microcomputer is better suited to a relatively simple application, circumventing the need to obtain a detailed understanding of mainframe operating procedures. Secondarily, there are areas, particularly in the field of instrument interfacing where links to a mainframe are impractical, expensive or otherwise difficult. Conversely, a more complex application might requires a rather extensive knowledge of the design and operation of the microcomputer. For this reason, an overview of the structure of microcomputer systems is given below.

2.2. *Hardware*

2.2.1. *The microprocessor*

About 150 different microprocessors can be identified currently which demonstrate a wide range of abilities and functions [3]. It is generally true to say that the nature of the microprocessor is unimportant to the microcomputer user, except if certain special or processor-specific facilities are required. Certain properties may, however, be of relevance in the choice of a system, one of these being the word length. All microprocessors act upon data that is presented as a word consisting of a number of binary digits (bits). The majority of the current generation of microprocessors exhibit a word length of 8 bits (also referred to as a byte) but newer devices have word lengths of 16 bits or even 32 bits giving greater flexibility, processing speed and programming convenience. Such considerations become important where high rates of data manipulation or extreme mathematical precision are required. The 16-bit microcomputers have not, as yet found extensive application in the desktop 'personal computer', a niche admirably and amply filled by their 8-bit ancestors.

A second property of the microprocessor that might need to be considered is the capacity for addressing individual memory elements. The microprocessor stores or retrieves data from individual elements by specifying a unique number as an address. Most 8-bit microprocessors possess a 16-bit address line which can take the unique binary numbers from 0000 0000 0000 0000 (decimal 0) to 1111 1111 1111 1111 (decimal 65535), a total of 65536 possible addresses (2^{16}). Thus, if each memory element could store an 8 bit number the total addressing capacity of an 8-bit micromputer is 65536 bytes. In computer circles it is customary to refer to 1024(2^{10}) bytes as 1 Kbyte, so that the 16-bit address line could reference 64 Kbytes of memory. This would normally represent the maximum memory capacity of an 8-bit microcomputer although it is now possible to extend the memory range further [3]. Memory address limitations are unlikely to be a serious restriction for 16-bit microprocessors, which can typically address between 1 and 8 megabytes of memory!

The hardware restrictions that are referred to above are less likely to be important than software considerations. Certain desirable software packages are restricted to one type of microprocessor or microcomputer, and can have a strong influence on the final choice of system. Further information may be found in Section 2.3.

2.2.2. Microcomputer memory

Semiconductor storage or memory is needed to store the sequence of instructions to the microprocessor and to store the data upon which the program will act. The memory is organised as a series of locations, the number of which is dictated by the address range; the size of each location is dictated by the word length of the microprocessor. Thus, an 8-bit microprocessor with a 16-bit address line can read from or write to 2^{16} bytes or 2^{19} bits of information (equivalent to approximately 16 pages of text in this book).

Two features of microcomputer memory that should be considered are volatility and programmability. A volatile memory is one that loses the data stored within it when the power is interrupted. Programmability refers to the ease with which data may be stored in a memory device.

A major division of memory separates random access memory (RAM) from read-only memory (ROM). RAM is volatile memory that is readily programmed to store data or programs and is the only type of memory that can be changed readily by the user. The amount of storage in a microcomputer, whether for programs or data is indicated by the amount of RAM that is provided. ROM, on the other hand is non-volatile and the information stored within it can be considered as invariant. The data in ROM is read many more times than it was written, although the write process must occur at least once (ROM is thus a misnomer). Because ROM is non-volatile the data stored within it is always immediately available when the microcomputer is switched on. Programs committed to ROM tend, therefore to be those that perform basic function such as providing a monitor or interpreter (see Section 2.3). In dedicated applications it is feasible to store the application software in ROM, obviating a requirement for transfer of the program from some permanent storage. Variants of ROM are PROM (or programmable ROM) and EPROM (or erasable PROM) which find particular application in 'one-off' or experimental applications.

2.2.3. Mass storage

The programmability of the microcomputer implies a need for some type of permanent storage of programs or data that can be read into the microcomputer RAM. The current generation of machines meet this need by recording the data on magnetic tape or disk. By far the cheapest solution is to use ordinary audio magnetic tape encoding the data as a binary stream where one frequency represents binary 1 and a second frequency represents binary Ø. Appropriate decoding hardware and software can accept this stream of audio frequency information and convert it to a microcomputer compatible form but the use of magnetic tape has two major disadvantages. Firstly, the quality of the magnetic coating and the design tolerances of the record/playback mechanism mean that the maximum practicable data transfer rate is approximately 1,000 bits/second. To transfer data from tape to a 32 Kbyte area of memory would take about 4 minutes using such a system, and transfer rates of 300 bits/second are found in some systems. The delay in transfer of programs or data is obstructive to program development or file handling. Secondly, the magnetic tape stores information in a sequential form, meaning that the tape must be searched from the start to find a specific block of data. This lack of 'random access', where any block may be accessed individually is a primary disincentive to the use of tape storage.

It should be pointed out that cassettes of magnetic tape have been developed for use at higher transfer rates and sometimes permit rapid scanning for a specified block of information but these have not, in general, been implemented in microcomputer systems.

As an alternative to magnetic tape, the 'floppy disk' has gained wide acceptance. The disk, either 5¼″ or 8″ in diameter is coated with a magnetically sensitive surface and rotates under a read/write head. Data is stored as a series of concentric rings on the disk such that each ring is accessed serially as the disk rotates but each ring may be referenced individually as the read/write head moves radially over the disk. Thus, random access to the stored information is perfectly feasible. The rate of data transfer is typically 40,000 bits/second and the capacity of the disk can vary from approximately 70 Kbytes for a 5¼″ to 500 Kbytes for an 8″ disk and within this range exist a variety of recording formats and operating software that make many disks unreadable on other systems. Disadvantages of disks are the cost of the drives, the sensitivity to corruption and limited lifespan of the disks and the additional RAM or ROM needed to contain a disk operating system. However, any project involving extensive software development or manipulation or large volumes of data would be practically impossible without disk storage. For extremely large applications 3–10 Mbyte 'hard disks' (using a single fixed disk in a sealed unit) are available for most microcomputers, although the fixed disk generates problems of data security and backup.

2.2.4 Input/output devices

Stand alone microcomputers are usually supplied with a keyboard for input of data and a video signal for output of data on a monitor (or with the addition of an RF modulator, a domestic television). Alternatively, a connection can be provided for linking a basic microcomputer to a visual display unit, containing a keyboard and a cathode ray tube. Whilst this form of input and output are of greatest value for entry and display of alphanumeric data there are instances where the data to be processed cannot readily be expressed in a form suitable for keyboard entry and the need for prior translation of such data into an alphanumeric format may be avoided by the use of an input device that generates data from an unusual source. One of the most often used of these devices is the digitiser or 'graphics tablet'. This consists of an active surface that is sensitive to the position of a stylus moving over the surface or over a diagram or photograph placed on a surface. The stylus position can be converted into a stream of Cartesian coordinates which become available for manipulation by the users' software. Typical applications include the analysis of image data from photomicrographs of subcellular structures [4,5] or the input of recorder traces, although the last application might be better fulfilled by analogue to digital conversion in a more direct fashion (see Section 3).

For graphics programs the joystick or trackball input devices can provide two channels of analogue input that can be interpreted and acted upon by the software resident in the computer, to alter the rotation or size of an image, for example.

The most common alternative output device to the visual display on a monitor is 'hard copy' from a printer. The quality of the printer can vary enormously, from a crude dot matrix printer in which each character is made up of a pattern of dots in a 5×7 matrix to a correspondence quality daisywheel printer [6]. Some matrix printers have the ability to generate graphical output at the same resolution as the character matrix. The typical printing speeds of 60–150 characters/second are normally acceptable for microcomputer applications.

Higher resolution graphical or alphanumeric output can be obtained by the use of a digital plotter. The most primitive plotter uses simple commands to control movement in the directions $+X$, $-X$, $+Y$, $-Y$, the four diagonals and $-Z$, $+Z$ (pen up/down). All of the instructions for vector generation (line drawing) and character generation must be encoded in the users software. The more 'intelligent' plotters possess these functions as an intrinsic part of their software and permit commands specifying moves such as 'dashed line from X, Y to X', Y'' or 'circle, radius 10 at X,Y'. Other output devices are less likely to be utilised to any great extent, but for process control, for example, a single bit output could be used to change the position of a solenoid controlled valve. Sound output has limited value, although computerised speech generation is approaching the stage where a microcomputer may possess a 100 word, highly intelligible vocabulary for such facilities as prompting or error messages!

2.3. *Software*

2.3.1. *High and low level software*

Sofware that has been written for use in microcomputer systems can be classified in several ways. The programs that permit the user to perform primitive functions such as memory access, input/output or the storage or recovery of data from a mass storage device are classed in the same category as the software that provides an interpreter for a 'language'. This type of software is primarily supplied by the manufacturer of the system and can be referred to as systems software. Applications software refers to the programs that are written to perform a specific function on a microcomputer for a limited number of users. Applications software such as non-linear curve fitting would not be required by all users of a particular system but most of them would require systems software such as an interpreter for BASIC.

Systems software requires the manipulation of binary data in a raw form and the programs must be able to handle such data at the level of the microprocessor. All microprocessors possess an instruction set that manipulate data at this level and the machine code program that is produced is dictated by the instruction set of the particular device. Programming at the level of machine code requires a detailed knowledge of the architecture of the microcomputer and of the microprocessor but produces fast and memory-efficient programs. Machine code programming is facilitated by use of an assembler, which replaces machine code instructions by mnemonics during generation of the program. When the program is complete it may then be assembled into pure machine code (or object code). The initial source code is retained as a documented and much more readable form of the program. Machine code or assembler programming are difficult because of the need for a detailed understanding of data processing by the microcomputer. The data is restricted to n bit integers unless floating point software is written and 'simple' operations such as displaying information on the screen can require considerable effort. For this reason the development of computer languages has aimed to remove many of these 'low level' considerations and allow the programmer to define a program as a series of 'pseudo-natural language' statements at a 'high' level with the objective of optimising legibility. All of the high level languages have not as yet been implemented on microcomputers but a significant sub-set are available. High level programming is suited to applications software although speed or other considerations may occasion the need for a low level machine code 'patch' or utility routine. The feasibility of this type of hybrid programming is primarily a function of the abilities and knowledge of the user.

2.3.2. *The monitor*

Practically all microcomputers are supplied with a program called the monitor that resides permanently in ROM. The monitor is a small (typically 2–4 Kbyte) program that handles primitive functions such as entry of data into memory, display of data and storage or recovery of data from magnetic tape. More sophisticated monitors may provide facilities for testing, debugging and disassembly of machine code programs. The users own programs may call subroutines in the monitor, for example to display the result of a calculation. One of the major functions of a monitor is to permit the loading of more sophisticated machine code programs such as interpreters or assemblers, and in many cases the former is loaded automatically when power is supplied to the microcomputer.

2.3.3. Interpreters

Interpreters are machine code programs that allow the user to enter and run high level programs in an interactive fashion, the most common interpreter being for one of the many dialects of the BASIC language [7,8]. This interpreter is often seen as a standard feature of the microcomputer to the extent that it is encoded permanently in ROM and is therefore always present. The monitor routines are then transparent to the user as the interpreter performs their functions in a high level language. Interpreters for BASIC show a considerable variety, ranging from integer-only interpreters needing 4 Kbytes of memory to 12 digit precision, floating point interpreters with enhancements for graphics and file handling that may occupy up to 20 Kbytes of memory.

The lines of high level source code (such as BASIC) are stored in RAM as they are entered via the keyboard in a coded format, using a small section of the interpreter that may perform limited syntax checking on the statements. At any time, the user may call for the execution of the statements that have been entered into RAM. The interpreter takes each statement, reduces it to a sequence of high level actions and then reduces each action to a series of machine code instructions, before execution can proceed. This ‘interpretation’ of the high level statement occurs every time the statement is executed, making interpreted languages slow in execution. Their primary advantage is the ease with which programs or sections of programs may be entered, checked and run. Editing of program statements is variable, ranging from simple replacement of a whole line to specific alterations within a statement or even, with sophisticated editors, global replacement of one sequence of characters by another. The most basic editing facilities will however be adequate for program development, but it is as well to appreciate the range of facilities that can be obtained.

Interpreters are provided by some suppliers for languages other than BASIC, the most interesting of which is the P-code interpreter, a part of the Pascal language system. Pascal is a high level language that generates an object code intermediate between the source code and machine code. The P-code, as it is called is not compiled further to machine code but is interpreted by a program called a P-code interpreter. Because the P-code is well standardised the Pascal language can be executed on any microcomputer, once the P-code interpreter has been written, and the interpretation of a low level P-code means that program execution is more rapid than if the source code were interpreted directly. Development of Pascal programs will, of course, require a compiler to generate the P-code [9,10,11].

Finally, other languages such as Forth [12] and Lisp [13] can be implemented on microcomputers using interpreters or compilers.

2.3.4. *Compilers*

A compiler is different to an interpreter because the source code is only scanned to generate an object code (such as pure machine code or P-code), the object code is subsequently executed with a considerable speed advantage. The more sophisticated microcomputers are able to support compilers for several languages, although there are restrictions on the type of machine that is used, the nature of the disk operating system and the availability of sufficient memory (usually requiring most of the full 64 Kbytes). Compilers are available for Fortran 77, Algol 60, Pascal, APL, C, COBOL and BASIC. The BASIC compiler is claimed to increase speed over interpreted BASIC programs by a factor of between 3 and 10, and may prove valuable in some time-critical applications. In particular, development and debugging of the BASIC program can be performed in interpretive mode before the object code is used in the final compiled version.

2.3.5. *Disk operating systems*

Flexible disks are able to reduce program development time and are practically essential for handling files of data. The software that performs primitive functions such as starting/stopping the disk drive, moving the head and controlling the storage or retrieval of data is referred to as the Disk Operating System or DOS. The DOS tends to be large (10–20 Kbytes of RAM) and is normally provided with the disk system. In general, the choice of DOS is unimportant if the user wishes to use the disk drive for local applications only but the selection of a system becomes critical if externally created software is to be used, especially at a low level.

Information is stored on disks in a wide range of formats such that most individual DOS software is totally incompatible with other systems. The one major exception is CP/M or Control Program for Microprocessors [14,15] a ubiquitous operating system that is restricted to the 8080 and Z80 family of microprocessors. CP/M is implemented on practically all of the microcomputers that use either of these processors and a vast range of systems software such as compilers, or utility and applications software is available in CP/M compatible form. However, the most popular microcomputers at present use the 6502 microprocessor, so that in terms of absolute numbers, most microcomputer systems do not implement CP/M.

3. COMMUNICATING WITH THE MICROCOMPUTER

3.1. *Introduction*

To use the microcomputer effectively it is essential to have appropriate means for communication with external devices. The multiplicity of ways in which such communications may be established means that only a limited number of all the possibilities are likely to be provided with or available for any one system. Accordingly, the selection and availability of certain interfaces may be a critical factor in the final decision as to the best system to acquire.

The two major uses of interfaces are (1) communication with the computer-type external devices, and (2) communication with other equipment which has a primary function not associated with computing. In the first category are included printers, digitisers or even other computers, whilst the second class can be taken to include a wide domain of analytical equipment such as spectrophotometers or scintillation counters but additionally including cell-culture vessels, for example. The distinction between the two categories is becoming progressively more blurred as manufacturers of equipment in the second class provide their own local data handling facilities as an in-built feature. To illustrate, it is now possible to purchase spectrophotometers that will automatically store and correct spectral information from enzyme kinetics, gel scanning or the analysis of thermal transitions of nucleic acids. This level of sophistication almost totally removes any need for additional computational facilities unless the microcomputer is to be used for archival storage of data or for more sophisticated numerical analysis.

On a simpler level, manufacturers have often made provision for the presentation of data at a connector that is accessible to the user. Some equipment handles data primarily as discrete integral values (e.g., scintillation counters) which can be presented readily at a standard computer interface such as a serial connection using RS 232 protocols (see Section 3.2.1). It is then a relatively simple matter to make the correct electrical connections between the RS 232 output and a suitable microcomputer interface.

Many items of analytical equipment present data that is of a non-discrete, continuously variable, analogue form which must be converted into a digital representation before it can be subjected to microcomputer analysis. The currently fashionable practice of using 7-segment light emitting diode (LED) displays for analogue quantities means that the signal has already been digitised and it is sometimes possible to capitalise upon this fact.

In addition to the choice of interface between the microcomputer and other equipment there are other factors that should be noted. The ease of making the appropriate connection and the accuracy and precision with which the signal is presented to the microcomputer should all be given careful consideration. This section will attempt to provide a brief summary of the many approaches to interfacing microcomputers with equipment. The general importance of analogue to digital conversion is reflected in the slightly more detailed treatment of this process. The reader may find references [16] and [17] useful for additional information.

3.2. Communicating with digital devices

3.2.1. Serial interfaces

Serial transmission of data is often used to communicate with peripherals such as printers, mass storage devices or modems (for long distance communication). The binary data encoded as a series of pulses specifying '1' or 'Ø' can be carried using a pair of conductors. Because each word (for example, a byte) has to be converted to a stream of bits the transmission rate is relatively low, typically ranging from 110 bits/second to 9600 bits/second. The most familiar serial interface is that used for encoding binary data as a sequence of audio frequencies for storage on audio tape or for the related process of storage on floppy disks [1].

Additionally, and perhaps more relevantly, many equipment manufacturers are making provision for the presentation of serial encoded data in a semi standardised format. The format is called RS 232 (C) and has found extensive application in communications, printer interfaces and now in equipment interfaces. Most microcomputers can be provided with facilities for sending and receiving data according to this standard, but the opportunities for variation in software, but not hardware within this standard are large [18]. There are some 150 subtle variations of RS 232 data transmission protocols [16] so that it is essential to ascertain whether the equipment and/or the microcomputer interfaces are sufficiently versatile to cope with the idiosyncrasies of design variations.

To a certain extent, the interconnection of two RS 232 devices has been facilitated by the development of integrated circuits called "Universal Asynchronous Receiver/Transmitter" (UARTs). These devices convert parallel data into a serial format (or vice versa) and have been made sufficiently general to allow most of the RS 232 possibilities to be selected by software. Thus, if the microcomputer or interface uses a UART to communicate serial data then connection to an RS 232 originating or receiving device should be relatively straightforward.

The remaining obstacle is to develop software to accept the bit serial data and convert it to real or integer numbers, character data or control codes and act upon this data in an appropriate fashion. Software to perform these low level functions cannot normally be implemented in a high level language, either because of the low speed of an interpreted language or because of the need for data manipulation at a fairly primitive level. Suitable machine code programming for handling serial data may prove to be a formidable task and microcomputer or interface manufacturers' information should be scrutinised to ascertain whether suitable software has been written.

3.2.2. Parallel interfaces

The connection between the microcomputer and the external device could clearly be faster if all the bits of the data word were transmitted simultaneously along a parallel series of conductors, numbering at least as many as there are bits in the word. Parallel transmission is best suited to situations where the data path is short as there is a greater susceptibility to corruption caused by interference between the adjacent data lines. Normally, additional lines are required to provide a 'handshaking' sequence between originating and receiving devices, these ensure that data is transmitted only when it can be received. For example, the time taken for the print head of a printer to return to the beginning of a line represents an interval where data cannot be printed, hence the printer must send a 'not ready' signal to the computer until it is able to commence printing again. Because such handshaking is common in parallel interfaces there is less need to specify the speed of data transmission so that the user need not consider such factors as bit rate and parity. The most common parallel interface is the 'Centronics' standard mainly intended for printers but also of value as a general 8 bit interface for A/D converters, etc.

Implementation of 8 bit parallel interfaces on some microcomputers (e.g., MCS 6502) is relatively simple due to concept of memory mapped input/output (I/O). This means that 8 bit data may be handled as if it were transmitted from or received into a single byte memory location which can theoretically be any one of the 64 K possible locations. Other microcomputers (e.g., Intel 8080 and Zilog Z80) avoid the problem of possible memory conflicts between I/O locations and programs or data by using the concept of output ports [19] which can number up to 256 separate, 8 bit wide data paths. Finally, manipulation of I/O is facilitated by the availability of large scale integrated circuits providing much of the hardware for the control of 8 bit data paths [3].

3.2.3. *Binary coded decimal*

The trend towards LED or LCD displays on many items of analogue equipment means that the process of digitising the signal has been accomplished as an internal function. Additionally, many manufacturers provide an external connector for taking off this digital information, e.g. for a printer suggesting the possibility of using this connector in a linkage to a microcomputer. The data will normally be present as a binary coded decimal (BCD) format in which the decimal digits are coded by groups of four bits (Table 1). Thus, the decimal equivalent of BCD 0101 0011 is 53. A four digit seven segment display could be used to provide microcomputer data using two input ports or memory locations, accessible from high or low level software. Problems might include decimal point manipulation and the interface which may need to be designed locally. Furthermore, most digital displays do not provide the same resolution as is provided by the unconverted analogue signal (via a chart recorder, for example). The practical solution may in fact involve bypassing

TABLE 1

Binary coded decimal representation of decimal quantities. The binary representation is given for comparison. Note that 4 bit numbers greater than 1001 are illegal in BCD.

Decimal	Binary coded decimal	Binary
0	0000 0000	0000 0000
1	0000 0001	0000 0001
2	0000 0010	0000 0010
5	0000 0101	0000 0101
9	0000 1001	0000 1001
10	0001 0000	0000 1010
11	0001 0001	0000 1011
19	0001 1001	0001 0011
20	0010 0000	0001 0100
56	0101 0110	0011 1000
99	1001 1001	0110 0011

this digital signal, and using an analogue to digital convertor that better meets the need of the experimenter.

3.2.4. *The IEEE-488 bus*

The IEE-488 bus was designed to allow connection of several devices (up to 15) on a common series of conductors [17,20,21]. Communication between the devices, which may be computer, peripherals or laboratory equipment, is in bit parallel, byte serial format at relatively high data rates (up to 500 Kbytes/s). Some microcomputer systems use this bus for peripheral interfacing whilst additional interface cards for other systems convert the microcomputer address and data configuration to IEEE 488 protocol. Most microcomputer systems and peripherals do not, however, provide this rather specialised feature as standard.

3.3. *Communicating with analogue devices*

3.3.1. *Introduction*

The biologist is presented with much of his experimental data in analogue form, limited only by the resolution of his measuring devices and subsequent analysis of this data can be performed in one of two ways. Firstly, the analogue data can be processed in its raw form (e.g. differentiation or integration), using complex analogue circuitry that must demonstrate linearity and noise immunity. Alternatively, the analogue data can be digitised and if needed, stored by the microcomputer; subsequent processing is then performed by software manipulation of the digitised information. The second approach has a number of advantages such as the versatility of analysis attainable by reprogramming of the microcomputer and the ability to conduct repeated analyses on the same stored data. Furthermore, the configuration of the equipment may be simplified with the shift to computer analysis of the data. One of the best examples of this consequence is in scanning spectrophotometers where, in the latest generation of machines, the ability to store a baseline or reference scan prior to a sample scan means that single beam instruments can be used, giving greater light energy and often, better performance than their double beam counterparts.

Facilities for digital data processing are becoming standard on many types of analytical equipment complete with inbuilt analytical software, usually in ROM. The software is therefore fixed and the user has little opportunity to alter or optimise the software to his own purposes. As an alternative, the user may be tempted to consider linking the piece of equipment to a microcomputer and producing software that is tailored perfectly to his needs. The first stage is the generation of microcomputer readable digital data from the analogue output from the experimental equipment. Thus a knowledge of analogue to digital (A/D) and to a lesser extent, the digital to analogue (D/A) conversion is essential if a suitable interface is to be built or acquired. The remainder of this section will discuss some of the principles and practice of A/D and D/A conversion.

3.3.2. *Digital to analogue conversion*

D/A converters generate an analogue voltage or current that is specified by a binary data word, and which is proportional to a reference supply; for this reason they are multiplying devices [22,23]. The reference supply can be external to the integrated circuit that performs the conversion (the term 'multiplying device' is usually restricted to this type) or alternatively may be provided internally as part of the function of the circuit.

The proportion of the reference supply that is generated by the D/A converter is dictated by the magnitude of the binary word that is presented on the data lines. For example, an 8 bit D/A converter, supplied with a reference voltage of 1 volt would be capable of generating a minimum voltage of 0 volts (data byte = 0000 0000) and a maximum voltage of $255/256 \times 1$ volts (data byte = 1111 1111) = 0.996 volts. As the data word is incremented by 1 bit the voltage increases by 1/256 volts = 0.039 volts. The maximum voltage that can be generated is always less than the reference value because one of the 256 steps is used to specify a voltage of Ø. In general terms, for an n-bit D/A converter the step size of the output is $V\text{ref}/2^n$ and the maximum voltage generated is equal to $V\text{ref}.\ (2^n - 1)/2^n$.

Applications for D/A converters are unlikely to be as extensive as for A/D converters. The primary uses would probably lie in the area of signal reconstruction where a digital representation of experimental or simulated data is used to generate an analogue output which could then be used to drive oscilloscope displays or generate 'hard copy' output on an X–Y or X–t plotter. The speed of D/A converters, typically in the region of 1 MHz makes simulation of rapid processes quite feasible, although the possibility of a solution using digital techniques should be investigated.

3.3.3. Analogue to digital conversion

Microcomputer-mediated analysis of analogue information from experimental equipment represents one of the more intriguing of potential applications, taking many precedents from the use of mini- or mainframe computers in this area [24,25]. The overall process can be separated into data capture, where the analogue data is converted to a digital format and data reduction, where the digitised information is processed according to the needs of the user. The second phase is totally dependent upon software, which, if written in a modular form can provide a versatile package that will be able to evolve with changing needs.

Since the quality of the data analysis is only as good as the data that is used, the signal acquisition rate, linearity of the conversion process and precision of the conversion must all be considered in the design of a microcomputer based system for analogue data processing.

The conversion of analogue information to digital data has been refined to the extent where the electronic circuitry has been relegated to a single integrated circuit. There are many methods for A/D conversion differing in resolution, speed and ease of implementation in microcomputer systems [22]. Some of these provide a means of sensing the position of a level, spindle or some other mechanical device, usually by translating position to a resistance through a potentiometer. The potentiometer can then be used to alter the behaviour of digital circuitry that can in turn be sensed by the microcomputer and translated to a variable, useable by high or

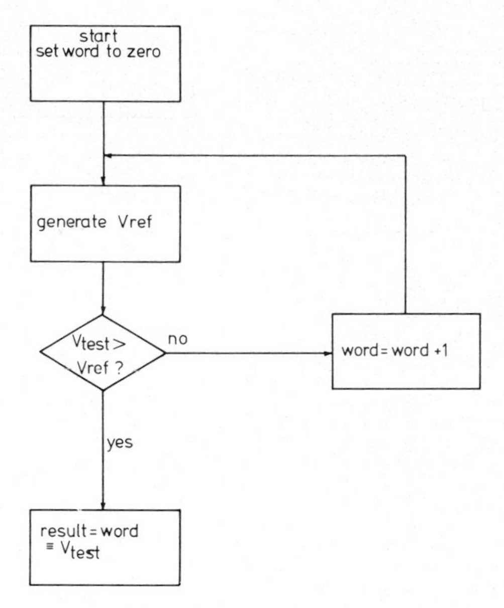

Fig. 2. A flowchart of the digitisation of an analogue voltage using the binary ramp or 'servo' method of A/D conversion.

low level software [26]. The most obvious application of this type of interface is in the 'game controls' used in many microcomputer systems or microprocessor controlled video games which can also function as relatively cheap analogue inputs.

More commonly, the signal to be digitised by the A/D interface is a continuously variable potential. In this case the user is presented with a bewildering array of complete A/D converters or, if a home made system is being considered, of isolated integrated circuits that perform the same basic function. The most popular A/D converters actually employ a D/A converter and a comparator that tests the input voltage against an internally generated reference voltage from the D/A converter. Depending upon the result of the comparison the binary word that was supplied to the D/A section is considered to be the closest binary value to V test or not, in the latter case the conversion process continues. The binary ramp (or servo or linear search) A/D converter uses this technique summarised in the flowchart in Fig. 2 (The whole of this process is conducted by a single integrated circuit).

After initialisation, the binary word is set to zero and the output of the D/A converter (V ref) is compared with V test. If V ref is less than V test the binary word is incremented and tested by one bit per cycle until V ref $\geqslant$ V test, at which time the binary word, the closest approximation to V test, can be sampled by the microcomputer. Additional signals will, of course, be required to indicate the start and end of the conversion process, to ensure that the data is valid.

By starting with V ref $= \emptyset$ and ramping upwards until V test is reached, this type of converter can take a variable time, with a resolution of n bits, up to 2^n cycles, to make a valid digitisation. A more sophisticated approach towards the D/A step can

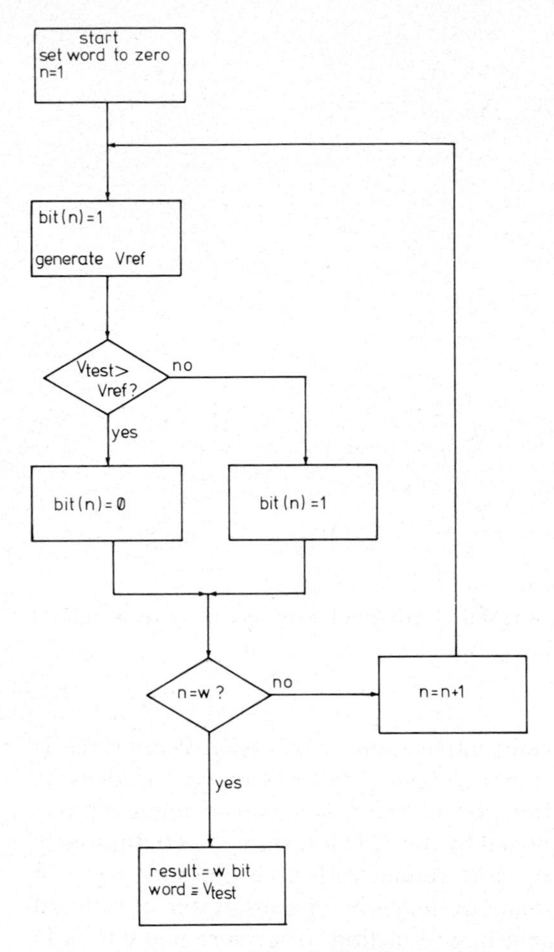

Fig. 3. A flowchart of the digitisation of an analogue voltage using the 'successive approximation' method of A/D conversion.

reduce this overhead by a considerable margin, (especially as n becomes larger) making the conversion time constant at $N + 1$ cycles. The successive approximation A/D converter achieves this improvement by using a logical trial and error method of determining the value of V test [27]. At initialisation, the binary word is set to zero and at the next cycle the most significant bit (msb) is set to 1 (e.g., for a 4 bit converter from 0000 to 1000). V test is then compared with V ref, and if it is greater, the msb remains at 1, if it is lower the msb is set to 0. Next, msb-2 is altered to 1 and the same comparison is made. When the process has been repeated for all n bits of data, the status of each bit of the word is the digitised value of V test. A flowchart of the process is shown in Fig. 3, and the sequence involved in digitising a voltage to 8 bits resolution is demonstrated in Table 2. Finally, the successive voltages assumed by V ref during this process are shown in Fig. 4.

The two most important considerations in the selection of an A/D interface for a

TABLE 2

The 8-bit digitisation of an analogue voltage by successive approximation techniques. The input test voltage was 0.43 and the A/D converter was adjusted such that binary 1000 0000 produced exactly 0.5 volts as *V* ref.

Test	Word before	*V* ref (volts)	*V* test > *V* ref?	Word after
1	1000 0000	0.5	no	0000 0000
2	0100 0000	0.25	yes	0100 0000
3	0110 0000	0.375	yes	0110 0000
4	0111 0000	0.4375	no	0110 0000
5	0110 1000	0.40625	yes	0110 1000
6	0110 1100	0.421875	yes	0110 1100
7	0110 1110	0.4296875	yes	0110 1110
8	0110 1111	0.4335737	no	0110 1110

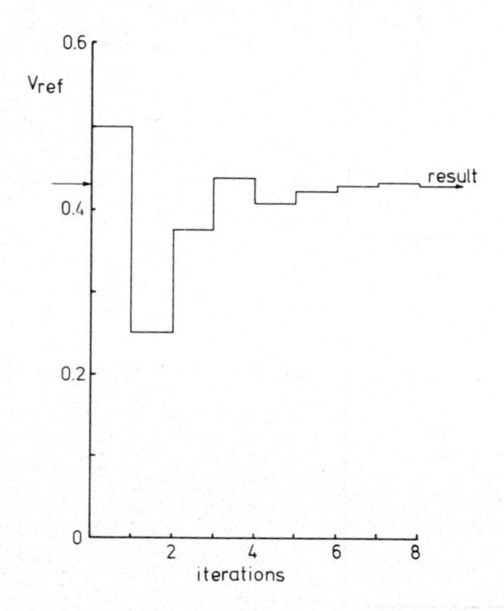

Fig. 4. Alteration of the test voltage generated by an 8-bit successive approximation analogue to digital converter. The voltages are the same as those in Table 2.

microcomputer system are resolution and sampling rate. The resolution of any converter is dictated by the number of bits that are presented to the D/A circuitry as this decides the size of the step in *V* ref. Clearly, the greater the number of steps assumed by *V* ref the greater is the precision with which the signal can be digitised. In general, *n*-bit analogue to digital converter will be able to discriminate signals that are separated by $100/2^n\%$ of the full scale signal. A summary of these values for different bit lengths is given in Table 3 and an illustration of the effect of resolution is shown in Fig. 5. Whilst it may appear superficially that one should choose the greatest value of *n* that is available for the system, many other factors come into play. Firstly, irrespective of the method that used, the time required for digitisation of a

TABLE 3

The effect of word length on the resolution of D/A or A/D converters.

Word length (bits)	Range	Resolution (%)
4	0–15	6.2500
6	0–63	1.5625
8	0–255	0.3906
10	0–1023	0.0977
12	0–4095	0.0244
14	0–16384	0.0061
16	0–65535	0.0015

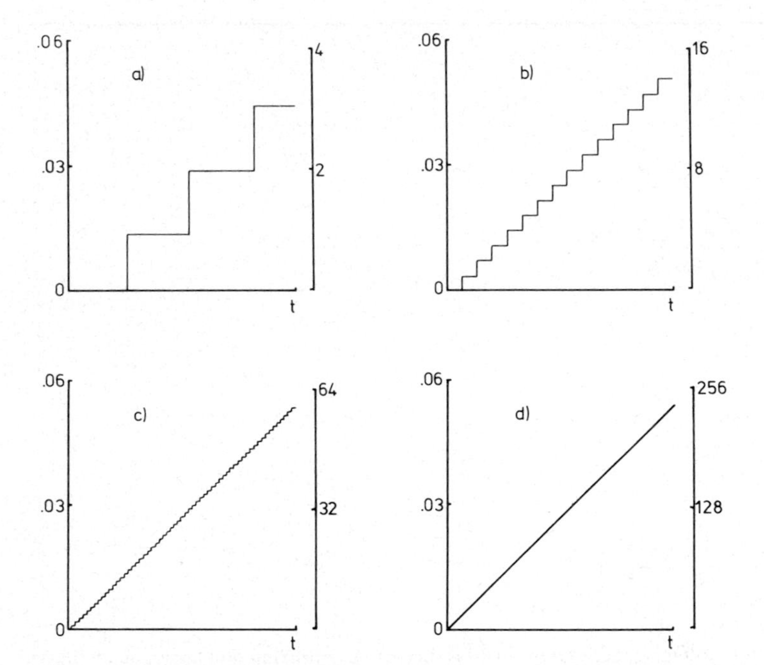

Fig. 5. The effect of word length on the resolution of A/D converters. To illustrate the effect it is assumed that in each case the maximum binary number has been set to an analogue fsd of 1.0. As an imaginary signal increases from 0, the decimal value of the digitised word increases in a stepwise fashion according to the right-hand scale. The left-hand scale shows the proportion of the analogue signal that has been covered. Conversion has been plotted at a resolution of: (a) 6 bits; (b) 8 bits; (c) 10 bits; and (d) 12 bits.

voltage increases as the value of n increases, although much more severely with binary ramp circuits than successive approximation circuits. Accordingly, the sampling rate will be reduced. Secondly, the current generation of microcomputers, being primarily 8 bit machines, handle data most conveniently and most rapidly as 8-bit words. Increasing the resolution above 8 bits means that the digitised value must be stored in two bytes of memory with attendant overheads in software retrieval of the value. Finally, the resolution of the A/D converter should be considered with respect to the quality of the analogue signal. All analogue signals contain random noise and if a piece of equipment is generating a signal that contains noise at a level of 1% there is little value in using a 16-bit A/D converter since any resolution greater than 8 bits is merely providing an accurate record of the noise component without generating additional information about the noise-free signal. If on the other hand, the noise is periodic or reproducible, sampling at high resolution may permit a subsequent signal refinement step in which the noise component is removed by software.

Related to, and to some extent compromised by the objective of high resolution conversion is the rate at which sampling can be conducted. Since high speed A/D converters are costly, it is important to determine a realistic rate of conversion commensurate with accurate data sampling. For signals that vary in either direction with time and which may be considered as a Fourier series, the Nyquist Theorem states that the rate of conversion should be at least twice the frequency of the highest component of the series [26,28]. For unidirectionally varying signals the compromise between speed and accuracy can be determined by calculating a parameter called 'aperture time' [22] which will then allow the user to specify the minimum operating conditions of the converter.

Associated with the question of sampling rate is the problem raised by the need to simultaneously convert information from more than one channel. The simplest approach is to use a separate A/D converter for each signal but this may create interfacing difficulties and the speed of conversion may not be met by the software for sampling each channel successively. The alternative approach is to use the same A/D converter to sample and present the data for each channel successively. Multiplexing of A/D channels in this way does not degrade the speed performance in proportion to the number of inputs so that a very fast converter can be made to look like a series of slower ones [26].

Finally, free running A/D converters that simply sample and store data must be managed carefully. An 8-bit converter operating at a perfectly feasible rate of 1000 samples/second will fill 32 Kbytes of memory in 33 seconds! If the process of interest is, relative to these rates, much slower there is limited value in generating such large quantities of data that must be processed or stored before the next phase of data acquisition can begin. Data from ten experiments such as the one above, each of 30 seconds duration, would, when saved on disk bring the user to the stage where new disks would need to be inserted to save the next sequence of experiments. Furthermore, processing of 32,000 data points in a high level language is not a trivial computational task even with a compiled language, and the speed of evaluation of some mathematical functions could delay analysis by large periods (see Section 4).

If the specifications of the data acquisition system exceed the requirements of the user it is a simple matter to delay the converter by incorporting timing loops. The delay may even be sufficient to use this 'wait' time to analyse the data previously collected, provided that these routines can be interrupted as new data becomes available. Whilst this approach makes best use of the capabilities of the microcomputer it is unlikely that such facilities could be readily implemented in a high level language.

A final consideration in the application of data acquisition to laboratory equipment relates to the need to synchronise the sampling process with the phenomenon of interest. For example, in interfacing a microcomputer to a scanning spectrophotometer is is essential that the A/D conversion process and the wavelength drive are started simultaneously, especially for overlaid scans, if errors in the registration between wavelength and data in memory are to be avoided. This suggests that control signals must be passed between the microcomputer and the equipment one simple way to do this is to use switches that can be tested as single bit inputs by the microcomputer to specify 'start of scan' and 'end of scan'. The A/D converter could then 'idle' until the status of the 'start of scan' bit changed.

Related to this problem is the need to obtain an accurate measure of the time dependence of the signal. Many microcomputers do not have internal clocks that count in seconds or submultiples thereof. It is possible to implement such a clock in software since all microcomputers are controlled by crystal oscillators, alternatively, a more costly approach is to obtain an external clock that will interface to the microcomputer. In both cases it should be perfectly feasible to arrange, via clock generated pulses, for data acquisition to occur at specified time intervals, allowing kinetic processes to be expressed in standard time units.

4. LIMITATIONS OF THE MICROCOMPUTER

4.1. *Software*

Whether because of hardware restriction or because of pricing policy, microcomputer software is not, as a rule, sophisticated. Basic systems are often supplied with one high level language (usually BASIC), a restricted monitor for limited machine code programming and limited editing facilities. Thus, the user who transfers his attention to microcomputers from mainframes may be surprised at the limited facilities offered by the basic software. Expandability may be an important criterion in the selection of a system; this applies equally well to software as hardware. One major feature of the microcomputer world is the appearance of companies that specialise in microcomputer software. For example, for one particular system it is possible to purchase enhanced editors for program development, a compression utility that reduced the amount of RAM needed by a program (and which, incidentally makes programs unreadably compact!) and a compiler that converts an interpreted BASIC program into a machine code equivalent. These utilities originated from different software suppliers, some of them as 'plug in and go' ROM, and could not be considered (by software standards) to be expensive. The movement of 'personal computing' into the consumer market has been valuable in maintaining software costs at reasonable levels. Additionally, the popularity of the CP/M operating system has resulted in the production of good quality software, including interpreters or compilers for many high level languages [14].

Because of the diversity in software supplies there is no straight-forward means to acquire information on such products. Manufacturers of the more popular specific systems are beginning to collate information on 'approved' software but these products will be highly machine-specific.

As a rule, software written for one type of microcomputer will not run on another type, either because of differences in the instruction set of the microprocessor, or even if the microprocessor is the same, because of differences in the architecture of the microcomputer and especially in the use of memory.

Additional problems arise from the role of BASIC as the most popular microcomputer language. The ease of use of an interpretive system and the simplicity of the language mean that the newcomer to programming can rapidly acquire a knowledge of the processes and concepts that are involved. However, BASIC interpreters come in many different forms to capitalise upon the special features of different systems and as such, portability of programs from one system to the next is practically difficult. (For a comparison of several current BASIC dialects, see [29].) More serious objections to BASIC are that it discourages programming that follows logical and clear paths, is unreadable because of limited variable names, lack of indentation and formatting and is restricted in data types. To some extent recent enhancements of the language have provided partial solutions but it is clear that BASIC should be replaced by a language that will combine its good features with a code that conforms to, or even enforces the principles of structured programming. At present Pascal, an Algol-like language is the prime contender for this position although it is more difficult for a newcomer to learn. The data types in Pascal permit, for example, the definition of a data set such as 'type bases (Adenine, Guanosine, Thymine, Cytosine)' or 'type base (A,G,T,C)'. Programming constructs would then allow the user to manipulate the new data type 'base' which can only assume A, G, T or C as acceptable values [11]. In BASIC, using string (character) handling statements the programmer is restricted to statements such as BA$ = 'A' which make a large program less readable. Such advantages in programming practice have not yet been fully realised and it seems likely that BASIC will remain the primary high level language of the microcomputer for some time [30].

For the biologist as in other specialist areas, the limitations of software extend to applications programs. The highly specialised nature of many research applications means that a high volume market for software does not exist, leaving the user with three options; find an alternative source of the software from the scientific literature, write the programs himself or commission a software company to write the programs to his specifications. Selection between these options is dictated by cost, ability and the availability of suitable software, the last being dependent upon the appropriate means of communication with other users[1]. The chances of finding another user employing the same microcomputer for the same task must be small, such that the most likely solution is the production of new code.

Other software of general utility can often be modified or adapted to suit a new purpose (see Section 5). A general database can, for example be modified to suit diverse applications as reference retrieval for a local bibliography or maintenance of animal records. In these cases commercially available software should be readily available for most makes of microcomputer.

4.2. Arithmetic precision and range

All computers must convert real decimal numbers to a binary form before manipulating them in arithmetic routines. The binary representation of a number is restricted in the numbers of bits that are used and thus, the precision with which numbers may be stored is limited. Calculations involving real numbers may require rounding or truncation of the least significant bit in the number. When the binary number is subsequently converted to a floating point representation the rounding error can become apparent.

Rounding errors become significant when the course of a series of computations takes the value of a real to sufficiently high or low values that rounding of the last bit affects the result. This is not only a function of the computation but is also dependent on the numeric range of the microcomputer. Most microcomputers store real numbers in exponential format, for example using 4 bytes (32 bits) for the mantissa and 1 byte for the exponent giving 9 significant digits and a range of $\pm 1 \times 10^{38}$. Zero is defined as any number less than approximately 2.9×10^{-39} with this particular system. Some systems restrict real numbers to 6 digits precision but others now offer the possibility of double precision representation of reals to, for example, 17 significant digits.

In practice, it is only individual applications that dictate the need for a particular degree of precision, but this is a feature that should be investigated in the selection of a system. The arithmetic precision of microcomputers does not approach the digit precision of the mainframe counterparts. For example, the local mainframe running Fortran offers up to 20 digit precision in real numbers in a range of 10^{-76} to 10^{76} approximately. Whether this level of precision is ever needed must be established for individual applications.

4.3. Processing speed

The rate at which the microprocessor runs is dictated by the frequency of the master oscillator and is typically from 1 to 4 MHz. Each cycle triggers an operation within the microprocessor and a single machine code instruction will consist of several (2–8) such operations, giving a range from 0.5 μs to 8 μs. Execution times are highly dependent upon the nature of the instruction and also upon the oscillator frequency, both of these factors are invariant in any one microcomputer.

High level instructions consist of series of machine code instructions meaning that the execution speed of the former is necessarily much lower that the latter. This tends to be especially true with software routines that perform high level functions such as exponentiation, vector calculation in graphics or the evaluation of trigonometric functions (Table 4). Whilst 24 ms may not seem unworkable in isolation, the repeated calculation of an exponential function may soon add a heavy burden to the total time of a computational task (by comparison, mainframe times for such calculations are in the order of 80 μs).

TABLE 4

A comparison of the execution times of various floating point instructions on a microcomputer[2]. Figures are presented for the routines executed in software and by interfacing to a hardware floating point processor[3] and were calculated with n and p equal to positive, real numbers.

Function	Execution time (ms)	
	Software	Hardware[3]
$\sqrt{n}$	47.6	1.1
$ln(n)$	20.6	2.9
n^p	47.7	5.6
e^p	24.3	2.9
$n*p$	2.9	1.2
$1/n$	3.6	1.6
sin (n)	24.3	2.8

Because of the interpretive nature of BASIC in most microcomputer systems considerable speed overheads may develop in large programs. Additionally, the handling of variable and statement references can delay program execution. Most current implementations of BASIC reserve memory locations for variables as they are 'discovered' during the interpretation of the program. When the variable is subsequently referenced the interpreter looks through the table of variables in the order in which they were reserved, such that the contents of a variable at the end of the table may take considerably longer to find than for the first variable in the table.

Execution of BASIC programs can also generate delays in the handling of line references. If a program executes the statement 'GO TO 6700' the interpreter is obliged to look through all of the line references (even all 6699 of them) before finding the correct destination. Clearly then, a portion of code that is executed repeatedly should be given low line numbers at the beginning of the program.

In the belief that most microcomputers are still being programmed in BASIC, the following example will serve to illustrate how careful consideration can have a significant effect on program execution time. The non-linear curve fitting program described in Section 5 has been used in the analysis of multicomponent cDNA hybridisation curves, by fitting an equation of the form

$$Y = P_7 + P_2\,(1-(1+\frac{x}{P_1})^{-1}) + P_4\,(1-(1+\frac{x}{P_3})^{-1}) + P_6(1-(1+\frac{x}{P_5})^{-1})$$

where X and Y are the independent and dependent variables and $P_1 \ldots P_7$ are the parameters to be optimised by the program [31]. The program was originally written in modular form (as far as BASIC would permit) and occupies some 700 lines of code. As originally coded, the program was laborious in its calculations, requiring approximately 6 seconds to evaluate the error sum (sum of the squares of the residuals) for 20 data points (Fig. 6). At this point, several options were available to the author: (1) to compile the program into machine code[4]; (2) to use a hardware floating point processor[3]; or (3) to restructure the program to make the most effective use of the BASIC interpreter. Compilation of the BASIC source code was a simple process but produced an object code requiring 25 Kbytes of memory as

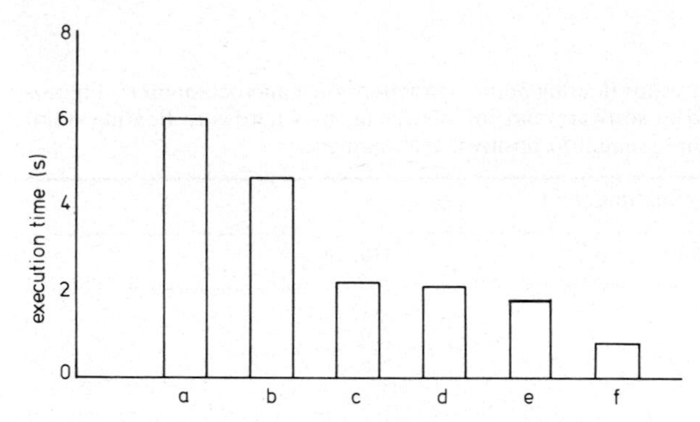

Fig. 6. Optimisation of a BASIC program for speed. The data show the execution times for evaluation of an error sum over 20 data points (see text). (a) Initial program; (b) compiled version of a; (c) a running with a hardware floating point processor; (d) an optimised version of a; (e) d with the hardware processor; and (f) a compiled version of d.

compared with 12 Kbytes for the original program. In addition, compilation required the deletion of some lines because the source file was too large as originally written and would have required division of the program into linked segments. The execution time for a single evaluation was reduced to 4.6 seconds for the compiled program, a relatively minor advantage.

Closer inspection of the original BASIC program revealed that a large proportion of the execution time was occupied by calculation of the expressions n^{-1} and m^2 in the evaluation of the error sum. Inspection of the data in Table 4 demonstrated two possible solutions. Firstly, using a hardware floating point processor would give approximately a 10-fold increase in the execution time for these two expressions. Alternatively, replacement of the expressions by $1/n$ and $m * m$ would also give a considerable speed advantage which the hardware processor would do little to improve. Initially, the fitting routines (which were executed frequently) were placed at high line numbers in the sequences, the modular structure of the program made it a simple matter to move those routines to low line numbers, improving the speed of the program further. Finally the variables that were used repeatedly by the fitting routines were declared at an early point in the program, reducing the time needed to access their values. The initial program (taking 6 seconds/evaluation) was improved when run in the presence of the hardware floating processor, mainly because of the increased rate of calculation of n^{-1} and m^2, bringing the evaluation time down by 63% to 2.2 seconds. However, when the program was optimised by altering variable declarations, line numbers and by replacing n^{-1} by $1/n$ and m^2 by $m * m$ a similar advantage was attained in the absence of the hardware processor, the evaluation time being 2.1 seconds. For this version of the program the floating point hardware made little difference (1.8 seconds/evaluation), as expected from the marginal advantage that it contributed to reciprocation or multiplication (Table 4). Compilation of the optimised version (Fig. 6d) was even more effective, reducing the execution time to 0.81 seconds (Fig. 6f). From the original version of the program it was possible to speed up the evaluation of the function by a factor of 7.5, a highly significant advantage that was attained primarily by an understanding of the modus operandii of this particular microcomputer. Since the overall processing speed is low, such knowledge may be essential for optimisation of BASIC programs, although compiled high level languages may not be amendable to refinement to a similar extent.

5. SPECIFIC MICROCOMPUTER APPLICATIONS

5.1. *Curve fitting*

There are a large number of circumstances where the experimenter wishes to know the relationship of his real-world data to a theoretical model of the system that he is studying. The model, expressed as a mathematical function will define a curve that will represent the behaviour of perfect data. Since real experiments are never perfect the problem becomes one of assessing the suitability of the model, or in other words, evaluating the fitness of the curve to describe the data. The behaviour of the curve is defined both by the form of the equation (invariant) and the values of constants that may be inserted into the equation. The curve fitting problem then develops into the need to obtain the values of the constants (or parameters) which, when inserted into the equation, give the best fit of the theoretical line to the experimental data. There are numerous approaches to curve fitting but the absence of straightforward methods has in most cases meant that this type of analysis is ignored. Graphical methods for finding the values of the parameters 'by eye' are often used and give acceptable results when the error in the data is small. If, however, the scatter is large then the subjective nature of this approach can generate systematic or random errors in the parameter estimates [32,33].

One simple approach to curve fitting relies on transformation of the equation to a straight line, followed by linear regression to calculate the slope and intercept of the line. Because linear regression can be conducted using simple graphical procedures these functions have often been provided as standard on pocket calculators and many BASIC programs have been written for this purpose; e.g. [34].

At a more sophisticated level, fitting a straight line by this procedure is only suitable for a limited number of problems, specifically those in which the dependent variable is linearly related to the parameters to be optimised. Thus, a straight line, $y = m.x + c$ has two parameters, m and c. It is clear that y is proportional to m or c in a linear fashion. In contrast, a hyperbolic curve of the form $y = p.x/(q + x)$, with parameters p and q is not a linear equation since y does not vary linearly with q (although y is a linear function of p). Because the hyperbola is non-linear, transformation of the equation to a straight line such as $1/y = q/p.x + 1/p$ introduces systematic bias into the data that gives erroneous estimates of the parameter (many readers will recognise this transformation in the Lineweaver–Burke transformation; further discussion of this and related problems may be found in [32,35,36]).

The alternative approach which avoids the errors caused by transformation of the data is to use non-linear curve fitting, methods which, as their name implies, especially applicable to estimation of the parameters non linear equations. Because of the difficulties in obtaining algebraic solutions for non-linear functions these solutions generally use iterative procedures where the estimates of the parameters are modified according to the relative success or failure of previous estimates; the repetitive nature of these calculations is ideally suited to the computer. In order to ascertain the success or failure of a set of parameter estimates the criterion of improved fit must be defined. In a large number of curve fitting methods the criterion is the same as in linear regression and is defined as a 'least squares' minimisation. The curve fitting program attempts to improve the fit of the line to the data by reducing the error sum, defined as the sum of the squares of the differences between the observed and theoretical values of the dependent variable at each value of the independent variable [32]. Least squares optimisation is easy to implement and has many precedents in the field of curve fitting although certain assumptions are inherent in the adoption of this criterion.

Taking the previous example of the hyperbola, a non linear curve fitting procedure must supply estimates of the two parameters p and q (in simple enzyme kinetics, equivalent to V_{max} and K_m) and calculate the error sum (ES) over all of the data points (x_1 to x_n) using the untransformed equation:

$$ES = \sum (y_i - p.x_i/(q + x_i)^2$$

where y_i is the ith experimental value of the dependent variable. The success of the current estimates of p and q can thus be related to the error sum of a previous set of estimates and future values computed accordingly, until an acceptable fit has been attained (convergence).

Within this general approach to non-linear curve fitting there are a large number of algorithms that vary widely in robustness, efficiency and ease of implementation [37,38,39]. Two large categories of algorithms can be identified [40]: (1) those that only require the function that is to be fitted to the data (*direct search* methods); and (2) those that require the first or higher partial derivatives of the function (*gradient methods*). The latter category are more efficient but require a knowledge of analytical differentiation and are best suited to matrix manipulations which may create problems of 'ill conditioning' if the initial estimates of the parameters are not reasonably close to the final values [41]. This problem is more prevalent on machines with a limited numerical range, a category largely populated by microcomputers.

Direct search methods have the advantage of being relatively immune to limitations in numerical accuracy and are therefore termed 'robust'. Although more forgiving of initial guesses, the approach to convergence tends to be less efficient than with gradient methods.

The large number of mainframe programs for non-linear curve fitting have not in general been widely employed in the biological sciences. However, there is increasing interest in the application of microcomputer programs to the analysis of a

TABLE 5

A summary of BASIC curve fitting programs suitable for implementation on microcomputers. The column headed 'Parameters' gives the maximum number optimised in the citation but does not necessarily imply that this number is a theoretical upper limit.

Program	Parameters	Comments	Reference
COMPT	6	Gradient method	[54]
NON-LINEAR REGRESSION	2	Gradient method Permits alternative weighting of errors	[33]
R'VOL	6	Direct search method Interactive initial parameter estimation	[55]
–	3	Polynomial fit Notes on extension to > 1 dimension	[56]
ESTRIP	8	Exponential 'stripping' method for initial parameter estimates in pharmacokinetics	[57]
PATTERN-SEARCH	7	Direct search method Interactive with graphical presentation	Appendix (7)
LINREG	2	Linear regression	[34]
ENZKIN	2	Hyperbolic curve fitting Non parametric estimation	Appendix (6)
–	–	Summary of several programs Unpublished or commercial	[51]

range of biological processes and several algorithms have been coded in BASIC (Table 5). One method that has proved valuable in the authors laboratory is based on the Patternsearch algorithm [42] a direct search method that is intuitively pleasing and robust [32,43]. A BASIC program[5] using this algorithm has been used successfully for fitting a number of different functions in such diverse areas as enzyme kinetics, toxin-induced cell lysis [44] and nucleic acid hybridisation [45].

As with most curve fitting programs the primary objective is to minimize the error sum by systematic perturbation of the initial estimates of the parameters. The program conducts local explorations around each parameter in turn, calculating the error sum each time. When each parameter has been perturbed, the new improved estimates are used to form the base point of a new exploration. One of the strengths of the Patternsearch algorithm lies in the ability to accelerate the approach to the best fit where appropriate and to reduce the size of the perturbation when convergence is attained. Further details of this algorithm may be found in [32,37,46].

The implementation of Patternsearch on a microcomputer system was based on two primary objectives. Firstly, the program was designed to be versatile but to require little computer experience on the part of the user. Secondly, the graphics facilities of the microcomputer were utilised to permit plotting of the data and function before, during or after fitting had taken place, enabling reasonable starting guesses for parameters to be made and encouraging a familiarity with the form of the

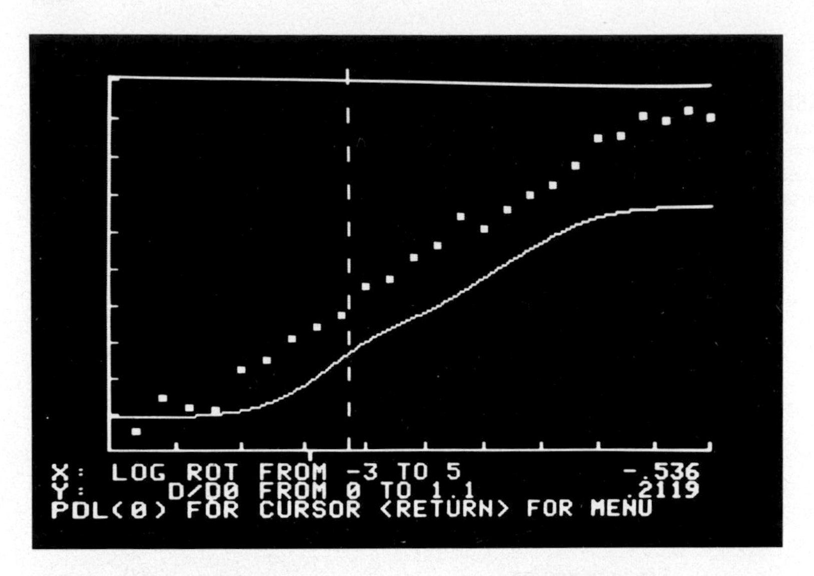

Fig. 7. The microcomputer screen display of the Patternsearch program before fitting. The estimates of the 7 parameters in this equation were adjusted by trial and error to give a reasonable approximation to the experimental data. The resolution of the screen display (280 × 192) is sufficient to convey the impression of 'goodness of fit'.

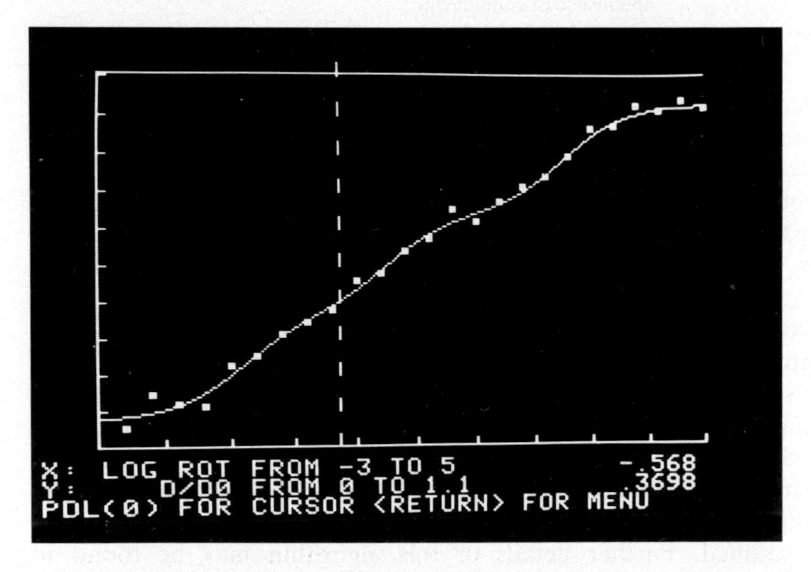

Fig. 8. The sreeen display after the fitting process is complete.

function. Figures 7 and 8 show the screen plot before and after fitting of a three component curve for nucleic acid hybridisation [45]. This function contains seven parameters and the program requires approximately two hours to bring all of the parameters to within 0.1% of their best fit values (for 45 data points). The results

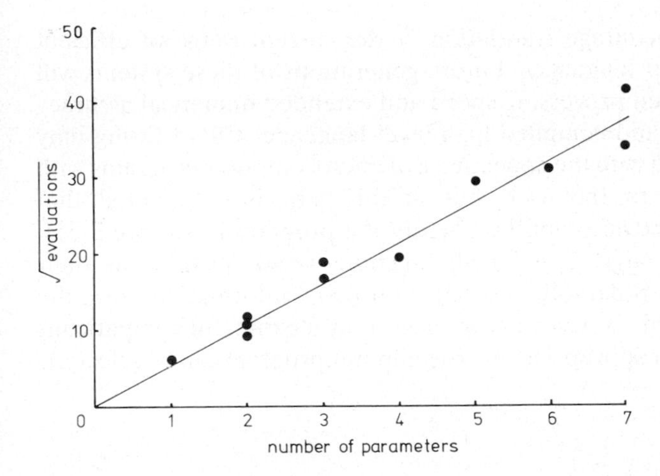

Fig. 9. The dependence of the Patternsearch fitting process on the number of parameters. The data are collected from several types of function and represent the mean of at least twenty separate experiments.

obtained by the Patternsearch method were compared with those from a PL/1 program running of an IBM 4341 and using a gradient method for optimisation [31]. In all cases the final values attained by both programs were within 2% of each other although, as might be expected, the run time of the mainframe program was considerably shorter. For other functions with less parameters the run time on the microcomputer was lower, exhibiting an approximately linear dependence upon the square of the number of parameters (Fig. 9).It should be pointed out that the BASIC program can be automatically executed from 'power-up'; the user requires *no* knowledge of the albeit primitive microcomputer operating system needing only to follow the instructions given by the program. Alteration of the program to fit another function requires a knowledge of BASIC but involves alteration of just 7 of the 700 lines of code to produce a dedicated program that asks for parameter estimates and data points by name, e.g., 'Substrate (7)?' rather than '*X* (7)?'. (It is the author's opinion that such an approach is essential if microcomputers are to gain wide acceptance.)

One major objection to the least squares optimisation methods is that the nature of the error in data is rarely studied, and is therefore unknown [47]. Least squares analyses assume that the data is normally distributed, an assumption that has in certain cases been shown to be unjustified [48]. Non-parametric statistical methods that made no assumptions concerning the error in data have acquired increasing relevance in biological systems and for simple systems, such as hyperbolic saturation curves, some non-parametric methods have been coded as microcomputer[6] programs, or described in the literature [19]. For complex functions containing several parameters, non-parametric methods are difficult to implement although weighting functions, in which the error sum is adjusted to alter the relative effect of different data points, may prove valuable [33,50].

Microcomputer curve fitting procedures have not as yet attained the same level of sophistication as with larger computers. It is possible that the extensive research that has been conducted may encourage translation of the current range of efficient programs into microcomputer languages. Future generations of these systems will undoubtedly provide increased processing speed and extended numerical assuracy probably using a structured and compiled high level language. Curve fitting may become more widely accepted with the appearance of microcomputer programs with an acceptable presentation to the user, but in this case it is essential that documentation and testing are accomplished before the program is accepted [51]. Lastly, the selection of one algorithm for all functions is not justified as their performance can vary considerably [52,53]. Further information on the suitability of different methods is necessary in order that meaningful comparisons may be made, so that the most appropriate microcomputer program can be selected.

5.2. Molecular graphics

Few people can fail to be impressed by the computer generated views of molecular structures that are becoming increasingly popular in education[7] and research [58,59]. The ability to perform scaling and rotation or to generate views containing perspective information have quite reasonably been seen as restricted to fast and powerful mainframe machines running complex software. However, the availability of limited graphic features on microcomputers and an appreciation of the information that can be conveyed even by simple representations of the structures have prompted some investigators to look at the feasibility of microcomputer generated molecular graphics. Because mainframe resources are not available to the microcomputer user it is necessary to consider several aspects before a realistic appraisal of this application can be made.

The quality of the screen image that can be presented is primarily a function of the resolution of the graphics software and the construction of the cathode ray tube. The majority of microcomputers use a raster-scan display similar to domestic television receivers where the electron beam sweeps in a systematic pattern across the whole of the display face. The intensity of the electron beam at any point dictates whether the phosphor coating is excited and consequently, the brightness of the spot at that position. In VDU type displays (including microcomputers) the screen is restricted to an array of dots that are illuminated in patterns to make up characters or graphics images. The size of the array is variable but displays offering greater than 512 points square are exceeding the display capabilities of even the better television monitors. A realistic microcomputer graphics system would normally offer a display of approximately 256 × 256 individually addressable points or ‘pixels’.

Practically all of the current generation of microcomputers use a ‘memory mapped’ display in which a binary image of the screen display is retained in RAM; appropriate circuitry reads the contents of this RAM and translates the bit pattern into a sequence of pulses that dictate whether the pixel is illuminated or not, the whole process being repeated at a rate that gives the impression of a steady display (typically 25 frames/second). The primary advantage of memory mapping is that alteration of the display is readily and rapidly accomplished by manipulation of the binary image in RAM, a simple matter using either machine code or high level software.

The amount of RAM needed to map a display can vary considerably. Simple textual displays do not demand large sections of RAM because the number of dot patterns that can be displayed are severely limited (usually to a character set of 256 or less). Additionally, the defined pattern of dots is restricted to specific lines forming character cells. Memory requirements for this type of display are usually limited to 1 byte/cell and a maximum of 256 bytes for the character set (often permanently coded in ROM). A display of 25 lines of 80 characters/line thus requires 2000 bytes to map display. The memory saving that is achieved is at a considerable sacrifice in the number of dot patterns that can be displayed, it has become impossible to specify the status ('on' or 'off') of each individual pixel [16].

Full 'dot resolution' graphics are essential for displaying graphs (see Fig. 7) or molecular structures to the maximum resolution of the screen. Clearly, the memory map in this instance will be larger because each point must have its status mapped in RAM by at least 1 bit. A display of 256 by 256 pixels would thus require a minimum of 65536 bits or 8192 bytes (8 Kbytes) of RAM. If the resolution is increased to 512 by 512 pixels the RAM needed is increased to 262,144 bits (32 Kbytes). Even 8 Kbytes of RAM represents a major incursion into the normal addressing range of an 8-bit microprocessor and unlimited resolution in a screen display is not possible unless special display techniques and extended memory addressing are available – neither of which have yet become widely available. The memory requirement increase further if colour information is to be added; a four colour display would require all 64 K RAM of a fully expanded microcomputer leaving no memory for display software! Microcomputers are therefore not particularly suited to extremely high resolution displays.

Additional problems arise from the relatively low processor speed. The higher the resolution of the display the greater the computational effort in calculating and displaying lines or points on the screen. The system used by the author requires an average time of approximately 2.2 ms to plot a vector on the screen at a resolution of 280 by 192 points. Thus, to draw a two-dimensional representation of the α-carbon atoms of for example lactate dehydrogenase would take 0.73 s representing an upper maximum for the rate of animation that could be attained in a high level, interpreted language. Although machine code would be faster the example illustrates the problems of using a relatively slow processor, even at limited resolution.

At present, the only feasible way of representing molecular structures in microcomputer graphics is as 'stick' structures where each atom centre of the object is represented as a single point with bond represented as lines joining the vertices. The amount of information that can be communicated by this type of representation is limited as individual atoms cannot be represented or identified and no space-filling information can be given. However, some success has been achieved using line representations and whilst the displays may have limited value they indicate the feasibility of using microcomputers for simple molecular graphics. The major computational overhead in molecular graphics lies in taking the information relating the eye position of the viewer to the three-dimensional coordinates of the structure and transforming this information into two-dimensional coordinates for display purposes [60,61]. Two common ways of representing the structure either use a

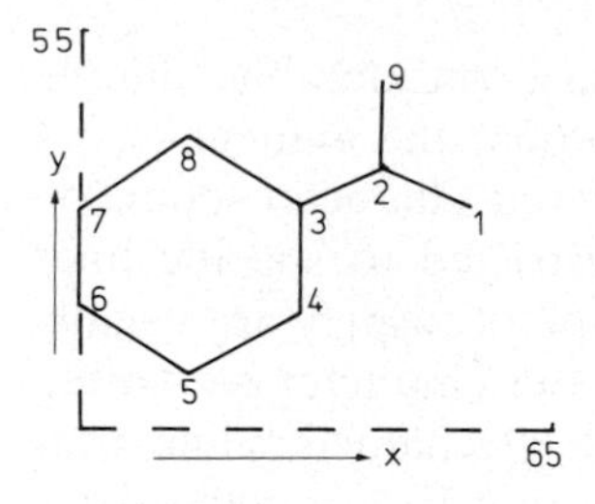

(a)

Points	X	Y	Lines
(1)	53,	31	(1) 1–2
(2)	42,	36	(2) 2–3
(3)	30,	30	(3) 3–4
(4)	30,	16	(4) 4–5
(5)	15,	7	(5) 5–6
(6)	0,	16	(6) 6–7
(7)	0,	30	(7) 7–8
(8)	15,	40	(8) 8–3
(9)	42,	48	(9) 2–9

(b)

Line		X	Y
(1)	MOVE TO	53,	31
(2)	DRAW TO	42,	36
(3)	DRAW TO	30,	30
(4)	DRAW TO	30,	16
(5)	DRAW TO	15,	7
(6)	DRAW TO	0,	16
(7)	DRAW TO	0,	30
(8)	DRAW TO	15,	40
(9)	DRAW TO	30,	30
(10)	MOVE TO	42,	36
(11)	DRAW TO	42,	48

Fig. 10. Two ways of representing molecular structures for graphical manipulation. For simplicity the object is shown as a two dimensional shape. Both methods have been used in microcomputer software suited to simple molecular graphics[8,9,10].

series of lines or vectors and include the information that specifies whether the line must 'draw to' or 'move to' the end coordinate or represent the object as a series of points with a second set of information indicating which joints are to be joined (Fig. 10). In either case, each three-dimensional coordinate must be transformed, scaled and rotated to generate a correct two-dimensional view. In addition, some programs permit the generation of perspective rather than isometric views (compare Figs. 11 and 12). Although Fig. 11 is plotted at approximately twice the resolution of Fig. 12, the lower resolution figure has the added advantage of perspective and conveys more information about the shape of the molecule.

Speed considerations acquire the greatest significance when an animated display is required to project successively transformed views of a line structure. To provide an acceptably smooth display a projection rate of better than 25 Hz is desirable requiring the transformation and display of an object in less than 40 ms. The transformation routines require extensive multiplication and division, in addition to trigonometric functions, inspection of the typical figures in Table 4 indicate the near impossibility of calculation at these rates using floating point, interpreted arithmetic. However, floating point precision is unnecessary with the current generation of graphics displays. Intger machine code arithmetic of 8 or 16 bit precision can give ranges of +127 to −127 or +32767 to −32767 for coordinates of vertices which are more than adequate for simple application. Two commercially available packages use integer arithmetic (calculating a 16 bit sin(x) in approx 50 μs) and achieve display rates with for example, haem (Fig. 11, Fig. 12) of 5 display/se-

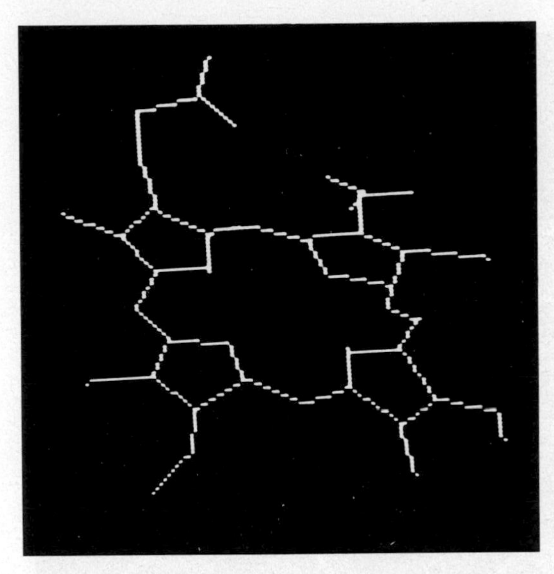

Fig. 11. Screen output of haem by a microcomputer graphics system[8]. In real life this structure was rotating about both the X and Y cartesian coordinates simultaneously at rates of approximately 5 projections/second.

cond and 1 display/second.[8,9] In contrast, a BASIC program[10] performing the same task requires 15 s but presents the transformation at a much higher accuracy, suitable for plotter output (Fig. 13).

Because of the sequential transformation of each point in simple systems, the object is drawn on the screen as the calculations proceed. When the final line is displayed the requirement for maximum display rate means that the screen will be erased immediately and the next transformation commenced. Thus, the finished view is only seen transiently, whilst the first vector remains on the screen practically continuously. A straightforward solution that improves the appearance of animated displays is that of 'double buffering' or 'page flipping'. Briefly, two areas (pages) of RAM are used for display purposes. The first is used for the generation of a screen view of the completed while the second is being written to with the next transformation. When the calculations are complete the roles are reversed and page 2 is displayed whilst the next transformation is stored in page 1. Using this technique with a rapid page switching method means that the complete object is presented on the screen at all times, differing only according to the transformation requirements.

Whilst such refinements can offer considerable improvements in the quality of microcomputer graphics the software that is currently available[8–11] can only describe objects as 'stick' structures. Hidden line removal [62] or 'space-filling' drawing arc not available and indeed, it may not even be practicable to include such features although graphics software for microcomputers is becoming increasingly sophisticated.

Most of the software described in this section was not written specifically for molecular graphics but for computer games or architectural applications, although the underlying principles are the same in all cases. The microcomputer user may benefit

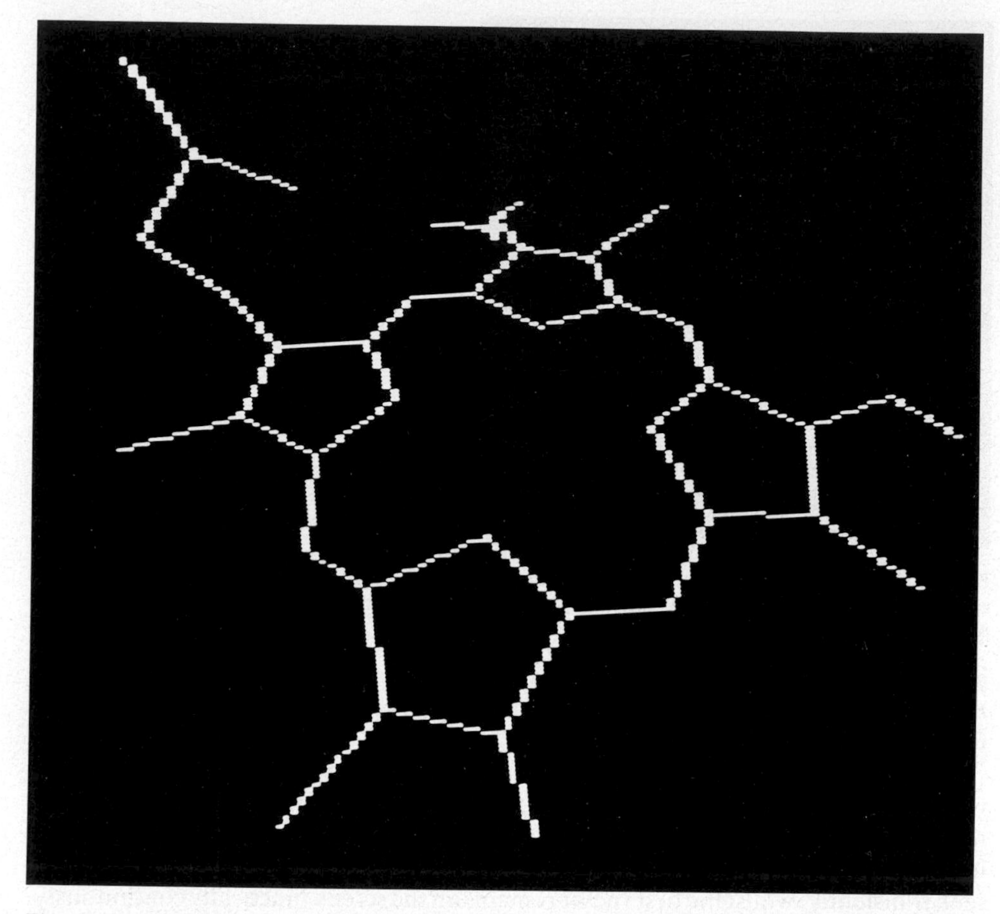

Fig. 12. Screen output of haem using a different graphics package[9]. Although the structure is displayed at a lower resolution than in Fig. 11, the additional information of perspective has been added.

even further from the impetus behind software development for domestic or other large volume markets.

5.3. *On-line interfacing*

When it is interfaced to an item of laboratory equipment the microcomputer can become a powerful tool for the acquisition and subsequent analysis of experimental data. The advantages include increased productivity, extended data analysis and the reduction of subjective errors. More and more equipment manufacturers are providing microprocessor-based data handling facilities as standard features such that interfacing a 'stand alone' microcomputer to laboratory equipment can be seen from one point of view as a temporary solution that capitalises upon existing facilities. However, the equipment that contains intrinsic computating facilities usually has the operating and analytical software installed in ROM, giving non

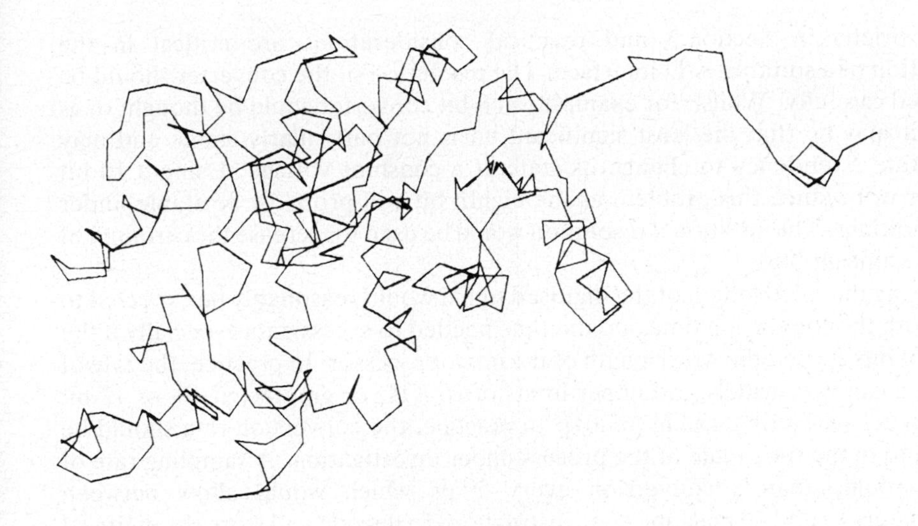

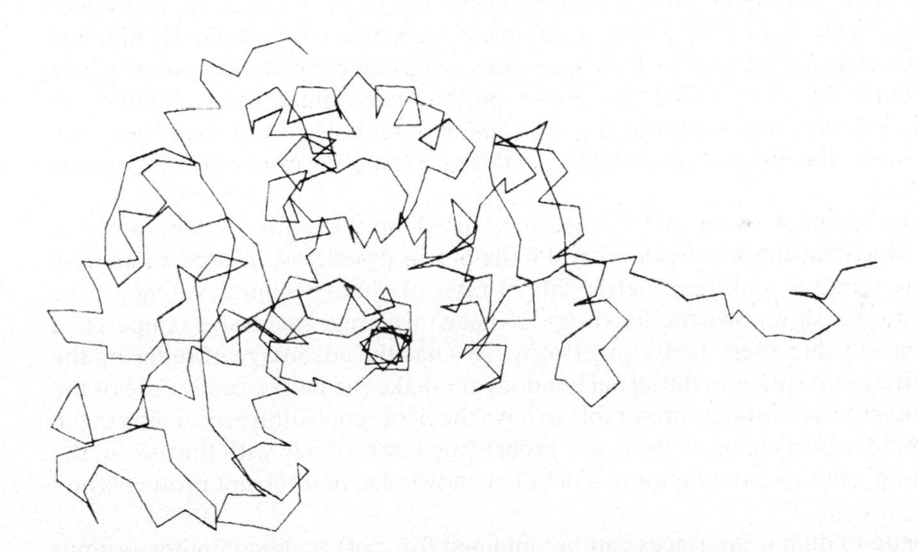

Fig. 13. Higher resolution plotter output of lactate dehydrogenase[10]. This view of the α-carbon atoms was one of an infinite range of projections that could have been produced. The lower projection has been rotated through 45°C about the x-Cartesian axis.

volatile software for data analysis at the expense of flexibility. Until equipment such as the 'programmable spectrophotometer' becomes generally available the flexible approach given by use of a microcomputer must be considered as a significant advantage, although the investment of effort will inevitably be considerable.

The acquisition of the analytical data dictates the quality of the subsequent analysis of the digitised signal. The principles of A/D conversion have been covered briefly in Section 3 and practical considerations are critical in the specification of a suitable A/D interface. The resolution of the converter should be considered carefully. Whilst, for example, an 8-bit converter could be thought of as suitable it may be that the least significant bit is not particularly stable and may demonstrate a tendency to change its state at a constant voltage. Using a 10 bit converter will reduce this problem as the eighth bit will probably be stable under these conditions. The additional resolution would be used to increase the certainty of the less significant bits.

Increasing the word length of the digitised signal would reasonably be expected to extend both the conversion time and the time needed to access data especially if the number of bits exceed the word length of the microprocessor. In practice, the rate of conversion can vary widely and upper limits of 20 KHz or greater, at 10- or 12-bit resolution are perfectly feasible [63,64]. In practice, the conversion rate should be appropriate to the time scale of the process under investigation. A sampling rate of 40 KHz would mean a conversion every 50 μs which would allow between approximately 10 and 40 machine code instructions to decode and store the digitised value in RAM and then update the pointer to the next available memory location in the RAM buffer. An 8-bit converter would fill 32 Kbytes of RAM buffer area in under a second, so that the choice of sampling rate should be realistic. For analysis of chromatographic data [65] where a complete separation may take 10 min the restriction of sampling rate to 1 Hz generates a reasonable 600 data points whilst 1 KHz would have exceeded the RAM buffer many times over. Analysis of neurophysiological data such as compound action potentials may involve frequency components of the order of several KHz, in this case sampling rate of 10 KHz may be needed [62].

A single channel, 8-bit A/D converter[11] used in the author's laboratory is capable of generating a sample every 140 μs but in practice is needed to analyse spectrophotometric and fluorimetric data at rates of about 1 sample/s. One of the simplest ways to slow down the converter has been to average successive samples and use the mean value every 1000 conversions. This has the advantage of reducing the effect of transient spikes in the signal but does not make the most effective use of the microprocessor. It would be preferable to have the data acquisition proceeding in the 'background' whilst signal analysis was proceeding interactively with the user in the 'foreground' but this would require a detailed knowledge of interrupt protocols and software.

Analogue to digital interfaces can be obtained for most microcomputer systems but often provide facilities that are not required such as multiple data channels. The alternative approach, of designing an interface locally has been the course of choice for many microcomputer users e.g. [63,66,67] using a few of the A/D devices that are available. Particularly in the case of locally designed interfaces but also important for a commercially produced circuit it is important that the correct testing procedures are applied. This can only be performed correctly when the interface is installed and connected, as some parts of the composite system may interact and generate problems [68].

Once the data has been digitised and stored in RAM or on magnetic media the data reduction software can be utilised. Software for integration, analysis of gradients, curve fitting and smoothing can be written in high level languages and developed and changed as the needs of the users alter [69,70]. The ability to perform repeated analyses on the same data is an important point but the use of a microcomputer system for archival storage of data is an area where the software should be exhaustively tested before commission, ensuring the provision of feature that perform data validation, 'hard copy' dumps and which enforce frequent copying of data.

It is difficult to provide many specific examples of microcomputer interfacing applications because much of the work has remained unpublished. The use of minicomputers has created many precedents [71,72] it will be interesting to observe future developments in this area. It is possible that 'computerised' equipment will remove the pressure to develop microcomputer-based systems for the analysis of analytical data.

5.4. *Other microcomputer applications*

Microcomputers can be usefully employed in almost any area of biological science, functioning either as a cheap source of local computing power or as a ready assembled system dedicated to a single application. Although the computational abilities of microcomputers do not resemble those of mainframe systems the processing speed is nonetheless sufficient for many repetitive tasks. For example, the comparison of two sequences of nucleotides or amino acids is not readily achieved by visual inspection of the character sequences representing such structures but can be conducted readily by a microcomputer. A program written by the author in an interpreted high level language for translation of a messenger RNA sequence is much faster than manual translation and the consequences of shifted reading frames are readily observed. Whilst this program is extremely simple it is feasible to include many of the features of more sophisticated software for nucleic acid analysis [73] in a microcomputer based system.

Graphical methods for sequence comparison such as the 'diagram' or diagonal plot method for polypeptides [74] are readily programmed provided that the output device of the system can present the data at sufficiently high resolution. For example, the comparison of two ferredoxin sequences (Fig. 14) was produced in the author's laboratory using an inexpensive digital plotter driven by a short BASIC program. The maximum resolution would have permitted the comparison of two sequences up to 700 residues long. Additionally, the non-parametric statistics for comparison of sequences or compositions [75] can be programmed readily for microcomputers, and may find value in teaching as well as in research.

The current generation of programs written for microcomputers in life sciences reflect the popularity of certain systems and the most common applications of numerical data analysis and on-line interfacing, whilst areas such as modelling and molecular graphics might be expected to develop further. Some of the more novel applications that have been published include karotype analysis [76], predictive modelling [77] and monitoring of the activity of dystrophic mice [78]. Program

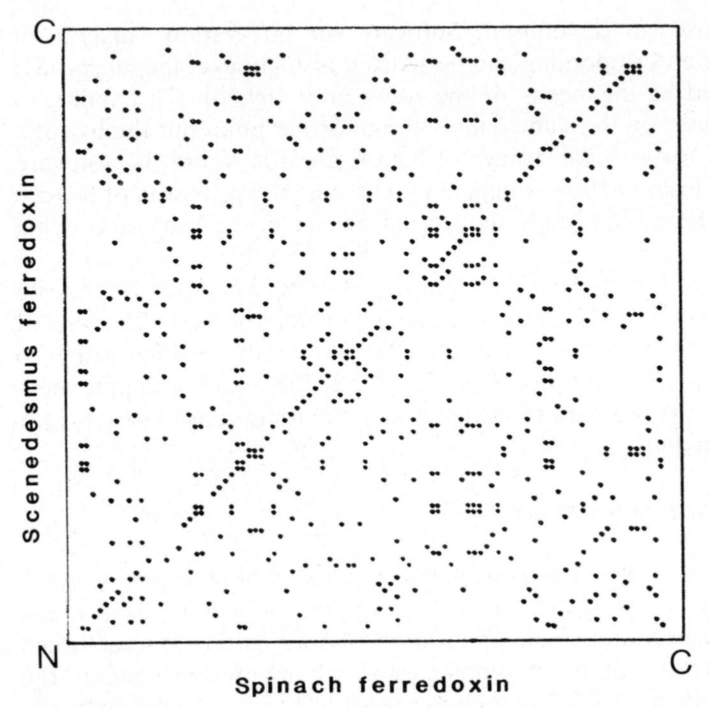

Fig. 14. Microcomputer output of the diagonal plot for comparison of the polypeptide sequences of two ferredoxins.

development would seem to be limited primarily by the ingenuity of the user, not by the microcomputer system. However, despite the publication of methods as diverse as these, there is a definite need for publication of novel or traditional microcomputer applications to prevent duplication of effort. Some communication channels that have been established may provide a partial solution[1] but the newcomer planning a new project is advised to make exhaustive enquiries before embarking upon programme development.

6. CONCLUSIONS

The availability of the microcomputer has caused an almost explosive increase in awareness of digital computing. Microprocessor-based systems are being installed in sites where practically any form of machine computation was previously unknown, providing a new opportunity for entry into the domain of large scale computing. Perhaps of all the epithets, the 'approachability' of the microcomputer is perhaps its greatest advantage, the application of power 'brings up' a system with a high level language giving simple control over screen display, mass storage or other forms of input/output. The interpretive nature of BASIC makes it ideally suited to reduction of the barrier between the user and machine, to the extent that a novice may reasonably expect to enter and run a small program within a few hours of exposure to the system.

However, by divorcing himself from the restrictions and sophistications of the mainframe computer the microcomputer user is making a number of sacrifices. Numeric accuracy, speed, range of languages and advice are all limited in the microcomputer world, to the extent that microcomputing must be seen as a different activity to traditional mainframe computing. Computational speed is considerably lower on the microcomputer, even using machine code, so there are restrictions on the types of manipulations that are feasible. On the other hand, to claim the superiority of a 2 s mainframe job over a similar microcomputer task requiring 2 min is unrealistic. Submission of a task to a larger computer is more protracted, and often involves moving to a different building. In this case, the microcomputer sitting in the laboratory can have a distinct advantage. Even with tasks requiring 1–2 h of microcomputer time we have found a general willingness to accept that interval rather than submit to the lengthy process of establishing a mainframe job. Additionally, the ability to prepare 'load and go' software means that users need have no knowledge of even the primitive operating systems of microcomputers. Loading a tape or disk and switching on a system can hardly be claimed to be difficult procedures. Although there are undoubtedly many tasks for which it is unsuitable the relative differences between the microcomputer and the mainframe are distorted by simple performance calculations.

Admittedly, the adoption of BASIC as the standard microcomputer language has done little to further the cause of structured programming. BASIC does not prevent the application of structured programming techniques but none the less allows programs to be written that are subsequently difficult to follow, modify or correct. This problem is amplified by the lack of a standard for microcomputer BASIC interpreters which range from simple, integer only versions to a large complicated language with file handling and structured constructs. To a certain extent this reflects the differences in the microcomputers themselves — enhancements for graphics will require a set of BASIC commands to control this facility. Consequently, the chances of a BASIC program written for one machine running on another are remote.

Given the differences between the species in the current generation of microcomputers there is little point in generating software than can be implemented on any machine. Such software would be rather dull and avoid using the machine specific features of any type. To make the most effective use of microcomputers the user is well advised to study and understand one system and capitalise upon its special features. The algorithm underlying any program can be readily translated into another dialect or even language, perhaps even more efficiently than entering and correcting the source code of a machine specific program. There are in excess of 250 different species of microcomputer currently available, no one person could be expected to have a deep understanding of more than one, or maybe two of these.

Attaining a deep understanding of even one system is not an easy process. Unlike many mainframe users, the microcomputer user often needs a greater understanding of the system software or hardware. This information is unfortunately lacking in most manufacturers' literature, providing no easy means of support. However, the scientist using microcomputer '*X*' is only one of a large body of people who are using the same system for diverse purposes such as for business or even as a hobby. The collective experiences of such a group, the majority of whom are highly motivated to delve into the depths of their system, represent a massive resource that should be used as fully as possible. User groups, machine-specific magazines or the popular microcomputer publications all provide a pool of knowledge that should not be dismissed lightly. The application of a machine code routine for purpose '*a*' (e.g., a games program) does not debar its use in scientific application '*b*'. In addition, attendant upon the diverse popularity of microcomputers is the growth of software firms that market at reasonable cost a wide range of potentially valuable software packages – where possible the cost effectiveness of writing a microcomputer program should be closely compared to the alternative approach of purchasing a 'ready-made' program.

It is only by virtue of considerable expenditure of effort that the most effective use of any microcomputer system can be realised. In addition to the obvious returns for this effort such as local data processing with a straightforward machine, there are other less tangible advantages that may accrue.

Firstly, the 'approachability' of the microcomputer means that there is a considerable incentive to develop novel and application-oriented software. The first generation of such software may not be elegant but the exercise of translating an objective into a computer language will have constrained the author to approach the problem in a different way. The increased 'computer literacy' can only be of value in the future development of biological research. Additionally, the strong motivation to produce microcomputer programs may hold valuable lessons for the adoption of future, more powerful computer systems [79].

A second advantage of the use of microcomputer in as local an environment as a laboratory lies in the 'democratization' of computing. The user can now write and/or enter a program, subunit his own data and collect results from a single system functioning as a laboratory tool. It has been suggested that such a scheme allows the user to assume greater responsibility for the analysis of his data and forces him to consider carefully such factors as experimental error [55].

At present, it may seem as though the tangibility of the disadvantages of using a microcomputer exceeds that of the advantages. The very real nature of the current problems does however imply that solutions will be found and the rate of development of this area means that the solutions will appear rapidly. The benefits extend beyond the additional facilities for data processing to include the development of computer awareness amongst biological scientists, it remains to be seen whether any further benefits are consequential upon this. Furthermore, this chapter has not covered areas such as information handling [89,81] or biological education [82,83] where the microcomputer is beginning to make an impact. With the pace of current technological advantages it is difficult to predict the future but it seems to the author that the microcomputer will be instrumental in demoting the computer to a valuable research tool that is used by many people to assist in the solution of many problems. The potential of the microcomputer lies with the systems of the future but the learning process must begin now.

ACKNOWLEDGEMENTS

I am grateful to G. Place, S. Green, N. Kelly and T. Best for assistance with the development of individual programs described in this chapter. The atomic coordinates for myoglobin and lactate dehydrogenase were kindly provided by Dr. S. Bellard. Thanks are due to Miss G. Devine for typing the manuscript and to Miss E. Aspinall for preparation of the illustrations.

REFERENCES

1. Ogdin, C.A. (1978) Microcomputer Design, Prentice-Hall Inc., Englewood Cliffs, N.J.
2. Givone, D.D. and Roesser, R.P. (1980) Microprocessors/Microcomputers, An Introduction, McGraw-Hill, New York.
3. ERA Technology (1979) Microprocessors – Their Development and Application, Leatherhead, U.K.
4. Green, R.J., Perkins, W.J., Piper, E.A. and Stenning, B.F. (1979) J. Biomed. Eng. 1, 240–246.
5. Tenny, J.R., Long, J.W., McFarland, W.D., Vorbeck, M.L., Townsend, J.F. and Martin, A.P. (1980) Computer Programs in Biomedicine 12, 1–6.
6. Printer Survey (1981) Pract. Comput. 4, 147–158.
7. Worland, B.B. (1979) Introduction to BASIC Programming – A Structured Approach, Houghton-Mifflin Co.
8. Coan, J.S. (1977) Advanced BASIC, Applications and Problems, Hayden Book Co., New Jersey.
9. Jensen, K. and Wirth, N. (1974) Pascal User Manual and Report, 2nd Edn., Springer-Verlag, New York.
10. Miller, A.R. (1981) Pascal Programs for Scientists and Engineers, Sybex Inc.
11. Atkinson, L. (1980) Pascal Programming, Wiley-Interscience, Chichester, U.K.
12. James, J.S. (1980) Byte 5, 100–126.
13. Levitan, S.P. and Bonar, J.G. (1981) Byte 6, 338–412.
14. Zaks, R. (1980) The CP/M Handbook, Sybex Inc.
15. Murtha, S.M. and Waite, M. (1980) CP/M Primer. Prentice Hall Inc., Englewood Cliffs, New Jersey.
16. Witten, I.H. (1980) Communicating with Microcomputers, Academic Press, London.
17. Leventhal, L.A. (1978) Introduction to Microprocessors, Prentice-Hall Inc., Englewood Cliffs, New Jersey.
18. Camp, R.C., Smay, T.A. and Triska, C.J. (1978) Microcomputer Systems Principles, Featuring the 6502/KIM, Matrix Publishers Inc., Portland, Oregon.
19. Barden,W. (1977) How to Program Microcomputers, Sams and Co. Inc., Indianapolis, U.S.A.
20. Hewlett Packard (1975) Condensed Description of the Hewlett Packard Interface Bus, HP Part No 59401 90030.
21. Fisher, E. and Jensen, C.W. (1980) Pet and the IEEE 488 BUS (GPIB), Osborne-McGraw Hill.
22. Carr, J.J. (1980) Microcomputer Interfacing Handbook, A/D and D/A, Tab Books Inc., Blue Ridge Summit, PA.
23. Garrett, P.H. (1981) Analog I/O Design – Acquisition, Conversion, Recovery, Reston Publishing Co., Reston, Virginia.
24. Page, C.F. (1979) Clin. lab. Haemat. 1, 153–164.
25. Werner, J. and Graener, R. (1980) Meth. Inform. Med. 19, 69–74.
26. Chamberlain, H. (1980) Musical Applications of Microprocessors, Hayden Book Co., New Jersey.
27. Garrett, P.H. (1978) Analog Systems for Microprocessors and Minicomputers, Reston Publishing Co., Reston, Virginia.
28. Hamming, R.W. (1977) Digital Filters, Prentice Hall Inc., Englewood Cliffs, New Jersey.
29. Li, T. (1981) Byte 6, 318–327.
30. Morris, R.A. (1980) Byte 5, 128–139.
31. Monahan, J.J., Harris, S.E. and O'Malley, B.W. (1977) In: Receptors and Hormone Action (B.W. Malley and L. Bitnbaumer, Eds.) 1, 297–329, Academic Press, New York.
32. Colquohn, D. (1971) Lectures in Biostatistics, Clarendon Press, Oxford.
33. Duggleby, R.G. (1981) Anal. Biochem. 110, 9–18.
34. Lee, J.D., Beech, G. and Lee, T.D. (1978) Computer Programs that Work, Sigma Technical Press, U.K.
35. Eisenthal, R. and Cornish-Bowden, A. (1974) Biochem. J. 139. 715–720.
36. Cornish-Bowden, A. (1979) Fundamentals of Enzyme Kinetics. Butterworth.
37. Dixon, L.C.W. (1972) Non-Linear Optimisation, English Universities Press, London.

38. Lam, C.F. and Cross, A.P. (1979) Comput. Biol. Med. 9, 145–153.
39. Nguyen, T.C. (1981) Byte 6, 435–446.
40. Box, M.J., Davies, D. and Swann, W.H. (1969) Non-Linear Optimization Techniques, Oliver and Boyd, Edinburgh.
41. Murray, W. (1972) In: Numerical Methods for Unconstrained Optimization (W. Murray, Ed.), pp. 107–122, Academic Press, London.
42. Hooke, R. and Jeeves, R.T. (1961) J. Ass. Comp. Mach. 8, 212–219.
43. Wilde, D.J. (1964) Optimum Seeking Methods, Prentice-Hall Inc., New Jersey.
44. Martin, D.F., Padilla, G.M. and Dessent, T.A. (1973) Anal. Biochem. 51, 32–41.
45. Green, S., Field, J.K., Green, C.D. and Beynon, R.J. (1982) Nucl. Acids Res. 11, 1411–1421.
46. Swann, W.H. (1972) In: Numerical Methods for Unconstrained Optimization (W. Murray, Ed.) pp. 13–28, Academic Press, London.
47. Eisenthal, R. and Cornish-Bowden, A. (1974) Biochem. J. 139, 715–720.
48. Nimmo, I.A. and Mabood, S.F. (1979) Anal. Biochem. 94, 265–269.
49. Nimmo, J.A. and Atkins, G.L. (1979) Anal. Biochem. 94, 270–273.
50. Cornish-Bowden, A. and Endrenyi, L. (1981) Biochem. J. 193, 1005–1008.
51. Peck, C.C. and Barrett, B.B. (1979) J. Pharmacokinet. Biopharm. 7, 537–541.
52. Wijnand, H.P. and Timmer, C.J. (1979) J. Pharmacokinet, Biopharm. 7, 685–687.
53. Muir, K.T. and Riegelman, S. (1979) J. Pharmacokinet. Biopharm. 7, 685–687.
54. Pfeffer, M. (1973) J. Pharmacokinet. Biopharm. 1, 137–163.
55. Koeppe, P. and Hamann, C. (1980) Comput. Programs Biomed. 12, 121–128.
56. Ruckdeschel, F.R. (1979) Byte 4, 150–160.
57. Brown, R.D. and Manno, M.E. (1978) J. Pharm. Sci. 67, 1687–1691.
58. Feldmann, R.J. (1976) Ann. Rev. Biophys. Bioeng. 5, 477–510.
59. Wipke, W.T., Heller, S.R., Feldmann, R.J. and Hyde, E (Eds.) (1974) Computer Representation and Manipulation of Chemical Information, Wiley, New York.
60. Rogers, D.F. and Adams, J.A. (1977) An Introduction to Computer Graphics, McGraw-Hill Inc.
61. Ryan, D.L. (1979) Computer Aided Graphics and Design, Marcel-Dekker Inc. New York.
62. Crow, F.D. (1981) Byte 6, 54–82.
63. Hallgren, R.C. (1980) IEEE Trans Biomed. Eng. 27, 161–164.
64. Mike, R.W. (1981) Byte 6, 312–316.
65. Reese, C.E. (1980) J. Chromatog. Sci. 18, 201–206.
66. Kirkpatrick, C.T. (1979) Br. J. Parmacol. 67, 497.
67. De Sieno, R.P. (1981) Byte 6, 274–278.
68. Lenz, J.E. and Kelly, E.F. (1980) IEEE Trans. Biomed. Eng., 27, 668–669.
69. Reese, C.A. (1980) J. Chromatog. Sci. 18, 249–257.
70. Linkens, D.A. (1979) Comput. Programs Biomed. 10, 114–124.
71. Kleinfield, A.M., Pandiscie, A.A. and Solomon, A.K. (1979) Anal. Biochem. 94, 65–74.
72. Fox, J.E. and Wilkinson, J.M. (1976) Anal. Biochem. 76, 387–391.
73. Osterburg, G. and Sommer, R. (1981) Comput. Programs Biomed. 13, 101–109.
74. Gibbs, A.J. and McIntyre, G.A. (1970) Eur. J. Biochem. 16, 1–11.
75. Cornish-Bowden, A. (1980) Anal. Biochem. 105, 233–238.
76. Green, D.M., Bogart, J.P. Anthony, E.H. and Genner, D.L. (1980) Comput. Biol. Med. 10, 219–227.
77. Powers, W.F., Abbrecht, P.H. and Covell, D.G. (1980) IEEE Trans. Biomed. Eng. 27, 520–523.
78. Nakatsu, K. and Owen, J.A. (1980) J. Pharmacol. Methods, 3, 71–82.
79. Beynon, R.J. (1980) Trends Biochem. Sci. 6, vi–vii.
80. Bertrand, D. and Bader, C.R. (1980) Int. J. Biomed. Computing, 11, 285–293.
81. Giles, I.T. (1981) Int. J. Biochem. 13, 673–680.
82. Smythe, R. and Lovatt, K. F. (1979) J. Biol. Ed. 13, 207–220.
83. Cunningham, P. (1979) Biochem. Educ. 7, 83–85.

Appendix notes

1. A Biochemistry Microcomputer Group has been established to encourage communication between microcomputer users. Further details may be obtained from the author.
2. The author now uses an Apple II microcomputer to almost complete

exclusion of all others and is only too aware of the tendency of user to claim the superiority of their system. It is hoped that the reader will not interpret the bias of the chapter as an unqualified recommendation of the system; there are many factors that must be considered in the light of individual needs.

3. Hardware floating point processor with ROM patch into Apple BASIC interpreter. California Computer Systems, 250 Caribbean Drive, Sunnyvale CA 94086, U.S.A.
4. Compiler for floating point, BASIC, Apple software. 'Expediter' Hayden Software Ltd.
5. Biochemistry Microcomputer Group contributed Program no. 31 'PATTERN-search' (R.J. Beynon).
6. Biochemistry Microcomputer Group contributed Program no. 10 'ENZKIN' (C.L. Bashford).
7. 'The Structure and Function of Haemoglobin' Computer generated 3-D film. University of London Audio-Visual Centre, 11 Bedford Square, London, WC1B 3RA, U.K.
8. 'Bill Budges Games Tool' for the Apple II Microcomputer California Pacific Computer Co., 1623 Fifth Street, Suite B, Davis, CA 95616, U.S.A.
9. 'A2–3D1' Graphics Software for the Apple II. Sublogic Inc., Box V, Savoy, Il 61874, U.S.A.
10. 'ASK' J. Swift, Blackpool and Fylde College of Further and Higher Education, Palatine Road, Blackpool, RY1 4DW, U.K.
11. Biochemistry Microcomputer Group Contributed Program no. 25 'FIGURE GROW' (M.R. Kibby).
12. Model 2200 A/D Converter for the Apple II Microcomputer. Merton Electronics, 8, Rutlish Road, London, SW19, U.K.

Bibliography

Brown, P.B. (Ed.), Computer technology in neuroscience, Halstead Press, 1976.

Cavill, I. *et al.*, Computers haematology, Butterworths, 1976.

Cooper, J.W., Minicomputer in the laboratory with examples using the PDP–11, Wiley, 1977.

Davies, R.G., Computer programming in quantitative biology, 1971. Academic Press.

Enlander, D. (Ed.) Computers in laboratory medicine, Academic Press, 1975.

Finkel, J., Computer aided experimentation: interfacing to minicomputers, Wiley, 1975.

Fu, K.S., Digital pattern recognition. Second edition, Springer, 1980.

Hawkes, P.W., Computer processing of electron microscopy images. (Topics in Current Physics 13.) Springer, 1980.

Howe, W.J., Milne, M.M. and Pennell, A. F. (Eds.), Retrieval of medicinal chemical information. (Advances in Chemistry, 84.)

Institute of Electrical Engineers *Conference Publication No. 79* Computer for analysis and control in medical and biological research conference proceedings, Dawson, 1971.

Klopfensein, C.E. and Wilkins, C.L., Computers in chemical and biochemical research, Vol. 1, 1972; Vol. 2, Academic Press, 1974.

La Fara, R.L., Computer methods for science and engineering, Hayden, 1973.

Lewis, R. (Ed.), Computers in the life sciences: applications in research and education, Kroom Helm Ltd., 1979.

Lewis, T.G. and Smith, B.J., Computer principle of modelling and simulation, Houghton Mifflin, 1979.

Lindsay, R.D., Computer analysis of neuronal structures, Plenum, 1977.

Mattson, I. *et al.* (Eds.), Computers in polymer science, Dekker, 1976 (Computers in science and instrumentation *6*).

Mayzner, M.S. and Dolan, T.R., Minicomputers in sensory and information processing research, Wiley, 1978.

Meisel, W.S., Computer operated approaches to pattern recognition, Academic Press 1972, (Mathematics in science and engineering somes 83.)

Pankhurst, R.J. (Ed.), Biological identification with computers, (Proceedings of a meeting held at King's College, Cambridge, 1973). Academic Press, 1975.
Preston, K. and Onoe, M., Digital processing of biomedical images, Plenum, 1976.
Rosenfeld, A., Digital picture analysis, Springer, 1970.
Siemazzko, F., Computing in clinical laboratories, Pitman, 1978.
Siler, W. and Linberg, D.A.B. (Eds.), Computers in life science research. Plenum, 1975.
Sims, G.E., Automation of a biochemical laboratory, Butterworths, 1972.
Soucek, B. and Carlson, A.D., Computers in neurobiology and behaviour (Sic). Wiley, 1976.
Spencer, D.D., Computer and programming guide for scientists and engineers, 2nd Ed., Sams, 1980.
Stacey, R.W. and Waxman, B., Computers in biomedical research, 4 vols. Academic Press, 1965–1964.
Sterling, T.D. and Pollack, S.V., Computers and the life sciences, Columbia University Press, 1965.
Stuper, A.J. et al., Computer assisted studies of chemical structure and biological function. Wiley, 1979.
Watkin, R.V., Computer technology for technicians and technical engineers, Longman, 1976.

Subject Index

Entries in upper case are programs and program languages